Forest Hydrology: Processes, Management and Assessment, 2nd Edition

Forest Hydrology: Processes, Management and Assessment, 2nd Edition

Edited by

Leon J. Bren
Faculty of Science, The University of Melbourne, Creswick, Victoria, Australia

Devendra M. Amatya
Santee Experimental Forest, Southern Research Station, USDA Forest Service Cordesville, South Carolina, USA

Thomas M. Williams
Clemson University Clemson, South Carolina, USA

Carmen de Jong
University of Strasbourg, Strasbourg, France

Ge Sun
Eastern Forest Environmental Threat Assessment Center, USDA Forest Service, Research Triangle Park, North Carolina, USA

CABI

CABI is a trading name of CAB International

CABI
Nosworthy Way
Wallingford
Oxfordshire OX10 8DE
UK

Tel: +44 (0)1491 832111
E-mail: info@cabi.org
Website: www.cabi.org

CABI
200 Portland Street
Boston
MA 02114
USA

T: +1 (617)682-9015
E-mail: cabi-nao@cabi.org

A catalogue record for this book is available from the British Library, London, UK.

ISBN-13: 9781800625235 (hardback)
9781800625303 (ePDF)
9781800625310 (ePub)

DOI: 10.1079/9781800625310.0000

Commissioning Editor: Ward Cooper
Editorial Assistant: Emma McCann
Production Editor: James Bishop

Typeset by Exeter Premedia Services Pvt Ltd, Chennai, India
Printed and bound in the UK by CPI Group (UK) Ltd, Croydon, CR0 4YY

Forest Hydrology: Processes, Management and Assessment, 2nd Edition

Edited by

Leon J. Bren
Faculty of Science, The University of Melbourne, Creswick, Victoria, Australia

Devendra M. Amatya
Santee Experimental Forest, Southern Research Station, USDA Forest Service Cordesville, South Carolina, USA

Thomas M. Williams
Clemson University Clemson, South Carolina, USA

Carmen de Jong
University of Strasbourg, Strasbourg, France

Ge Sun
Eastern Forest Environmental Threat Assessment Center, USDA Forest Service, Research Triangle Park, North Carolina, USA

CABI

CABI is a trading name of CAB International

CABI
Nosworthy Way
Wallingford
Oxfordshire OX10 8DE
UK

Tel: +44 (0)1491 832111
E-mail: info@cabi.org
Website: www.cabi.org

CABI
200 Portland Street
Boston
MA 02114
USA

T: +1 (617)682-9015
E-mail: cabi-nao@cabi.org

A catalogue record for this book is available from the British Library, London, UK.

ISBN-13: 9781800625235 (hardback)
9781800625303 (ePDF)
9781800625310 (ePub)

DOI: 10.1079/9781800625310.0000

Commissioning Editor: Ward Cooper
Editorial Assistant: Emma McCann
Production Editor: James Bishop

Typeset by Exeter Premedia Services Pvt Ltd, Chennai, India
Printed and bound in the UK by CPI Group (UK) Ltd, Croydon, CR0 4YY

Contents

List of Contributors vii

Preface to the Second Edition xi

Acknowledgements xiii

1 **An Introduction to Forest Hydrology** 1
L.J. Bren

2 **Forest Runoff Processes** 19
T.M. Williams

3 **Forest Evapotranspiration: Processes and Quantification** 35
G. Sun, J.-C. Domec and D.M. Amatya

4 **Forest Hydrology and Climate Change** 54
S. Liu, M. Zhang and G. Sun

5 **Trends and Characteristics of Hydroclimatic Extremes on Reference Watersheds in Experimental Forests in the USA** 72
D.M. Amatya, S. Mukherjee, E. Keppeler, P. Caldwell, S.L. Johnson, S. Sebestyen, J. Campbell, B. Rau, K. Elder, D. Misra and P. Wohlgemuth

6 **Forest Cover Changes and Hydrology in Large Watersheds** 97
X. Wei, Q. Li, M. Zhang, W. Liu, H. Fan and Y. Hou

7 **Hydrology of Cold-region Montane Forested Watersheds** 109
M. Dobre, A. Srivastava, W.J. Elliot, K.J. Elder, T.E. Link and E.S. Brooks

8 **Hydrology of Flooded and Wetland Forests** 124
L.J. Bren, T.M. Williams and D.M. Amatya

9 **Tropical Forest Hydrology** 143
T. Kumagai, H. Kanamori and N.A. Chappell

10 **Hydrology of Forests After Wildfire** 160
P.R. Robichaud

11 Hydrology of Taiga Forests at Northern High Latitudes 176
T. Burenina, A. Onuchin, I. Danilova, M. Korets and A. Musokhranova

12 European Perspectives on Forest Hydrology 196
C. de Jong

13 Forest Hydrology in China 218
K. Duan, M. Liu, X. Liu, Z. Zhang, Y. Wang and G. Sun

14 Forests for Municipal Water Supply Watersheds 234
L.J. Bren and A.A. Webb

15 Forest Hydrology Modeling: Processes, Methods and Example Models 258
H.E. Golden, D.M. Amatya and G. Sun

16 Geospatial Technology Applications in Forest Hydrology 287
S.S. Panda, E. Masson, M.S. Rana, S. Sen and H.W. Kim

17 Future Directions in Forest Hydrology 314
J. Nettles, L.J. Bren, D.M. Amatya, G. Sun, T.M. Williams and C. De Jong

Index 327

List of Contributors

Devendra M. Amatya, Santee Experimental Forest, Southern Research Station, USDA Forest Service, Cordesville, South Carolina, USA. Email: devendra.m.amatya@usda.gov

Leon J. Bren, The University of Melbourne, Creswick, Victoria 3363, Australia. Email: l.bren@unimelb.edu.au

Erin S. Brooks, University of Idaho, Moscow, Idaho, USA. Email: ebrooks@uidaho.edu

Tamara Burenina, V.N. Sukachev Institute of Forest Siberian Branch of the Russian Academy of Sciences, Krasnoyarsk, Russia. Email: burenina@ksc.krasn.ru

Peter Caldwell, Center for Integrated Forest Science, Southern Research Station, USDA Forest Service, Otto, North Carolina, USA. Email: peter.v.caldwell@usda.gov

John Campbell, Hubbard Brook Experimental Forest, Northern Research Station, USDA Forest Service, Durham, New Hampshire, USA. Email: jlcampbell@usda.gov

Nick A. Chappell, Lancaster Environment Centre, Lancaster University, Lancaster, UK. Email: n.chappell@lancaster.ac.uk

Irena Danilova, V.N. Sukachev Institute of Forest Siberian Branch of the Russian Academy of Sciences, Krasnoyarsk, Russia. Email: tiv80@mail.ru

Carmen de Jong, University of Strasbourg, Strasbourg, France. Email: carmen.dejong@live-cnrs.unistra.fr

Mariana Dobre, University of Idaho, Moscow, Idaho, USA. Email: mdobre@uidaho.edu

Jean-Christopher Domec, Bordeaux Sciences Agro, UMR1391 ISPA INRA, 1 Cours du général de Gaulle 33175, Gradignan Cedex, France; Nicholas School of the Environment, Duke University, 9 Circuit Dr, Durham, NC 27710, USA. Email: jc.domec@agro-bordeaux.fr

Kai Duan, School of Civil Engineering, Sun Yat-Sen University, Guangzhou, China. Email: duank6@mail.sysu.edu.cn

Kelly J. Elder, Fraser Experimental Forest, Rocky Mountain Research Station, USDA Forest Service, Fort Collins, Colorado, USA. Email: kelly.j.elder@usda.com

William J. Elliot, Forestry Sciences Laboratory, Rocky Mountains Research Station, USDA Forest Service, Moscow, Idaho, USA. Email: william.j.elliot@usda.gov

Houbao Fan, Institute of Ecology and Environmental Science, Nanchang Institute of Technology, Nanchang, China. Email: hbfan@jxau.edu.cn

Heather E. Golden, Office of Research and Development, Center for Environmental Measurement and Modeling, US Environmental Protection Agency, Cincinnati, Ohio, USA. Email: golden.heather@epa.gov

Yiping Hou, Jiangxi Agriculture University, Nanchang, China. Email: yiping.hou@ubc.ca

Sherri Johnson, HJ Andrews Experimental Forest, Pacific Northwest Research Station, USDA Forest Service, Corvallis, Oregon, USA. Email: sherri.johnson@oregonstate.edu

Hironari Kanamori, Kobe Gakuin University, Kobe, Japan.

Elizabeth Keppeler, Caspar Creek Experimental Watershed, Pacific Southwest Research Station, USDA Forest Service, Fort Bragg, California, USA. Email: elizabeth.keppeler@usda.gov

Hyun Woo Kim, Department of Environmental & Energy Engineering, Anyang University, Anyang, Gyeonggi-do, Korea. Email: kim4ecohydro@anyang.ac.kr

Mikhail Korets, V.N. Sukachev Institute of Forest Siberian Branch of the Russian Academy of Sciences, Krasnoyarsk, Russia. Email: mik@ksc.krasn.ru

Tomo'omi Kumagai, University of Tokyo, Yayoi, Bunkyo-ku, Tokyo 113-8657, Japan; University of Hawaii, Honolulu, Hawaii, USA. Email: tomoomikumagai@gmail.com

Tim E. Link, University of Idaho, Moscow, Idaho. Email: tlink@uidaho.edu

Qiang Li, College of Forestry, Northwest A&F University, Yangling, China. Email: qiang.li@nwafu.edu.cn

Meixian Liu, School of Civil Engineering, Sun Yat-Sen University, Guangzhou, China. Email: liumx56@mail.sysu.edu.com

Shirong Liu, Key Laboratory of Forest Ecology and Environment of National Forestry and Grassland Administration, Ecology and Nature Conservation Institute, Chinese Academy of Forestry, Beijing 100091, China. Email: liusr@caf.ac.cn

Wenfei Liu, Institute of Ecology and Environmental Science, Nanchang Institute of Technology, Nanchang, China. Email: liuwf729@126.com

Xiaodong Liu, College of Forestry and Landscape Architecture, South China Agricultural University, Guangzhou, China. Email: ljyymy@vip.sina.com

Eric Masson, Université Lille 1 Sciences and Technologies, Geography Department, Cedex, France. Email: eric.masson@univ-lille.fr

Debasmita Misra, University of Alaska-Fairbanks, Caribou-Poker Experimental Forest, Fairbanks, Alaska, USA; Email: debu.misra@alaska.edu.

Sourav Mukherjee, Oak Ridge Institute of Science and Education, Santee Experimental Forest, Southern Research Station, USDA Forest Service, Cordesville, South Carolina, USA. Email: sourav.mukherjee@usda.gov

Anastasia Musokhranova, V.N. Sukachev Institute of Forest Siberian Branch of the Russian Academy of Sciences, Krasnoyarsk, Russia. Email: nastya.krasn@mail.ru

Jami Nettles, Weyerhaeuser Company, Columbus, Mississippi, USA. Email: jami.nettles@weyerhaeuser.com

Alexander Onuchin, V.N. Sukachev Institute of Forest Siberian Branch of the Russian Academy of Sciences, Krasnoyarsk, Russia. Email: onuchin@ksc.krasn.ru

Sudhanshu S. Panda, Institute of Environmental Spatial Analysis, University of Georgia, Gainesville, Georgia, USA. Email: sudhanshu.panda@ung.edu

Manish Sing Rana, Civil Engineering Department, Indian Institute of Technology, Guwahati, Assam, India. Email: manishsingh@iitg.ac.in

Benjamin Rau, Fernow Experimental Forest, Northern Research Station, USDA Forest Service, Parsons, West Virginia, USA. Email: benjamin.rau@usda.gov

Peter R. Robichaud, Rocky Mountain Research Station, USDA Forest Service, Moscow, Idaho, USA. Email: peter.robichaud@usda.gov

Stephen Sebestyen, Marcell Experimental Forest, Northern Research Station, USDA Forest Service, Grand Rapids, Minnesota, USA. Email: stephen.sebestyen@usda.gov

Sumit Sen, Department of Hydrology, Indian Institute of Technology, Uttarakhand, India. Email: sumit.sen@hy.iitr.ac.in

Anurag Srivastava, University of Idaho, Moscow, Idaho, USA. Email: srivanu@uidaho.edu

Ge Sun, Eastern Forest Environmental Threat Assessment Center, USDA Forest Service, Research Triangle Park, NC 27709, USA. Email: ge.sun@usda.gov

Ashley A. Webb, WaterNSW, PO Box 1251, Tamworth, NSW PO Box 1251, Australia. Email: ashley.webb@waternsw.gov.au

Xiaohua Wei, Department of Earth, Environmental and Geographic Sciences, University of British Columbia (Okanagan campus), Kelowna, Canada. Email: adam.wei@ubc.ca

Thomas M. Williams, Clemson University, Clemson, South Carolina, USA. Email: tmwllms@clemson.edu

P. Wohlgemuth, Formerly with San Dimas Experimental Forest, Pacific Southwest Research Station, USDA Forest Service, Riverside, California, USA. Email: pjdnw@yahoo.com

Yanhui Wang, Ecology and Nature Conservation Institute, Chinese Academy of Forestry, Beijing, China. Email: wangyh@forestry.ac.cn

Minfang Zhang, School of Resources and Environment, University of Electronic Science and Technology of China, Chengdu 611731, China. Email: mingfangzhang@uestc.edu.cn

Zhiqiang Zhang, School of Soil and Water Conservation, Beijing Forestry University, Beijing, China. Email: zhqzhang@bjfu.edu.cn

Preface to the Second Edition

It has been nine years since we produced the first edition of this book. This new edition has many commonalities with the first edition but also many differences. Every chapter from the first edition has been thoroughly revised and/or updated with new information. We have lost some authors due to retirement but introduced new chapters, materials and authors. We welcome our new editor, Dr Ge Sun, to the existing editorial team. Our material is sourced from North America, Europe, Oceania and Asia, so we believe that we have covered the expanding global footprint of forest hydrology research.

The first edition in 2016 was commissioned by CABI to gather together the vast amount of "new information" (at that time) on forest hydrology. This brought together world-renowned professors, scientists, engineers, managers and researchers with long experience in forest hydrology (and related disciplines) to contribute. The contributions covered the hydrology of forests of diverse types and the techniques used in their investigations. This included reviews and synthesis on the past and likely future of forest hydrology. We were pleased that the first edition was translated into the Chinese language by a team led by Dr Kai Duan, in China, so that it could be accessed by a wider readership.

The second edition uses a similar approach and is complementary to various introductory texts in the same field (see the Preface of the first edition for a history of these). Since the first edition, literature on forest hydrology has grown exponentially, reflecting the fact that forests are increasingly valued for their ecosystem services. A 2024 "scientometric" analysis showed that there are about 1300 new refereed scientific papers per year in forest hydrology (roughly four new papers per day if you take Sunday off). This can be totally overwhelming, even for scientists with years of experience in the field, and our hope is that this book will help readers navigate the large and growing ocean of knowledge in the field. The contributions in this new edition cover the hydrology of more forest types, newer techniques used in their investigations, approaches to management and the past, issues of climate change (including extreme events) and the likely future of forest hydrology. We hope that it will be a valuable addition to the library of any worker involved in forest and watershed management.

The new edition features entirely new chapters on climate change (led by Shirong Liu), development and application of forest hydrology in China (led by Kai Duan) and municipal watersheds and their management (Leon J. Bren). The new addition includes a greater focus on forest evapotranspiration and its components (interception, transpiration and forest floor evapotranspiration). This reflects the fact that the process is (after precipitation) the largest component of the hydrological cycle and probably the major source of "feedback" in the hydrological system. The book includes a new look at the hydrological processes and extreme events of reference watersheds in experimental forests using the large collection of data from US Forest Service Experimental Watersheds.

The five editors of the book have worked hard to invite previous and new accomplished authors for contributions, preparing the book, liaising with authors, referees and CABI editors and making this second edition a reality. We trust that the book will provide a good understanding of the basic principles of forest hydrology and research techniques for graduate students, land managers, researchers and practitioners. Hopefully, some readers will be enthused enough to join the ranks of researchers in the discipline; it is not always easy, but it can be a useful and rewarding field.

In this revision, we feel that we have addressed (at least to some extent) the controversies of the age. The contribution of forests to water storage in watersheds directly affects hundreds of millions of people around the world. "Climate change" can be omnipresent in our media. Drought and flooding remain vast concerns across the world. This is especially important in populous developing countries such as China. The effects of forest fires have increasing hydrological consequences on a worldwide basis and their issues remain as problematic as ever. The twinned, perverse issues of floods and droughts are as present in the world now as at any time in the past. New advances in remote-sensing data monitoring, data mining and modeling are making the world of forest hydrology pregnant with new opportunities, and we hope that the reader can come away enthused by these. We are reaping the benefit of years of small watershed experimentation and tens of thousands of research papers. The accumulation of data at multiple scales allows resolution of age-old questions about forests and water. Forest hydrologists are also developing a growing knowledge of how to "scale up" and "scale down" these results to watersheds of differing sizes. We can't pretend that we have answered all questions, but hopefully the reader will leave with a greater knowledge of the issues including the contemporary ones and their complexities and potential solutions – and hopefully some creative (and perhaps startling) new approaches to the discipline.

In preparing the book, we have been struck by the excellent work done in the past and, although it may not be referred to much by more modern work, it has certainly underpinned these more recent studies. Much (or most) of this work was done in the days when computations were done by hand or using log-tables, and graphs were plotted manually. We hope that this "pioneering work" will not be forgotten by modern practitioners; we also hope that our work will be viewed as well if someone looks back many decades hence.

Finally, we would like to thank the many people who contributed to our production of this book. A major thanks goes to the staff at CABI who gave us the opportunity to create a second edition of the book – particularly Managing Editor Kin Yunn Ho and Commissioning Editor Ward Cooper. Dr Jami Nettles and her employer, the Weyerhaeuser Company, contributed in many ways to this book. A third thanks goes to the various authors (old and new) who contributed to the chapters. A fourth thanks are to the many colleagues who critiqued chapters. And finally, thanks to the partners and colleagues who put up with the editors (and authors) being ensconced behind a barricade of computers, books and papers during the preparation of this volume.

Leon J. Bren
Devendra M. Amatya
Thomas M. Williams
Carmen de Jong
Ge Sun

Acknowledgements

During the preparation of this publication, the Forest Hydrology and Biogeochemistry community lost a beloved colleague, our coauthor, Dr. John L. Campbell. With a positive attitude and gracious smile, John made significant contributions to our understanding of Northern Forest ecosystems.

We acknowledge the help given in the production of this volume by the CABI production staff, particularly Commissioning Editor Ward Cooper and Managing Editor Kin Yunn Ho. The following scientists assisted in refereeing or discussing points:

- Dr. Ning Sun, Pacific Northwest National Laboratory, Richland, Washington, USA
- Dr. Steve McNulty, USDA Forest Service, North Carolina, USA
- Dr. Peter C. Caldwell, USDA Forest Service, Otto, North Carolina, USA
- Dr. Stephen Sebestyen, USDA Forest Service, Grand Rapids, Minnesota, USA
- Dr. Safeeq Khan, University of California, Merced, California, USA
- Dr John Turner, Principal Consultant, ForSci Consulting, Sydney, New South Wales Australia
- Professor Patrick Lane, University of Melbourne, Melbourne, Victoria, Australia
- Professor Jim Gregory, North Carolina State University, Raleigh, North Carolina, USA
- Professor William Conner, Clemson University, Clemson, South Carolina, USA
- Dr. Eva-Verena Müller, Forest Monitoring and Environmental Protection Department, Germany

We would also like to thank the efforts of partners, pets, etc. who put up with the turmoil of assembling a book like this.

1 An Introduction to Forest Hydrology

L.J. Bren*
The University of Melbourne, Creswick, Victoria, Australia

Abstract
Forest hydrology is the science of the relationship between forests and water. The chapter explores the development of scientific knowledge in this discipline. In the 20th century, this passed from philosophic discussion to field measurement experimentation. Although not without problems, the results showed the responsiveness of streamflow to forest manipulation and the large amount of water stored in watershed slopes. In recent years, there has been a large growth in published work involving new concepts, techniques and measurement devices. The accumulation of data from long-term paired-watershed experiments is allowing many concepts to be examined scientifically for the first time. This ability is enhanced by a new generation of remote sensing devices ranging from personal aerial vehicles to satellites. We believe that the discipline is in excellent shape to meet the challenges of the coming century; these challenges include demands for more water and those imposed by climate change.

1.1 What Is Forest Hydrology?

Forest hydrology is the study of the hydrological processes in forested watersheds and the influence of forest ecosystem characteristics and silvicultural practices on those processes. In its purest form, forest hydrology is a quantitative discipline underpinned by conservation of mass and energy in connected, continuous media. The application of such 'pure' theories is rendered difficult by the variations of inputs across space and time and in the properties of materials comprising the watersheds. Such difficulties are the challenges of forest hydrology.

In writing this overview, one is struck by the vastness of the publications across 'forest hydrology' (and if you doubt this, see Farooqi *et al.*, 2024). These encompass theory, observations, methodologies, processes, results and political advocacy. The scale of work ranges from molecular to effectively the size of the Earth. Interests may be in the science, economics or politics of land-use management. Forest hydrology is graded into the wider disciplines of meteorology, geology, hydrology, ecohydrology, forestry, soil science and plant physiology. There is a diverse, voluminous, worldwide literature in the discipline. This has answered many questions from the past, but newer and more difficult questions are being raised, particularly concerning the long-term future of our planet rather than a particular watershed. Issues of forest hydrology and 'climate change' are particularly difficult to resolve.

Interest in forest hydrology dates back to three sources. The first of these was intellectual curiosity about the way the world works (dating back to the earliest writings of humanity). The second was the observation of landowners that actions such as clearing forests often generated consistent, observable and (in hindsight at least) predictable results in streamflow and sediment load. The third was an age-old concern about the 'sustainability' (as we would now define it)

*Corresponding author: l.bren@unimelb.edu.au

DOI: 10.1079/9781800625310.0001

of land uses and rainfall. Underpinning this was, and is, the socio-economic importance of streamflow to the survival of communities, and some harsh experiences when rainfall and consequent streamflow was either extremely low or extremely high.

1.1.1 Role of forest hydrology in resolving conflicts

Wood and water are fundamental economic goods, and from time to time there are inherent conflicts. Demand for resolution of these conflicts has driven the development of the discipline. Points of conflict include the following:

1. Impacts of logging and/or deforestation on water quality and quantity (e.g. Benyon *et al.*, 2023).
2. Impacts of afforestation on water quantity (e.g. Kaptein *et al.*, 2024).
3. Water use of different species of trees and water use of forests compared with agriculture, and the impacts of these on downstream water users (e.g. Zhang *et al.*, 2001).
4. 'Reversing' land-use changes to return areas to an earlier land-use hydrology (known as restoration conservation) (e.g. Jones *et al.*, 2022).

Forestry hydrology has evolved to both help resolve such conflicts and to avoid future conflicts.

1.2 Development of Forest Hydrology

1.2.1 Historical antecedents

The history of hydrology dates back to the earliest civilizations such as ancient Rome, as they had the ability to measures flows and to manage water with canals, quanats, drainage tunnels and dams. Scientific historians note the growth of hydrological science for many centuries but usually denote the starting point as the work of Frenchmen Pierre Perrault (1608–1680) and Edme Marriotte (1620–1684) in the period 1670–1680. They showed that rainfall in the Seine Basin was entirely adequate to sustain the flow of the river (Biswas, 1970). Around 1700, the brilliant English polymath Edward Halley advanced the field further by providing the first quantitative estimates of what we would now call the hydrological cycle (Hubbart, 2011). Unfortunately, there seems to be little information on who first formulated that key complementary idea to the rainfall – the watershed. McCulloch and Robinson (1993) suggested that the concept has been used for millennia. The well-known scientist Cayley (1859) refined the concept of contours and slope lines and might well be viewed as an early scientific user. Examination of early dam-building projects in Australia, at least, suggest that dam size was usually based on the size of the river feeding the dam, and that determination of the size and properties of the watershed usually came (much) later.

The emergence of forest hydrology as a subdiscipline of hydrology appears to owe much to the French Revolution (Andréassian, 2004). This led to an unparalleled expansion of land clearing in France as 'the King's Forests' were cleared for settlement. Landholders then encountered many of the same problems – erosion, flooding, streams drying up, landslips or other forms of mass erosion, and sedimentation – now encountered in developing countries. At the time, France was probably the most technically advanced country in the world. The ills and possible remedies caused much discussion in intellectual circles of post-revolutionary France, although by modern standards the discussion was philosophical rather than scientific. Out of this came a view of the forested watershed as being something analogous to a 'sponge' (sometimes called the Law of Dausse, after Dausse, 1842) and this oversimplification still underpins the view of non-technical citizens.

Among other things, Dausse (1842) argued that 'Rain is formed when a warm and humid wind comes in contact with strata of cold air; and since the air of forests is colder and more humid than that of the open, rain must fall there in greater abundance.' The view was then expressed that the forests constitute 'a vast condensing apparatus'. This message became codified into 'trees bring rain', which was a worldwide catchcry of a century ago.

Interestingly, satellite measurement of air temperatures has confirmed that the air of forests is colder than surrounding agricultural land because of heat loss associated with greater transpiration (Mildrexler *et al.*, 2011), but the link to greater rainfall and condensation appears elusive. Other chapters in this book will explore some of the same ideas.

1.2.2 The era of hydro-mythology

In the latter part of the 19th century, views concerning the role of forests in hydrology began to become accepted and, indeed, were viewed as 'conventional wisdom'. These include 'trees bring rain', forests modify flooding, forests provide 'healthier water', forests providing increased dry-season flows, and forests reducing erosion. A century and a half later, such statements would be viewed as 'partly true', 'generalizations', 'sweeping statements' or 'unproven' but are still commonly held. In this period, data started to be collected to 'prove' such statements; the concepts of experimental design, rigorous measurement and hypothesis testing were yet to arrive in the world of forest hydrology.

By the start of the 20th century, there was a body of advanced thought on the role of forests in protecting watersheds and some skilled observation but little that we would now recognize as 'science'. Some authors (e.g. Marsh, 1864; Zon, 1912) were far ahead of their time and contemporaries in examining the beneficial effects of the presence of large forests on streamflow. In retrospect, their work was a seminal contribution to the developing field of forest hydrology and watershed science. With the development of forestry science, stable forest management organizations, and the advent of sophisticated and reliable instruments (e.g. water level, precipitation, air temperature and solar radiation recorders), the discipline was ripe for development – and that development continues today.

The 20th century also brought the concept of the 'hydrological cycle' as we know it today (see Linton, 2008, for a history and sociological critique). This is taught in schools and underpins our thoughts. As presented, it is incomplete in both scientific and social criteria (Bren, 2023) but is undeniably useful. 'Climate-change' concerns have also raised questions about the stability of the hydrological cycle (e.g. Koutsoyiannis, 2020). Traditionally forest hydrology was viewed as a subset of the hydrological cycle, but questions as to whether forest treatments may perturb the global cycle are sometimes raised.

1.2.3 The era of small-watershed measurement

Around the middle of the 19th century, the value of hydrological data was realized. In general, these data took the form of periodic readings of major river levels. Although informative, it was found that with this approach it was impossible to link rainfall and streamflow except in the crudest sense, and that large rivers were both difficult to measure and too complex for simple water-balance studies. This led to the first true 'watershed study' in the Bernese Emmental region of Switzerland in 1906. In this study, the hydrological response of two watersheds of 0.6 km^2 was compared. These had different distributions of land use. Inferences on the hydrology of the slopes were drawn by comparison. In general, the results showed a moderating influence of the presence of forests on peak flows and a slower summertime recession from the forested watersheds (reflecting better slope storage). Measurement at Emmental still continues and the data set is a valuable asset for climate-change researchers; Hegg *et al.* (2006) provided an overview of this project, while Stähli *et al.* (2021) looked at the utility of the data in terms of later research.

By contemporary standards, the early Emmental project was far from perfect. It relied on correlation between land use and outputs rather than experimental manipulation, the data were sometimes discontinuous, and the project appears to have had a somewhat tenuous political existence. From this writer's distant viewpoint (in space and time), one has to admire the work and the people that made it happen – going out to the field on horseback or on foot, measuring in wet and cold conditions, countless hours of tedious manual calculations using logarithmic tables or slide rules, laborious hand-plotting of graphs, the constant struggle to maintain and upgrade equipment, and the ever-present demand from administrators of 'What is more data going to show you that you don't already know?' But the project did set the scene for the big advance in forest hydrology – paired-watershed experiments.

1.2.4 That great leap forward: paired-watershed experiments

The European experiences were not lost on a generation of US settlers, with massive efforts directed at controlling large rivers. The value of forests in protecting watersheds was recognized explicitly by the formation of the National Forest Service in 1891. However, there was no clear basis of information beyond the earlier observations of Marsh (1864) – a deficiency clearly evident to the early forestry scientists.

In 1910, the Wagon Wheel Gap experiment was commenced in Arizona by the US Forest Service (Bates and Henry, 1921, 1928). This was the first formal examination of the effects of forest denudation on streamflow and sediment yield. This study ran until 1926 and was the prototype of hundreds of paired-watershed experiments around the world; arguably, this has been the most successful forest hydrology technique. In this, a 'to-be-treated' stream is 'calibrated' against a 'control' or reference stream. The forest on the first watershed is then altered, and the effect on streamflow is determined by comparison with the flow in the 'control' stream. Their conclusions were based on mean values of study variables without the benefit of a statistical treatment of year-to-year variability.

Van Haveren (1988) revisited the data set produced by Bates and Henry (1928) to ascertain whether a more sophisticated 'modern' approach (including covariance and regression analysis) would give the same result as that of the older work. Table 1.1 summarizes his findings.

The results of this analysis showed that 'many of the original conclusions stated by Bates and Henry (1928) are statistically supportable. A few of their conclusions could not be supported statistically.' These finding underline the discipline of the early researchers doing what is now viewed as 'computationally intensive' work in the pre-computer days. Study of the Bates and Henry (1928) work also shows tentative first steps in 'hydrograph analysis' – relating specific characteristics of the flow record to the land use or land-use change. This continues to be something of a specialty area in the discipline of forest hydrology today.

Table 1.1. A comparison between the results of the van Haveren (1988) examination of the Wagon Wheel Gap experimental and the conclusions reached by Bates and Henry (1928).

Hydrograph parameter	Original conclusion	Re-evaluation result
Average annual water yield	Increased 24 mm	Increased 25 mm
Annual max. daily flow	Increased 50%	Increased an average of 50%
Date of the annual maximum flow	Advanced 3 days	Advanced 6 days (NS)
Starting date of snowmelt	Advanced 12 days	Advanced 5 days (NS)

NS, not significant.

1.2.5 Proliferating paired-watershed experiments

The success of the Wagon Wheel Gap experiment led to a large increase in paired-watershed projects around the world. These were generally classed as 'deforestation experiments' in which the effects of forest harvesting were studied, or 'afforestation experiments' in which the effects of plantation formation were measured. Fig. 1.1 shows an example of such a project in which the native forest was cleared and replaced with radiata pine in Australia; this project is continuing. There are comprehensive lists of past projects (e.g. Brown *et al.*, 2005), and perusal of the literature suggests that the list of projects is growing steadily with time. The technique took root in the USA, but examples can now be found in all corners of the world. In general, the data sets of matched streamflow and rainfall records have been invaluable in the development of modelling, testing of specific hypothesis and

Fig. 1.1. The experimental phase. A small watershed was converted from native eucalypt forest to a radiata pine plantation as part of a paired-watershed project in north-eastern Victoria, Australia, in 1980. The watershed is now on its second rotation of pine, and measurement is continuing.

estimation of the effects of climate change. This approach might be classed as 'bottom up', as the amalgamation of results is producing testable hypotheses.

A large body of experience has developed with this technique, demonstrating the following:

1. There has been a rapid build-up of hydrological knowledge by the experimenters, with much information gained that is peripheral to the main aims of the experiment (e.g. Hewlett *et al.*, 1969; for a quantitative example, see Bren and Lane, 2014).
2. The projects serve as a great 'teaching tool' (usually self-education) for forest hydrologists (Hewlett and Pienaar, 1973).
3. The results of the experiments are usually respected by courts and similar bodies as being 'trustworthy' and are not often attacked in courts. The author contends that this is partly due to the 'visual, tangible' nature of the experiment – people can see and visit the areas, and the concepts being explored are understandable.
4. The experiments involve a substantial capital cost and organizational commitment to get established. Once established, they are relatively inexpensive to maintain. This maintenance fits well with the routine of research organizations (Bren and McGuire, 2011).
5. Reflecting the nature of forests, the project may take many decades to bring to completion. In societies in which there are constant rearrangements of (or, worse, no) land-management agencies, long-term management may prove difficult.
6. Although individual watersheds may have 'different' properties, there is a corpus of

scientific knowledge that applies to them, and that encompasses most sources of variation.

7. The accumulation of a century or more of data allows interesting hypothesis to be explored as never before. For an example, see the analysis by Nippgen *et al.* (2016) of the 'memory' of small watersheds for past periods of rainfall.

The major (technical) disadvantage of the technique is that the watersheds are small and that scaling up of results to regional watersheds can be problematic; this is discussed by Bren (2022).

Paired-watershed experimentation and experimental science?

Given that forest hydrology is a science, concepts should be able to be quantified and hypotheses tested. Paired-watershed projects are a specialized form of experiment in which a time-variant effect is measured relative to a reference state ('control'). From the scientific point of view, there are some difficulties:

1. Replication of treatments is rarely feasible in economic or geographical terms. There are many differences between neighbouring watersheds, and these can be viewed to some extent as uncontrolled differences, so perhaps a true 'replication' is impossible? In any case, few organizations would be prepared to replicate the treatment because of time, cost and space requirements.
2. Concepts of blinding in the experimental design (in which the analyst has no specific knowledge of treatments of individuals concerned, or even no knowledge of what has been done to a watershed) have been little investigated. This partly reflects the fact that usually the experimenter ends up doing the necessary analysis. This is necessary in practical terms but does not conform to the best scientific practice in more mature (particularly medical) sciences.
3. Statistical analysis of data can be rendered difficult by the non-normality of residuals from many models, thereby limiting appropriate testing. Hewlett and Pienaar (1973) noted that hydrologists were divided on the importance of this, and this schism still exists today.
4. Other than a null hypothesis (i.e. that there is no effect), development of testable hypothesis can be difficult. To date, many of the paired-catchment experiments have been 'exploratory' in the sense that the aim was to examine the deviations in stream properties from the 'normal' ones.

Although the concept is not perfect, the development of paired-watershed projects over the years has provided increasing rigour and sophistication of experimental design, newer methods of dealing with non-normality of data residuals in testing, and use of hourly, daily or monthly data rather than annual data as a means of increasing degrees of freedom in the data (at the expense of autocorrelation). Over the years, there has been a constant improvement in the rigour of data examination and the use of sophisticated analytical software (e.g. Pagliari *et al.*, 2021).

1.2.6 'Closing' the water balance

Fundamental to paired-watershed experiments are the measurement of rainfall and other precipitation entering the watershed and the measurement of water (or vapour) leaving the watershed. It is an axiom of forest hydrology that water entering and leaving a watershed can be quantified to give a 'water balance' or water budget. Thus, over any period, the water entering the watershed equals the volume of water leaving the watershed plus the change in water stored in the watershed. It follows that, by careful measurement of the processes and summation over a suitable period of time, one can compare inflows and outflows. Differences are a measure of error. This is called 'closing the water balance' and is commonly an aim of forest hydrology research (Waichler and Wemple, 2005; Scott, 2010).

Some difficulties are implicit in this scheme, leading to experimental uncertainty (Fisher *et al.*, 2005). All field-based measurement schemes have proven costly and laborious to maintain for years on end. Some variables such as rainfall and streamflow are relatively easy to measure. Others, such as evapotranspiration have proven elusive, laborious and difficult to measure. Many parameters necessary to describe the hydrology have a wide stochastic variability. Formation of the water balance also depends on the key assumption that the actual watershed boundaries coincide with the surface boundaries; this is usually viewed as an axiom rather than a testable hypothesis as we have no explicit way to test this premise (particularly at substantial depths below the surface). Similarly, leakage into or out of the watershed is assumed to be zero or, at most, a small, relatively constant value. Review of recent history does show that we have growing sophistication in our methods and that the issues of the past are fading away.

1.2.7 The search for experimental alternatives

A strength and weakness of the paired-watershed approach is the sequential nature of the study. If development of the forest takes a century, then a paired-watershed project following the full life of the forest will take at least this time. This is usually too long for most researchers or research organizations. In addition, the concept of a 'control watershed' remaining 'stationary' (i.e. unchanged for a century) is problematic and difficult to overcome – particularly as some projects are close to a century in age.

One approach to speeding up the process has been the omission of a 'calibration period' before treatment. This has the disadvantage that it is difficult to set any statistical error limits (or, indeed, sometimes to form a view of just what is the treatment effect). An alternative approach is the use of plots to measure hydrological variables of interest. The experimenter is not bound by the sequential nature of measurement but, rather, can have many plots in different age classes. Sophisticated plot designs have the disadvantages that the plots are difficult to sustain for long time periods and that the usual variable of interest – streamflow – is not commonly directly measurable. Thus, streamflow effects must be inferred by water-balance differencing. It is arguable whether this is as satisfactory as a direct measurement of streamflow. With good statistical design, the errors involved can be quantified. An excellent example of plot use is the work of Kaptein *et al.* (2024) in assessing the water use of hybrid *Eucalyptus* trees in Zululand, South Africa. This work showed that the use of these trees gave benefits in wood productivity but would have water resource implications.

1.2.8 Coming to grips with the dynamics of watershed flows

As forest hydrology knowledge grew, there was a concomitant increase in knowledge in other fields of watershed hydrology. Some of this impetus came from the brilliant theoretical work in soil physics by Buckingham (1907); see Philip (1974) for an incisive review of this and how it did not sit well with the hydrology establishment of the time. This provided the model of the slopes having a continuum of energy levels of water, manifesting themselves in saturated and unsaturated zones. Although the model was applicable to agricultural and forest soils, the inhomogeneity of forest soils made it more difficult to apply.

Hydrology on agricultural land was substantially predicated on increases in flow during and after rain ('stormflow') being due to overland flow from an infiltrating surface. Groundwater was not usually considered a contributing agent to stormflow (or even streamflow). The infiltrated water was considered to pass downwards though the pore structure of the soil, with a portion reaching the aquifer to support 'baseflow'. A substantial body of theory linking these processes developed (e.g. Horton, 1945). Attempts to apply these formulations to forested watersheds were (and still are) unsatisfactory. Forest hydrologists found little evidence of overland flow, nor could correct infiltrations values be developed from sieved soil samples due to rocks, root material and 'macropores' (inhomogeneities) found in forest soils. The presence of such fundamental forest materials in samples was, in fact, viewed as 'errors' or 'imperfections' in the forest soils.

The differences between agricultural and forested slopes were the subject of much research (and some acrimonious debates or arbitrary paper rejections) in the 1950s to 1970s. Slope studies faded in academic prominence, being viewed as difficult and laborious, but recent measurement techniques in characterizing energy potential in unsaturated soils have suddenly made large gains possible. In doing this research, the network of paired-watershed projects provided both invaluable sites and data. The results of this research can be summarized as follows:

1. Commonly forested slopes have a high infiltration capacity, and true 'overland flow' is rarely generated. An account of where it did occur in massive rainfalls is given by Orr (1973), who noted large amounts of litter movement but little actual erosion.
2. Infiltrated water moves both through 'macropores' (holes) in the soil and through the soil matrix (see Auberton, 1971, for a classic description, or Kan *et al.*, 2023, for a more modern account). Because of the action of roots and forest biota, the soil is constantly 'turning over'. It is difficult to characterize the hydraulic properties of such media using simple models. Stochastic variation in soil texture, structure and pore space geometry, and anisotropy and air compression effects make water behaviour complex. Isotope work has shown that trees may preferentially remove water from the smaller pores only, leading to a 'two water worlds' model (Brooks *et al.*, 2009; McDonnell, 2014). Thus, water stored in larger pores or fractures in rock generates streamflow and the hydrograph response, while water stored in the soil pore matrix is 'tapped' by tree roots.
3. The watershed 'soil' is a complex medium composed of rocks, mineral soil and organic matter. Often, forest soils are better viewed as decomposing rock ('saprolites'). Simple models based on agricultural soils are difficult to apply. Šamonil *et al.* (2020) provided a good account of regolith development in forests; a classic difficulty is whether the deeper material is better classed as 'rotten rock' or 'soil'.
4. The hydrological response of the watershed may sometimes be generated many metres (or tens of metres) below the soil surface. Surface soil is generally a superconductive layer, which may transmit water to substantial depths or to the stream. In general, relating surface-soil properties in a forest to stream hydrographs is difficult.
5. Most slopes are characterized by a water-table aquifer at some depth below the surface. Elements of the behaviour can be approximated using groundwater theory (Troch *et al.*, 2003). However, as shown by workers such as Loague and Freeze (1985), this is never simple. Reconciliation of forest watershed data and 'classic groundwater theory' commonly hinges on subtle points of definition. Other issues include stochastic variation, anisotropy, discontinuities, difficulty in specifying initial and boundary conditions, and how the presence of macropores and air flow may be incorporated (Morel-Seytoux, 1973). Somers and McKenzie (2020) provided an excellent discussion of the issues of groundwater in mountainous watersheds.
6. The role of small pores holding water at high tensions in watershed slopes is not much explored. The advent of new instrumentation at the time of writing is likely to change this.

Notwithstanding the difficulties of quantification, the research has given a clear picture of the watershed slopes recharging and discharging, with the presence of trees providing a highly conductive surface layer and maintaining infiltration pathways to the subsurface water stores. The effect of the forest and understorey vegetation is to, by transpiration, deplete the soil water content in the slope. This leads to reduced storm responses during the next period of rainfall. The watershed behaviour at depth in soils is likely to be an area of research for the foreseeable future (Šamonil *et al.*, 2020). Not uncommonly, one can show deep movement of water in the slopes, but the volumetric flux is small compared with the upper layers (Bren, 1979). This basic knowledge provides an excellent foundation for the next generation of forest hydrologists.

1.2.9 Hydrograph analysis: the last refuge of the desperate hydrologist

The hydrograph is the record of outflow of a watershed over time; ideally, this is collected in conjunction with a hyetograph (a record of rainfall intensity over time). The conventional (and still fundamental) approach to such records is to integrate them over a year to obtain the volume or depth of both annual rainfall and annual streamflow. Integration smooths errors and often makes long-term relationships apparent.

An alternative approach is to use the data directly or even to differentiate with respect to time; the latter process enhances both variability and errors in the data (Whittaker and Robinson, 1924). Generically, such operations come under the category of hydrograph analysis, and this comes and goes in fashion. The doyen of forest hydrology, John Hewlett (1922–2004), is reputed to have quipped at a conference that hydrograph analysis was the last refuge of the desperate hydrologist. The technique (loved by this author) has provided much information on the dynamic behaviour of forest stream systems but usually shows that streams emanating from forested watersheds have complex dynamic behaviour that cannot be encapsulated by simple equations or simple explanations. Thus, elegant formulations may explain some of the behaviour but cannot reproduce all facets of it.

The plethora of paired-watershed experiments and associated data has made excellent sequences of rainfall and streamflow data available for testing and model calibration. An early approach was the 'quickflow separation' technique proposed by Hibbert and Cunningham (1966) and illustrated in Fig. 1.2. As initially envisaged, 'stormflow' (also known as 'quickflow') was delineated by an upward sloping line. Storm rainfall was the rainfall occurring between the initiation of the line and the intersection of the line with the receding hydrograph. The concept worked well for small-storm hydrographs. However, difficulties quickly manifested themselves:

1. By 1966, it was known that stormflow from forested watersheds was substantially a groundwater response. The concept of 'quickflow' as delineating a particular and discrete process could not be sustained; rather, it arbitrarily partitioned a part of a longer-lived slope response.
2. The process used the dependent variable (the storm hydrograph) to define the independent

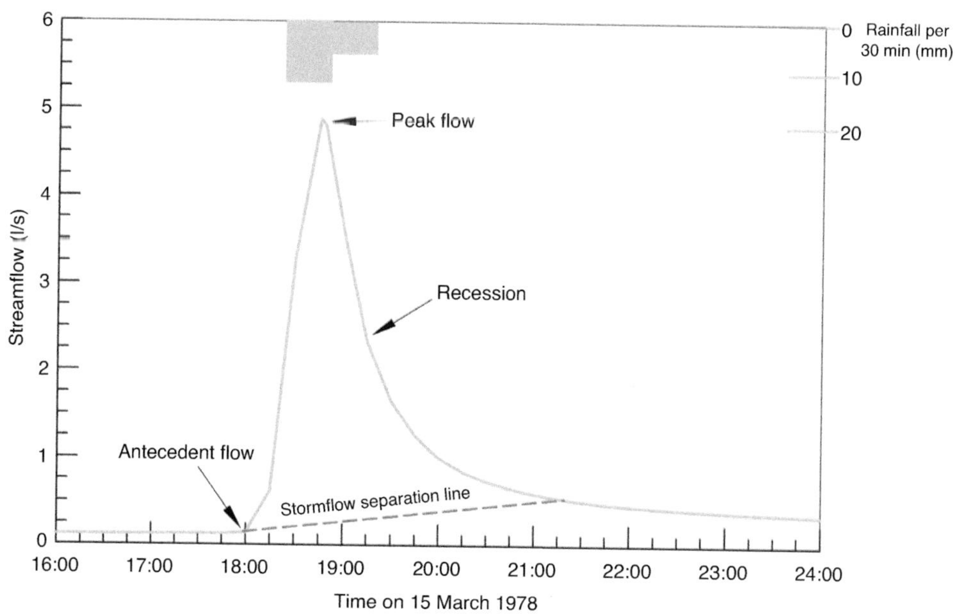

Fig. 1.2. Hydrograph terminology and the application of a stormflow separation procedure to a small hydrograph from a native forest watershed in north-eastern Victoria, Australia.

variable (storm rainfall). This is dubious in a statistical sense.

3. For large storms, the 'quickflow' separation line could take many days or weeks to intercept the receding hydrograph. Given this, the concept of 'quickflow' becomes a confusing misnomer not applicable to storm flows many days (or weeks) after the causal rainfall.

Alternative hydrograph separation procedures (sometimes generically known as 'baseflow–stormflow separation') suffer from similar issues. It is usually unclear what was being delineated in hydrological terms. The method did produce a national (Hewlett *et al.*, 1977) and an international data set (Hewlett *et al.*, 1984) that allowed hypotheses arising from non-forest hydrology to be at least partly tested using data from forested watersheds. These showed that maximum short-term intensities had little impact on the depth or volume of storm flow, and that the depth of storm rainfall received was the best predictor of the stormflow arising from a forested watershed. Subsequent work (including Bren *et al.*, 1987) showed that rainfall intensity was, indeed, a factor in stormflow generation, but that simple measures such as maximum 15 min, 30 min or 1 hour intensity were inadequate. Some 30 years after this was published, Howard *et al.* (2010) revisited this discussion using data from a watershed subject to very high-intensity rainfall in a tropical zone and suggested that the matter is still not resolved. The concept of 'baseflow' is enduringly popular with hydrologists (e.g. Chen and Ruan, 2023).

Hydrograph analysis and hydrograph separation still occupies an interesting place in forest hydrology and fluctuates in fashion. This partly reflects the difficulties of obtaining good, matched data sets of rainfall and volumetric streamflow. The technique is very demanding in time.

'Old' and 'new' water

During the 1970s, application of isotope-tagged water to forested slopes engendered a fast hydrological response, but the water entering the stream from the slopes was water that had been stored in the slopes for long periods and was commonly not the same water that 'rained' on the slopes. Thus, the 'new' water was pushing out the 'old' water. Effectively, this suggested an orderly process of replacement of slope water. Residence time could be weeks to months (Sklash and Farvolden, 1979; Pearce *et al.*, 1986).

At the time of this discovery, it was thought that stormflow separation based on hydrograph analysis would give new information on slope hydrology processes. However, as discussed by Burns (2002), this has not been the case; indeed, the status of such studies was downgraded to 'just one more tool'. Difficulties relate to models of mixing, homogeneity of the slopes and, as always, the role of macropores in providing preferential flow paths.

The 'variable source area' concept

An important – but somewhat ethereal and enigmatic – advance in forest hydrology was the concept of the 'variable source area' (VSA); Hibbert and Troendle, 1988) presented an account of this and some of the passions that went with it. The concept originated from work at the Coweeta Research Laboratory led by Hewlett and Nutter (1969). This had its origins in dissatisfaction with existing hydrological theory based on low rates of infiltration and predicted overland flow across the watershed surface. This theory stated that storm runoff was due to rainfall infiltrating into the watershed slopes near the stream (Fig. 1.3). This area would become saturated and quickly contribute runoff to the stream. In heavy rainfall, the source area would expand, while in dryer periods it would contract – hence the term VSA. In very large storms, the source area would expand to occupy the whole watershed (Orr, 1973).

The concept has been verified to some extent by studies in small watersheds. The VSA was and is a useful qualitative concept, but it is an abstraction of a more complex reality. McDonnell (2003) revisited this model some 40 years after Hewlett and colleagues articulated it. He noted that mathematical models of small-watershed behaviour usually implicitly use a structure based on VSA concepts but noted a disconnect between modellers and field investigators, which has slowed down attempts to link numerical modelling and VSA concepts. Modern explorations of the concept include Yeakley *et al.* (2020), Costa *et al.* (2020) and Kim (2022).

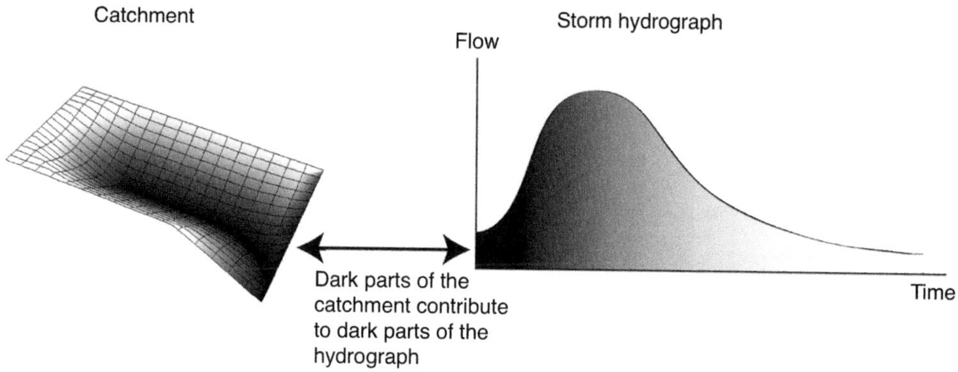

Fig. 1.3. The author's perception of the VSA model; the interpretation is that the 'blacker' parts of the watershed have a higher probability of contributing to the 'blacker' parts of the hydrograph than the lighter colours. From Bren (2022).

1.2.10 Forest fires and watershed hydrology

Forest fires have been around for as long as forests, but their effect on landscape formation has not been appreciated until relatively recently. In recent decades, Australia, USA and Canada have experienced 'megafires' – large, destructive forest fires hardly known on a recent historical scale (with the area measured in hundreds or thousands of square kilometres). Some of these fires were possibly fostered by the cumulative effect of fire-suppression policies, unprecedented fuel loads, insect and disease infestations, and drier conditions linked to climate change. The impact of these high-intensity wildfires on the hydrology of forested landscapes has been of great importance. In general, the results can be summarized as follows:

1. Fires change the forest age class and/or type, which may have long-term consequences on the hydrological regime. Thus, in Australia, the water use of the key commercial species, mountain ash (*Eucalyptus regnans*), appears to vary with forest age and changes in forest density (see Benyon *et al.*, 2023, for a concatenation of results on this point). Mountain ash forests are killed by wildfire, and a new, even-aged forest regenerates. Major fires in these forests introduce a long-term change in the water yield (relative to annual rainfall). Because of the economic importance of water from these forests, this is a major management concern.
2. After fires, 'spike hydrographs' in which very high rates of streamflow are generated for a short time are common (Brown, 1972). These high runoff rates have very high erosive power. Fig. 1.4 shows an example of this after fire burnt the experimental watershed in Fig. 1.1.
3. Fire-induced erosion can have major consequences in degradation of important watersheds and appears to be altering the hydrology of large watersheds (Smith *et al.*, 2011). The relative importance of this in large watersheds, which have many other agents of change, is an important (and difficult) field.

A study of one such burnt watershed showed that it took about 3 years to recover (Bren, 2012). The removal of undergrowth by burning showed many erosion features, which appear to be associated with past burns dating back for unknown time periods. In this environment, on northern slopes at least, it is likely that the combination of fire and hydrology has accelerated the development of the Australian mountain landscapes (see, for instance, Nyman *et al.*, 2011, and Inbar, 2017). Definitive work on these processes is currently being done in many countries around the world.

Fig. 1.4. Fire-induced erosion fills a weir at the Croppers Creek hydrological project in Victoria, Australia.

1.2.11 The era of integration and the age of Budyko formulations

Integration of data has always been an effective way of subduing the influence of errors in data. Paired-watershed projects produced annual volumes of input (rainfall) and output (streamflow), and tabulated versions of these were readily available. Additionally, the dynamic water-yield behaviour of small watersheds has rarely been of interest to water-supply managers compared with the annual outflow volumes of larger watersheds. Hence, the use of integrated values over a year made sense. The 'year' was often a water year (from summer to summer) to avoid 'change of storage' effects. In Australia, this was often from May to April because, at the end of April, the soil moisture and groundwater status of the watershed was predictably and consistently low.

Russian scientist Mihail Budyko developed an energy balance of the Earth's climate (Budyko, 1982). Aspects of the methodology have direct relevance to forest hydrology and have been widely applied in climate-change modelling. In turn, the excellent small-watershed data from paired-watershed experiments have proven to be ideal for testing the theories of climate change. A widely used outcome of this approach in forest hydrology has been the work of Zhang *et al.* (2001) in producing generalized evaporation (or runoff) curves as a function of mean annual rainfall (Fig. 1.5). Such curves have been used to make coarse comparisons between the hydrology of mature forests and pasture.

More recent work has relied heavily on small-watershed data to give estimates of ET/P where ET is annual evapotranspiration and P is annual precipitation. These use aspects of the Budyko model to characterize regional hydrology (e.g. Donohue *et al.*, 2012). This has allowed characterization of regional and global trends in watershed hydrology, and linking of regional work with local work. An example is Fig. 1.6

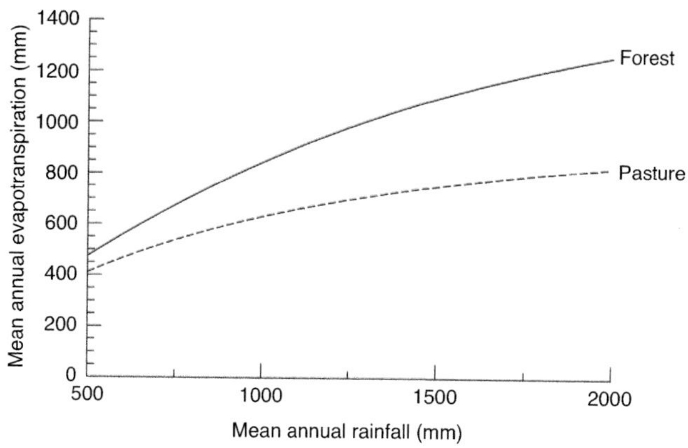

Fig. 1.5. The curves of Zhang *et al.* (2001) used to estimate evapotranspiration from grassland and forest.

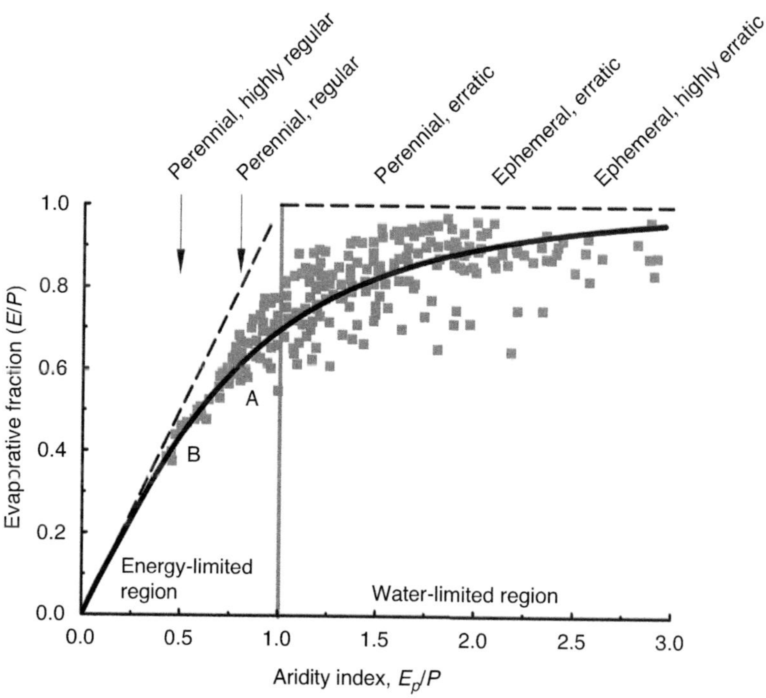

Fig. 1.6. Budyko plot showing catchments in the Murray–Darling Basin. From Trancoso *et al.* (2016) and their classification of streams (using Dunne, 1983) based on the aridity index.

(from Trancoso *et al.*, 2016) for Australia's Murray–Darling Basin. This links the Budyko framework with the more traditional approach to forest hydrology of Dunne (1983). The growth of this theory holds promise for a direct linkage to climate-change research.

1.2.12 More integrated modelling

The glittering prospect was that if one knew enough about plant–water–atmosphere relationships, the physics of water behaviour in watershed materials, and the movement of water vapour in the atmosphere, and if one had a large enough computer, then 'integrated modelling' would be entirely adequate to resolve any forest hydrology issue. This concept has been around for decades, but accomplishment seems to move further away with time. An example of moving towards this was the inclusion of hydrology modules in forest growth/plant physiology models such as 3PG (e.g. Feikema *et al.*, 2010). Similarly, GIS packages could include spatially distributed hydrology modules, allowing accurate prediction, and evapotranspiration from watersheds could be linked to climate models to allow answering of the long-standing question of whether transpiration is really water lost to the local forest.

To date, noticeable success in this direction has not been achieved, and this conclusion has not changed since the publication of the first edition of this book. The concept of measuring parameters directly and inserting these into reliable 'integrated' models to estimate hydrological behaviour seems as distant as ever. In their review, Meresa *et al.* (2023) outlined many difficulties including adequate model structure and the quality of input data, with the aim 'to identify and reduce uncertainty in projecting hydrological extremes'.

1.3 Where Are We at in 2025?

A good place to end an introduction to forest hydrology is to review the status of the discipline at the time of writing by looking at recent papers, and to see if we can discern some trends. This review is helped by the timely 'scientometric' analysis of Farooqi *et al.* (2024), who carried out a systematic analysis on some 17,000 articles retrieved from the Web of Science. The following observations can be made:

1. Studies in the discipline are burgeoning. This includes both 'bottom-up' (field studies) and 'top-down' (remote sensing and modelling work) studies. Farooqi *et al.* (2024) noted a remarkable surge in publications between 2000 and 2022. On average, 787 publications emerged per year, with the current rate (2024) being around 1200 per year. These were published in 751 leading journals and encompassed 112 countries.
2. Using the number of research articles, the USA has the largest research activity, followed by the People's Republic of China, Canada, Germany and Australia.
3. Studies include new takes on 'old' problems (e.g. Kaptein *et al.*, 2024, on the water use of eucalypts in South Africa) and new issues (e.g. how long does a Chilean catchment take to return to 'normality' after plantation removal; Lara *et al.*, 2021; Balocchi *et al.*, 2023).
4. New models combining forestry and hydrology principles are being developed (Speich *et al.*, 2020; Benyon *et al.*, 2023).
5. 'Personally controlled' aerial vehicles (drones) with sophisticated sensors are feasible, allowing a new approach to field measurement (Hill *et al.*, 2020; Vélez-Nicolás *et al.*, 2021).
6. Modelling is increasing the timespan of our studies to many thousands of years (Inbar, 2017; Inbar *et al.*, 2020; Aakala *et al.*, 2023).
7. The research workforce is becoming more gender inclusive (Rustad *et al.*, 2023).
8. The large body of paired-watershed experimental data is proving a fruitful field for reanalysis (Nippgen *et al.*, 2016) or even repeating 'classic' field experiments (Lee *et al.*, 2019).

We conclude that forest hydrology is thriving, awash with new challenges and techniques, and facing up to a new range of issues from those dealt with in the past.

References

Aakala, T., Remy, C.C., Arseneault, D., Morin, H., Girardin, M.P. *et al*. (2023) Millennial-scale disturbance history of the boreal zone. In: Girona, M.M., Morin, H., Gauthier, S. and Bergeron, Y. (eds) *Boreal Forests in the Face of Climate Change: Sustainable Management*. Springer International Publishing, Cham, Switzerland, pp. 53–87.

Andréassian, V. (2004) Waters and forests: From historical controversy to scientific debate. *Journal of Hydrology* 291(1–2), 1–27. DOI: 10.1016/j.jhydrol.2003.12.015.

Auberton, G.M. (1971) *Nature and Extent of Macropores in Forest Soils and Their Influence on Subsurface Water Movement*, Vol. 192. North-Eastern Forest Experiment Station, US Forest Service, Upper Darby, Pennsylvania.

Balocchi, F., Galleguillos, M., Rivera, D., Stehr, A. and Arumi, J.L. (2023) Forest hydrology in Chile: Past, present, and future. *Journal of Hydrology* 616, 128681. DOI: 10.1016/j.jhydrol.2022.128681.

Bates, C.G. and Henry, A.J. (1921) Streamflow at Wagon Wheel Gap, Colorado. *Monthly Weather Review* 49, 637–650.

Bates, C.G. and Henry, A.J. (1928) Second phase of streamflow experiment at Wagon Wheel Gap, Colo. *Monthly Weather Review* 56(3), 79–80. DOI: 10.1175/1520-0493(1928)56<79:SPOSEA>2.0.CO;2.

Benyon, R.G., Inbar, A., Sheridan, G.J., Lyell, C.S. and Lane, P.N.J. (2023) Variable self-thinning explains hydrological responses to stand replacement in even-aged forests. *Journal of Hydrology* 618, 129157. DOI: 10.1016/j.jhydrol.2023.129157.

Biswas, A.K. (1970) *History of Hydrology*. North-Holland Publishing Company, Amsterdam/London.

Bren, L.J. (1979) The numeric prediction of hydrologic processes on a small, forested catchment. PhD thesis, The University of Melbourne, Creswick, Victoria, Australia.

Bren, L.J. (2022) *Forest Hydrology and Catchment Management: An Australian Perspective*, 2nd edn. Springer.

Bren, L.J. (2023) Our great benefactor – The hydrologic cycle. *Science Victoria* 3, 25–27.

Bren, L.J. (*et al*2012) Hydrologic impact of fire on the croppers creek paired catchment experiment. In: Webb, A.A., Bonell, M., Bren, L., Lane, P.N.J., McGuire, D. et al. (eds) *Revisiting Experimental Catchment Studies in Forest Hydrology. Proceedings of a Workshop held during the XXV IUGG General Assembly in Melbourne 2011*, IAHS-AISH Publication, pp. 154–165 (Vol. 353).

Bren, L.J. and Lane, P.N.J. (2014) Optimal development of calibration equations for paired catchment projects. *Journal of Hydrology* 519, 720–731. DOI: 10.1016/j.jhydrol.2014.07.059.

Bren, L.J., Farrell, P.W. and Leitch, C.J. (1987) Use of weighted integral variables to determine the relation between rainfall intensity and storm flow and peak flow generation. *Water Resources Research* 23(7), 1320–1326. DOI: 10.1029/WR023i007p01320.

Bren, L.J. and McGuire, D. (2011) Paired catchment experiments and forestry politics in Australia. In: Webb, A.A., Bonell, M., Bren, L., Lane, P.N.J. and McGuire, D. (eds) *Revisiting Experimental Catchment Studies in Forest Hydrology. Proceedings of a Workshop held during the XXV IUGG General Assembly in Melbourne*, IAHS-AISH Publication, pp. 154–165 (Vol. 353).

Brooks, J.R., Barnard, H.R., Coulombe, R. and McDonald, J.J. (2009) Ecohydrologic separation of water between trees and streams in a mediterranean climate. *Nature Geoscience* 3, 100–104.

Brown, J.A.H. (1972) Hydrologic effects of a bushfire in southeastern new South Wales. *Journal of Hydrology* 15, 72–96.

Brown, A.E., Zhang, L., McMahon, T.A., Western, A.W. and Vertessy, R.A. (2005) A review of paired catchment studies for determining changes in water yield resulting from alterations in vegetation. *Journal of Hydrology* 310(1–4), 28–61. DOI: 10.1016/j.jhydrol.2004.12.010.

Buckingham, E. (1907) *Studies on the Movement of Soil Moisture*. USDA Bureau of Soils Bulletin 38, United States Department of Agriculture, Washington, DC.

Budyko, M.I. (1982) *The Earth's Climate: Past and Future*. Academic Press, New York.

Burns, D.A. (2002) Stormflow-hydrograph separation based on isotopes: The thrill is gone — what's next? *Hydrological Processes* 16(7), 1515–1517. DOI: 10.1002/hyp.5008.

Cayley, A. (1859) On contours and slope lines. *Philosophical Magazine Series* 18, 264–268.

Chen, S. and Ruan, X. (2023) A hybrid budyko-type regression framework for estimating baseflow from climate and catchment attributes. *Journal of Hydrology* 618, 129118. DOI: 10.1016/j.jhydrol.2023.129118.

Costa, D., Shook, K., Spence, C., Elliott, J., Baulch, H. *et al.* (2020) Predicting variable contributing areas, hydrological connectivity, and solute transport pathways for a Canadian Prairie basin. *Water Resources Research* 56(12), e2020WR027984. DOI: 10.1029/2020WR027984.

Dausse, M. (1842) De la pluie et de l'influence des forets sur la course d'eau. *Annales Des Ponts et Chaussees March-April* 14–209.

Donohue, R.J., Roderick, M.L. and McVicar, T.R. (2012) Roots, storms and soil pores: Incorporating key ecohydrological processes into Budyko's hydrological model. *Journal of Hydrology* 436–437, 35–50. DOI: 10.1016/j.jhydrol.2012.02.033.

Dunne, T. (1983) Relation of field studies and modeling in the prediction of storm runoff. *Journal of Hydrology* 65(1–3), 25–48. DOI: 10.1016/0022-1694(83)90209-3.

Farooqi, T.J.A., Portela, R., Xu, Z., Pan, S., Irfan, M. *et al.* (2024) Advancing forest hydrological research: Exploring global research trends and future directions through scientometric analysis. *Journal of Forestry Research* 35(1), 128. DOI: 10.1007/s11676-024-01771-1.

Feikema, P.M., Morris, J.D., Beverly, C.R., Collopy, J.J., Baker, T.G. *et al.* (2010) Validation of plantation transpiration in South-Eastern Australia estimated using the 3PG+ forest growth model. *Forest Ecology and Management* 260(5), 663–678. DOI: 10.1016/j.foreco.2010.05.022.

Fisher, J.B., DeBiase, T.A., Qi, Y., Xu, M. and Goldstein, A.H. (2005) Evapotranspiration models compared on a Sierra Nevada forest ecosystem. *Environmental Modelling & Software* 20(6), 783–796. DOI: 10.1016/j.envsoft.2004.04.009.

Hegg, C., McArdell, B.W. and Badoux, A. (2006) One hundred years of mountain hydrology in Switzerland by the WSL. *Hydrological Processes* 20(2), 371–376. DOI: 10.1002/hyp.6055.

Hewlett, J.D. and Nutter, W.L. (1969) *An Outline of Forest Hydrology*. University of Georgia, Athens, Georgia.

Hewlett, J.D. and Pienaar, L. (1973) Design and analysis of the catchment experiment. In: *Use of Small Watersheds in Determining Effects of Forest Land Use on Water Quality Conference, University of Kentucky, Lexington, Kentucky*.

Hewlett, J.D., Lull, H.W. and Reinhart, K.G. (1969) In defense of experimental watersheds. *Water Resources Research* 5(1), 306–316. DOI: 10.1029/WR005i001p00306.

Hewlett, J.D., Fortson, J.C. and Cunningham, G.B. (1977) The effect of rainfall intensity on storm flow and peak discharge from forest land. *Water Resources Research* 13(2), 259–266. DOI: 10.1029/WR013i002p00259.

Hewlett, J.D., Fortson, J.C. and Cunningham, G.B. (1984) Additional tests on the effect of rainfall intensity on storm flow and peak discharge from wild-land basins. *Water Resources Research* 20, 8985–8999.

Hibbert, A.R. and Cunningham, G.B. (1966) Streamflow data processing opportunities and application. In: Sopper, W.E. and Lull, H.W. (eds) *Forest Hydrology: Proceedings of an International Symposium*, Pergamon Press, Oxford, UK, Pennsylvania State University, Pennsylvania.

Hibbert, A.R. and Troendle, C.A. (1988) Streamflow generation by variable source area. In: Swank, W. and Crossley, D.A. (eds) *Forest Hydrology and Ecology at Coweeta*. Springer, New York, pp. 111–127.

Hill, D.J., Pypker, T.G. and Church, J. (2020) Applications of unpiloted aerial vehicles (UAVs) in forest hydrology. In: Levia, D.F., Carlyle-Moses, D.E., Iida, S., Michalzik, B. and Nanko, K.*et al* (eds) *Forest–Water Interactions*. Springer, Cham, Switzerland, pp. 55–85.

Horton, R.E. (1945) Erosional development of streams and their drainage basins; hydrophysical approach to quantitative morphology. *Geological Society of America Bulletin* 56(3), 275. DOI: 10.1130/0016-7606(1945)56[275:EDOSAT]2.0.CO;2.

Howard, A.J., Bonell, M., Gilmour, D. and Cassells, D. (2010) Is rainfall intensity significant in the rainfall–runoff process within tropical rainforests of northeast queensland? The hewlett regression analyses revisited. *Hydrological Processes* 24(18), 2520–2537. DOI: 10.1002/hyp.7694.

Hubbart, J.A. (2011) Origins of quantitative hydrology: Pierre perrault, edme mariotte, and edmund halley. *Journal of the American Water Resources Association* 13, 15–17.

Inbar, A. (2017) The role of fire in the coevolution of vegetation, soil, and landscapes in South-Eastern Australia. PhD thesis, The University of Melbourne, Creswick, Victoria, Australia.

Inbar, A., Nyman, P., Lane, P.N. and Sheridan, G.J. (2020) The role of fire in the co-evolution of soils and temperate forests. *Water Resources Research* 56, e2019WR026005.

Jones, J., Ellison, D., Ferraz, S., Lara, A., Wei, X. *et al.* (2022) Forest restoration and hydrology. *Forest Ecology and Management* 520, 120342. DOI: 10.1016/j.foreco.2022.120342.

Kan, X., Zheng, W., Cheng, J., Zhangzhong, L., Li, J. *et al.* (2023) Investigating soil pore network connectivity in varied vegetation types using X-ray tomography. *Water* 15(21), 3823. DOI: 10.3390/w15213823.

Kaptein, N.D., Toucher, M.L., Clulow, A.D., Everson, C.S. and Germishuizen, I. (2024) Water use and potential hydrological implications of fast-growing *Eucalyptus grandis* × *Eucalyptus urophylla* hybrid in Northern Zululand, South Africa. *Water SA* 50, 57–68.

Kim, J. (2022) Equivalent Horton's ratios of channel network within the framework of variable source basin area. *Journal of Hydrology: Regional Studies* 39, 100994.

Koutsoyiannis, D. (2020) Revisiting the global hydrological cycle: Is it intensifying? *Hydrology and Earth System Sciences* 24(8), 3899–3932. DOI: 10.5194/hess-24-3899-2020.

Lara, A., Jones, J., Little, C. and Vergara, N. (2021) streamflow response to native forest restoration in former *eucalyptus* plantations in South Central Chile . *Hydrological Processes* 35(8), e14270. DOI: 10.1002/hyp.14270.

Lee, R.M., McGuire, K.J., Strahm, B.D., Knoepp, J.D., Jackson, C.R. *et al.* (2019) Revisiting the Hewlett and Hibbert (1963) hillslope drainage experiment and modelling effects of decadal pedogenic processes and leaky soil boundary conditions. *Water Resources Research* 56, e2019WR025090.

Linton, J. (2008) Is the hydrologic cycle sustainable? A historical–geographical critique of a modern concept. *Annals of the Association of American Geographers* 98(3), 630–649. DOI: 10.1080/00045600802046619.

Loague, K.M. and Freeze, R.A. (1985) A comparison of rainfall-runoff modeling techniques on small upland catchments. *Water Resources Research* 21(2), 229–248. DOI: 10.1029/WR021i002p00229.

Marsh, G.P. (1864) *Man and Nature: Or, Physical Geography as Modified by Human Action*. 1965 reprint with introduction by David Lowenthal. Belknap Press of Harvard University, Cambridge, Massachusetts.

McCulloch, J.S.G. and Robinson, M. (1993) History of forest hydrology. *Journal of Hydrology* 150(2–4), 189–216. DOI: 10.1016/0022-1694(93)90111-L.

McDonnell, J.J. (2003) Where does water go when it rains? Moving beyond the variable source area concept of rainfall-runoff response. *Hydrological Processes* 17(9), 1869–1875. DOI: 10.1002/hyp.5132.

McDonnell, J.J. (2014) The two water worlds hypothesis: Ecohydrological separation of water between streams and trees? *WIREs Water* 1(4), 323–329. DOI: 10.1002/wat2.1027.

Meresa, H., Zhang, Y., Tian, J., Ma, N., Zhang, X. *et al.* (2023) An integrated modeling framework in projections of hydrological extremes. *Surveys in Geophysics* 44(2), 277–322. DOI: 10.1007/s10712-022-09737-w.

Mildrexler, D.J., Zhao, M. and Running, S.W. (2011) A global comparison between station air temperatures and MODIS land surface temperatures reveals the cooling role of forests. *Journal of Geophysical Research* 116(G3), G03025. DOI: 10.1029/2010JG001486.

Morel-Seytoux, H.J. (1973) Two-phase flows in porous media. *Advances in Hydroscience* 9, 119–202.

Nippgen, F., McGlynn, B.L., Emanuel, R.E. and Vose, J.M. (2016) Watershed memory at the coweeta hydrologic laboratory: The effect of past precipitation D and storage on hydrologic response. *Water Resources Research* 52, 1673–1695.

Nyman, P., Sheridan, G.J., Smith, H.G. and Lane, P.N.J. (2011) Evidence of debris flow occurrence after wildfire in upland catchments of South-East Australia. *Geomorphology* 125(3), 383–401. DOI: 10.1016/j.geomorph.2010.10.016.

Orr, H.K. (1973) The Black Hills (South Dakota) Flood of June 1972: Impacts and Implications. General Technical Report RM2. Washington, D.C: USDA Forest Service.

Pagliari, P.H., Ranaivoson, A.Z. and Strock, J.S. (2021) Options for statistical analysis of unreplicated paired design drainage experiments. *Agricultural Water Management* 244, 106604. DOI: 10.1016/j.agwat.2020.106604.

Pearce, A.J., Stewart, M.K. and Sklash, M.G. (1986) Storm runoff generation in humid headwater catchments: 1. Where does the water come from? *Water Resources Research* 22(8), 1263–1272. DOI: 10.1029/WR022i008p01263.

Philip, J.R. (1974) Fifty years progress in soil physics. *Geoderma* 12(4), 265–280. DOI: 10.1016/0016-7061(74)90021-4.

Rustad, L., Adams, M.B., Dymond, S.F., Gregory, M. and Miniat, C.F. (2023) Perspectives on the contributions of women to the hydrologic sciences and their changing demographics at USDA forest service experimental forests and ranges. *Journal of Hydrology* 621, 129469. DOI: 10.1016/j.jhydrol.2023.129469.

Šamonil, P., Phillips, J., Daněk, P., Beneš, V. and Pawlik, L. (2020) Soil, regolith, and weathered rock: Theoretical concepts and evolution in old-growth temperate forests, central europe. *Geoderma* 368, 114261. DOI: 10.1016/j.geoderma.2020.114261.

Scott, R.L. (2010) Using watershed water balance to evaluate the accuracy of eddy covariance evaporation measurements for three semiarid ecosystems. *Agricultural and Forest Meteorology* 150(2), 219–225. DOI: 10.1016/j.agrformet.2009.11.002.

Sklash, M.G. and Farvolden, R.N. (1979) The role of groundwater in storm runoff. *Developments in Water Science* 12, 45–65.

Smith, H.G., Sheridan, G.J., Lane, P.N.J., Nyman, P. and Haydon, S. (2011) Wildfire effects on water quality in forest catchments: A review with implications for water supply. *Journal of Hydrology* 396(1–2), 170–192. DOI: 10.1016/j.jhydrol.2010.10.043.

Somers, L.D. and McKenzie, J.M. (2020) A review of groundwater in high mountain environments. *WIREs Water* 7(6), e1475. DOI: 10.1002/wat2.1475.

Speich, M.J.R., Zappa, M., Scherstjanoi, M. and Lischke, H. (2020) FORests and HYdrology under climate change in Switzerland v1.0: A spatially distributed model combining hydrology and forest dynamics. *Geoscientific Model Development* 13(2), 537–564. DOI: 10.5194/gmd-13-537-2020.

Stähli, M., Seibert, J., Kirchner, J.W., Freyberg, J. and Meerveld, I. (2021) Hydrological trends and the evolution of catchment research in the Alptal valley, Central Switzerland. *Hydrological Processes* 35, e14113.

Trancoso, R., Larsen, J.R., McAlpine, C., McVicar, T.R. and Phinn, S. (2016) Linking the Budyko framework and the dunne diagram. *Journal of Hydrology* 535, 581–597. DOI: 10.1016/j.jhydrol.2016.02.017.

Troch, P.A., Paniconi, C. and Emiel van Loon, E. (2003) Hillslope-storage Boussinesq model for subsurface flow and variable source areas along complex hillslopes: 1. Formulation and characteristic response. *Water Resources Research* 39(11), 1316. DOI: 10.1029/2002WR001728.

Van Haveren, B.P. (1988) A reevaluation of the Wagon Wheel Gap forest watershed experiment. *Forest Science* 34(1), 208–214. DOI: 10.1093/forestscience/34.1.208.

Vélez-Nicolás, M., García-López, S., Barbero, L., Ruiz-Ortiz, V. and Sánchez-Bellón, Á. (2021) Applications of unmanned aerial systems (UASs) in hydrology: A review. *Remote Sensing* 13(7), 1359. DOI: 10.3390/rs13071359.

Waichler, S.R. and Wemple, B.C. (2005) Simulation of water balance and forest treatment effects at the H.J. Andrews Experimental Forest. *Hydrological Processes* 19(16), 3177–3199. DOI: 10.1002/hyp.5841.

Whittaker, E.T. and Robinson, G. (1924) *The Calculus of Observations*. Blackie, London.

Yeakley, A., Swank, W.T., Hornberger, G.M. and Shugart, H.H. (2020) *A Method of Modeling Source Area Response to Climate Variability*. UMBC Geography and Environmental Systems Department Collection.

Zhang, L., Dawes, W.R. and Walker, G.R. (2001) Response of mean annual evapotranspiration to vegetation changes at catchment scale. *Water Resources Research* 37(3), 701–708. DOI: 10.1029/2000WR900325.

Zon, R. (1912) *Forests and Water in the Light of Scientific Investigation*. US Government Printing Office, Washington, DC, Senate Document No. 469, 62nd Congress, reprinted in 1927 with a revised bibliography.

2 Forest Runoff Processes

T. M. Williams*
Clemson University, Clemson, South Carolina, USA

Abstract

In the years since the first version of this book, our understanding of forest runoff processes has matured. Eddy-covariance and sap-flow monitoring have allowed measurement of evaporation and transpiration as separate interdependent processes producing a burgeoning science of ecohydrology. Variation in topography, vegetation, soil, climate and geology create conditions that result in runoff in several surface and subsurface flow paths. These flow paths can be explained as connections among a group of various-sized storage reservoirs that exist between solid bedrock and the top of the canopy. Reservoirs fill from direct rainfall or spill from a reservoir at higher elevation. Reservoir capacities are determined by a balance between gravity and vapor-pressure deficit. Vapor-pressure deficit drives evaporation or transpiration and is much stronger than gravity. Although the concept is simple, spatial and temporal variation in climate and vegetation combine with spatial variation in soil and geology to make prediction highly complex.

2.1 How Did We Get to Our Current Knowledge?

This chapter reviews and updates the information presented in the first edition. In the 10 years since that edition was compiled, ecohydrology has become a maturing field of study. Afforestation is likely to become an important part of our efforts to mitigate climate change by capture of carbon (Nunes *et al.*, 2020). As carbon assimilation and transpiration are closely linked (Sun *et al.*, 2016), efforts at carbon capture are likely to have profound effects on hydrology as well. Essentially, we are concerned with the inverse of the 19th-century controversies about forest cutting in France, which lead to the development of the paired-watershed method (Andreassian, 2004).

The paired-watershed experimental design has been the backbone of forest hydrology research (Neary *et al.*, 2012) and still has much to offer for the future (Richardson *et al.*, 2023). Much of our understanding of runoff production in forests has come from study of the reference watersheds in paired experiments. Proper design of watershed pairs (Reinhart, 1967) includes similar (ideally identical but this is unattainable) soils, vegetation, slope, aspect and geology. The vegetation on the watersheds should be stable prior to treatment. Streamflow should be measured at a point constrained, naturally or by a structure, such that all water leaving the watershed will flow through a measurement device. These constraints limit where paired-watershed studies have been conducted. To assure that the measurement device does not leak, many watersheds are on relatively shallow soil over a confining layer that has limited fractures. In order to define the contributing area, the topography should be clear and reflect confining layer topography. Due to these constraints, relatively steep watersheds on crystalline bedrock have been

*Corresponding author: tmwllms@clemson.edu

 Forest Hydrology: Processes, Management and Assessment (eds Leon J. Bren *et al.*)
DOI: 10.1079/9781800625310.0002

the favored location of paired-watershed forest hydrology experiments (Brown *et al.*, 2005). A discrete channel is a stream, and water traveling along the channel is the streamflow (also called runoff). The area that can conceivably contribute water to the cross-section within the stream is called the watershed of that stream.

2.2 Review of Runoff Processes

2.2.1 Infiltration excess versus variable source area

A recap of early research discoveries illustrates how local watershed characteristics biased early work on runoff production mechanisms. Horton (1933) offered the compelling 'infiltration-excess overland flow' mechanism. If the precipitation rate exceeded the infiltration rate, the excess rainfall would run off from the surface to a nearby stream. Not only was this proposal easily verified empirically by observation of a suburban yard or farm field, but it also could be reproduced by long-understood soil physical process (Green and Ampt, 1911) and the (then) latest theoretical formulations (Richards, 1931). It is the mechanism that occurs on most human-altered systems across the world. However, as with earlier French observations, Hursh (1944) noted that surface flows were seldom found in forests at the Coweeta Hydrologic Laboratory in the Appalachian Mountains of western North Carolina, USA. He used the term 'subsurface stormflow' to describe streamflow in the peak portion of a hydrograph from subsurface sources. Hewlett and Hibbert (1963) proposed that subsurface stormflow came from lateral flow from saturated zones near an expanding stream network that they later termed a 'variable source area' (VSA; Hewlett and Hibbert, 1967). The VSA concept, however, is not exclusive to the channel-expansion process they described, which is particular to the deep permeable regolith and fractured bedrock found at Coweeta (Nippgen *et al.*, 2016).

Infiltration-excess overland flow has been considered to be a process that only occurs in disturbed forests (DeBano, 2000; Lang *et al.*, 2015). This contention has been challenged by expanded research in tropical and semi-arid forest systems (Bonnel, 1993) due to sparse vegetation or water repellency following long dry periods (Miyata *et al.*, 2019). Bonnel (1993) observed that the bulk of research, on which many of our theories of runoff processes were developed, occurred in the humid temperate zone, which does not experience the large, high-intensity storms commonly found in the tropics.

2.2.2 Controversy over processes

Dunne *et al.* (1975) mapped source areas in north-eastern USA and south-eastern Canada where thin soils overlie glacially eroded bedrock, or compacted tills. These slopes are less steep than those of Coweeta. They suggested that, where slopes are gentler, surface runoff will occur from wider areas of concave topography along streams, and they termed this 'saturation-excess overland flow.' At the same time, in the Ouachita Mountains in Arkansas, USA, Beasley (1976) found subsurface stormflow originating from discrete channels within the soil along existing stream channels. There, the soils had developed a fragipan layer, which inhibited vertical root growth and turned root channels parallel to the slope above the fragipan. At the Maimai watersheds in New Zealand, with thin soil and steep slopes, Mosley (1979) described similar flow within subsoil macropores as 'macropore flow.'

2.2.3 Proposed and "discarded" mechanisms: "because it doesn't work here it must be wrong!"

The Maimai watersheds provided a focused effort to explain runoff production over several decades, with researchers testing new sets of hypotheses. Pearce *et al.* (1986) used oxygen isotopes to show that water flowing into a stream was chemically more akin to water stored in the soil rather than rainwater. That put an end to the simple hypothesis that rain traveled rapidly to the stream in soil macropores or pipes. Sklash *et al.* (1986) argued for a fast groundwater mechanism they called 'groundwater ridging.' They argued that the capillary fringe of high water tables near the stream remained saturated because the negative hydrostatic pressure was insufficient to cause air to enter the soil, known as tension-saturated soil (Abdul and Gillham, 1989). A small increase in hydrostatic pressure caused by upslope infiltration caused these saturated soils to release water, resulting in a rapid increase in streambank seepage. Buttle (1998) argued against the widespread adoption of this mechanism, and suggested that the conditions needed were large areas of tension-saturated soils close to the streams, and these did not generally occur. Katsura *et al.* (2014) found a ridging mechanism with shallow soil over a weathered and cracked granite with no indication of a capillary fringe. Rapid water-table rise produced a ridge of groundwater in the weathered granite only with rainfall greater than 160 mm. Wenninger *et al.* (2004) also found a fast groundwater input in conjunction with a gravel layer that was exposed in the upper slopes and intersected the stream and adjacent wetlands. Although 'groundwater ridging' may not be a common process, it occurs when conditions create high infiltration at some point in the watershed, which results in a rapid increase in groundwater seepage into a stream channel. Although tension-saturated soil is not common, it may not be necessary to produce a rapid increase in seepage rates.

During the late 1980s, McDonnell *et al.* (1991) examined an old hypothesis of Hursh and Brater (1941) of a mechanism best described as 'fast forest floor flow.' This produced rapid stormflow at Maimai. They did find fast flow within the forest floor at Maimai but found that water entered macropore channels within a few meters. McDonnell *et al.* (1991) dismissed 'fast forest floor flow' as a local phenomenon rather than a watershed-wide runoff mechanism. Nevertheless, there have been occasional descriptions of this mechanism in a variety of locations. Timmons *et al.* (1977) measured runoff in the forest floor and ascribed 97% of snowmelt runoff and 57% of annual runoff to this mechanism in glacial till soils in north-central Minnesota. Kendall *et al.* (1999) used end-member mixing of dissolved organic carbon and cations to demonstrate that snowmelt runoff was dominated by 'O horizon overland flow' in a till-covered watershed in Vermont. Kim *et al.* (2005) found that rapid flow in the forest floor was more prevalent in wet antecedent conditions and extended for distances > of more than 36m in wet conditions in steep watersheds with glacial till and colluvium soils in British Columbia. They noted that shallow flow occurred despite the underlying soil being unsaturated. They also concluded, like McDonnell *et al.* (1991), that flow was limited to short distances and was not a significant contributor to stormflow. In contrast, Terajima and Moriizumi (2013) showed large contributions of stormflow originating from 'biomat flow' during wet periods or high rainfall on steep terrain in Japan. Likewise, Crespo *et al.* (2012) found that flow was dominated by near-surface flow in the organic horizon during wet periods in the montane cloud forests of Ecuador.

Bishop *et al.* (2011) used the term 'transmissivity feedback' to describe the rapid increase in streamflow as the water table rose within highly permeable surface soils in Sweden. They noted that saturation-excess overland flow was occasionally seen, but flow within the permeable organic layers was prevalent. Skaggs *et al.* (2011) reported that apparent soil hydraulic conductivities in drained forest plantations were twofold higher than comparable drained agricultural fields in coastal North Carolina. However, the higher conductivity disappeared when forest floors were destroyed during bedding for reforestation (Skaggs *et al.*, 2006). Amatya *et al.* (2022) found a fast flow component was needed to predict stormflow from a coastal South Carolina watershed when the water table was above a critical depth near the soil surface. It appears that 'fast forest floor flow' occurs widely when antecedent moisture is high but is not important for stormflow generation in conditions where preferential flow to deeper macropores or soil pipes is prevalent. In some high water table conditions, saturation-excess overland flow may occur within the forest floor at velocities that do not move litter.

2.2.4 Macropore flow and soil piping

After over 20 years and many studies, an explanation of the flow process at Maimai was settled with an understanding of the complex interactions found there (McGlynn *et al.*, 2002). Across the watershed, any lateral water movement moves into soil macropores within a few meters, and travels to the lower confining layer. Undulations in the surface of that confining layer determine whether the infiltrating water is stored or moves into 'soil pipes' at the lower soil interface. Flow in macropores is common to forests (Gao *et al.*, 2018) and can result in flow to deeper layers well before the soil matrix is saturated, making the Richards equation invalid (Beven and Germann, 1983). Weiler and McDonnell (2007) suggested that vertical preferential flow channels should be called macropores, and horizontal or downslope conduits be called soil pipes. This provides a separation useful for hydrological processes but differs from the geomorphic distinction proposed by Jones (2010), where a macropore becomes a pipe only when velocity within the pipe can move soil particles. Flow within pipes and macropores has been the subject of research with a variety of techniques ranging from plot, trench and watershed tracers (dyes, chemistry, isotopes) to multiple model formulations, two- and three-dimensional computer-aided tomography with X-ray, electrical resistance, radar, and seismic sampling (Beven and Germann, 2013).

The change from macropore to pipe flow depends on several factors. Soil wetness appears to be the key to initiation of pipe flow. Sidle *et al.* (2007) found relatively short macropores that organized into pipes and created two thresholds of runoff generation at rainfall depths of over 40 mm and over 110 mm in mountains of Japan. Negishi *et al.* (2007) found that pipes in Malaysia near the regolith surface flowed quickly after light rains, but pipes at shallower depths responded only after heavier rains, and individual pipes had unique response thresholds. Trenching studies often found soil pipes concentrated near an interface where hydraulic conductivity decreases. In well-developed soils, an interface often occurs between the zone of leaching (where salts or colloids are removed by water) and the zone of illuviation (where salts or colloids are added by water) and a second zone near the soil–bedrock interface. Pipes at the bedrock interface are more likely over fine-grained rock like shale than more permeable sandstone (Tang *et al.*, 2020).

Slope is also an important factor in determining whether macropore flow enters soil pipes. Jackson *et al.* (2014) noted that saturation at the top of the B horizon was not sufficient to initiate pipe flow on gentle slopes. As watershed relief decreases, vertical macropore flow to a regional water table is more likely than lateral pipe flow. On watersheds where relief is minimal (microtopography <10% of total relief), macropore flow is entirely vertical. On such watersheds, when precipitation is much greater than evapotranspiration, saturation excess can occur across the entire watershed (Amatya *et al.*, 2020).

Downslope pipes generally empty into a saturated zone near the stream edge, resulting in enhanced groundwater flow into the streambank vicinity or saturation-excess overland flow in the riparian zone. Pipes may also exit the soil in areas where the soil thins, such as windthrow pits, or over bedrock outcrops to produce 'return flow,' as named by Dunne *et al.* (1975). Pipe outlets in concave landscape positions may also form ephemeral streams in what have been called '0' order basins (Sidle *et al.*, 2007). Pipes or macropores may also deliver water to weathered or cracked bedrock, contributing to the fracture flow system within the bedrock system (Gabrielli *et al.*, 2012). Rapid groundwater flow may also originate at the bedrock interface without identifiable pipes in areas where glacial erosion has removed the regolith (Buttle and McDonald, 2002).

Flow in soil pipes becomes both a geomorphic and hydrological process where lateral flow is rapid enough to move soil particles (Jones, 2010). As a geomorphic process, soil pipes are found in all parts of the world and not restricted to forested landscapes (Bernatek-Jakiel and Poesen, 2018). Soil pipes have been identified in very humid (>2000 mm rain; Bhagyalekshmi *et al.*, 2015) and quite dry (300 mm; Faulkner *et al.*, 2008) landscapes. Pipes may be a significant portion of small stream sediment delivery (20–50%) associated with larger rainfall amounts or higher rainfall intensity, which may cause minor wall or roof collapses (Sayer *et al.*, 2006). Pipe collapse can form topography similar to that of 'epikarst.' This is a perched aquifer found in the upper part of karst areas where flow in fractured limestone forms discontinuous streams and sinkholes (Faulkner *et al.*, 2008). Discontinuous pipes and pipe outlets may also lead to channel expansion (Zimmermann *et al.*, 2014), even in tropical watersheds where infiltration excess can also be found.

2.2.5 Saturated zone processes

The last two runoff processes are associated with permanent groundwater and may not always produce a rapid response to rainfall. Carbonate rocks commonly have large reservoirs of groundwater and are a source of water for about 25% of the world population (Ford and Williams, 2007). Large conduits are often aligned with faulting of the carbonate rocks and can be observed by remote sensing and other geotechnical analyses (Redhaounia *et al.*, 2015). The rate and timing of flow from these aquifers is generally determined by the conditions of the epikarst, the weathered and cracked upper surface of the carbonate rock, which varies up to 10m thick (Williams, 2008). In karst systems, different species of trees have developed roots that withdraw water from within the soil profile, the epikarst and the deeper cavities (Geekiyanage *et al.*, 2019). Epikarst acts in ways similar to macrospores and pipes in soil systems. Epikarst often produces runoff in a similar manner but may also have vertical connections to deeper 'cavity flow systems' (Williams, 2008). Hydrograph and isotope analysis, dye tracing, and chemical end-member analysis have been used to analyze epikarst and karst systems in ways similar to those used in other types of runoff process research (Branka, 2007; Zhongcheng *et al.*, 2024). In general, epikarst behaves similar to cracked and weathered silicate bedrocks, being recharged by preferential flow, discharging after a rainfall threshold is reached and providing a conduit to the permanent saturated zone. The primary difference is that the enhanced solution of preferential flow paths in epikarst leads to greater positive feedback. A major difference is in the role of surface topography. Conduit flow is driven by hydraulic gradients and subsurface factors (faults and fractures) that may be interpretable from surface topography but not necessarily driven by it. Probably the greatest difficulty is in determining the area contributing to a spring or spring-fed stream.

In limited circumstances, silicate materials can also result in karst-like topography and runoff processes. Soils over quartzite sandstone may also produce preferential flow and karst-like surfaces (Weary and Doctor, 2014). Migon *et al.* (2023) studied sandstone mesas in Eastern Czechia and western Poland and found that 89% of flow from caprock areas was in subsurface pathways. Solution of cementing materials in the sandstone was evident. 'Sapping' is a process of subsurface erosion due to saturated aquifer flows (Dunne, 1990). It is found where a saturated permeable layer is located above a less-permeable layer, such as sand over clay or sandstone over shale (Schumm and Rea, 1995). Sapping represents an end point of the processes of runoff generation. In this case, rainfall is not transported laterally by any process described previously and percolates to a water-table aquifer where it moves laterally to a point where the water table intersects the land surface. Sapping results from positive feedback of heterogeneity in the aquifer material forming preferential flow paths and discrete springs (Higgins, 1984). This differs from piping, as erosion occurs primarily at the spring, and channel development is by slumping, creep and 'backward head-cutting' rather than subsurface erosion and pipe collapse. The process produces a particular landscape characterized by long narrow valleys, few short tributaries that intersect at high angles and steep open amphitheater-like valley heads (Howard *et al.*, 1988). The landform also allows sapping to be examined by remote sensing and numeric landscape analysis. Luo (2000) showed that sapping had statistics of their hypsometric curve (graph of relative area versus relative digital elevation model (DEM) height) that significantly differed from the general fluvial watershed. The landform is found in both the arid Colorado Plateau and the humid windward side of the Hawaiian Islands (Howard *et al.*, 1988). Depending on rainfall and cohesiveness of the aquifer material, valley progression may be slow, taking millions of years (Howard, 1982), or fast, measured in thousands of years with small gully formation as fast as 30 m/day (Micallef *et al.*, 2021). Rapid expansion occurred after rainfall intensity exceeding 40 mm/day.

2.3 Thresholds and Connectivity

Runoff from forests is produced by a variety of processes that depend on climate, geology and forest stand factors. It is little wonder that hydrologists in the late 20th century searched vainly for a general theory of runoff production from forested land. From their careful descriptions, we have been able to recognize those disparate processes, although we still struggle to describe them with consistent terminology (McMillan, 2022). One important commonality we have recognized in the last two decades is the importance of 'thresholds' in the response of forested watersheds to rainfall (Ross *et al.*, 2021). A threshold is a magnitude or intensity that must be exceeded for a certain result to occur. Scaife and Band (2017) outlined a series of articles documenting threshold behavior in differing forest types and runoff processes. They noted that earlier studies, like those cited in the first edition (Williams, 2016), had identified a threshold by inspection of plotted data rather than quantitative evaluation by piecewise regression analysis (Toms and Lesperance, 2003). Much of the earlier work also used data from unique events as defined by Hibbert and Cunningham (1967). Using such a definition reduces the number of events that can be analyzed and precludes data from multiple rainstorms that are often responsible for exceeding runoff thresholds. Beiter *et al.* (2020) used a numerical definition based on runoff quantity and timing, allowing inclusion of the entire record.

Earlier work tended to identify a single threshold at a site and associate a threshold with the predominant process at that site. For example, the predominant process at Panola Mountain was hill-slope pipe and regolith activation after 55 mm of rain (Tromp-van Meerveld and McDonnell, 2006), or 316 mm of cumulative rainfall at Hubbard Brook where saturation excess was the primary process (Detty and McGuire, 2010). Current research reveals that thresholds are not unique values at a site but functionally depend on the input components considered (rainfall rates, season, prior moisture content) (e.g. Ross *et al.*, 2021), or differ with variation in season, altitude, slope and tree species (Scaife and Band, 2017). These studies suggest that runoff process threshold behavior is not as simple as was originally assumed.

The basic explanation of threshold behavior was a concept of connectivity; before water from any part of the watershed becomes runoff, it must be connected to the stream, first described for infiltration-excess overland flow (Betson, 1964). Connections of overland flow are easily seen and a threshold amount of excess rainfall (alternatively called wetting volume, initial abstraction or depression storage) is generally incorporated into most hydrological models based on infiltration-excess overland flow. Thresholds and connections may be as vital for invisible subsurface processes of forested watersheds (Ali *et al.*, 2013; McDonnell, 2013). The idea that some storage amount must be filled before any runoff process begins was termed 'fill and spill' by Spence and Woo (2003) for a visible process of saturation excess flow in northern Canada and by Tromp-van Meerveld and McDonnell (2006) for a subsurface process of flow at the soil–bedrock interface. They argued that threshold responses are driven by filling the storage reservoir resulting in overflow to a conduit to the stream. As the theory has developed (McDonnell *et al.*, 2021), the watershed can be viewed as a series of storages that must fill until water can flow to another storage compartment or the stream. While each watershed is a unique expression of climate, parent material, geomorphology, soil and vegetation, these 'hydro-ecosystems' are all realizations of 'filling and spilling' at differing scales, locations, quantities and timings that can be related to climate, parent material, geomorphology, soil and vegetation.

2.4 "Fill and Spill" on a Generalized Watershed

Figures 2.1–2.3 represent an imaginary headwaters watershed where all the described runoff mechanisms occur, a 'fill-and-spill' schematic of the flows, and a hypothetical hydrograph, respectively. These allow an explanation of the interaction of storages, connections and runoff processes across the variety of forested systems. These simplified schematic drawings do require some imagination from the reader.

In Fig. 2.1, tree caricatures represent the various types of tree canopies that influence interception and stemflow. The generalized sediments and varying depths of rooting are drawn to represent some of the various possibilities of runoff generation. Soil macropores and soil pipes are assumed to be associated with roots and soil

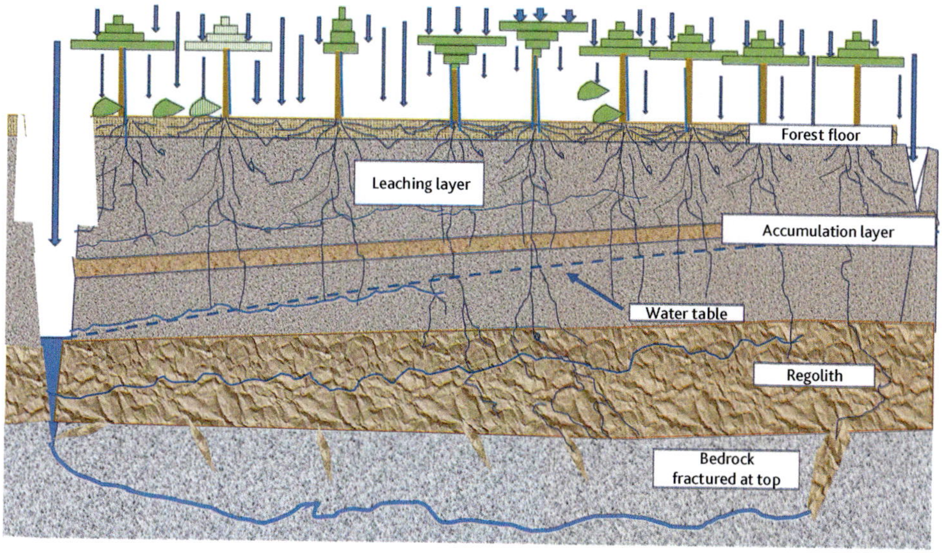

Fig. 2.1. Generalized forested cross-section without slopes. Various caricatures of trees represent differing crown sizes and shapes that can influence how much water is intercepted. Tree roots (black) are assumed to include macropores into the profile. Water flows are depicted in blue. Soil pipes may alternatively be in leached soil, at the regolith boundary or within the regolith. Solution channels may occur in the bedrock.

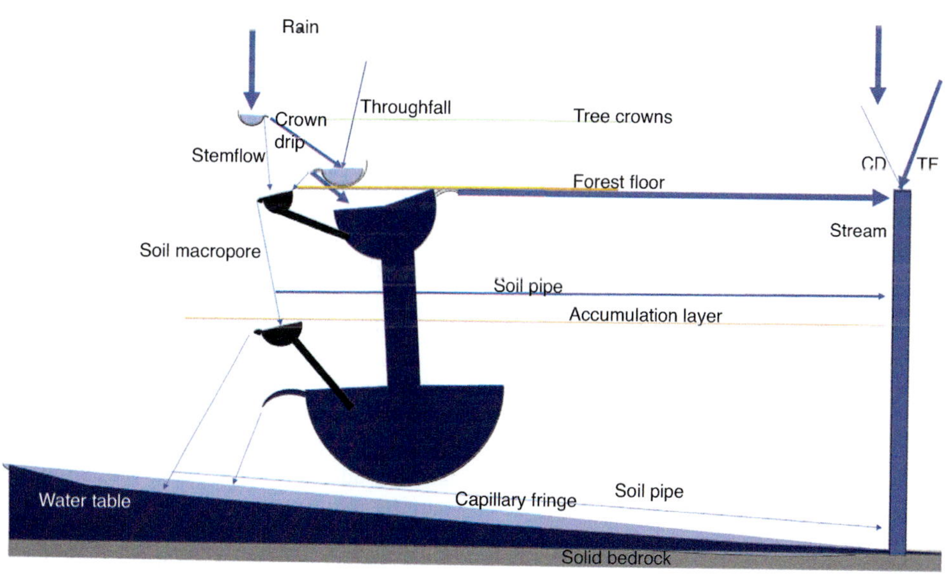

Fig. 2.2. A schematic of fill-and-spill reservoirs approximating the layers depicted in the generalized cross-section of Fig. 2.1. CD, canopy drip; TF, throughfall.

fauna, which are aerobic and concentrated in the unsaturated region. Some tree species can maintain healthy roots in the saturated zone and into cracked bedrock. In this figure, preferential flow paths are depicted by the fine black lines generally associated with tree roots. Soil pipes

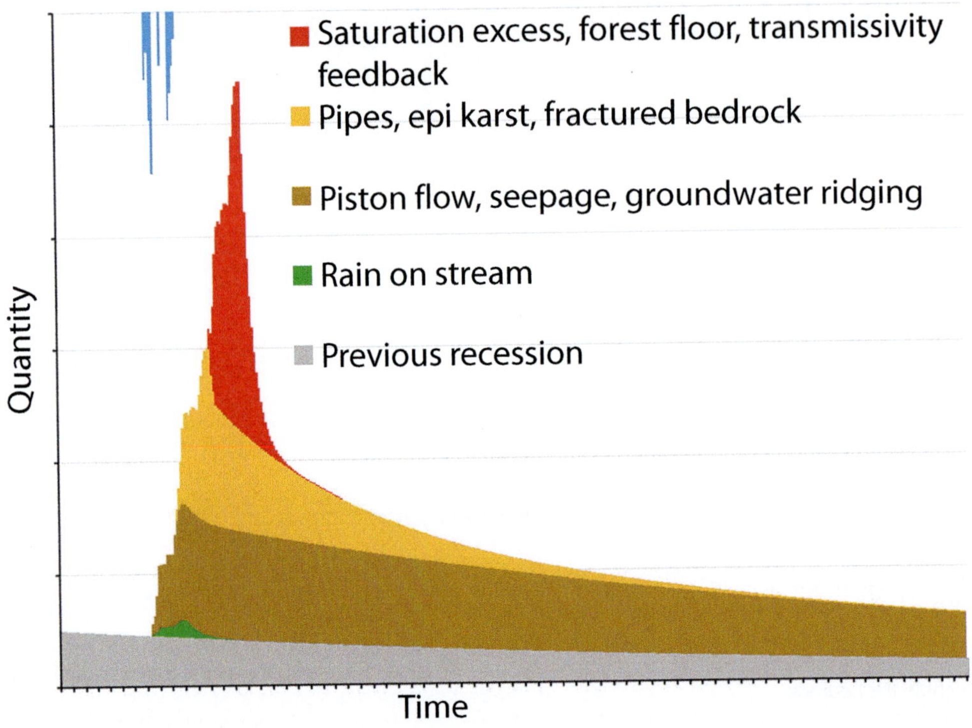

Fig. 2.3. A hypothetical separation of a hydrograph depicting the size and delay of runoff from differing hypothesized runoff process depicted in Figs 2.1 and 2.2.

are represented by blue lines that are nearly horizontal. Finally, there is one solution channel in the bedrock shown as a thick blue line. Real watersheds have complex slopes, the underlying rock is anisotropic and the surfaces also have microtopographic features. Likewise, real sediment profiles will have different layer depths, often with some of the depicted layers absent, and complex variations in soil particle and pore sizes. It is suggested that the reader's imagination can visualize real forested watersheds from this simplistic caricature. Fig. 2.2 is a 'fill-and-spill' schematic of the same cross-section, listing interactions that these processes would entail. Fig. 2.3 is a hydrograph separation that might reflect the results of these same runoff processes on the watershed for a hypothetical rainfall.

Rainfall intercepted by the forest canopy is the first reservoir to be filled. Total interception is determined by several physical factors such a frequency of rain, humidity between rains, wind speed and rain intensity. Biological factors are also important and include canopy height, canopy roughness, leaf area, leaf shape, tree shape, leaf retention and the structure of the layered canopy (Fig. 2.1). Although interception can be a significant factor in overall hydrology, generally representing 10–50% of incoming rain (Roth *et al.*, 2007), it does not represent a large storage potential in an individual storm. Iida *et al.* (2017) made measurements of individual branches to find storage of between 5.3 and 7.2 mm that lowered to 4.4–5.4 mm if the branches were shaken to simulate winds. Pineda and Rigo (2017) found that 90% of lightning-caused wildfires occurred with rainfall of less than 10 mm. Interception storage is a rather small storage to fill. Once filled, there is very little change to rainfall intensity at the ground, suggesting that the rate of spill approximates the precipitation rate.

Spilling is usually described as throughfall or stemflow. Throughfall may have two components that are slightly different in effect on the watershed. Rain may fall through canopy gaps without interception, or intercepted water may drip from the canopy. Only the canopy drip component and stemflow will behave as spill from the interception storage. Characteristics of the tree crowns can greatly influence the distribution of incoming rain into gap throughfall, canopy drip or stemflow (Crockford and Richardson, 2000). Many of these crown characteristics are represented graphically in Fig. 2.1. Tree density and crown size determine canopy gap throughfall. Branch angle is very important to partitioning between canopy drip and stemflow. Narrow pyramidal crowns with downward-sloping branch angles favoring crown drip, while branches that slope steeply upwards favor stemflow. Stem flow of *Acacia* spp. can be more than four times greater than the normal 1–4% of rainfall (Pressland, 1973). Larger stemflow values may be characteristic of species in more arid regions.

In watersheds with a perennial (intermittent) stream, filling interception storage results in the first runoff mechanism – rain on the stream surface (Fig. 2.2). Canopy gap throughfall begins simultaneously with the rain event, while canopy drip begins after the small interception storage is filled. Water in the stream can be treated as another fill-and-spill reservoir, which can be quantified relatively easily. Each millimeter of rain adds 10 m^3/ha of stream surface. The reservoir spills at a rate determined by water surface slope, hydraulic radius and bed friction, such as expressed by the familiar Manning's equation for open-channel flow in smooth channels. The ephemeral stream depicted in Fig. 2.1 would alter the linear relation of fill rate to rainfall rate. With ephemerals, as rainfall increases both the rate of addition to the stream area and an increase in stream area would occur. As depicted, a small increase in the height of the water table would cause water to flow in the ephemeral channel, resulting in an increase in stream surface area. Hewlett (1982) saw channel expansion as an important cause of the non-linear response of forested watersheds. For most headwater watersheds, rain on the stream is not a large component of streamflow, begins to contribute very quickly after rain begins, and declines rapidly after rainfall ceases, quickly moving downstream due to gravity.

Stemflow is generally not a large percentage of rainfall, except for some tree species associated with arid regions, but is delivered at the root collar (Levia and Germer, 2015). In addition to increasing soil moisture near the tree, stemflow can facilitate movement into preferential flow channels. As a proportion of rainfall, stemflow increases with both intensity and volume of rain. It is also subject to wide eco-hydrological variability as both branching and rooting characteristics of different tree species. These differences will impact the quantity and timing of stemflow, as well as which subsurface storage component it will fill. If we assume roots produce the majority of preferential flow paths, then stemflow will follow the paths of tree roots into soil macropores (Fig. 2.2).

Rainfall that falls through or drips from the canopy reaches the forest floor of decomposing leaf litter and fine roots. The forest floor can be of two types, mor with distinct organic horizons lying on the mineral soil, or mull with greater mixing of organic and the upper mineral soil. Soil faunae are probably more important in the mull-type forest floor, which is usually associated with hardwood forests, while fungal hyphae, decomposed litter and fine roots are more likely in coniferous forests. High porosity and rapid permeability are the defining aspects of the forest floor. Hydraulic conductivity of over 1000 mm/h (Torres *et al.*, 1998, measured rates near 95 mm/min), ensures that the forest floor will transmit even intense rain to the mineral soil surface. The high conductivity of the forest floor suggests that it will generally be a small storage volume and will spill rather quickly. McDonnell *et al.* (1991) found that near-surface flow quickly moved into preferential flow paths to become part of subsurface runoff systems. Rapid production of preferential flow appears to be common, with connection of water in the forest floor to the stream occurring only if the soil has very low conductivity, is very steep or the entire soil profile is saturated.

The boundary between the forest floor and the mineral soil also marks the boundary where the rate of spill is no longer determined by the rate of rainfall. For this discussion, 'soil' will refer to all porous material between the forest floor and solid bedrock. From the forest floor, the direction and rate of water movement is determined by conditions within the soil. These materials are either saturated or unsaturated and form the major reservoir that fills and spills. Aubertin (1971) argued water would only enter an empty soil pore if the water potential was greater than atmospheric pressure. Runoff does not occur unless gravity is stronger than the soil matric potential, hydrostatic (saturated) to a minimum tension of –0.1 bar in sand or –0.3 bar in clay. Plants can effectively remove water from soil to −15 bars. These values indicate that any boundary between saturated and unsaturated soil is a one-way gate leading only to the unsaturated soil. Water found in plants is unlike that found in runoff in both ^{2}H and ^{18}O (natural stable isotopes of water; Dusek and Vogel, 2024), leading McDonnell (2014) to conceive 'two water worlds', one of which sustains plants and the other of which leads to runoff generation. McDonnell (2013) suggested that Hortonian overland flow was only one manifestation of flow generation. Flow generation occurs when pore water potential due to gravity exceeds potential soil moisture tension at 'field capacity.' This happens when water is added faster than it can move into unsaturated soil. All flow generation is saturation excess occurring anywhere between solid bedrock and the soil surface. This generally happens at points within the soil profile where the conductivity of the soil matrix decreases abruptly.

In addition to the forest floor–mineral soil interface, soil conductivity often declines abruptly between the zone of leaching and the zone of illuviation, where soil development has occurred (Fig. 2.1). The interface of the regolith and solid bedrock will also produce an abrupt decline in conductivity. These same zones may also limit vertical root growth, creating lateral roots, and increase the probability of downslope soil pipes. Anywhere pore water moves due to gravity the downslope vector component will be equal to the tangent of the slope angle. Finally, the rate of flow will be proportional to the ratio of potential gradient divided by frictional loss. Except for very steep watersheds, the gradient into the profile will exceed the downslope component. Rapid slope parallel flow will occur in soil pipes when the gradient:friction ratio is less than the gradient:friction ratio in the macropore.

If there is a perennial stream, a water table will exist somewhere between the forest floor and solid bedrock. Flow to the stream will be determined by the potential:friction ratio (usually estimated using the well-known 'Darcy' equation). The ultimate source of streamflow will be water drained as the capillary fringe is lowered and the unsaturated storage reservoir expands. The expansion is called drainable porosity, defined as the pore space that desaturates when the water table drops a small distance. This is represented in Fig. 2.3 as the gray section that asymptotically approaches zero.

When it starts raining, stemflow begins as the interception reservoir begins to spill, and stemflow starts to move in the macropores around the roots. Crown drip and throughfall on the forest floor will fill another interception reservoir that will spill on to the mineral soil surface. The soil pore structure can be categorized as matrix or macropore, with macropore being pores that will drain by gravity within 2 days, a traditional definition of field capacity. In the two water worlds context, the matrix is the 'plant water world.' The distribution of forest floor spill that enters the matrix is described by the Richards equation and is inversely proportional to the soil moisture content. Forest floor flows to macropores occurs when spill from the forest floor exceeds fill into the matrix. If the slope is mild and rooting extends to the impervious bedrock, macropore flow will proceed to the water table after filling the small soil moisture reservoir formed by unsaturated soil surrounding the macropore. This reservoir is represented in Fig. 2.2 as two small reservoirs connected to the main unsaturated soil above and below the accumulation layer in a well-developed soil.

As water reaches the water table, the water potential there increases, causing the capillary fringe to fill unsaturated soil as fillable porosity, which will equal drainable porosity if soil moisture content is at field capacity. If soil moisture potential is less than field capacity, fillable porosity will include the additional water needed to saturate the drier soil. In most cases, fillable porosity will limit water-table rise, and the rate of groundwater flow to the stream will be small, and stormflow (the brown hydrograph in Fig. 2.3) will be in proportion to the rise of the water table. The soil–bedrock interface can be complex, including solution conduits, bedrock cracks and down-slope soil pipes. In addition to matrix groundwater flow, any of these other conduit types may produce rapid flow to the stream. Chang *et al.* (2015) even used pipe flow equations to model conduit flow in a karst system.

As macropore flow continues to raise the water table, soil pipes will begin to fill and provide a low-friction route to the stream. These lower soil pipes will function similarly to epikarst or bedrock cracks but, with more friction, will flow more slowly (the yellow hydrograph in Fig. 2.3). If the soil is young and shallow, matrix conductivity may be relatively uniform from the surface to the bedrock interface. In this case, soil pipes will be distributed in the zone of annual water-table elevations near the bottom of the soil. Pipe activation thresholds for these types of systems occur with accumulations of 20–70 mm of rainfall.

In areas where weathering has progressed to produce a mature soil, matrix flow from the soil surface carries dissolved and colloidal particles deeper into the soil profile, producing a zone of leaching. As evaporation and transpiration are withdrawing water from the soil, the thickness of this leaching zone will depend on the water balance of the site. The bottom of the leaching zone can be thought of as the threshold between the two water worlds. Above the transition, water moves with gravity every time it rains. Below this, gravity becomes less important and capillary forces are more likely to prevail. Colloidal particles are deposited, and oxidation–reduction reactions change the solubility of elements, most notably iron. A zone of accumulation of occurs in this region of the soil. The heterogeneity of this accumulation zone influences flow generation in humid warm temperate, subtropical and tropical climates. The uniform depth, thickness and orientation represented in Fig. 2.1 is only an abstraction of reality.

In areas where the zone of accumulation is thicker and more homogeneous, matrix conductivity will be smaller and connections of the unsaturated reservoir above and below the layer may equilibrate slowly. Flow into the lower unsaturated reservoir may come primarily from macropores that penetrate the accumulation layer. In this case, water may tend to pool at the top of the accumulation layer, and soil pipes may also form above the layer. Both evaporation and transpiration will tend to drain the upper section of unsaturated soil much faster, so this reservoir will tend to have greater deficits and will spill into these upper soil pipes less often (Fig. 2.2). Thresholds of flow from these upper pipes may occur with cumulative excess rain of 70–120 mm. These pipes near the surface may have less flow resistance resulting in larger and more rapid drainage.

When the upper soil unsaturated zone fills to the surface, it spills as saturation-excess overland flow. Saturation generally occurs in limited portions of the watershed and can often be associated with shallow soil pipe exits or along stream riparian zones. Saturated surface flows are visible during the rain, with connectivity being apparent. For the entire profile to become saturated, this flow occurs only after large amounts of excess rain, with thresholds as high as 150–300 mm. Surface channels have much less resistance than subsurface pipes so this flow will produce large rises and rapid declines in the hydrograph (Fig. 2.3).

2.5 Final Thoughts

Forest hydrology has been an applied science, concerned primarily with how various forest management activities change the quantity or quality of water in streams. Forest, as a land use, is generally restricted to regions too wet, dry, cold, hot, steep, rocky or sandy for agriculture. Forest management has also focused on a few commercially important forest species. As the science of hydrology developed, the process of runoff production was observed on lands devoted to infrastructure and agriculture. These processes could be adequately explained with a few mathematical relations. The Manning or Chezy equations could be used for surface flow, the Darcy equation for saturated media and the Richards equation for unsaturated media. Although runoff production from forests was recognized as unique (rain always disappeared into the forest floor) in the 19th century, researchers pointed out that runoff generation from forests could not be explained by theories that were developed for built-up and agricultural land. Yet it was recognized that forests produced water in streams and that the ratio of runoff to rainfall increased with larger storms.

In order to explain runoff generation from forests, data from control watersheds on many paired-watershed experiments were examined. Although examination of hydrograph recession analysis had been established early in the 20th century (Maillet, 1905), recession coefficients were not easily determined on the small watersheds characteristic of paired-watershed experiments. A simpler method was needed to determine runoff from rainfall events, and the straight line separation of base and stormflow was adopted. Unfortunately, that simplification limits analysis to only hydrographs from distinctly separate rainfall events. In the 1960s, calculations were limited to a small number of hydrographs by the effort required. With few data points and continuous distribution analysis, plots of rainfall and runoff could generally be fitted by an exponential model. Such data supported a VSA model with channel expansion and saturation-excess overland flow restricted to a riparian corridor (Hewlett, 1982). Between 1970 and 2000, the runoff process explanation produced a number of hot disputes as researchers found that the standard explanation did not occur in the particular watershed they were studying. Isotope analysis and end-member mixing techniques confirmed the inadequacy of the VSA model depending on channel expansion, and also revealed the importance of climate, soil, geology and vegetation, not only in the quantity of runoff but also in the very runoff mechanism involved. The controversies occurred mostly from limited provincial thinking.

Between 1990 and 2010, it would seem that we discovered most of the processes that produce runoff from forested watersheds. We are even beginning to understand the physical and biological factors that govern them. We now have field and numerical techniques to quantify the runoff production processes even in the unseen subsurface. This chapter proposes that these processes are governed by a balance between flow driven by gravity and water retention by adhesive and cohesive forces. Streamflow depends on two factors: (i) water must be added to the system faster than capillarity moves it into the unsaturated system; and (ii) a saturated path must lead from the incoming rain or the stream. The challenge to forest hydrology will be to understand where and how that happens in the wide variety of forests of today and what we will need in order to preserve a livable earth in the future.

References

Abdul, A.S. and Gillham, R.W. (1989) Field studies of the effects of the capillary fringe on streamflow generation. *Journal of Hydrology* 112, 1–18.

Ali, G., Oswald, C.J., Spence, C., Cammeraat, E.L.K., McGuire, K.J. *et al.* (2013) Towards a unified threshold-based hydrological theory: Necessary components and recurring challenges. *Hydrological Processes* 27, 313–318.

Amatya, D.M., Chescheir, G.M., Williams, T.M., Skaggs, R.W. and Tan, S. (2020) Long-term water table dynamics of forested wetlands: Drivers and their effects on wetland hydrology in the Southeastern Atlantic coastal plain. *Wetlands* 40, 65–79.

Amatya, D.M., Walega, A., Callahan, T.J., Morrison, A., Vulava, V. *et al.* (2022) Storm event analysis of four forested catchments on the Atlantic coastal plain using a modified SCS-CN rainfall-runoff model. *Journal of Hydrology* 608, 127772.

Andreassian, V. (2004) Water and forests: From historical controversy to scientific debate. *Journal of Hydrology* 291, 1–27.

Aubertin, G.M. (1971) *Nature and Extent of Macropores in Forest Soils and their Influence on Subsurface Water Movement.* USDA Forest Research Paper NE-192. USDA Forest Service, Northeastern Forest Experiment Station, Upper Darby, Pennsylvania.

Beasley, R.S. (1976) Contribution of subsurface flow from the upper slopes of forested watersheds to channel flow. *Soil Science Society of America Journal* 40, 955–957.

Beiter, D., Weiler, M. and Blume, T. (2020) Characterizing hillslope-stream connectivity with a joint event analysis of stream and groundwater levels. *Hydrology and Earth System Sciences* 24, 5713–5744.

Bernatek-Jakiel, A. and Poesen, J. (2018) Subsurface erosion by soil piping: Significance and research needs. *Earth-Science Reviews* 185, 1107–1128.

Betson, R.P. (1964) What is watershed runoff? *Journal of Geophysical Research* 69, 1541–1552.

Beven, K.J. and Germann, P.F. (1983) Macropores and water flow in soils. *Water Resources Research* 18, 1311–1325.

Beven, K.J. and Germann, P.F. (2013) Macropores and water flow in soils revisited. *Water Resources Research* 49, 3071–3092.

Bhagyalekshmi, S., Mohan, M., Sreedharan, K. and Ajaykumar, B. (2015) Evaluation of the factors controlling spatial distribution of soil piping: A case study from the Southern Western Ghats, India. *Arab Journal of Geoscience* 8, 8055–8067.

Bishop, K., Seibert, J., Nyberg, L. and Rodhe, A. (2011) Water storage in a till catchment II: Implications of transmissivity feedback for flow paths and turnover times. *Hydrological Processes* 2, 3950–3959.

Bonnel, M. (1993) Progress in the understanding of runoff generation dynamics in forests. *Journal of Hydrology* 150, 217–275.

Branka, T. (2007) How can the epikarst zone influence the karst aquifer hydraulic behavior. *Environmental Geology* 51, 761–765.

Brown, A.E., Zhang, L., McMahon, T.A., Western, A.W. and Vertessy, R.A. (2005) A review of paired catchment studies for determining changes in water yield resulting from alterations in vegetation. *Journal of Hydrology* 310, 28–61.

Buttle, J.M. (1998) Fundamentals of small catchment hydrology. In: Kendall, C. and McDonnell, J.J. (eds) *Isotope Tracers in Catchment Hydrology*, Vol. 38. Elsevier, Amsterdam, The Netherlands, pp. 1–49.

Buttle, J.M. and McDonald, D.J. (2002) Coupled vertical and lateral preferential flow on a forested slope. *Water Resources Research* 38, 18–16.

Chang, Y., Wu, J. and Jiang, G. (2015) Modeling hydrological behavior of a karst spring using a non-linear reservoir–pipe model. *Hydrogeology Journal* 23, 901–914.

Crespo, P., Bucker, A., Feyen, J., Vache, K.B., Frede, H.G. *et al.* (2012) Preliminary evaluation of the runoff processes in a remote montane cloud forest basin using mixing model analysis and mean transit time. *Hydrological Processes* 26, 3896–3910.

Crockford, R.H. and Richardson, D.P. (2000) Partitioning of rainfall into throughfall, stemflow, and interception: Effect of forest type, ground cover, and climate. *Hydrological Processes* 14, 2903–2920.

DeBano, L.F. (2000) The role of fire and soil heating on water repellency in wildland environments: A review. *Journal of Hydrology* 231–232, 195–206.

Detty, J.M. and McGuire, K.J. (2010) Threshold changes in storm runoff generation at a till-mantled headwater catchment. *Water Resources Research* 46, W07525.

Dunne, T., Moore, T.R. and Taylor, C.H. (1975) Recognition and prediction of runoff-producing zones in humid regions. *Hydrological Sciences Bulletin* 20, 305–327.

Dunne, T. (1990) Hydrology, mechanics, and geomorphic implications of erosion by sub-surface flow. In: Higgins, C.G. and Coates, D.R. (eds) *Groundwater Geomorphology: The Role of Subsurface Water in Earth-Surface Processes and Landforms*. Special Paper 252. Geological Society of America, Boulder, Colorado, pp. 1–28.

Dusek, J. and Vogel, T. (2024) Isotopic signatures and soil water partitioning in a humid temperate catchment: Implications for the “two water worlds” hypothesis. *Journal of Hydrology* 632, 130893.

Faulkner, H., Alexander, R. and Zukowsky, R. (2008) Slope-channel coupling between pipes, gullies, and tributary channels in the mocatan catchment badlands, Southeast Spain. *Earth Surface Processes and Landforms* 33, 1242–1260.

Ford, D.C. and Williams, P.W. (2007) *Karst Hydrogeology and Geomorphology*. Wiley, Chichester, UK.

Gabrielli, C.P., McDonnell, J.J. and Jarvis, W.T. (2012) The role of bedrock groundwater in rainfall–runoff response at hillside and catchment scales. *Journal of Hydrology* 450–451, 117–133.

Gao, M., Li, H.Y., Liu, D., Tang, J., Chen, X.Y. *et al.* (2018) Identifying the dominant controls on macropore flow velocity in soils: A meta-analysis. *Journal of Hydrology* 567, 590–604.

Geekiyanage, N., Goodale, U.M., Cao, K. and Kitajima, K. (2019) Plant ecology of tropical and subtropical karst ecosystems. *Biotropica* 51, 626–640.

Green, W.H. and Ampt, G.A. (1911) Studies on soil physics. *Journal of Agricultural Science* 4, 1–24.

Hewlett, J.D. and Hibbert, A.R. (1963) Moisture and energy conditions within a sloping soil mass during drainage. *Journal of Geophysical Research* 68, 1081–1087.

Hewlett, J.D. (1982) *Principles of Forest Hydrology*. University of Georgia Press, Athens, Georgia.

Hewlett, J.D. and Hibbert, A.R. (1967) Factors affecting the response of small watersheds to precipitation in humid areas. In: Sopper, W.E. and Lull, H.W. (eds) *Forest Hydrology – Proceedings of an International Symposium, Pennsylvania State University, Pennsylvania*, Pergamon Press, Oxford, New York, pp. 275–290.

Hibbert, A.R. and Cunningham, G.B. (1967) Streamflow data processing opportunities and application. In: Sopper, W.E. and Lull, H.W. (eds) *Forest Hydrology – Proceedings of an International Symposium Pennsylvania State University, Pennsylvania*, Pergamon Press, Oxford, New York, pp. 725–736.

Higgins, C.G. (1984) Piping and sapping, development of landforms by ground-water outflow. In: LaFleur, R.G. (ed.) *Groundwater as a Geomorphic Agent*. Routledge, London, pp. 18–58.

Horton, R.E. (1933) The role of infiltration in the hydrologic cycle. *Eos* 14, 446–460.

Howard, A.D. (1982) Equilibrium and time scales in geomorphology: Application to sand-bed alluvial streams. *Earth Surface Processes and Landforms* 7, 303–325.

Howard, A.D., Kochel, R.C. and Holt, H.E. (1988) *Sapping Features of the Colorado Plateau*. NASA Special Publications 491. Scientific and Technical Information Office, National Aeronautics and Space Administration, Washington, DC.

Hursh, C.R. (1944) Appendix B: Report of sub-committee on subsurface flow. *Transactions of the American Geophysical Union* 25, 743–746.

Hursh, C.R. and Brater, E.F. (1941) Separating storm-hydrographs from small drainage-areas into surface- and subsurface-flow. *Eos* 22, 863–871.

Iida, S., Levia, D.F., Shimiza, A., Shimiza, T., Tamai, K. *et al.* (2017) Intrastorm rainfall interception dynamics in a mature coniferous forest. *Journal of Hydrology* 343, 770–783.

Jackson, C.R., Bitew, M. and Du, F. (2014) When interflow also percolates: Downslope travel distances and hillslope process zones. *Hydrological Processes* 28, 3195–3200.

Jones, J.A. (2010) Soil piping and catchment response. *Hydrological Processes* 24, 1548–1566.

Katsura, S., Kosugi, K., Yamakawa, Y. and Mizuyama, T. (2014) Field evidence of groundwater ridging in a slope of a granite watershed without the capillary fringe effect. *Journal of Hydrology* 511, 703–718.

Kendall, K.A., Stanley, J.B. and McDonnell, J.J. (1999) A hydrometric and geochemical approach to test the transmissivity feedback hypothesis during snowmelt. *Journal of Hydrology* 219, 188–205.

Kim, H.J., Sidle, R.C. and Moore, R.D. (2005) Shallow lateral flow from a forested hillslope: Influence of antecedent wetness. *Catena* 60, 293–306.

Lang, A.J., Aust, W.M., Bolding, M.C., Barrett, S.M., McGuire, K.J. *et al.* (2015) Streamside management zones compromised by stream crossings, legacy gullies, and over-harvest in the piedmont. *Journal of the American Water Resources Association* 51, 1153–1164.

Levia, D.F. and Germer, S. (2015) A review of stemflow generation dynamics and stemflow–environment interactions in forests and shrublands. *Review of Geophysics* 53, 673–714.

Luo, W. (2000) Quantifying groundwater-sapping landforms with a hypsometric technique. *Journal of Geophysical Research: Planets* 105, 1685–1694.

Maillet, E. (1905) *Essai d'Hydraulique Souterraine et Fluviale*. Librairie Scientifique. Hermann, Paris.

McDonnell, J.J. (2013) Are all runoff processes the same. *Hydrological Processes* 27, 4103–4111.

McDonnell, J.J. (2014) The two water worlds hypothesis: Ecohydrological separation of water between streams and trees. *Water* 1, 323–329.

McDonnell, J.J., Owens, I.F. and Stewart, M.K. (1991) A case study of shallow flow paths in a steep zero-order basin. *Journal of the American Water Resources Association* 27, 679–685.

McDonnell, J.J., Spence, C., Karran, D.J., Meerveld, H.J. and Harman, C.J. (2021) Fill-and-spill: A process description of runoff generation at the scale of the beholder. *Water Resources Research* 57, e2020WR027514.

McGlynn, B.L., McDonnell, J.J. and Brammer, D.D. (2002) A review of the evolving perceptual model of hillslope flow paths at the Maimai catchments, New Zealand. *Journal of Hydrology* 257, 1–26.

McMillan, H. (2022) A taxonomy of hydrological processes and watershed function. *Hydrological Processes* 36, e14537.

Micallef, A., Marchis, R., Saadatkhah, N., Pondthai, P., Everett, M.E. *et al.* (2021) Groundwater erosion of coastal gullies along the Canterbury coast (New Zealand): A rapid and episodic process controlled by rainfall intensity and substrate variability. *Earth Surface Dynamics* 9, 1–18.

Migon, P., Duszynski, F., Jancewicz, K., Kotowski, M. and Porebna, W. (2023) Subsurface–subsurface connectivity and the morphological evolution of sandstone-capped tabular hills – how much analogy to karst? *Geomorphology* 440, 108884.

Miyata, S., Gomi, T., Sidle, R.C., Hiraoka, M., Onda, Y. *et al.* (2019) Assessing spatially distributed infiltration capacity to evaluate storm runoff in forested catchments: Implications for hydrological connectivity. *The Science of the Total Environment* 669, 148–159. DOI: 10.1016/j.scitotenv.2019.02.453.

Mosley, M.P. (1979) Streamflow generation in a forested watershed. *New Zealand Journal of Hydrology* 15, 795–806.

Neary, D., Hayes, D., Rustad, L., Vose, J., Gottfried, G. *et al.* (2012) US forest service experimental forests and ranges network: A continental research platform for catchment-scale research. In: Webb, A.A., Bonell, M., Bren, L., Lane, P.N.J. and McGuire, D. (eds) *Revisiting Experimental Catchment Studies in Forest Hydrology. Proceedings of a Workshop held during the XXV IUGG General Assembly in Melbourne, June–July 2011*, IAHS Publication 353. IAHS, Wallingford, UK, pp. 49–57.

Negishi, J.N., Nogushi, S., Sidle, R.C., Ziegler, A.D. and Nik, A.R. (2007) Stormflow generation involving pipe flow in a zero-order basin of Peninsular Malaysia. *Hydrological Processes* 21, 789–806.

Nippgen, F., McGlynn, B.L., Emanuel, R.E. and Vose, J.M. (2016) Watershed memory at the Coweeta Hydrologic Laboratory: The effect of past precipitation and storage on hydrologic response. *Water Resources Research* 52, 1673–1695.

Nunes, L.J.R., Meireles, C.I.R., Pinto Gomes, C.J. and Almeida Ribeiro, N.M.C. (2020) Forest contribution to climate change mitigation: Management oriented to carbon capture and storage. *Climate* 8, 21.

Pearce, A.J., Stewart, M.K. and Sklash, M.G. (1986) Storm runoff generation in humid headwater catchments: 1. Where does the water come from? *Water Resources Research* 22, 1263–1272.

Pineda, N. and Rigo, T. (2017) The rainfall factor in lightning-ignited wildfires in Catalonia. *Agricultural and Forest Meteorology* 239, 249–263.

Pressland, A.J. (1973) Rainfall partitioning by an arid woodland (*Acacia anuera* f. Muell.) in South-Western Queensland. *Australian Journal of Botany* 21, 235–245.

Redhaounia, B., Aktatakci, H., Ilondo, B.O., Gabtni, H., Khomsi, S. *et al.* (2015) Hydro-geophysical interpretation of fractured and karstified limestone reservoirs: A case study from Amdoun region (NW Tunisia) using electrical resistivity tomography, digital elevation model (DEM) and hydrogeochemical approaches. *Journal of African Earth Sciences* 112, 328–338.

Reinhart, K.G. (1967) Watershed calibration methods. In: Sopper, W.E. and Lull, H.W. (eds) *Forest Hydrology – Proceedings of an International Symposium, Pennsylvania State University, Pennsylvania*, Pergamon Press, Oxford, New York, pp. 715–723.

Richards, L.A. (1931) Capillary conduction of liquids in porous mediums. *Physics* 1, 318–333.

Richardson, P.W., Cafferata, P.H., Dymond, S.F., Keppeler, E.T., Wagenbrenner, J.W. *et al.* (2023) Past and future roles of paired watersheds: A North American inventory and anecdotes from the Caspar Creek Experimental Watersheds. *Frontiers in Forest and Global Change* 6, 1275392.

Ross, C.A., Ali, G., Spence, C. and Courchesne, F. (2021) Evaluating the ubiquity of thresholds in rainfall-runoff response across contrasting environments. *Water Resources Research* 57, e2020WR027498.

Roth, B.E., Slatton, K.C. and Cohen, M.J. (2007) On the potential for high-resolution lidar to improve rainfall interception estimates in forest ecosystems. *Frontiers in Ecology and the Environment* 5, 421–428.

Sayer, A.M., Walsh, R.P.D. and Bidin, K. (2006) Pipe flow suspended sediment dynamics and their contribution to sediment budgets in small rainforest catchments, Sabah, Malaysia. *Forest Ecology and Management* 224, 119–130.

Scaife, C.I. and Band, L.E. (2017) Nonstationarity in threshold response of stormflow in Southern Appalachian headwater catchments. *Water Resources Research* 53, 6579–6596.

Schumm, S.A. and Rea, D.K. (1995) Sediment yield from disturbed earth systems. *Geology* 23, 391–394.

Sidle, R.C., Hirano, T., Gomi, T. and Terajima, T. (2007) Hortonian overland flow from Japanese forest plantations – An aberration, the real thing, or something in between? *Hydrological Processes: An International Journal* 21, 3237–3247.

Skaggs, R.W., Amatya, D.M., Chescheir, G.M. and Blanton, C.D. (2006) *Soil Property Changes during Loblolly Pine Production*. ASABE Paper No. 068206. American Society of Agricultural and Biological Engineers, St. Joseph, Michigan.

Skaggs, R.W., Chescheir, G.M., Fernandez, G.P., Amatya, D.M. and Diggs, J. (2011) Effects of land use on soil properties and hydrology of drained coastal plain watersheds. *Transactions of the ASABE* 54, 1357–1365.

Sklash, M.G., Stewart, M.K. and Pearce, A.J. (1986) Storm runoff generation in humid headwater catchments 2. A case study of hillslope and low order stream response. *Water Resources Research* 2, 1723–1282.

Spence, C. and Woo, M.K. (2003) Hydrology of subarctic Canadian shield: Soil-filled valleys. *Journal of Hydrology* 279, 151–166.

Sun, G., Domec, J.C. and Amatya, D.A. (2016) Forest evapotranspiration: Measurement and modeling at multiple scales. In: Amatya, D.M., Williams, T.M., Bren, L. and Jong, C. (eds) *Forest Hydrology: Processes, Management, and Applications*. CABI International, Wallingford, UK, pp. 32–90.

Tang, Q., Duncan, J.M., Guo, L., Lin, H., Xiao, D.M. *et al*. (2020) On controls of preferential flows in soils of differing hillslope position and lithological origin. *Hydrological Processes* 34, 4295–4305.

Terajima, T. and Moriizumi, M. (2013) Temporal and spatial changes in dissolved organic carbon concentration and fluorescence intensity of fulvic acid like materials in mountainous headwater catchments. *Journal of Hydrology* 478, 1–12.

Timmons, D.R., Verry, E.S., Burwell, R.E. and Holt, R.F. (1977) Nutrient transport in surface runoff and interflow from an aspen-birch forest. *Journal of Environmental Quality* 6, 188–192.

Toms, J.D. and Lesperance, M.L. (2003) Piecewise regression: A tool for identifying ecological thresholds. *Ecology* 84, 2034–2041.

Torres, R., Dietrich, W.E., Montgomery, D.R., Anderson, S.P. and Loague, K. (1998) Unsaturated zone processes and the hydrologic response of a steep, unchanneled catchment. *Water Resources Research* 34, 1865–1879.

Tromp-van Meerveld, H.J. and McDonnell, J.J. (2006) Threshold relations in subsurface stormflow: 1. A 147-storm analysis of the panola hillslope. *Water Resources Research* 42, W02410.

Weary, D.J. and Doctor, D.H. (2014) *Karst in the United States. A Digital Map Compilation and Database.* USGS Open File Report No.2014-1156. US Geological Survey, Reston, Virginia.

Weiler, M. and McDonnell, J.J. (2007) Conceptualizing lateral preferential flow and flow networks and simulating the effects on gauged and ungauged hillslopes. *Water Resources Research* 43, W03403.

Wenninger, J., Uhlenbrook, S., Tilch, N. and Libengut, C. (2004) Experimental evidence of fast groundwater responses in a hillslope/floodplain area in the Black Forest Mountains, Germany. *Hydrological Processes* 18, 5305–3322.

Williams, P.W. (2008) The role of epikarst in karst and cave hydrogeology: A review. *International Journal of Speleology* 37, 1–10.

Williams, T.M. (2016) Forest runoff processes. In: Amatya, D.M., Williams, T.M., Bren, L. and Jong, C. (eds) *Forest Hydrology: Processes, Management, and Applications*. CABI International, Wallingford, UK, pp. 17–31.

Zhongcheng, J., Zhenwei, L.I., Weiqun, L., Funing, L. and Zeyan, W.U. (2024) Research progress and prospect of soil and water leakage in karst areas of Southwest China. *Science of Soil and Water Conservation* 22, 1–11.

Zimmermann, B., Zimmermann, A., Turner, B.L. and Francke, E.H. (2014) Connectivity of overland flow by drainage network expansion in a rain forest catchment. *Water Resources Research* 50, 1457–1473.

3 Forest Evapotranspiration: Processes and Quantification

G. Sun[1]*, J.-C. Domec[2,3] and D.M. Amatya[4]

[1]Eastern Forest Environmental Threat Assessment Center, USDA Forest Service, Research Triangle Park, North Carolina, USA; [2]Bordeaux Sciences Agro, Gradignan Cedex, France; [3]Nicholas School of the Environment, Duke University, Durham, North Carolina, USA; [4]Santee Experimental Forest, Southern Research Station, USDA Forest Service, Cordesville, South Carolina, USA

Abstract

Next to precipitation, evapotranspiration (ET) is the second largest flux in forest hydrological cycles. Forest ET rates are generally higher than other vegetated ecosystems under the same climatic conditions. ET serves as the nexus of the carbon, water and energy balances in forests. Understanding ET dynamics is critical for quantifying forest–water relationships and ecosystem services (e.g. carbon sequestration, climate and streamflow flow moderation, biodiversity) at multiple scales. This chapter describes the basic ET processes, which include forest canopy and litter interception, tree transpiration and soil evaporation. Finally, practical methods for measuring and modeling ET at tree, watershed and regional levels are presented to show the challenges in accurately quantifying ET under a changing environment.

3.1 Introduction

Forest evapotranspiration (ET) refers to the sum of evaporation from the soil, canopy and litter interception and plant transpiration, and is critical to understanding of effects of forest change (Bosch and Hewlett, 1982) and the energy, water and biogeochemical cycles in forests (Yi *et al.*, 2024). Compared with traditional hydrology, the science of forest hydrology is rooted in understanding the 'forest influences' on microclimate and has a relatively long history of studying the effects of forest vegetation change in regulating the water cycles through ET (Sun *et al.*, 2016). The importance of ET in understanding forest ecohydrological processes, quantifying ecosystem services and developing 'nature-based solutions' for climate change mitigation and adaptation has been increasingly recognized (Hao *et al.*, 2023; Yang *et al.*, 2023).

The linkage among energy, water and carbon balances at a forest stand level over a long time period (Fig. 3.1), in which ET plays a key role, can be described conceptually in the following interlinked formulae (Sun *et al.*, 2011a):

Water balance:

$$P = ET + Q \pm \Delta S \tag{3.1}$$

where ET = litter and canopy interception + evaporation from the soil + plant transpiration.

Energy balance:

*Corresponding author: ge.sun@usda.gov

 Forest Hydrology: Processes, Management and Assessment (eds Leon J. Bren *et al.*)
DOI: 10.1079/9781800625310.0003

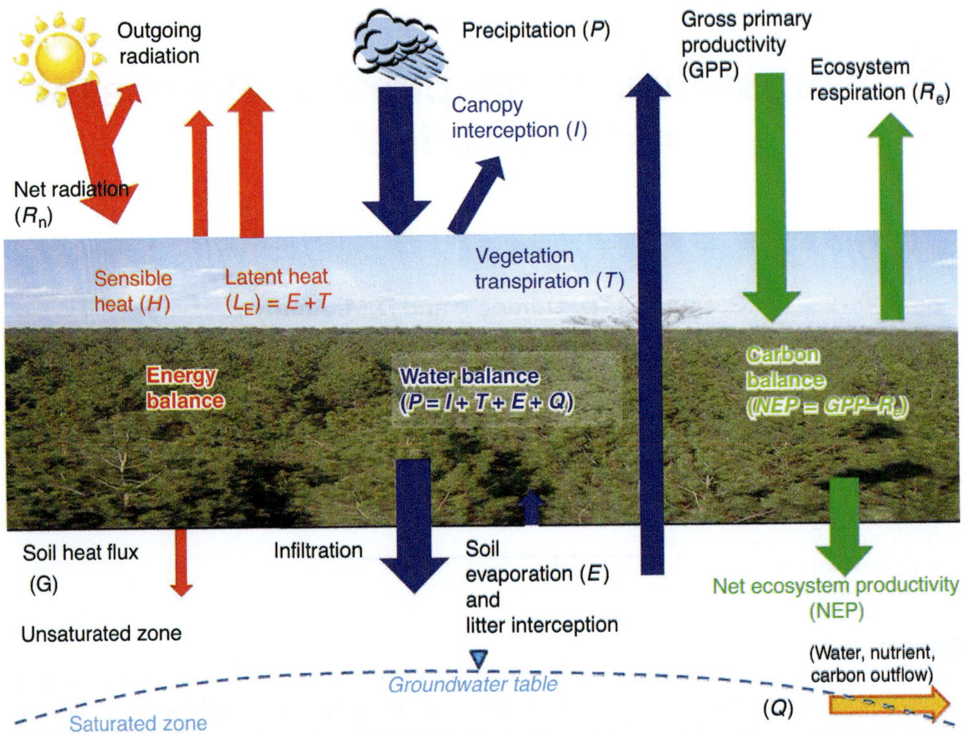

Fig. 3.1. Linkages among energy, water and carbon cycles in a forest ecosystem on the lower coastal plain of North Carolina in the USA. Note that net radiation (R_n) is a result of total incoming minus reflected shortwave radiation, along with the absorbed minus emitted longwave radiations.

$$R_n = L_E + H = ET \times L + H \quad (3.2)$$

Carbon balance:

$$\begin{aligned} NEP &= GPP - R_e - L_c \\ &= E \times WUE - R_e - L_c \end{aligned} \quad (3.3)$$

In the above, P is precipitation (mm), Q is runoff (mm), E is given as mm or kg or m^3 of water/m^2, ΔS is change in soil water storage (mm), R_n is net radiation (W/m^2; 1 W = 1 Joule per second, J/s), L_E is latent heat (W/m^2) representing the energy needed to evaporate the amount of water, assuming a constant conversion factor called latent heat of vaporization of water ($L = 2.256 \times 10^6$ J/kg H_2O) and H is sensible heat (W/m^2) representing the loss of energy to heat the air within/near the forest canopies. The net ecosystem productivity (*NEP*, g carbon (C)/m^2) is the carbon balance between carbon gain by gross ecosystem productivity (i.e. plant photosynthesis) and carbon loss by ecosystem respiration (R_e, g C/m^2) and by lateral stream runoff (L_c, g C/m^2). The magnitude of both gross primary productivity (*GPP*, g C m^2) and R_e are much larger than *NEP* and L_c, and all four variables are influenced by stand soil moisture and the hydrology. ET explains most of the seasonal variability of GPP for all ecosystems (Sun *et al.*, 2011a). The ratio of GPP:ET is termed water-use efficiency (WUE) and has been used as an important variable to understand water–carbon coupling and for quantifying the trade-offs between water yield and carbon sequestration (Zhang *et al.*, 2023).

ET is a key variable linking meteorology (Baldocchi *et al.*, 2001), hydrology (Asbjornsen *et al.*, 2011; Ukkola and Prentice, 2013) and ecosystem sciences (Sun *et al.*, 2017). ET is the only variable that links hydrology and biological processes in many ecosystem models. ET is also highly linked to ecosystem productivity (Beer *et al.*, 2007) and net ecosystem exchange of CO_2

because both photosynthesis and ecosystem respiration are controlled by soil water availability (Jackson *et al.*, 2005; Sun *et al.*, 2011a). Most of the water lost through tree transpiration is used to cool leaves when air temperature increases (Taborski *et al.*, 2022), and thus the ET processes are closely linked to energy partitioning and micro- and macro-climate systems (Fig. 3.1) (Bonan, 2008) and climate change (Jackson *et al.*, 2008). Global climate change, in turn, affects the local forest water resources directly (e.g. precipitation and runoff forms and patterns) (Amatya *et al.*, 2024) or indirectly (e.g. ecosystem structure) by affecting all ET components (Duarte *et al.*, 2024). ET has long been regarded as an index to represent the available environmental water and energy resources and ecosystem productivity by bioclimatologists (Stephenson, 1998). Thus, ET has been used to explain the large regional variations in plant and animal distribution and species richness (Currie and Paquin, 1987; Currie, 1991; Stephenson, 1998; Hawkins *et al.*, 2003).

3.2 ET Processes

Both physiological (e.g. stomata control) and physical (e.g. water potential control) processes influence water movements from the plant organs of roots, xylem, and leaves to stands (Domec *et al.*, 2012a) and landscapes (i.e. watersheds). Soil evaporation can be a large part of the total ET for young plantation forests or forests with sparse canopies (Younger *et al.*, 2023). Soil evaporation is minor in closed canopy forests (Domec *et al.*, 2012b) so this chapter focuses on the processes that control canopy and litter interception and transpiration. Partitioning ET to soil evaporation, canopy and litter interception and transpiration is important to understanding the ecosystem functional response such as carbon sequestration to climate change and how plants adapt to novel soil moisture and atmospheric regimes (Wang and Dickinson, 2012; Wei *et al.*, 2017; Stoy *et al.*, 2019).

3.2.1 Canopy and litter interception

There is a long history of studying forest canopy interception in forest hydrology (Fischer *et al.*, 2023). The quantity of canopy and litter interception in forests can be a large component of the ET and water balances, depending on forest stand age (Sadeghi *et al.*, 2024), structure characteristics such as leaf area index (LAI), canopy holding capacity, and the amount of litter and litter water holding capacity (Deguchi *et al.*, 2006), as well as soil moisture. In addition, the frequency of storms and the drying and wetting cycles affect total canopy and litter interception (Lian *et al.*, 2022).

The earliest studies by Horton (1919) showed that forest interception rates are highly variable between and across species, with the spruce/fir/hemlock forest type the highest, followed by pines and then hardwoods. Helvey (1971) reported annual canopy interception of 17% for red pine (*Pinus resinosa* Ait.), 16% for ponderosa pine (*Pinus ponderosa* Dougl. ex. Laws.), 19% for eastern white pine and 28% for the spruce/fir/hemlock forest type. Summer rates of deciduous forests in the south-eastern USA ranged from 8% to 33%, with a mean of 17%, and winter rates ranged from 5% to 22%, with a mean of 12% (Helvey and Patric, 1965). Annual canopy interception rates were 18% for wetland sites, 20% for hardwood sites and a longleaf pine (*Pinus palustus* Mill.) plantation, and 23% for pine-dominated forests in the south-eastern USA (Bryant *et al.*, 2005). Interception rates vary from 10% to 35% and from 5% to 25% for unthinned and thinned loblolly pine stands, respectively (Gavazzi *et al.*, 2016). In the USA, reported annual litter interception rate values for eastern forests varied by about 2–5%, generally less than 50 mm per year (Helvey and Patric, 1965). However, litter interception may be higher than canopy interception in other forest ecosystems (Gerrits *et al.*, 2007). The mean canopy and litter interception interception rates for pine forests decrease as precipitation increases at both event and longer timescale (Fig. 3.2) (Gavazzi *et al.*, 2016).

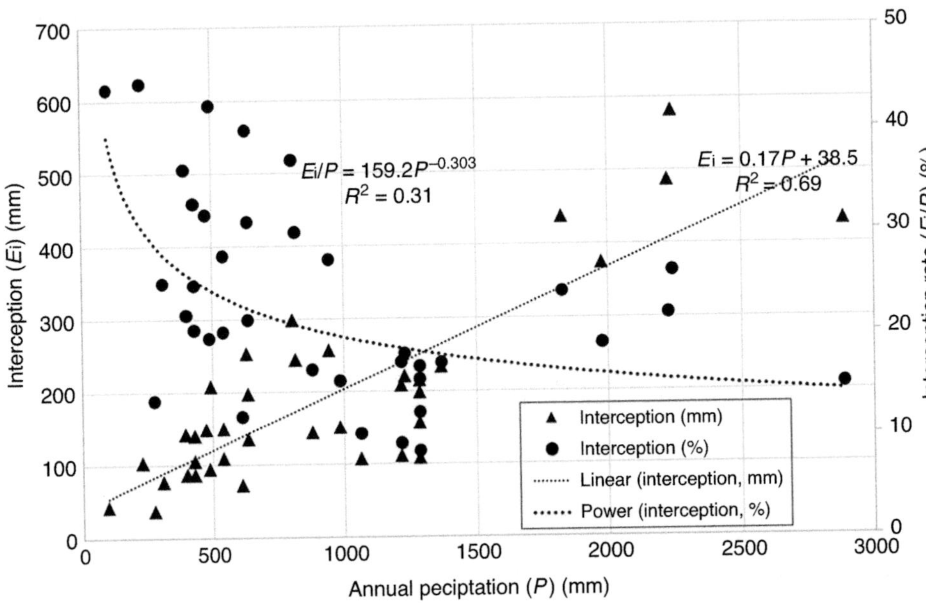

Fig. 3.2. Large variability in canopy interception rates among pine forests. In general, annual accumulative pine forest interception rates decrease with an increase in precipitation. Each dot represents one global study site.

3.2.2 Transpiration

The transpiration process represents water uptake from the root zone to its loss through the leaf stomata, the tiny openings found on one or both sides of the tree leaves in the canopies (Canny, 1998). In most plants, transpiration is controlled largely by the vapor pressure deficit of the atmosphere and soil moisture. Because transpiration is an inevitable consequence of CO_2 assimilation by plants through photosynthesis and maintaining leaf tissue turgidity and water uptake, together with soil evaporation, transpiration represents an ecosystem water loss and thus is a 'necessary evil' (Curtis, 1926) that benefits plants from cooling leaf surfaces to nutrient transport from the root zone to the top of the canopies. In fact, less than 5% of the water taken by up roots is used to assimilate CO_2 and used for growth of the trees, with the remaining 95% returning to the atmosphere through the stomata (Steudle, 2002). A global synthesis indicated that transpiration accounts for 61 ± 15% of ET, and returns approximately 39 ± 10% of incident precipitation to the atmosphere, playing an important role in the global water cycle (Schlesinger and Jasechko, 2014), which is consistent with a study by Good *et al.* (2015) who reported that globally transpiration is 64 ± 13% (mean ± 1 standard deviation) of evapotranspiration. The transpiration:ET ratios are highest in tropical rainforests (70 ± 14%) and lowest in steppes, shrub lands and deserts (51 ± 15%). Thus, biophysical parameters such as stomatal conductance and leaf area that control transpiration have important implications for total ET (Jasechko *et al.*, 2013). Therefore, changes in transpiration due to increasing CO_2 concentrations, land-use changes, shifting ecozones, air pollution, climate warming and drought may have significant impacts on ET and water balances (Schlesinger and Jasechko, 2014).

Transpiration is an active water translocation process that occurs only when water exists continuously along the soil–root–stem–branch–leaf–stomata flow pathways (Kumagai, 2011). However, transpiration rates differ tremendously among different tree species and ages (Fig. 3.3). For example, *Quercus rubra* trees

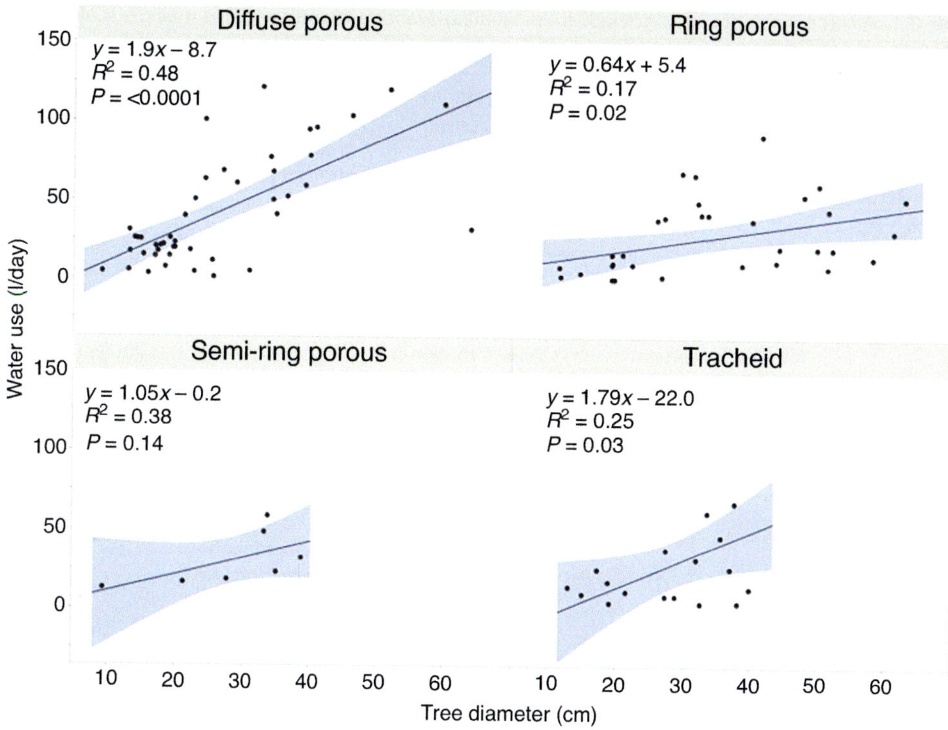

Fig. 3.3. Measured growing season daily mean sapflow rates (tree transpiration) grouped by xylem anatomy and tree diameter at breast height for major species in the USA. The shaded area depictsthe 95% confidence intervals for the trend. Diffuse porous species: red maple (*Acer rubrum*), sugar maple (*Acer saccharum*), birch (*Betula* spp.), hackberry (*Celtis laevigata*), dogwood (*Cornus florida*), American beech (*Fagus grandifolia*), European beech (*Fagus sylvatica*), sweetgum (*Liquidambar styraciflua*), tulip poplar (*Liriodendron tulipifera*), blackgum (*Nyssa sylvatica*), hop hornbeam (*Ostrya virginiana*), sycamore (*Platanus occidentalis*), American basswood (*Tilia americana*) and catalpa (*Catalpa speciosa*); ring porose species: green ash (*Fraxinus pennsylvanica*) and oak (*Quercus* spp.); semi- porous species: hickory (*Carya* spp.), aspen (*Populus* spp.) and black cherry (*Prunus serotina*); tracheid-bearing species: balsam fir (*Abies balsamea*), oneseed juniper (*Juniperus monosperma*), white spruce (*Picea glauca*), pine (*Pinus* spp.), northern white cedar (*Thuja occidentalis*) and eastern hemlock (*Tsuga canadensis*).

with a 50 cm diameter transpire an average of 30 kg/day, but *Betula lenta* can transpire as much as 110 kg/day under the same climate in the southern Appalachians of the eastern USA (Vose *et al.*, 2011). A review of 52 whole-tree water-use studies for 67 tree species worldwide using different techniques concluded that maximum daily water-use rates for trees averaging 21 m in height were within 10–200 kg/day (Wullschleger *et al.*, 1998).

Tree transpiration rates are controlled by numerous biophysical factors such as microclimate, atmospheric CO_2 concentration, soil water potential, stand characteristics (e.g. leaf area, sapwood area, species compositions, tree density) and hydraulic transport properties of plant tissues (Table 3.1) (Domec *et al.*, 2010, 2012a, b). The species composition changes over space and time due to natural regeneration or the response to climatic change or human activities such as silviculture (i.e. reforestation, afforestation). In addition, the forest ecosystem structure may change dramatically over time in both above- and belowground characteristics including leaf (i.e. leaf biomass, LAI, morphology), stem (i.e. sapwood area) (Domec *et al.*, 2012a; Komatsu and Kume, 2015), and belowground factors (i.e. root biomass).

Table 3.1. A comparison of major methods for estimating evapotranspiration (ET) at multiple scales

	Method	Strength	Weakness	Source
Direct field-based	Porometer and cuvette	Leaf level physiological process	Difficult to scale up due to uncertainty of the influence of boundary layers and variability of leaf age, radiation and humidity	Olbrich (1991)
	Weighing lysimeter	Single whole-tree water use	High maintenance cost for small trees	Müller and Bolte (2009)
	Heat balance/heat dissipation sapflow	Allows routine unsupervised measurement accurately at single plant scale	Large-scale measurement errors are determined by the sample size and variability of samples	Granier (1987)
	Eddy covariance	Measuring fluxes continuously, offering high temporal resolution data	High cost of instrumentation, gap filling required, energy imbalance problems	Baldocchi *et al.* (2001)
	Bowen ratio	Low cost; works for both crops and natural vegetation	Relies on several assumptions; errors associated with low gradients	Vogt and Jaeger (1990)
	Surface renewal	Low cost; works for both crops and natural vegetation, including forests; can be applied close to the canopy surface	Requires extensive fetch requirements	Fischer *et al.* (2023)
	Catchment water balance	Easy to measure; low cost	Only long-term average is reliable	Ukkola and Prentice (2013)
Remote sensing	MODIS, OpenET	Provides spatial, continuous and temporal scale high-resolution data	Uncertainties due to errors generated by measurement of sparse canopies; data mostly from clear-sky conditions	Melton *et al.* (2022)
Mathematical modeling for ET alone or the full hydrological cycles	Theoretical models (e.g. Penman–Monteith equation)	Widely tested, including all conditions; low cost	Requires site-specific parameters; not easy to apply to data-poor regions	McMahon *et al.* (2013)

Continued

Table 3.1. Continued

	Method	Strength	Weakness	Source
Isotopes	Stable isotopes of hydrogen (H) and oxygen (O)	Process-based understanding of water source of ET; partitioning evaporation and transpiration	Cost and scaling up to stand level	Good *et al.* (2015)

Forest canopies have a complex vertical structure, and the understorey vegetation may be an important component of a forest stand by intercepting and transpiring a significant amount of water. For example, over 20% of the total ET for a 17-year-old pine plantation was from understories (Domec *et al.*, 2012b) and as much as one-third of the total ET during the growing season in temperate and boreal forests (Balandier *et al.*, 2022). Emergent understorey vegetation soon after harvest in the humid coastal plain was shown to have substantial LAI, potentially affecting water balance for 4–5 years until the planted pine seedlings dominated the understorey (Sampson *et al.*, 2011).

3.2.3 Hydraulic redistribution by roots: exchange of water at the soil–root interface

Plants can reduce water stress by extracting water from deeper and moister soil layers through their roots and storing it in the upper, drier soil layers for use by shallow roots. The bidirectional (upward and downward) process is termed 'hydraulic redistribution' (HR) (Burgess *et al.*, 1998). The HR process occurs widely in all water-limited vegetated environments (Neumann and Cardon, 2012). HR is a passive process that depends on the soil suction head and the root distribution within the soil column. HR by roots acts as a large water capacitor, increasing the efficiency of whole-plant water transport, buffering the seasonality of ET against water stress during seasonal water deficits, and representing 20–40% of whole-stand water use (Domec *et al.*, 2010; Asensio *et al.*, 2020). Even when HR only represents a relatively small amount of ecosystem water use (<0.5 mm/day) and just a fraction (5–10%) of total ET during dry period, the partial recharge of upper soil moisture by HR is important to slow down the decline of soil water content and thus maintain water availability in the top soil horizons (Warren *et al.*, 2007). The influx of soil water maintains root water uptake capacity and extends root functioning later into the drought period, influencing forest productivity (Domec *et al.*, 2010).

3.2.4 Total ET

The ET rates at the ecosystem, watershed or regional level are controlled mainly by energy and water availability (Zhang *et al.*, 2001; Aguilos *et al.*, 2024), but are also influenced by other management factors such as chemical fertilization, CO_2 'fertilization effects' and nitrogen deposition tree genetic improvement, species conversion, artificial drainage and irrigation (Tian *et al.*, 2012). During the forest stand development, site-level energy and water availability also vary, resulting in dramatic seasonal changes in total ET and the partitioning of ET. For example, ET of a loblolly pine plantation stand on the lower coastal plain of the southeastern USA increased dramatically prior to age 10 but stabilized around years 10–15 (Fig. 3.4).

Forested watershed ET generally decreases and water yield increases soon after tree removal by harvesting (Bosch and Hewlett, 1982). Other natural disturbances such as hurricanes (Jayakaran *et al.*, 2014), wildfires (Boden *et al.*, 2023; Mankin and Patel, 2023), invasive species, defoliation due to insects and diseases (Sidhu *et al.*, 2024), wind and snowstorms, and urbanization also alter ET as a result of reduced canopy interception and transpiration and soil disturbances (Diem *et al.*, 2023). However, the hydrological responses vary depending on season and local conditions, and ET generally tends to increase soon after afforestation or reforestation and after natural

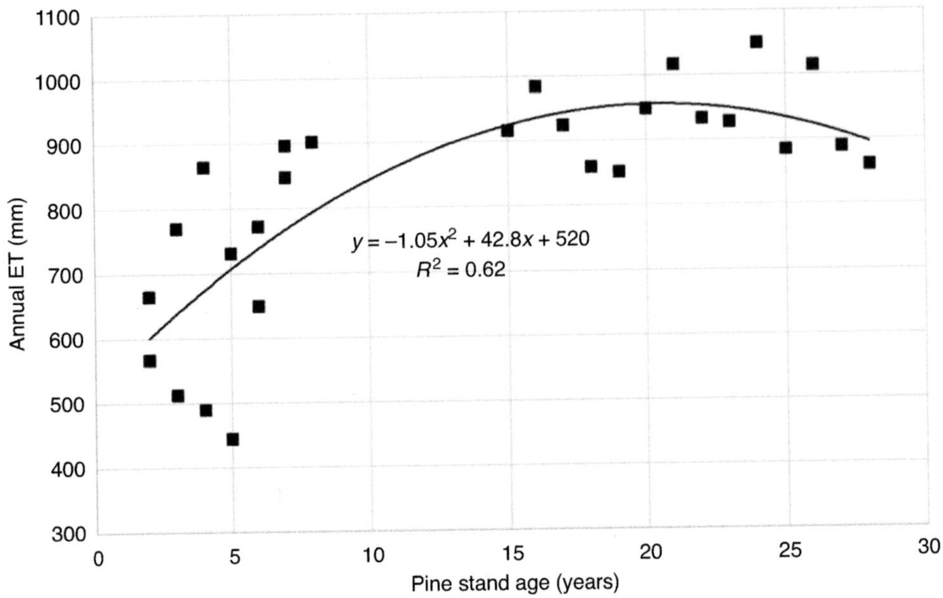

Fig. 3.4. A general curve showing that evapotranspiration (ET) rates of a loblolly pine (*Pinus taeda*) plantation increase with stand age, especially from seedlings to 15 years old. The ET data (2005–2018) were collected using an eddy flux method for stands with multiple ages in the lower coastal plain of North Carolina in the south-eastern USA (Domec *et al.*, 2012b; Aguilos *et al.*, 2024).

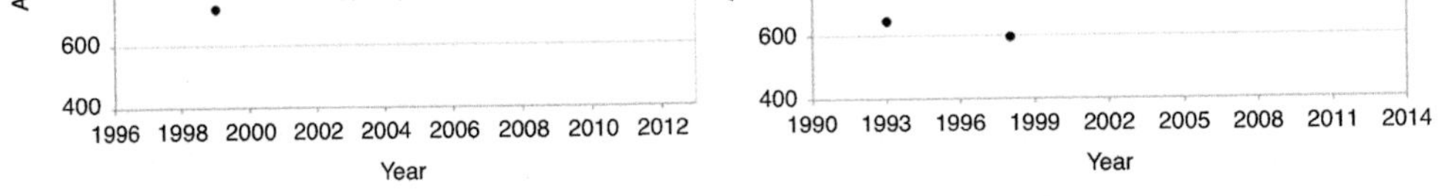

Fig. 3.5. Annual watershed forest ecosystem evapotranspiration (ET) calculated as the differences between measured precipitation and measured streamflow for an experimental watershed recovering gradually after disturbances: (a) following harvesting in 1995 and replanting with loblolly pine in 1997 in Carteret County, coastal North Carolina, USA, and (b) a mixed forest naturally regenerated after the impact of Hurricane Hugo in 1989 at Santee Experimental Forest in coastal South Carolina, USA.

regeneration from disturbances. Such changes in ET are often estimated as the differences between measured precipitation and streamflow runoff, assuming the change in soil water storage is negligible (Fig. 3.5). The inter-annual variability of ET is a result of variability of both the precipitation and potential ET (Ukkola and Prentice, 2013). Direct ET measurements using the eddy covariance over a full stand rotation (Aguilos *et al.*, 2024) and remote sensing-based estimates (Yang *et al.*, 2017) provide insights of the ET dynamics over space and time.

Forest ET rates also vary dramatically across a heterogeneous terrain with significant topographical gradients that modify natural water and energy availability (Boggs *et al.*, 2021). For example, ET rates of a forest stand are higher on the sunny side and/or near the ridges on a mountain watershed due to more available solar radiation (Emanuel *et al.*, 2010). Forest thinning practices reduce forest biomass, and thus canopy interception and transpiration from the remaining trees, but do not necessarily reduce total ET until the thinning treatment exceeds a threshold (del Campo *et al.*, 2022). The long-term effects of forest disturbances on ET and water yield by wildland fires and insects and diseases appear to be rather variable in the western USA (Goeking and Tarboton, 2022).

3.3 Direct ET Measurement

Forest ET processes can be quantified at multiple temporal (from subhourly to annual) and spatial scales from leaf to watershed, and even to a global scale using various methods from hand-held cuvette to remote sensing (Table 3.1). Because canopy and litter interception are important components of total forest ET, throughfall measurements under forest canopies are essential. For tree transpiration, the porometer method has been used to measure leaf water use for understanding the environmental control of gas exchange (CO_2 and H_2O) at the leaf level (Olbrich, 1991). Other methods to quantity tree transpiration include ventilated chambers (Denmead *et al.*, 1993), large lysimeters (Müller, 2009), sap flux density based on thermal dissipation and heat transport theories (Granier *et al.*, 1996).

The sapflow technique has the advantage of being economical, easy to install, and applicable to forests with multiple tree species and on heterogenous landscapes (Ford *et al.*, 2007; Boggs *et al.*, 2021). The ET of forest understories can be determined by direct measurements with the eddy covariance method with a short tower or by calculating the differences between total ET measured by the eddy covariance method above the tree canopy and tree sapflow (Domec *et al.*, 2012a). Sapflow measurements provide a powerful tool for quantifying plant water use and the physiological responses of plants to environmental change. SAPFLUXNET (Poyatos *et al.*, 2021) is a global database that compiles global sapflow data for understanding environmental and biological controls on water use across species and extreme climates such as drought and heatwaves (Ahongshangbam *et al.*, 2023). In contrast, the eddy covariance technique measures forest ET by calculating the covariance between fluctuations in vertical eddy velocity and the specific water vapor content above forest canopies (Baldocchi and Ryu, 2011). The method is designed to understand the gas exchange at the boundary layer between vegetation and the atmosphere and to answer landscape-scale (the footprint of the flux tower) questions (Baldocchi *et al.*, 2001). Global participation in the flux measurements through FLUXNET (over 900 sites) or other regional networks in different countries since the 2000s has been a major driving force for advancing ET science. However, the recent literature suggests that the eddy covariance measurement method may result in a large underestimation of ET for forests (Volk *et al.*, 2024). Most of the discrepancy can be explained by an underestimation of turbulent fluxes, at which data are most affected during rainfall conditions, as well as insufficient closure correction during interception episodes (Fischer *et al.*, 2023; Pluntke *et al.*, 2023).

The Bowen ratio energy balance method and, more recently, the surface renewal biometeorological technique have been used for quantifying ET in croplands and forests, and have similar accuracy to the eddy flux methods (Vogt and Jaeger, 1990; Fisher *et al.*, 2023). These methods estimate ET based on the ratio of sensible heat to latent heat with measurements of air temperature, humidity gradients, net radiation and soil heat flux. It is relatively inexpensive but relies on several assumptions such as an extensive fetch over a homogeneous surface (Barr *et al.*, 1994).

Stable isotopes have been used as tracers for identifying the sources of water uptake in ecosystems and for evaluating quantitatively the relationships among water, energy and isotopic budgets. For example, effects of drought on forest carbon balance can be detected by analyzing the isotopic budget of $\delta^{13}C$ and $\delta^{18}O$ (Lin et al., 2019). The deuterium:hydrogen isotope ratios of continental runoff and ET have been used to determine the transpired fraction of ET as 56–74% (25th to 75th percentiles), with a median of 65% and mean of 64% (Good *et al.*, 2015). Furthermore, studies across an ecosystem gradient in the USA and Mexico provided evidence of ecohydrological separation, whereby different subsurface compartmentalized pools of water supply either plant transpiration fluxes or the combined fluxes of groundwater and streamflow (Evaristo *et al.*, 2015).

Satellite remote-sensing technology has been used to estimate landscape-scale ET since the 1980s (Zhang *et al.*, 2016), and recently in an urban environment (Qiu *et al.*, 2023). Coupling energy balance models with remotely sensed land-surface temperature retrieved from thermal infrared imagery provides proxy information of the surface moisture and vegetation cover growth status (Anderson *et al.*, 2021). Models such as the regional atmosphere–land exchange inverse (ALEXI) model and the associated flux disaggregation model (DisALEXI) are based on the two-source energy balance (TSEB) land-surface representations (Yang *et al.*, 2017). An application of these modeling systems in a lower coastal plain in North Carolina showed promise to map high-resolution ET (e.g. daily, 30 m) for a landscape with mixed land uses with natural wetland forests, drained pine forest with multiple stand ages, and croplands (Yang *et al.*, 2017). Large uncertainty remains among various remote-sensing ET products, and thus an OpenET platform has been developed to include an ensemble of multiple remote-sensing models to increase access to field-scale (30 m) ET data for the contiguous USA (Volk *et al.*, 2024).

Long-term and watershed-scale ET rates are estimated as the difference in measured precipitation and streamflow, assuming a negligible change in storage (Wilson *et al.*, 2001). Watershed-scale ET is dependent on its land use, vegetation cover and dominant soils, in addition to climate (Ukkola and Prentice, 2013; Amatya *et al.*, 2015).

A few ET method comparison studies at the watershed scale found that each approach had its own limitations (Wilson *et al.*, 2001; Ford *et al.*, 2007; Domec *et al.*, 2012b; Shimizu *et al.*, 2015; Tie *et al.*, 2018; Pluntke *et al.*, 2023). The eddy covariance method measures fluxes continuously, offering high temporal resolution data series, but data availability is limited by site topography, the cost of instrumentation, systematic errors and gap-filling issues. In addition, the eddy covariance method may underestimate ET by as much as 20% due to a lack of energy balance closure (Wilson *et al.*, 2002), especially on wet days, because the sonic anemometer and infrared gas analyzer must be dry to function properly (Fischer *et al.*, 2023).

3.4 Indirect ET Estimates

3.4.1 PET-based methods

Due to the high cost of measuring ET at large scales, mathematical modeling has been widely used to estimate ET (McMahon *et al.*, 2013; Komatsu and Kume, 2020). ET models can be roughly divided into two groups: biophysical (theoretical) and empirical models. Many theoretical models have evolved around the famous Penman–Monteith model (Monteith, 1965), which represents the most advanced process-based ET model for a single canopy.

In practice, it is often rather difficult to parameterize the process-based ET models to estimate actual ET, which is influenced by many complex factors other than meteorology that can be routinely measured. To simplify calculations, the concept of potential ET (PET) was introduced in the 1940s and represents the potential maximum ecosystem water loss when the water supply is not limiting. Although the ambiguous concept of PET is increasingly challenged, it remains used by practitioners. Actual ET is scaled down from the hypothetical PET by limiting canopy conductance and soil moisture and correlates to pan evaporation (Fisher *et al.*, 2017). Such PET models are often embedded in hydrological models that can simulate the dynamics of soil moisture, a major control of soil evaporation and transpiration (Tian *et al.*, 2012).

Many PET models (>50) exist and can be classified into five groups (Lu *et al.*, 2005) based on the theories behind the algorithms: (i) water budget; (ii) mass-transfer; (iii) combination; (iv) radiation; and (v) temperature based. These models give inconsistent values due to their different assumptions and input data requirements, or because they were often developed for specific climatic regions (McMahon *et al.*, 2013).

Because of the large uncertainties of different PET methods (Amatya *et al.*, 1995), the agricultural ET community suggests not using the term PET (Xiang *et al.*, 2020), and efforts have been made to develop standardized methods to better estimate ET for different crops. One standard crop reference ET method has been adopted by the Food and Agriculture Organization of the United Nations (FAO), formally in the FAO Irrigation and Drainage Paper No. 56 (FAO-56; Allen *et al.*, 1998). ET_0 is defined as 'the rate of evapotranspiration from a hypothetic crop with an assumed crop height (12 cm) and a fixed surface resistance (70 s/m) and albedo (0.23), which would closely resemble evapotranspiration from an extensive surface of green grass cover of uniform height, actively growing, completely shading the ground and not short of water.' ET_0 is calculated using the Penman–Monteith model with the specific parameters for a standard grass with a LAI of 4.8 (Sun *et al.*, 2011a):

$$ET_0 = \frac{0.408\Delta(R_n - G) + \gamma \frac{C}{T+273.16} U_2 (e_s - e_a)}{\Delta + \gamma(1 + 0.34U_2)} \quad (3.4)$$

where ET_0 is the daily grass reference ET (mm), Δ is the slope of the saturation water vapor pressure at air temperature T (kPa/°C), R_n is net radiation (MJ/m^2), G is soil heat flux (MJ/m^2), γ is the psychrometric constant (kPa/°C); e_s is saturation vapor pressure (kPa), e_a is actual vapor pressure (kPa), U_2 is mean wind speed (m/s) at 2m height and C is the unit conversion factor with a value of 900.

Once ET_0 is calculated, actual ET for a particular ecosystem type can be estimated by simply multiplying by a crop coefficient (K_c) developed for that crop using field-measured ET. The K_c method works well in irrigation agriculture for various croplands that have uniform phonology. However, for forests, this method can be problematic given the large variability of species composition of a forest, leaf biomass dynamics throughout the season, and the stand age, structure and density effects on tree biomass and hydraulic properties (e.g. canopy conductance, sapwood area). A casual use of ET_0 as the maximum ET in a hydrological model may result in underestimation of ET for a forest-dominated landscape (Amatya *et al.*, 1995; Amatya and Harrison, 2016), where the actual forest ET rates in a humid climate often exceed ET_0. A study based on global eddy covariance data suggests that K_c for any forest type may vary tremendously, and that latitude, precipitation and LAI are the best predictors of K_c (Liu *et al.*, 2017). Forests generally have higher K_c values than other ecosystem types (Fig. 3.6).

3.4.2 Empirical ET models

Empirical ET models are derived from direct ET measurements at the ecosystem scale and may be best used as a first-order approximation under mean climatic conditions. The following model was derived from field data collected at 13 sites using a variety of methods (Sun *et al.*, 2011b). The model estimates monthly ET for any land-cover types as a function of LAI, ET_0 (mm/month) and precipitation (P; mm/month):

$$ET = 11.94 + 4.76 \times \text{LAI} + ET_0(0.032 \times \text{LAI} + 0.0026 \times P + 0.15) \quad (3.5)$$

where ET_0 is the FAO-56 grass reference ET as discussed above.

Other forms of the ET model use Hamon's PET instead of the more data-demanding FAO reference ET method (Sun *et al.*, 2011a):

$$ET = 0.174 \times P + 0.502 \times PET + 5.31 \times \text{LAI} + 0.0222 \times PET \times \text{LAI} \quad (3.6)$$

Using a similar concept and a 250 FLUXNET synthesis data set, Fang *et al.* (2016) developed the following two types of monthly ET models requiring different input variables:

$$ET = 0.42 + 0.74 \times PET - 2.73 \times \text{VPD} + 0.10 \times R_n \quad (R^2 = 0.73, \text{RMSE} = 17.0\text{mm/month}) \quad (3.7)$$

where PET is monthly potential ET (mm) calculated by Hamon's method, VPD is vapor pressure deficit (hundreds of Pascals) that can be estimated from relative air humidity, R^2 is the coefficient of determination and RMSE is the root mean squared error. As R_n is rarely available at the regional scale, another model

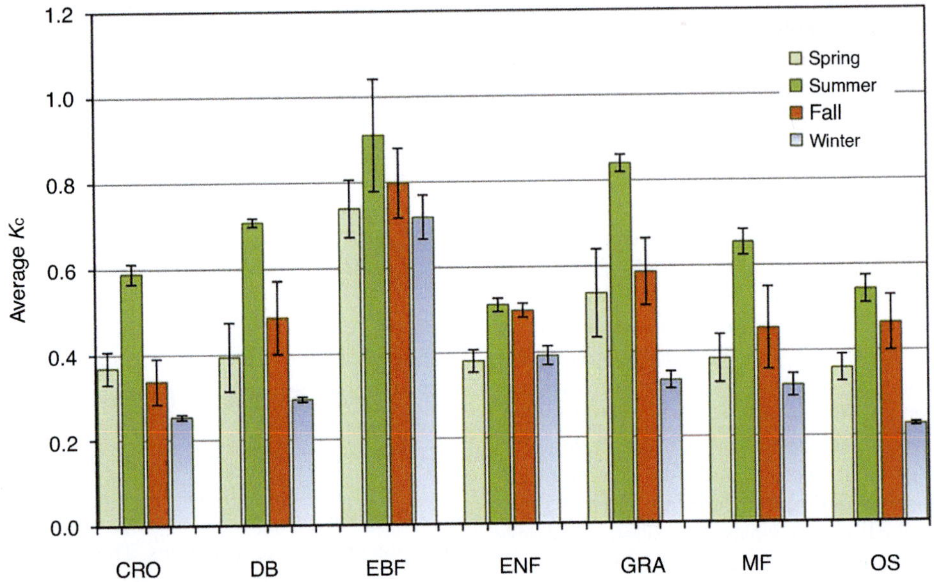

Fig. 3.6. A comparison of the seasonal mean crop coefficient (K_c) calculated from global eddy flux measurements for cropland (CRO), deciduous broad-leafed forest (DB), evergreen broad-leafed forest (EBF), evergreen needle-leaf forest (ENF), grassland (GRA), mixed forest (MF), and open shrubland (OS). K_c is estimated as measured ET divided by the grass reference ET (ET_o) calculated by the standard FAO-56 method (Allen *et al.*, 1998). Error bars represent standard deviation.

Table 3.2. Empirical evapotranspiration (ET) models by land-cover type developed using three commonly measured biophysical variables.

Land-cover type	Model	RMSE	R^2	*n*
Shrubland	$ET = -3.11 + 0.39 \times \text{PET} + 0.09 \times P + 11.127 \times \text{LAI}$	12.5	0.80	193
Cropland	$ET = -8.15 + 0.86 \times \text{PET} + 0.01 \times P + 9.54 \times \text{LAI}$	20.9	0.70	653
Grassland	$ET = -1.36 + 0.70 \times \text{PET} + 0.04 \times P + 6.56 \times \text{LAI}$	16.8	0.66	803
Deciduous forest	$ET = -14.82 + 0.98 \times \text{PET} + 2.72 \times \text{LAI}$	23.7	0.74	754
Evergreen needle leaf forest	$ET = 0.10 + 0.64 \times PET + 0.04 \times P + 3.53 \times \text{LAI}$	17.8	0.68	1382
Evergreen broad leaf forest	$ET = 7.71 + 0.74 \times PET + 1.85 \times \text{LAI}$	16.8	0.76	233
Mixed forest	$ET = -8.763 + 0.95 \times PET$	13.1	0.79	259
Savannas	$ET = -5.66 + 0.18 \times \text{PET} + 0.10 \times P + 44.63 \times \text{LAI}$	11.1	0.68	36

ET and *P* are measured in mm/month; *PET* is estimated by Hamon's method (mm/month)
LAI, leaf area index (m²/m²);R², coefficient of determination;RMSE, root mean square error;*n*, sample size;

that uses commonly available weather data was developed:

$$ET = -4.79 + 0.75 \times PET + 3.92 \times \text{LAI} + 0.04 \times P (R^2 = 0.68, \text{RMSE} = 18.1\text{mm/month}) \quad (3.8)$$

A series of ecosystem-specific monthly scale ET models were also developed using the global eddy flux data (Table 3.2) (Fang *et al.*, 2016). An empirical annual ET model was developed by combining a water balance method with a climate and land-cover regression equation to estimate mean annual ET across the conterminous USA (Sanford and Selnick, 2013). The climate variables included mean annual daily, maximum daily and minimum daily temperatures, and mean annual precipitation. The land-cover types included developed forest, shrubland, grassland, agriculture and marsh.

The long-term mean ET in a region is controlled mainly by water availability (precipitation) and atmosphere demand (PET), and this relationship is well described in the Budyko framework (Budyko *et al.*, 1962; Zhang *et al.*, 2004; Gan *et al.*, 2021). Using the same concept, Zhang *et al.* (2001) analyzed watershed balance data for over 250 catchments worldwide and developed a simple two-parameter ET model. The model offers a practical tool that can readily be used for assessing the long-term average effect of vegetation changes on catchment evapotranspiration:

$$ET = P \times \frac{1 + w\frac{PET}{P}}{1 + w\frac{PET}{P} + \frac{P}{PET}} \quad (3.9)$$

where w is the plant-available water coefficient representing the relative difference in plant water use for transpiration and P is annual precipitation. PET can be estimated by the Priestley and Taylor (1972) model. The best-fitted values for forest and grassland are 2.0 and 0.5, respectively, when PET is estimated using the Priestley and Taylor (1972) model (Zhang *et al.*, 2001).

The Budyko framework has been widely used to explain the mean spatial patterns of ET under land-cover change (Zhou *et al.*, 2015; Gan *et al.*, 2021) and climate change (Li and Quiring, 2021). The model has been also been extended to finer temporal scales such as monthly to fully capture the dynamics of ET over time (Zhang *et al.*, 2008).

3.4.3 Emerging ET estimation methods

Emerging novel ET methods based on the maximum entropy production (MEP) theory, offer new ways to estimate ET with limited meteorological data (Wang and Bras, 2011). For example, the MEP model provides simultaneous solution of latent, sensible and surface medium heat fluxes using only three input variables: net radiation, surface temperature and specific humidity, without the need for surface conductance parameters used in the traditional Penman–Monteith equation.

Two inexpensive micrometeorological methods, flux variance method and surface renewal analysis, are available as an alternative to estimate ET. Both methods estimate sensible heat flux derived from high frequency (~10 Hz) air temperature measurement, using a fine-wire thermocouple (Shapland *et al.*, 2014; Gray *et al.*, 2022; Fisher *et al.*, 2023) and require net radiation and soil heat flux to estimate ET.

A new generation of ecohydrological models that combine the effects of CO_2 on ET processes and couple the physical and biological processes such as soil moisture redistribution, hydraulic distribution, photosynthesis, canopy conductance and tree growth is available to fully understand the atmosphere–vegetation–soil processes mechanistically (Cheng *et al.*, 2017). Such models can provide climate models with better ET information for quantifying the feedback of forest cover change to regional and global climate systems (Fisher *et al.*, 2017).

3.5 Conclusions

ET has been increasingly recognized as an important measure in many academic disciplines (Yi *et al.*, 2024) and resource management concerned with tree species, density, stand age, prescribed burning, fertilization and thinning under a changing climate (Sun *et al.*, 2023). Much progress has been made in the past three decades toward measuring forest ET 'everywhere all the time' (Baldocchi and Ryu, 2011; Jung *et al.*, 2019). However, there still are large data gaps in many parts of the world in applying existing ET models, resulting in difficulty in answering critical questions such as how reforestation and global greening up affect water supply and low flows (Yang *et al.*, 2023). Climate change and rising demands for forest water services require better models to reliably account for the role of forest management in regulating ET and streamflow and other ecosystem services (e.g. carbon sequestration). Regional land-surface models should be rigorously validated with field measurements of ET at multiple scales to be creditable for quantifying the benefits of forests in climate mitigation and adaptation (Sun *et al.*, 2024).

References

Aguilos, M., Sun, G., Liu, N., Zhang, Y. and Starr, G. (2024) Energy availability and leaf area dominate control of ecosystem evapotranspiration in the southeastern US. *Agricultural and Forest Meteorology* 349, 109960.

Ahongshangbam, J., Kulmala, L., Soininen, J., Frühauf, Y., Karvinen, E. *et al.* (2023) Sap flow and leaf gas exchange response to drought and heatwave in urban green spaces in a nordic city. *Biogeosciences Discussions* 20, 4455–4475.

Allen, R.G., Pereira, L.S., Raes, D. and Smith, M. (1998) *Crop Evapotranspiration – Guidelines for Computing Crop Water Requirements*. FAO Irrigation and Drainage Paper No. 56. FAO, Rome.

Amatya, D.M. and Harrison, C.A. (2016) Grass and forest potential evapotranspiration comparison using five methods in the atlantic coastal plain. *Journal of Hydrologic Engineering* 21, 05016007.

Amatya, D.M., Skaggs, R.W. and Gregory, J.D. (1995) Comparison of methods for estimating ref-ET. *Journal of Irrigation and Drainage Engineering* 121, 427–435.

Amatya, D.M., Sun, G., Rossi, C.G., Ssegane, H.S., Nettles, J.E. *et al.* (2015) Forests, land use change, and water. In: Zolin, C.A. and Rodrigues, R. (eds) *Impact of Climate Change on Water Resources in Agriculture*. CRC Press, Boca Raton, Florida, pp. 116–153.

Amatya, D.M., Callahan, T.J., Mukherjee, S., Harrison, C.A., Trettin, C.C. *et al.* (2024) Hydrometeorological trends in a low-gradient forested watershed on the southeastern Atlantic coastal plain in the USA. *Hydrology* 11, 31.

Anderson, M.C., Yang, Y., Xue, J., Knipper, K.R., Yang, Y. *et al.* (2021) Interoperability of ECOSTRESS and Landsat for mapping evapotranspiration time series at sub-field scales. *Remote Sensing of Environment* 252, 112189.

Asbjornsen, H., Goldsmith, G.R., Alvarado-Barrientos, M.S., Rebel, K., Van Osch, F.P. *et al.* (2011) Ecohydrological advances and applications in plant–water relations research: A review. *Journal of Plant Ecology* 4, 3–22.

Asensio, V., Domec, J.C., Nouvellon, Y., Laclau, J.P., Bouillet, J.P. *et al.* (2020) Potassium fertilization increases hydraulic redistribution and water use efficiency for stemwood production in *Eucalyptus grandis* plantations. *Environmental and Experimental Botany* 176, 104085.

Balandier, P., Gobin, R., Prévosto, B. and Korboulewsky, N. (2022) The contribution of understorey vegetation to ecosystem evapotranspiration in boreal and temperate forests: A literature review and analysis. *European Journal of Forest Research* 141, 979–997.

Baldocchi, D., Falge, E., Gu, L., Olson, R., Hollinger, D. *et al.* (2001) FLUXNET: A new tool to study the temporal and spatial variability of ecosystem-scale carbon dioxide, water vapor, and energy flux densities. *Bulletin of the American Meteorological Society* 82, 2415–2434.

Baldocchi, D. and Ryu, Y. (2011) A synthesis of forest evaporation fluxes – from days to years – as measured with eddy covariance. In: Levia, D.F., Carlyle-Moses, D. and Tanaka, T. (eds) *Forest Hydrology and Biogeochemistry: Synthesis of Past Research and Future Directions*. Springer, Dordrecht, Netherlands, pp. 101–116.

Barr, A.G., King, K.M., Gillespie, T.J., Hartog, G. and Neumann, H.H. (1994) A comparison of bowen ratio and eddy correlation sensible and latent heat flux measurements above deciduous forest. *Boundary-Layer Meteorology* 71, 21–41.

Beer, C., Reichstein, M., Ciais, P., Farquhar, G.D. and Papale, D. (2007) Mean annual GPP of Europe derived from its water balance. *Geophysical Research Letters* 34, L05401.

Boden, K., Philippus, D., Sytsma, A., Kurzweil, J., Randell, J. *et al.* (2023) A multi-scale assessment of forest treatment impacts on evapotranspiration and water yield in the Sierra Nevada. *Ecohydrology* 16, e2548.

Boggs, J.L., Sun, G., Domec, J.C. and McNulty, S.G. (2021) Variability of tree transpiration across three zones in a southeastern US piedmont watershed. *Hydrological Processes* 35, e14389.

Bonan, G.B. (2008) Forests and climate change: Forcings, feedbacks, and the climate benefits of forests. *Science* 320(5882), 1444–1449. DOI: 10.1126/science.1155121.

Bosch, J.M. and Hewlett, J.D. (1982) A review of catchment experiments to determine the effect of vegetation changes on water yield and evapotranspiration. *Journal of Hydrology* 55, 3–23.

Bryant, M.L., Bhat, S. and Jacobs, J.M. (2005) Measurements and modeling of throughfall variability for five forest communities in the southeastern US. *Journal of Hydrology* 312, 95–108.

Budyko, M.I., Yefimova, N.A., Aubenok, L.I. and Strokina, L.A. (1962) The heat-balance of the surface of the earth. *Soviet Geography Review and Translation* 3, 3–16.

Burgess, S.S.O., Adams, M.A., Turner, N.C. and Ong, C.K. (1998) The redistribution of soil water by tree root systems. *Oecologia* 115(3), 306–311. DOI: 10.1007/s004420050521.

Canny, M.J. (1998) Transporting water in plants. *American Scientist* 86, 152–159.

Cheng, L., Zhang, L., Wang, Y.-P., Canadell, J.G., Chiew, F.H.S. *et al.* (2017) Recent increases in terrestrial carbon uptake at little cost to the water cycle. *Nature Communications* 8(1), 110. DOI: 10.1038/s41467-017-00114-5.

Currie, D.J. (1991) Energy and large-scale patterns of animal-species and plant-species richness. *American Naturalist* 137, 27–49.

Currie, D.J. and Paquin, V. (1987) Large-scale biogeographical patterns of species richness of trees. *Nature* 329, 326–327.

Curtis, O.F. (1926) WHAT is the significance of transpiration? *Science* 63(1628), 267–271. DOI: 10.1126/science.63.1628.267.

Deguchi, A., Hattori, S. and Park, H.T. (2006) The influence of seasonal changes in canopy structure on interception loss: Application of the revised gash model. *Journal of Hydrology* 318, 80–102.

del Campo, A.D., Otsuki, K., Serengil, Y., Blanco, J.A., Yousefpour, R. *et al.* (2022) A global synthesis on the effects of thinning on hydrological processes: Implications for forest management. *Forest Ecology and Management* 519, 120324.

Denmead, O.T., Dunin, F.X., Wong, S.C. and Greenwood, E.A.N. (1993) Measuring water use efficiency of Eucalypt trees with chambers and micrometeorological techniques. *Journal of Hydrology* 150, 649–664.

Diem, J.E., Carlton, D.K. and Pangle, L.A. (2023) Evapotranspiration from developed land and urban watersheds in a humid subtropical climate. *Water Resources Research* 59, e2023WR035276.

Domec, J.-C., King, J.S., Noormets, A., Treasure, E., Gavazzi, M.J. *et al.* (2010) Hydraulic redistribution of soil water by roots affects whole-stand evapotranspiration and net ecosystem carbon exchange. *The New Phytologist* 187(1), 171–183. DOI: 10.1111/j.1469-8137.2010.03245.x.

Domec, J.C., Lachenbruch, B., Pruyn, M.L. and Spicer, R. (2012a) Effects of age-related increases in sapwood area, leaf area, and xylem conductivity on height-related hydraulic costs in two contrasting coniferous species. *Annals of Forest Science* 69, 17–27.

Domec, J.C., Sun, G., Noormets, A., Gavazzi, M.J., Treasure, E.A. *et al.* (2012b) A comparison of three methods to estimate evapotranspiration in two contrasting loblolly pine plantations: Age-related changes in water use and drought sensitivity of evapotranspiration components. *Forest Science* 58, 497–512.

Duarte, H.F., Kim, J.B., Sun, G., McNulty, S.G. and Xiao, J. (2024) Climate and vegetation change impacts on future conterminous United States water yield. *Journal of Hydrology* 639, 131472.

Emanuel, R.E., Epstein, H.E., McGlynn, B.L., Welsch, D.L., Muth, D.J. *et al.* (2010) Spatial and temporal controls on watershed ecohydrology in the northern rocky mountains. *Water Resources Research* 46, W11553.

Evaristo, J., Jasechko, S. and McDonnell, J.J. (2015) Global separation of plant transpiration from groundwater and streamflow. *Nature* 525(7567), 91–94. DOI: 10.1038/nature14983.

Fang, Y., Sun, G., Caldwell, P., McNulty, S.G., Noormets, A. *et al.* (2016) Monthly land cover-specific evapotranspiration models derived from global eddy flux measurements and remote sensing data. *Ecohydrology* 9, 248–266.

Fischer, M., Katul, G., Noormets, A., Pozníková, G., Domec, J.C. *et al.* (2023) Merging flux-variance and surface renewal methods in the roughness sublayer and the atmospheric surface layer. *Agricultural and Forest Meteorology* 342, 109692.

Fischer, S., Moderow, U., Queck, R. and Bernhofer, C. (2023) Evaporation of intercepted rainfall – comparing canopy water budget and energy balance related long term measurements at a norway spruce site. *Agricultural and Forest Meteorology* 341, 109637.

Fisher, J.B., Melton, F., Middleton, E., Hain, C., Anderson, M. *et al.* (2017) The future of evapotranspiration: Global requirements for ecosystem functioning, carbon and climate feedbacks, agricultural management, and water resources. *Water Resources Research* 53, 2618–2626.

Fisher, J.B., Dohlen, M.B., Halverson, G.H., Collison, J.W., Pearson, C. *et al.* (2023) Remotely sensed terrestrial open water evaporation. *Scientific Reports* 13(1), 8174. DOI: 10.1038/s41598-023-34921-2.

Ford, C.R., Hubbard, R.M., Kloeppel, B.D. and Vose, J.M. (2007) A comparison of sap flux-based evapotranspiration estimates with catchment-scale water balance. *Agricultural and Forest Meteorology* 145, 176–185.

Gan, G., Liu, Y. and Sun, G. (2021) Understanding interactions among climate, water, and vegetation with the Budyko framework. *Earth-Science Reviews* 212, 103451.

Gavazzi, M.J., Sun, G., McNulty, S.G., Treasure, E.A. and Wightman, M.G. (2016) Canopy rainfall interception measured over ten years in a coastal plain loblolly pine (*Pinus taeda* L.) plantation. *Transactions of the ASABE* 59, 601–610.

Gerrits, A.M.J., Savenije, H.H.G., Hoffmann, L. and Pfister, L. (2007) New technique to measure forest floor interception – An application in a beech forest in Luxembourg. *Hydrology and Earth System Sciences* 11, 695–701.

Goeking, S.A. and Tarboton, D.G. (2022) Variable streamflow response to forest disturbance in the western US: A large-sample hydrology approach. *Water Resources Research* 58, e2021WR031575.

Good, S.P., Noone, D., Kurita, N., Benetti, M. and Bowen, G.J. (2015) D/h isotope ratios in the global hydrologic cycle. *Geophysical Research Letters* 42, 5042–5050.

Granier, A. (1987) Evaluation of transpiration in a douglas-fir stand by means of sap flow measurements. *Ree Physiology* 3, 309–320.

Granier, A., Huc, R. and Barigah, S.T. (1996) Transpiration of natural rain forest and its dependence on climatic factors. *Agricultural and Forest Meteorology* 78(1–2), 19–29.

Gray, B.A., Toucher, M.L., Savage, M.J. and Clulow, A.D. (2022) Seasonal evapotranspiration over an invader vegetation (*Pteridium aquilinum*) in a degraded montane grassland using surface renewal. *Journal of Hydrology Regional Studies* 40, 101012.

Hao, L., Sun, G., Huang, X., Tang, R., Jin, K. *et al.* (2023) Urbanization alters atmospheric dryness through land evapotranspiration. *Npj Climate and Atmospheric Science* 6, 149.

Hawkins, B.A., Field, R., Cornell, H.V., Currie, D.J., Guégan, J.F. *et al.* (2003) Energy, water, and broad-scale geographic patterns of species richness. *Ecology* 84, 3105–3117.

Helvey, J.D. (1971) A summary of rainfall interception by certain conifers on North America. In: *Proceedings of the 3rd International Seminar for Hydrology Professors, Purdue University West Lafayette, Indiana*, pp. 103–113.

Helvey, J.D. and Patric, J.H. (1965) Canopy and litter interception of rainfall by hardwoods of eastern United States. *Water Resources Research* 1, 193–206.

Horton, R.E. (1919) Rainfall interception. *Monthly Weather Review* 47, 603–623.

Jackson, R.B., Jobbágy, E.G., Avissar, R., Roy, S.B., Barrett, D.J. *et al.* (2005) Trading water for carbon with biological carbon sequestration. *Science*310, 1944–1947.

Jackson, R.B., Randerson, J.T., Canadell, J.G., Anderson, R.G., Avissar, R. *et al.* (2008) Protecting climate with forests. *Environmental Research Letters* 3, 044006.

Jasechko, S., Sharp, Z.D., Gibson, J.J., Birks, S.J., Yi, Y. *et al.* (2013) Terrestrial water fluxes dominated by transpiration. *Nature* 496(7445), 347–350. DOI: 10.1038/nature11983.

Jayakaran, A.D., Williams, T.M., Ssegane, H., Amatya, D.M., Song, B. *et al.* (2014) Hurricane impacts on a pair of coastal forested watersheds: Implications of selective hurricane damage to forest structure and streamflow dynamics. *Hydrology and Earth System Sciences* 18, 1151–1164.

Jung, M., Koirala, S., Weber, U., Ichii, K., Gans, F. *et al.* (2019) The FLUXCOM ensemble of global land-atmosphere energy fluxes. *Scientific Data* 6(1), 74. DOI: 10.1038/s41597-019-0076-8.

Komatsu, H. and Kume, T. (2015) Changes in the sapwood area of Japanese cedar and cypress plantations after thinning. *Journal of Forest Research* 20, 43–51.

Komatsu, H. and Kume, T. (2020) Modeling of evapotranspiration changes with forest management practices: A genealogical review. *Journal of Hydrology* 585, 124835.

Kumagai, T. (2011) Transpiration in forest ecosystems. In: Levia, D.F., Carlyle-Moses, D. and Tanaka, T. (eds) *Forest Hydrology and Biogeochemistry: Synthesis of Past Research and Future Directions*. Springer, Dordrecht, Netherlands, pp. 389–406.

Li, Z. and Quiring, S.M. (2021) Identifying the dominant drivers of hydrological change in the contiguous United States. *Water Resources Research* 57, e2021WR029738.

Lian, X., Zhao, W. and Gentine, P. (2022) Recent global decline in rainfall interception loss due to altered rainfall regimes. *Nature Communications* 13(1), 7642. DOI: 10.1038/s41467-022-35414-y.

Lin, W., Domec, J.-C., Ward, E.J., Marshall, J., King, J.S. *et al.* (2019) Using $\delta^{13}C$ and $\delta^{18}O$ to analyze loblolly pine (*Pinus taeda* L.) response to experimental drought and fertilization. *Tree Physiology* 39(12), 1984–1994. DOI: 10.1093/treephys/tpz096.

Liu, C., Sun, G., McNulty, S.G., Noormets, A. and Fang, Y. (2017) Environmental controls on seasonal ecosystem evapotranspiration/potential evapotranspiration ratio as determined by the global eddy flux measurements. *Hydrology and Earth System Sciences* 21(1), 311–322. DOI: 10.5194/hess-21-311-2017.

Lu, J.B., Sun, G., McNulty, S.G. and Amatya, D.M. (2005) A comparison of six potential evapotranspiration methods for regional use in the southeastern United States. *Journal of the American Water Resources Association* 41(3), 621–633. DOI: 10.1111/j.1752-1688.2005.tb03759.x.

Mankin, K.R. and Patel, R. (2023) Wildfire burn severity affects postfire shifts in evapotranspiration in subalpine forests. *Journal of Natural Resources and Agricultural Ecosystems* 1(1), 1–11. DOI: 10.13031/jnrae.15438.

McMahon, T.A., Peel, M.C., Lowe, L., Srikanthan, R. and McVicar, T.R. (2013) Estimating actual, potential, reference crop and pan evaporation using standard meteorological data: A pragmatic synthesis. *Hydrology and Earth System Sciences* 17(4), 1331–1363. DOI: 10.5194/hess-17-1331-2013.

Melton, F.S., Huntington, J., Grimm, R., Herring, J., Hall, M. *et al.* (2022) OpenET: Filling a critical data gap in water management for the western United States. *JAWRA Journal of the American Water Resources Association* 58(6), 971–994. DOI: 10.1111/1752-1688.12956.

Monteith, J.L. (1965) Evaporation and environment. *Symposia of the Society for Experimental Biology* 19, 205–234.

Müller, J. (2009) Forestry and water budget of the lowlands in Northeast Germany — consequences for the choice of tree species and for forest management. *Journal of Water and Land Development* 13a(1), 133–148. DOI: 10.2478/v10025-010-0024-7.

Müller, J. and Bolte, A. (2009) The use of lysimeters in forest hydrology research in north-east Germany. *Landbauforschung (Applied Agriculture and Forestry Research)* 59, 1–10.

Neumann, R.B. and Cardon, Z.G. (2012) The magnitude of hydraulic redistribution by plant roots: A review and synthesis of empirical and modeling studies. *The New Phytologist* 194(2), 337–352. DOI: 10.1111/j.1469-8137.2012.04088.x.

Olbrich, B.W. (1991) The verification of the heat pulse velocity technique for estimating sap flow in *Eucalyptus grandis*. *Canadian Journal of Forest Research* 21, 836–841. DOI: 10.1139/x91-117.

Pluntke, T., Bernhofer, C., Grünwald, T., Renner, M. and Prasse, H. (2023) Long term climatological and ecohydrological analysis of a paired catchment – flux tower observatory near Dresden (Germany). Is there evidence of climate change in local evapotranspiration? *Journal of Hydrology* 617, 128873. DOI: 10.1016/j.jhydrol.2022.128873.

Poyatos, R., Granda, V., Flo, V., Adams, M.A., Adorján, B. *et al.* (2021) Global transpiration data from sap flow measurements: The SAPFLUXNET database. *Earth System Science Data* 13(6), 2607–2649. DOI: 10.5194/essd-13-2607-2021.

Priestley, C.H.B. and Taylor, R.J. (1972) On the assessment of surface heat flux and evaporation using large-scale parameters. *Monthly Weather Review* 100(2), 81–92. DOI: 10.1175/1520-0493(1972)100<0081:OTAOSH>2.3.CO;2.

Qiu, G.Y., Yan, C. and Liu, Y. (2023) Urban evapotranspiration and its effects on water budget and energy balance: Review and perspectives. *Earth-Science Reviews* 246, 104577. DOI: 10.1016/j.earscirev.2023.104577.

Sadeghi, S.M.M., Epstein, J.M., Deljouei, A., Gorora, F.J. and Cohen, M.J. (2024) Stand age controls canopy and soil rainfall partitioning in slash pine forests. *Forest Ecology and Management* 572, 122307.

Sampson, D.A., Amatya, D.M., Lawson, C.D.B. and Skaggs, R.W. (2011) Leaf area index (LAI) of loblolly pine and emergent vegetation following a harvest. *Transactions of the ASABE* 54, 2057–2066.

Sanford, W.E. and Selnick, D.L. (2013) Estimation of evapotranspiration across the conterminous United States using a regression with climate and land-cover data. *Journal of the American Water Resources Association* 49, 217–230.

Schlesinger, W.H. and Jasechko, S. (2014) Transpiration in the global water cycle. *Agricultural and Forest Meteorology* 189, 115–117.

Shapland, T.M., Snyder, R.L., Paw, U.K.T. and McElrone, A.J. (2014) Thermocouple frequency response compensation leads to convergence of the surface renewal alpha calibration. *Agricultural and Forest Meteorology* 189–190, 36–47.

Shimizu, T., Kumagai, T.O., Kobayashi, M., Tamai, K., Iida, S. *et al.* (2015) Estimation of annual forest evapotranspiration from a coniferous plantation watershed in Japan. Comparison of eddy covariance, water budget and sap-flow plus interception loss. *Journal of Hydrology* 522, 250–264.

Sidhu, H.K., Kidd, K.A., Emilson, E.J.S., Stastny, M., Venier, L. *et al.* (2024) Increasing spruce budworm defoliation increases catchment discharge in conifer forests. *The Science of the Total Environment* 912, 168561. DOI: 10.1016/j.scitotenv.2023.168561.

Stephenson, N. (1998) Actual evapotranspiration and deficit: Biologically meaningful correlates of vegetation distribution across spatial scales. *Journal of Biogeography* 25, 855–870.

Steudle, E. (2002) Transport of water in plants. *Environmental Control in Biology* 40, 29–37.

Stoy, P.C., El-Madany, T.S., Fisher, J.B., Gentine, P., Gerken, T. *et al.* (2019) Reviews and syntheses: Turning the challenges of partitioning ecosystem evaporation and transpiration into opportunities. *Biogeosciences (Online)* 16, 3747–3775.

Sun, G., Caldwell, P., Noormets, A., McNulty, S.G., Cohen, E. *et al.* (2011a) Upscaling key ecosystem functions across the conterminous United States by a water-centric ecosystem model. *Journal of Geophysical Research: Biogeosciences* 116, G00J05.

Sun, G., Alstad, K., Chen, J., Chen, S., Ford, C.R. *et al.* (2011b) A general predictive model for estimating monthly ecosystem evapotranspiration. *Ecohydrology* 4, 245–255.

Sun, G., Amatya, D.M. and McNulty, S.G. (2016) Forest hydrology. In: Sing, V.P. (ed.) *Part 7: Systems Hydrology. Handbook of Applied Hydrology*, 2nd edn. McGraw Hill, New York, p. 85.

Sun, G., Hallema, D. and Asbjornsen, H. (2017) Ecohydrological processes and ecosystem services in the anthropocene: A review. *Ecological Processes* 6, 35.

Sun, G., Wei, X., Hao, L., Sanchis, M.G., Hou, Y. *et al.* (2023) Forest hydrology modeling tools for watershed management: A review. *Forest Ecology and Management* 530, 120755.

Sun, G., Tiwari, K.R., Hao, L., Amatya, D., Liu, N. *et al.* (2024) Climate change and forest hydrology in future forests. In: McNulty, S.G. (ed.) *Future Forests: Mitigation and Adaptation to Climate Change*. Elsevier, Amsterdam, Netherlands, pp. 95–124.

Taborski, T., Domec, J.C., Chipeaux, C., Lafont, S., Wingate, L. *et al.* (2022) Quantifying stomatal canopy conductance in a pine forest during drought from combined sap flow and canopy surface temperature measurements. *Agricultural and Forest Meteorology* 323, 108997.

Tian, S.Y., Youssef, M.A., Skaggs, R.W., Amatya, D.M. and Chescheir, G.M. (2012) Modeling water, carbon, and nitrogen dynamics for two drained pine plantations under intensive management practices. *Forest Ecology and Management* 264, 20–36.

Tie, Q., Hu, H., Tian, F. and Holbrook, N.M. (2018) Comparing different methods for determining forest evapotranspiration and its components at multiple temporal scales. *The Science of the Total Environment* 633, 12–29. DOI: 10.1016/j.scitotenv.2018.03.082.

Ukkola, A.M. and Prentice, I.C. (2013) A worldwide analysis of trends in water-balance evapotranspiration. *Hydrology and Earth System Sciences* 17, 4177–4187.

Vogt, R. and Jaeger, L. (1990) Evaporation from a pine forest-using the aerodynamic method and bowen ratio method. *Agricultural and Forest Meteorology* 50, 39–54.

Volk, J.M., Huntington, J.L., Melton, F.S., Allen, R., Anderson, M. *et al.* (2024) Assessing the accuracy of openet satellite-based evapotranspiration data to support water resource and land management applications. *Nature Water* 2, 193–205.

Vose, J.M., Sun, G., Ford, C.R., Bredemeier, M., Otsuki, K. *et al.* (2011) Forest ecohydrological research in the 21st century: What are the critical needs? *Ecohydrology* 4, 146–158.

Wang, J. and Bras, R.L. (2011) A model of evapotranspiration based on the theory of maximum entropy production. *Water Resources Research* 47, W03521.

Wang, K. and Dickinson, R.E. (2012) A review of global terrestrial evapotranspiration: Observation, modeling, climatology, and climatic variability. *Reviews of Geophysics* 50, RG2005.

Warren, J.M., Meinzer, F.C., Brooks, J.R., Domec, J.C. and Coulombe, R. (2007) Hydraulic redistribution of soil water in two old-growth coniferous forests: Quantifying patterns and controls. *The New Phytologist* 173(4), 753–765. DOI: 10.1111/j.1469-8137.2006.01963.x.

Wei, Z., Yoshimura, K., Wang, L., Miralles, D.G., Jasechko, S. *et al.* (2017) Revisiting the contribution of transpiration to global terrestrial evapotranspiration. *Geophysical Research Letters* 44, 2792–2801.

Wilson, K.B., Hanson, P.J., Mulholland, P.J., Baldocchi, D.D. and Wullschleger, S.D. (2001) A comparison of methods for determining forest evapotranspiration and its components: Sap-flow, soil water budget, eddy covariance and catchment water balance. *Agricultural and Forest Meteorology* 106, 153–168.

Wilson, K.B., Goldstein, A., Falge, E., Aubinet, M., Baldocchi, D.D. *et al.* (2002) Energy balance closure at FLUXNET sites. *Agricultural and Forest Meteorology* 113, 223–243.

Wullschleger, S.D., Meinzer, F.C. and Vertessy, R.A. (1998) A review of whole-plant water use studies in tree. *Tree Physiology* 18(8_9), 499–512. DOI: 10.1093/treephys/18.8-9.499.

Xiang, K., Li, Y., Horton, R. and Feng, H. (2020) Similarity and difference of potential evapotranspiration and reference crop evapotranspiration – A review. *Agricultural Water Management* 232, 106043.

Yang, Y., Anderson, M.C., Gao, F., Hain, C.R., Semmens, K.A. *et al.* (2017) Daily landsat-scale evapotranspiration estimation over a forested landscape in North Carolina, USA, using multi-satellite data fusion. *Hydrology and Earth System Sciences* 21, 1017–1037.

Yang, Y., Roderick, M.L., Guo, H., Miralles, D.G., Zhang, L. *et al.* (2023) Evapotranspiration on a greening earth. *Nature Reviews Earth & Environment* 4, 626–641.

Yi, K., Senay, G.B., Fisher, J.B., Wang, L., Suvočarev, K. *et al.* (2024) Challenges and future directions in quantifying terrestrial evapotranspiration. *Water Resources Research* 60, e2024WR037622.

Younger, S.E., Jackson, C.R., Dix, M.J., Caldwell, P.V. and Aubrey, D.P. (2023) Evapotranspiration partitioning of *Eucalyptus benthamii* and *Pinus taeda* during early stand development. *BioEnergy Research* 16, 2204–2218.

Zhang, K., Kimball, J.S. and Running, S.W. (2016) A review of remote sensing based actual evapotranspiration estimation. *Wiley Interdisciplinary Reviews: Water* 3, 834–853.

Zhang, L., Dawes, W.R. and Walker, G.R. (2001) Response of mean annual evapotranspiration to vegetation changes at catchment scale. *Water Resources Research* 37, 701–708.

Zhang, L., Hickel, K., Dawes, W.R., Chiew, F.H., Western, A.W. *et al.* (2004) A rational function approach for estimating mean annual evapotranspiration. *Water Resources Research* 40, W02502.

Zhang, L., Potter, N., Hickel, K., Zhang, Y.Q. and Shao, Q.X. (2008) Water balance modeling over variable time scales based on the Budyko framework – model development and testing. *Journal of Hydrology* 360, 117–131.

Zhang, Z., Zhang, L., Xu, H., Creed, I.F., Blanco, J.A. *et al.* (2023) Forest water-use efficiency: Effects of climate change and management on the coupling of carbon and water processes. *Forest Ecology and Management* 534, 120853.

Zhou, G., Wei, X., Chen, X., Zhou, P., Liu, X. *et al.* (2015) Global pattern for the effect of climate and land cover on water yield. *Nature Communications* 6, 5918. DOI: 10.1038/ncomms6918.

4 Forest Hydrology and Climate Change

S. Liu[1]*, M. Zhang[2] and G. Sun[3]

[1]Key Laboratory of Forest Ecology and Environment of National Forestry and Grassland Administration, Ecology and Nature Conservation Institute, Chinese Academy of Forestry, Beijing, China; [2]School of Resources and Environment, University of Electronic Science and Technology of China, Chengdu, China; [3]Eastern Forest Environmental Threat Assessment Center, USDA Forest Service, Research Triangle Park, North Carolina, USA

Abstract

Climate change affects hydrological processes directly by changing patterns of precipitation and evapotranspiration, and indirectly by impacting hydrology through altering forest growth rates and structure, and forest disturbances regimes. This chapter addresses two key aspects of forest hydrology under climate change: (i) climate change impacts on hydrology at the tree, stand and watershed levels; and (ii) interactive hydrological effects of climate and forest changes. Recommendations are made for region-specific management strategies of the forest–water nexus for climate change adaptation. These provide an essential scientific basis for adaptive forest management strategies to promote forest ecosystem resilience to climate change and to improve water-related ecosystem services.

4.1 Introduction

Forest hydrology is rooted in studying the influences of forest change on local climate (i.e. precipitation) and floods (Hewlett, 1982). Indeed, early works were largely written by forest micrometeorologists (Lee, 1980) who examined forest energy and water balances under human disturbances such as deforestation and acid rains assuming a stationary climate (McCulloch and Robinson, 1993). However, forest hydrologists started to pay more attention to climate change impacts on forest hydrological processes and management issues in the late early 1990s when the potential for anthropogenically caused global climate change was first being detected (Bren, 2023). This observation caused a surge in forest ecohydrological studies focused on illustrating the impacts of climate change on the forest water cycles at multiple scales beginning in the 2000s (Amatya *et al.*, 2006; Jones *et al.*, 2017; Sun *et al.*, 2023).

The global climate system has changed significantly due to elevated greenhouse gases and human activities since the 1850s (IPCC, 2021). Growing evidence suggests that increases in greenhouse gases are responsible for changes in precipitation and temperature, and associated sea-level rises have modified water cycles and forest dynamics at multiple scales (de Lombaerde *et al.*, 2021). More severe floods are now occurring in the Pacific Islands and North America, Europe, Australasia, and Central and South America, while large-scale droughts are

*Corresponding author: liusr@caf.ac.cn

 Forest Hydrology: Processes, Management and Assessment (eds Leon J. Bren *et al.*)

DOI: 10.1079/9781800625310.0004

projected to occur in Africa, South America and Europe, as well as in Australasia, Central and North America, and the Caribbean. These impacts are predicted to become worse in future decades (IPCC, 2021). Earlier and increased spring snowmelt peak flows and reduced summer flows due to global warming have also been well documented in snow-dominated regions (Berghuijs *et al.*, 2014).

Climate change induces hydrological change, directly or indirectly affecting forest–water relationships (Fig. 4.1) by altering precipitation, atmospheric water demand, forest water uses due to prolonging growing seasons and increasing forest growth potential (Adams *et al.*, 2020; Sun *et al.*, 2023), and forest leaf loss caused by climate change-related wildfires, insect outbreaks and windthrows (Jones *et al.*, 2012; Hallema *et al.*, 2018; Smith-Tripp *et al.*, 2021). Future shifts in tree species and forest community structure, and current land greening or forestation under climate change further complicate the forest–water nexus (Vose *et al.*, 2016; Zhang *et al.*, 2023).

Contemporary forest management has multiple objectives, including offsetting CO_2 emission as part of the nature-based solutions to mitigate climate change and its negative impacts on society (Sun and Vose, 2016; Hua *et al.*, 2022). Trade-offs and synergies among perceived ecosystem services of forest management practices exist. For example, large-scale forestation for soil erosion control and carbon sequestration may reduce streamflow and exacerbate water scarcity in areas with increasing droughts. In contrast, planting trees with high water consumption may help reduce flood risks in areas suffering from growing floods (Buechel *et al.*, 2022). Therefore, future forest management strategies must translate the understanding of the complex forest–water nexus in response to climate change for adaptive forest watershed management to mitigate anticipated climate change impacts and risks.

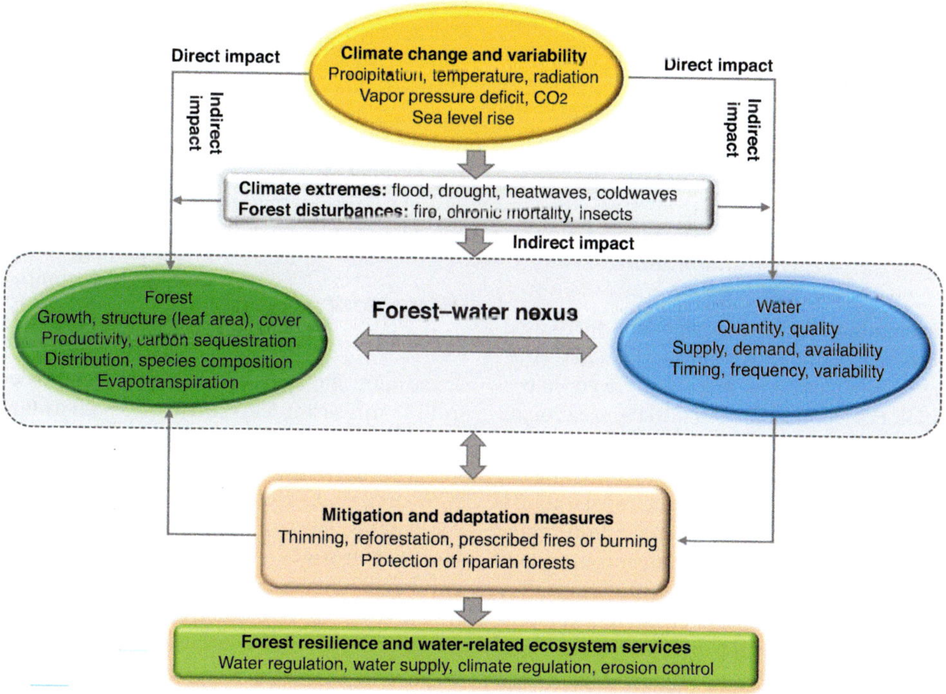

Fig. 4.1. Illustration of direct and indirect climate change impacts on the forest–water nexus, and potential interactions with mitigation and adaptation measures for forest resilience and water-related ecosystem services under climate change.

This chapter discusses the state-of-the-art knowledge about climate change impacts on the forest–water nexus and recommends management strategies for the forest–water nexus for climate change adaptation. Climate change, directly and indirectly, affects forest and water balances at multiple scales from trees to watersheds (Fig. 4.1). Conversely, forest feedback as a result of climate change affects forest hydrology interactively. Therefore, this chapter addresses two key aspects of forest hydrology under climate change: (i) climate change impacts on hydrology at the tree, stand and watershed levels; and (ii) interactive hydrological effects of climate and forest changes.

4.2 Principles of the Hydrological Response to Climate Change

Climate change affects water balance by directly altering precipitation, and by evapotranspiration (ET) through increasing atmospheric demand. However, the indirect effects of climatic change on ET through various biotic and abiotic processes can be more complex (Sun *et al.*, 2023; Zhang *et al.*, 2023). This section focuses on the impacts of climate change on key hydrological components including tree transpiration, ET and streamflow, and on drivers including water-use efficiency.

4.2.1 Transpiration

Transpiration is a large component of ET, especially for closed-canopy forests (Jasechko *et al.*, 2013), and variations in ET are driven by vegetation (Vadeboncoeur *et al.*, 2018; see Chapter 3, this volume). Future increases in climate variability and extremes can directly influence tree water use because tree-level transpiration is closely related to changes in temperature, vapor pressure deficit (VPD), CO_2 concentration, precipitation regimes and soil moisture availability. Climate extremes can impact the regeneration, distribution and mortality of trees, and can indirectly change tree transpiration rates (Brodribb *et al.*, 2020). However, patterns of tree-level water use in response to climate change can vary significantly among tree species and functional groups (Bryant *et al.*, 2021), and are influenced by the physiological adaptive strategies and acclimation potentials of the tree species (Thurman *et al.*, 2020).

Response of tree transpiration to warming

Future air temperature changes caused by global warming may prolong the growing season by either advancing the date of leaf flush or postponing the date of leaf senescence in temperate, boreal and subalpine forests. The number of days for plants to transpire water will be increased accordingly (Oishi *et al.*, 2018). However, trees have adaptive biophysical strategies such as accelerated transpiration for cooling to avoid thermal damage or mortality under high-temperature conditions during heatwaves (Aparecido *et al.*, 2020). This process depends mainly on soil water availability (Xu *et al.*, 2020). A general understanding is that tree transpiration in alpine and boreal forests is increased under a warming climate, where low air and soil temperatures are used to limit tree transpiration in the non-growing season (winter and early spring) (Wieser *et al.*, 2015). For example, experimental studies show that early-season transpiration and leaf water potential in alpine and boreal forests increase more significantly and quickly in heated plots than in unheated control and cooled plots. However, rapid temperature oscillation (i.e. freeze–thaw) in cold ecosystems can also induce xylem embolism and cavitation, as well as hydraulic failure, limiting tree transpiration increments (Harrison *et al.*, 2020).

Response of tree transpiration to increasing VPD

A warming climate is accompanied by increases in VPD. Increasing VPD can induce both instantaneous and long-term stomatal responses, and consequently, tree transpiration changes (Grossiord *et al.*, 2020). Tree-level transpiration could either increase or decrease with increasing VPD, depending on VPD ranges and other environmental variables such as soil moisture (Chen *et al.*, 2022). Some studies have found increases in transpiration rates in response to increasing VPD, although with decreasing leaf water potentials and stomatal conductance (Zhang *et al.*, 2023). The increase in transpiration can cause a more rapid depletion of soil moisture, thereby exacerbating drought stress, especially under conditions of high VPD and low precipitation (Duan *et al.*, 2014). In contrast, the transpiration of some tree species declines with high VPD (Whitley *et al.*, 2013). The adaptive strategies of transpiration for trees are related to their xylem sapwood characteristics (Zhang *et al.*, 2023). Trees with xylem susceptible to hydraulic failure require water storage in the sapwood (i.e. capacitance), and thus curtail transpiration via strict stomatal control (Meinzer *et al.*, 2008). In contrast, trees with xylems that are highly resistant to embolism, and thus are drought tolerant, can avoid hydraulic failure. They can maintain photosynthesis and transpiration with increasing VPD (Skelton *et al.*, 2015). In addition, the tree's transpiration response to VPD change can be affected by water availability in deep soils. Deep soil water sustains tree transpiration during rainless days, and thus modulates the transpiration response to VPD and droughts (Chen *et al.*, 2014).

Response of tree transpiration to elevated CO_2 concentration

The tree's transpiration response to elevated CO_2 concentration is complex and variable due to the feedback between plant biophysics and soil water, and differences in tree ages and species (Xu *et al.*, 2020). Some studies have reported a fertilization effect of elevated CO_2 concentration that increases photosynthesis rate and associated transpiration (Piao *et al.*, 2017), while others have revealed unchanged transpiration, ET and water yield due to CO_2 concentration enrichments (Ward *et al.*, 2018). However, tree growth may be limited by soil water deficit due to increasing tree transpiration, especially in dry environments or seasons (Jung *et al.*, 2010). Moreover, partial stomatal closure under elevated CO_2 can restrict water vapor diffusion out of leaves and restrain transpiration rates further (Swann *et al.*, 2016).

Response of tree transpiration to drought

Droughts are becoming severe and frequent globally. During droughts, trees generally decrease the stomatal aperture size to reduce water loss from transpiration. However, the stomatal regulation mechanism is highly variable across biomes and even among coexisting species (Chen *et al.*, 2014). Tree species with deep roots access water from deep soil layers, maintaining higher transpiration levels during droughts in comparison with shallow-rooted tree species (Zapater *et al.*, 2013). Coniferous species tend to be more hydrologically sensitive to drought than broadleaf species (Carnicer *et al.*, 2013). For example, the reduction in transpiration of *Pinus sylvestris* and *Pinus nigra* is greater than that of *Quercus faginea* in a continental Mediterranean mixed forest during summer drought (Grossiord *et al.*, 2015). Similarly, the coniferous species *Pinus strobus* in temperate mixed hardwood forest ecosystems in north-eastern USA also exhibits a greater stomatal sensitivity to moisture stress and a quick reduction in transpiration during droughts, consistent with a drought-avoidant (isohydric) strategy. In contrast, *Quercus rubra* significantly reduces its water use unless the drought increases significantly, consistent with drought-tolerant (anisohydric) behavior (Asbjornsen *et al.*, 2021). Also noteworthy is that *Q. rubra* is highly adaptive to moderate droughts rather than extreme droughts. Extreme droughts can push the species beyond its tolerance threshold. These changes in tree-level transpiration inevitably affect forest stand-level transpiration or watershed-level ET (Vadeboncoeur *et al.*, 2018).

Response of water-use efficiency to climate change

Forest water-use efficiency (WUE, defined as the amount of carbon assimilated as biomass produced per unit of water used by vegetation) is a useful metric for understanding the water–carbon coupling cycles of forest ecosystems under climate change. The elevated atmospheric CO_2 concentration can significantly affect WUE by altering CO_2 and H_2O exchanges via leaf stomata. Several studies have reported increases in forest WUE in recent decades (Adams *et al.*, 2020), but variations exist among forest and climate types. As indicated by a global study of intrinsic WUE (iWUE), the iWUE increment of evergreen needleleaf forests in response to rising atmospheric CO_2 ($\Delta WUE/\Delta c_a$, the rate of change in iWUE of trees relative to those in atmospheric $[CO_2]$ (c_a)) was significantly higher than that of deciduous broadleaf, deciduous needleleaf and evergreen broadleaf forests (Fig. 4.2a; Zhang

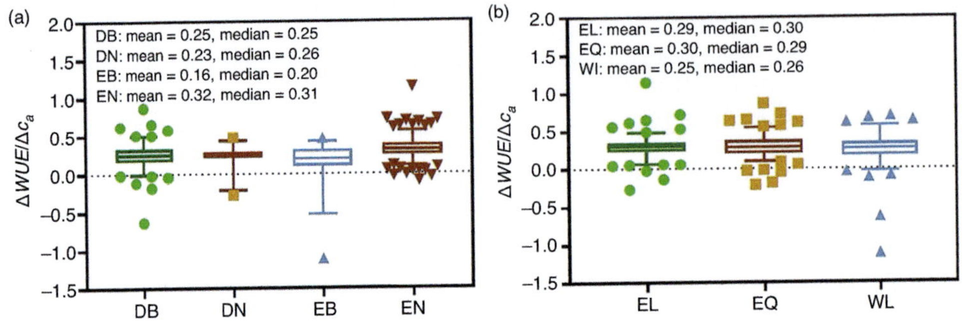

Fig. 4.2. Comparisons of $\Delta WUE/\Delta c_a$. Rates of change in intrinsic water-use efficiency (WUE) of trees relative to those in atmospheric [CO_2] (c_a) between different forest types (a) and climate types (b). DB, deciduous broadleaf forest; DN, deciduous coniferous forest, EB, evergreen broadleaf forest; EN, evergreen coniferous forest. EL, EQ and WL refer to energy-limited, equitant and water-limited environments respectively, which are classified by dryness index (Zhang *et al.*, 2017). Data from Adams *et al.* (2020).

et al., 2023). Meanwhile, the increase in iWUE of forests in energy-limited environments in response to elevated atmospheric CO_2 was significantly greater than that in water-limited environments (Fig. 4.2b).

However, the increase in WUE may not translate into enhanced tree growth due to limitations from other environmental factors. These include VPD, precipitation and temperature, which play their roles in the ratio of net photosynthesis and stomatal conductance. These factors can modulate the WUE response to rising CO_2, especially across functionally distinct plant groups with different wood anatomies and leaf morphologies (Rumman *et al.*, 2018). For example, increasing VPD was the major driver for the WUE increase in European forests during the 20th century according to the observations of tree rings. The VPD impact on WUE sometimes overrides the effect of CO_2 change (Xu *et al.*, 2018). Warming can also cause a strong increase in WUE by reducing stomatal conductance, mainly observed in trees from semi-arid regions (Szejner *et al.*, 2018) or in seasonal temperate forests during the dry season (Urrutia-Jalabert *et al.*, 2015).

4.2.2 Evapotranspiration

Global ET has been reported to be shifting from increasing (before 1998) to leveling off (Jung *et al.*, 2010) or decreasing (after 1998) (Zhang *et al.*, 2016). However, climate change impacts on ET vary among regions and over time (Yang *et al.*, 2015). Global warming is expected to increase ET in wet areas and decrease ET in dry areas because of the general pattern that wet areas are becoming wetter and dry areas are drier under climate change (IPCC, 2021). In addition, increasing humidity and higher CO_2 concentrations may reduce transpiration and counteract the warming effects on ET (Snyder *et al.*, 2011). In contrast, warming combined with wetting are likely to boost annual ET (Fig. 4.3). Concurrent warming (3.47°C) and rising annual precipitation (110 mm, +17%) was estimated to increase ET by 13% (41 mm) in Krycklan catchment, Sweden (87% of forest coverage) (Teutschbein *et al.*, 2015). Similarly, warming by 1°C, 2°C and 4°C increased ET by 3%, 6% and 15%, respectively, in the Upper Mississippi River Basin, while increased/decreased precipitation also has similar impacts to increase/reduce ET (Wu *et al.*, 2012). Moreover, under double CO_2 experiments, ET decreased by 11%, suggesting suppressed transpiration from stomatal closure due to elevated CO_2 concentrations (Wu *et al.*, 2012). Sometimes, warming alone can also enhance ET. As reported by Zhang *et al.* (2014), warming (+1.15°C, 6.97%) enhanced annual potential ET (+67 mm, 6.38%) in a densely forested watershed (70% of forest coverage) although with relatively stable rainfall from 1975 to 2009.

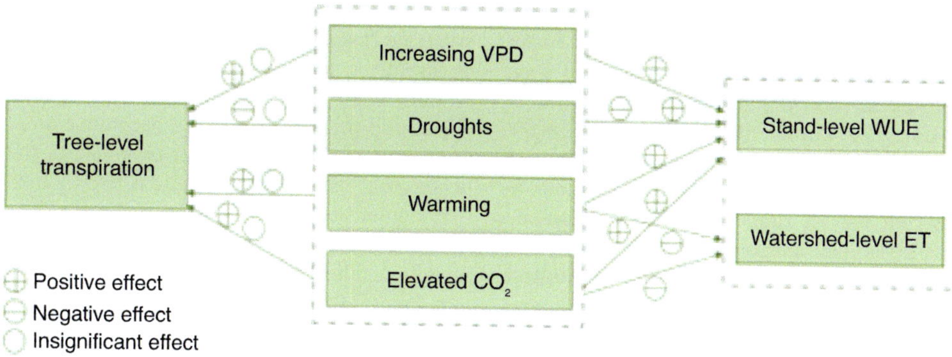

Fig. 4.3. Climate change impact on tree-level transpiration, stand-level water-use efficiency (WUE) and watershed-level evapotranspiration (ET).

4.2.3 Streamflow

Direct impact of climate change on streamflow

In forested watersheds, climate change can directly affect the magnitude, timing, duration, variability and frequency of streamflow by altering hydrological processes involving precipitation, evaporation, transpiration, soil water storage and then runoff generation. The impact of climate change on annual streamflow in a forested watershed can be positive or negative depending on the impact of climate change on precipitation and ET (Creed *et al.*, 2014). Climate change will increase streamflow in watersheds with a greater increment in annual precipitation than in ET (Lu *et al.*, 2013), and vice versa (Liu *et al.*, 2019). The seasonal streamflow response to climate change often varies with seasonal precipitation changes. Climate change is usually found to increase wet season streamflow and decrease dry season flow resulting from increased precipitation in the wet season and decreased in the dry season due to climate change in watersheds dominated by a monsoon climate, such as in Malaysia and China (Li *et al.*, 2021). The negative effects of climate change on streamflow are more widespread in water-limited environments such as in south-eastern Europe, the Mediterranean, South America and Africa (Ahmadalipour *et al.*, 2019). Climate change was predicted to cause 26% reductions in annual streamflow in a Mediterranean watershed dominated by shrubland (Senent-Aparicio *et al.*, 2018). Similar findings have been reported from the Ecuadorian Andes, where, despite increasing rainfall, afforestation and other land-use changes are associated with declining streamflow (Kayitesi *et al.*, 2022), and in the Gilgel Abay catchment in Lake Tana basin where climate change was projected to reduce seasonal and annual streamflow on average by 33% when the temperature increases by +2°C and rainfall decreases by 20% (Abdo *et al.*, 2010).

The effects of climate change on floods are mixed. A substantial increase in the magnitude of floods was projected in most areas below the 60°N line in Europe, even in the Mediterranean under a + 2°C global warming (Roudier *et al.*, 2016). However, based on the projected rainfall, the results showed that flow magnitudes of large floods are unlikely to increase in the future in two catchments in Queensland, although intensive land-use change coupled with climatic change has raised the concern on flood risk (Chen and Yu, 2015). Conversely, the magnitude of floods has even been projected to decrease due to a decline in snowpack in areas where most of the floods are caused by spring snowmelt and rainfall, mainly in regions above the 60°N line in Europe (Roudier *et al.*, 2016).

Climate change can significantly affect the timing and magnitude of streamflow in snow-dominated forested watersheds, because global warming can significantly alter water and energy balances in snow environments (Creed *et al.*, 2014). In the snow-dominated Rocky Mountain region in the western USA, global warming has been shown to cause early snowmelt and earlier arrival of floods (Foster *et al.*, 2016). Similar findings have also been reported in snow-dominated forest watersheds in northern Europe (Graham *et al.*, 2007). Climate change has increased precipitation in the north-eastern USA, the fraction of precipitation that falls as rain, and flooding (Marsooli *et al.*, 2019). The increase in air temperature and less snow-affected streamflow regimes suggest an increased role of vegetation in regulating flow regimes in the spring in north-eastern USA.

The streamflow response to climate change varies with climate types, hydrological regime and forest characteristics (e.g. tree species and ages). Generally, the streamflow response to climate change can be more pronounced in snow-dominated or water-limited watersheds, especially in watersheds with increasing droughts due to climate change and intensively managed plantations of non-native tree species (e.g. *Eucalyptus* plantations in Brazil, Chile, Uruguay and Australia) or with young plantations of coniferous species (Jones *et al.*, 2017). The tree species often plays a key role in determining the streamflow response to climate change. For example, water yield declined over 12 years under climate change in catchments with radiata pine and *Eucalyptus* plantations in Chile (Iroumé *et al.*, 2021). However, the baseflow steadily increased after the replacement of *Eucalyptus* plantations with the early stages of native forest restoration (Lara *et al.*, 2021). In addition, the response of streamflow to climate change can vary with forest ages in a watershed. For example, water yield during the dry summer season is lower in catchments with young (25–50 year-old) regenerating conifer forests than in catchments with mature and old forests in the Pacific Northwest of the USA. This effect is enhanced during dry years (Perry and Jones, 2017), implying that water yield from catchments dominated by young conifer forests may decline in future warmer, drier summers.

Indirect impact of climate change on streamflow

Streamflow can also be impacted indirectly through altered forest ecohydrological processes associated with changes in forest structure and composition, and by forest disturbances and succession in response to climate change (Winkler *et al.*, 2010). Climate change-related disturbances including fire, ice storms, hurricanes, insect and disease outbreaks, species shifts, and invasive plants are becoming more frequent and catastrophic. For example, in the 'water-rich' south-eastern USA, a rise in hurricane intensity is likely to increase flood risk in the coastal plains along with increasing drought severity and heatwaves. Global warming is likely to increase atmospheric evaporation demand, drought severity and frequency, resulting in more catastrophic fires and large-scale outbreaks of insects (Sun and Vose, 2016). Fires are likely to increase peak flow and total runoff and sediment loads to down streams, but the response intensities vary with climate and fire intensity (Hallema *et al.*, 2018). Similarly, the projected fire frequency and severity are very likely to grow with an increasing number of hot days in Afromontane Forest sites in Africa, resulting in more impacts on water resources in already water-stressed areas (Stehle, 2018). The impact of beetle infestation on streamflow can be limited. Studies suggest that streamflow responses to beetle infestation are less pronounced than to fire or logging in the Rocky Mountains of North America and British Columbia, Canada (Ren *et al.*, 2021).

Changes in forest structures and composition due to climate change can also impact streamflow. For instance, a 76-year monitoring study in the southern Appalachian Mountains of North Carolina, USA, showed that water yield decreased by 18% since the mid-1970s (Caldwell *et al.*, 2016) due to a shift in dominance from xerophytic oak and hickory species to several mesophytic species (i.e. mesophication) that consume relatively more water). Similarly, climate change leads to increasing birch (*Betula platyphylla* Sukaczev) and decreasing larch (*Larix gmelinii* Rupr.), and consequently a reduction of 10 mm in annual stream flow in a sub-boreal watershed in north-eastern China in recent 30 years (Yu *et al.*, 2019). In a montane forest in South Africa's Cape Fold Belt Mountains, climate change is projected to significantly increase alien invasive species that are more favorable for global warming (Holmes *et al.*, 2005). The potential increment of water consumption by invasive trees can be as much as millions m^3 (Le Maitre *et al.*, 2002), greatly reducing streamflow in the future. In addition, global warming can cause the upward shift of treelines in many alpine forested watersheds, which may have an unexpected impact on streamflow. For example, increased forest covers in the Swiss Alps due to rising treelines yielded limited effects on annual streamflow, but had a seasonally variable effect on evaporation and soil moisture (Köplin *et al.*, 2013). A recent study suggested that CO_2 fertilization has limited effects on water yield (Wei *et al.*, 2024)

Nevertheless, it is rather difficult to separate the direct and indirect effects of climate change on streamflow, given the close interactions and feedback among forests, climate and water. Past and present watershed disturbances and vegetation succession may neutralize the effects of climate change on streamflow in some forested sites over longer study periods. At long-term ecological research sites in forested headwater catchments in the USA with 20–60 year climate and streamflow records, streamflow trends were directly related to climate trends resulting from changes in ice and snow at only seven of 19 sites (Creed *et al.*, 2014). In contrast, at other sites, human and natural disturbances and vegetation succession obscured the effects of climate change on streamflow (Jones *et al.*, 2012). In general, the climate change effect on streamflow can be either negative (i.e. increasing water scarcity) or positive (i.e. increasing water availability). It is collectively determined by watershed climate conditions, the hydrological regime, vegetation change, human disturbances and the climate change trajectory, as well as interactions of these drivers.

4.3 Interactions and Feedback among Forest, Water and Climate

Forest change (e.g. deforestation and forestation including reforestation and afforestation) and climate variability play a co-equal role in shaping annual streamflow variations. Two climate change variables (temperature and precipitation) drive the forest hydrological processes that control water resources (Wei *et al.*, 2017; Zhou *et al.*, 2015). Interactions between forest and climate changes result in offsetting and amplifying hydrological effects and feedback (i.e. the impact of forest change on climate and/or vice versa and associated modifications of hydrological processes) (Ellison *et al.*, 2017). Thus, understanding the interactions and feedbacks among forests, water and climate is essential for future forest–water management under a changing climate.

4.3.1 Offsetting and amplifying effects

The offsetting and amplifying effects of forest and climate changes on streamflow depend on their individual strengths and direction of impact. Climate change can either have positive or negative effects on water yield. In comparison, forest change impacts on streamflow are more diverse and variable in their magnitudes and directions, which are likely related to forest change, climate and watershed characteristics (Andréassian, 2004; Zhang *et al.*, 2017; Zhang and Wei, 2021). When the influences of forest and climate changes move in opposite directions, the resulting offsetting effects tend to stabilize streamflow. In contrast, when the influences of climate change and forest change go in the same direction, they produce amplifying or cascading effects on streamflow, resulting in increased risks of floods or droughts. According to a study of 67 forested watersheds (with forest coverages > 30%), 51 watersheds exhibited cascading effects on annual streamflow, while 16 showed offsetting effects on annual streamflow. The offsetting or amplifying effects under future climate change are expected to intensify due to shifts in climate patterns and increasing forest changes (Feng *et al.*, 2019). In particular, the amplifying effects between climate change-induced extreme weather events such as storms, heat waves and large-scale forest disturbances such as severe wildfire may trigger additional risks (e.g. floods and landslides) (Hallema *et al.*, 2018). Significant uncertainty exists about the possible magnitude and frequency of these combined effects between extreme climate and severe forest disturbance (Creed and van Noordwijk, 2018).

4.3.2 Larger-scale forest–water–climate interactions and feedbacks

Forest–water interactions have been well studied within watershed boundaries. However, climate-related feedback to forest–water interactions can move across watershed boundaries. Forests act as reservoirs that store a large amount of water on terrestrial surfaces (Zhang and Wei, 2021). They also work as biological pumps that release a large proportion of stored or intercepted water into the atmosphere through ET (Ellison *et al.*, 2017). The water vapor evaporated or transpired by forests consistently returns as precipitation locally and in downwind locations. On average, about 80% of ET moves downwind to fall as precipitation in other watersheds (Creed and van Noordwijk, 2018). The local and regional precipitation recycling clearly demonstrate that forest–water interactions play a role in moving water across watersheds and regions, thereby affecting the geographical distribution of water resources available on terrestrial surfaces (Gebrehiwot *et al.*, 2019).

Forestation can change land surface roughness, albedo, sensible heat, and then ET rate, which further impact local and regional climate (e.g. air mass, cloud formation, surface temperature and precipitation cycling) and eventually alter available water resources (Ellison *et al.*, 2017). Various modeling studies or satellite observations suggest that the impact of large-scale forestation on precipitation is highly complex and variable among regions. Forestation can promote condensation by altering atmospheric temperature, moisture, wind speed and direction. It can also constrain convection by canopy interception and surface cooling. Whether forestation increases precipitation or not depends on these complex effects at regional scales. A continental-scale study by rain-gauge observations found that forestation of rain-fed agricultural land in Europe caused substantial precipitation changes (Meier *et al.*, 2021). Forestation was estimated to increase summer precipitation in most downwind regions. In a reforestation scenario, summer precipitation was predicted to increase by 7.6% ± 6.7% on average over Europe (0.13 ± 0.11 mm/day), potentially offsetting a substantial part of the projected precipitation decrease from climate change. However, the downwind effect in winter was positive in coastal areas but near neutral and negative in continental and northern Europe, respectively.

The precipitation change induced by forestation can be further transferred to surface-water availability. Forestation or greening was estimated to increase global water availability by a rate of 0.26 mm/year during 2001–2018, offsetting about 15% of the recently observed decline in global water availability (Cui *et al.*, 2022). The areas where greening is increasing water availability account for 53% of the global land surface, where the increment of precipitation input from upwind forest lands is higher than local ET loss by greening. These areas are mainly in north-east China, south-east China, Europe and West Africa. In contrast, greening has a negative impact on local and downwind water availability in 8% of the land surface, mostly located in water-limited or high-elevation regions (e.g. the southern Qinghai-Tibet Plateau, western Canada, and the northern Congo Basin) (Cui *et al.*, 2022).

4.4 Managing the Forest–Water Nexus for Climate Change Adaptation

4.4.1 Silvicultural options for enhancing forest hydrological resilience

Forest structure and species composition play significant roles in regulating ecohydrological processes across multiple spatial scales (Hu *et al.*, 2024). However, tree species and functional groups differ in their ecophysiological traits that influence water-use patterns. Thus, the hydrological impact of climate change can be either offset or enhanced by trees, depending on water consumption and the forest population's response to climate change (Vose *et al.*, 2016). Therefore, to cope with climate change, appropriate silvicultural options (e.g. selecting tree species with greater adaptive ability to climate change, selective logging, thinning and irrigation) are needed to enhance the resilience of the forest ecosystem, especially to extreme climate events (e.g. drought) (Table 4.1).

The selection of tree species relies on the site conditions, disturbance history, water demand, biophysical responses to climate change and potential climate change impact on local water resources. This could be particularly true for a water-limited environment with growing water stress under climate change. In semi-arid and arid environments or temperate monsoon

Table 4.1. Adaptive forest management options to manage the hydrological impact of climate change.

Environment	Hydrological impact	Risks to forest ecosystem and society	Adaptive forest management options
Semi-arid and arid watersheds; temperate watersheds with distinct dry seasons	ET (+), annual streamflow (–), low flow (–) and peak flow (–)	Water shortage and drying up of streams; Increasing soil moisture stress and hydrological droughts; Increasing invasive species and forest degradation; Loss of aquatic and flood-plain habitats	Thinning; Selective logging of plantations with high water use (e.g. young plantations, radiata pine or *Eucalyptus* plantations); Removal of invasive species; Planting of native species; Replacement of drought-avoidant species (e.g. *Pinus sylvestris*, *P. nigra* and *P. strobus*) with drought-tolerant species (e.g. *Quercus rubra and Q. faginea*); Increasing water-retention facilities (e.g.terraces and ponds) for water storage and irrigation
	ET (–), annual streamflow (+), low flow (–) and peak flow (+)	Increasing flood risks and sediment; Tree dieback or mortality; Increasing forest fires and insect infestations; Loss of aquatic habitats	Increasing water-retention facilities (e.g.terraces and ponds) for flood control, sediment control and dry-season water supply; Reducing tree stocking, removal of forest litter, increasing fire buffers, thinning forests and prescribed burning; Removal of infected or dead trees, and pest control by introducing predators; Restoration of fire-burnt forest floors; Maintaining and increasing forest riparian buffers
Alpine and boreal snow-dominated watersheds	ET (+), annual streamflow (–), low flow (–) and peak flow (–)	Water shortage; Increasing soil moisture stress and summer hydrological droughts; Forest structure and species composition changes (e.g.increasing invasive species or broadleaf species); Forest expansion with treeline shift; Increasing insect infestations; Loss of aquatic and flood-plain habitats	Thinning; Selective logging of invasive species or broadleaf species (e.g.birch) with high water consumption; More logging activities performed at higher elevations to synchronize snowmelt processes at both high and low elevations; Removal of infected or dead trees, and pest control by introducing predators; Increasing water-retention facilities (e.g.terraces and ponds) to guarantee water supply and restore aquatic and flood-plain habitats

Continued

Table 4.1. Continued

Environment	Hydrological impact	Risks to forest ecosystem and society	Adaptive forest management options
Subtropical and tropical rain-dominated watersheds	ET (–), annual streamflow (+), low flow (–) and peak flow (+)	Increasing flood risks and sediment; Loss of aquatic habitats	Restoring hydrological functions of natural forests; Increasing plantations; Redesigning logging roads (e.g.minimizing direct discharge of runoff from roads to streams) and installing larger culverts for flood control; Constraining or carefully designing logging activities (e.g.time, location, proportion and soil disturbance); Maintaining and increasing forest riparian buffers

ET, evapotranspiration; +, positive impact; –, negative impact;

environments with seasonal droughts, drought-tolerant native species could be the first option for silviculture, and tree species with high water consumption (e.g. *Pinus radiata* and *Eucalyptus globulus*) should be planted with caution (Table 4.1). For example, the drought-tolerant native species *Quercus rubra*, *Q. prinus*, and *Pinus palustris* in eastern USA (Klos *et al.*, 2009), and *Diospyros kaki* Thunb. and *Cotinus coggygria* Scop. in northern China (Chen *et al.*, 2019) could be selected for future silviculture to achieve climate change adaptation goals in these regions, given their adaptation to droughts.

In addition, tree species composition must also be carefully designed in silviculture to enhance the hydrological resilience of forests, given that climate change induces shifts in tree species composition and consequently affects hydrological processes (Wattenbach *et al.*, 2007). In semi-arid and arid regions where climate change reduces streamflow, the proportion of non-native species or species with higher water consumption in plantations must be determined by a robust assessment of their water consumption to identify the best tree species composition for mitigating negative climate change impact. In alpine and boreal forests where climate change leads to a degradation of coniferous species and an increase in broadleaf species, selective logging of broadleaf species (e.g. *Betula platyphylla* Sukaczev in north-east China) with high water consumption and replanting with coniferous species are necessary to recover the original tree species composition. This can help counteract the negative effect of climate change on water and maintain watershed hydrological resilience to climate change (Table 4.1). The appropriate approaches for enhancing the hydrological resilience of forests over different temporal and spatial scales are highly context dependent. Nevertheless, our current knowledge suggests that tree species selection and composition are crucial for strengthening the resilience of forests to climate change impacts. This strategy aligns well with nature-based solutions to confront the many emerging sustainability issues such as water shortages, urbanization and climate change (Springgay, 2019).

4.4.2 Watershed-level forest management for improving hydrological resilience

Forest management strategies to mitigate the hydrological impacts of climate change are essential for future watershed management. These strategies aim to promote and increase ecosystem resilience and restore the hydrological functions and services of forest ecosystems under changing climate conditions. Large-scale plantations of non-native tree species such as *P. radiata* and *Euc. globulus* or the expansion of invasive species with a higher water use may increase ET and thus reduce high flows and flood risks. However, densely planted fast-growing forests or increasing invasive species may aggravate drought and water shortages in dry seasons or in semi-arid and arid watersheds (Meli *et al.*, 2024). The replacement of native species with non-native species, the removal of invasive species, and increasing water-retention facilities for water storage and irrigation in dry environments should be adopted in forest management to increase the hydrological resilience of forested watersheds to climate change (Table 4.1). In contrast, in subtropical and tropical rain-dominated watersheds with increasing climate change-induced flood risks, forest management options such as restoring native forests including riparian forests and wetlands, redesigning logging roads and installing larger culverts are needed to mitigate the potential impacts of extreme precipitation (Table 4.1) (Mukherjee *et al.*, 2024a).

In addition, spatial distribution of forestry activities can also impact streamflow at a watershed level. Trees planted in riparian zones are found to have a greater influence on streamflow, especially in low-flow seasons. Similarly, plantations on upslopes often have less impact on streamflow than on downslopes in water-limited watersheds. In snow-dominated watersheds, logging at higher elevations can synchronize snowmelt processes at higher and lower elevations, resulting in increasing the magnitude and advancing of spring floods. Therefore, the spatial distribution of forest management practices can be designed to mitigate climate change impact on streamflow. For example, more logging activities are recommended to be performed at higher elevations in alpine watersheds if climate change decreases streamflow including peak flow due to less snowpack. These forestry practices can increase streamflow and help with regulating flow regimes, and can mitigate the negative hydrological impact of climate change.

Climate change leads to more frequent wildfires in dry watersheds or dry seasons, which can increase streamflow and flood risks, and degrade forest hydrological functions (e.g. burnt trees and hydrophobic soils) (Creed and van Noordwijk, 2018). Forest management options such as plantations of fire-tolerant tree species, reducing tree stocking, carefully designed fire buffers, prescribed fires, the removal of forest litter, thinning and restoring fire-burnt forest floors are recommended to mitigate climate change impact on water. In general, the risk and vulnerability of watershed-level forest management for mitigating extreme climate and associated forest disturbances (e.g. fires, insect outbreaks and droughts) must be explicitly evaluated within the site-specific context (Table 4.1).

Advanced decision support systems can help land managers and decision makers to manage water under a changing climate and other environmental changes (Sun *et al.*, 2023). Communicating forest hydrological research and climate change science to land managers and the public is needed for climate adaptation. One example is the USDA Forest Service 'Forests to Faucet' tool that assesses the effects of environmental stressors (e.g. climate change, land-use change, fire, and insects and diseases) on water quantity and quality, and ranks the importance of watershed for water supply under different land-use and climate scenarios (Mack *et al.*, 2022). More adaptive tools and indigenous knowledge are needed to evaluate the forest–water nexus in a complex context, especially for exploring unanticipated climate change impacts.

4.4.3 Integrating the forest–water nexus into adaptive forest management

As indicated in the sections above, climate change and its associated forest disturbances and extreme weather can alter forest composition and structure, distribution and growth, and thus forest hydrological functions and water-related ecosystem services. Current forest adaptation for climate change is often emphasized extensively in the forestry sector, with two basic options to prioritize carbon sequestration, and water-related forest ecosystem services that sustain carbon sequestration are hardly considered. Therefore, it is essential to integrate adaptive forest management into climate change mitigation and water resource management in the future (FAO/IUFRO/USDA, 2021).

The interactive hydrological effects between forest and climate changes need to be considered in designing climate change adaptation through forest management. Offsetting effects can further influence water supply stabilization, while amplifying effects can exacerbate floods or drought risk. Interactions between climate change-induced extreme weather events such as storms and heat waves, and large-scale forest disturbances such as severe wildfire may consequently trigger additional risks, such as floods, sediment and debris flow (Mukherjee *et al.*, 2024b) and landslides. Nevertheless, significant uncertainty exists about the possible geolocation, magnitude and frequency of interactive effects between extreme climate and severe forest disturbance.

Our current understanding of the forest–water nexus falls within the known impacts of a changing climate. This gained knowledge is either historical or based on forecasted data and modeling results based on historical ecosystem responses to climate change. However, climate change coupled with other environmental changes including air pollution, a changing wildland–urban interface, and invasive plant and animal species, complicates the forest ecosystem's responses. The complex interactions among these factors becomes increasingly difficult to anticipate (McNulty *et al.*, 2024). The previously unobserved impacts are growing and are poorly known under a changing environment, which poses great challenges for mitigating forest management for climate change.

4.5 Conclusions

Climate change can pose both direct and indirect effects on forest hydrology at multiple scales with variable influences. The impacts of climate change on the forest–water nexus are complex in watersheds due to differences in climate, watershed features and forest characteristics. The interactions and feedback among the forest, water and climate introduce more complexities in managing the forest–water nexus for the sustainable provision of water and other ecosystem services. Therefore, adaptive local forest management strategies are needed to promote forest ecosystem resilience to climate change and to improve water-related ecosystem services.

References

Abdo, K.S., Fiseha, B.M., Rientjes, T.H.M., Gieske, A.S.M. and Haile, A.T. (2010) Assessment of climate change impacts on the hydrology of Gilgel Abay catchment in Lake Tana basin, Ethiopia. *Hydrological Processes* 23, 3661–3669.

Adams, M.A., Buckley, T.N. and Turnbull, T.L. (2020) Diminishing CO_2-driven gains in water-use efficiency of global forests. *Nature Climate Change* 10, 466–471.

Ahmadalipour, A., Moradkhani, H., Castelletti, A. and Magliocca, N. (2019) Future drought risk in Africa: Integrating vulnerability, climate change, and population growth. *Science of The Total Environment* 662, 672–686.

Amatya, D.M., Sun, G., Skaggs, R.W., Chescheir, G.M. and Nettles, J.E. (2006) Hydrologic effects of global climate change on a large drained pine forest. In: *Proceedings of the ASABE-Weyerhaeuser sponsored International Conference on Hydrology and Management of Forested Wetlands, New Bern, North Carolina, 8–12 April 2006*, ASABE Publication No. 701P0406. American Society of Agricultural and Biological Engineers, St. Joseph, Michigan, pp. 383–394.

Andréassian, V. (2004) Waters and forests: From historical controversy to scientific debate. *Journal of Hydrology* 291, 1–27.

Aparecido, L.M.T., Woo, S., Suazo, C., Hultine, K.R. and Blonder, B. (2020) High water use in desert plants exposed to extreme heat. *Ecology Letters* 23, 1189–1200.

Asbjornsen, H., McIntire, C.D., Vadeboncoeur, M.A., Jennings, K.A., Coble, A.P. *et al.* (2021) Sensitivity and threshold dynamics of *Pinus strobus* and *Quercus* spp. in response to experimental and naturally occurring severe droughts. *Tree Physiology* 41, 1819–1835.

Berghuijs, W.R., Woods, R.A. and Hrachowitz, M. (2014) A precipitation shift from snow towards rain leads to a decrease in streamflow. *Nature Climate Change* 4, 583–586.

Bren, L. (2023) Climate change, drought, and forest hydrology. In: Bren, L. (ed.) *Forest Hydrology and Catchment Management: An Australian Perspective*, 2nd edn. Springer, Dordrecht, Netherlands, pp. 329–357.

Brodribb, T.J., Powers, J., Cochard, H. and Choat, B. (2020) Hanging by a thread? Forests and drought. *Science* 368, 261–266.

Bryant, K.N., Frederiksen, B.W. and Rosenthal, D.M. (2021) Ring- and diffuse-porous species exhibit a spectrum of hydraulic behaviors from isohydry to anisohydry in a temperate deciduous forest. *Trees* 36, 485–495.

Buechel, M., Slater, L. and Dadson, S. (2022) Hydrological impact of widespread afforestation in great britain using a large ensemble of modelled scenarios. *Communications Earth and Environment* 3, 6.

Caldwell, P.V., Miniat, C.F., Elliott, K.J., Swank, W.T., Brantley, S.T. *et al.* (2016) Declining water yield from forested mountain watersheds in response to climate change and forest mesophication. *Global Change Biology* 22, 2997–3012.

Carnicer, J., Barbeta, A., Sperlich, D., Coll, M. and Peñuelas, J. (2013) Contrasting trait syndromes in angiosperms and conifers are associated with different responses of tree growth to temperature on a large scale. *Frontiers in Plant Science* 4, 409.

Chen, Y.R. and Yu, B.F. (2015) Impact assessment of climatic and land-use changes on flood runoff in southeast Queensland. *Hydrological Sciences Journal* 60, 1759–1769.

Chen, L.X., Zhang, Z.Q., Zeppel, M., Liu, C.F., Guo, J.T. *et al.* (2014) Response of transpiration to rain pulses for two tree species in a semiarid plantation. *International Journal of Biometeorology* 58, 1569–1581.

Chen, S.N., Chen, Z.S.N., Xu, H., Kong, Z., Xu, Z.B. *et al.* (2022) Biophysical regulations of transpiration and water use strategy in a mature Chinese pine (*Pinus tabulaeformis*) forest in a semiarid urban environment. *Hydrological Processes* 36, e14485.

Chen, Z.C., Li, S., Luan, J.W., Zhang, Y.T., Zhu, S.D. *et al.* (2019) Prediction of temperate broadleaf tree species mortality in arid limestone habitats with stomatal safety margins. *Tree Physiology* 39, 1428–1437.

Creed, I.F., Spargo, A.T., Jones, J.A., Buttle, J.M., Adams, M.B. *et al.* (2014) Changing forest water yields in response to climate warming: Results from long-term experimental watershed sites across North America. *Global Change Biology* 20, 3191–3208.

Creed, I.F. and van Noordwijk, M. (eds) (2018) *Forest and Water on a Changing Planet: Vulnerability, Adaptation and Governance Opportunities*, World Series Volume 38. IUFRO, Vienna, Austria.

Cui, J.P., Lian, X., Huntingford, C., Gimeno, L., Wang, T. *et al.* (2022) Global water availability boosted by vegetation-driven changes in atmospheric moisture transport. *Nature Geoscience* 15, 982–988.

de Lombaerde, E., Vangansbeke, P., Lenoir, J., Meerbeek, K., Lembrechts, J. *et al.* (2021) Maintaining forest cover to enhance temperature buffering under future climate change. *Science of The Total Environment* 810, 151338.

Duan, H.L., Duursma, R.A., Huang, G.M., Smith, R.A., Choat, B. *et al.* (2014) Elevated [CO2] does not ameliorate the negative effects of elevated temperature on drought-induced mortality in *Eucalyptus radiata* seedlings. *Plant, Cell and Environment* 37, 1598–1613.

Ellison, D., Morris, C.E., Locatelli, B., Sheil, D., Cohen, J. *et al.* (2017) Trees, forests and water: Cool insights for a hot world. *Global Environmental Change* 43, 51–61.

FAO/IUFRO/USDA (2021) *A Guide to Forest-Water Management*, FAO Forestry Paper No.185. FAO, Rome.

Feng, X., Thompson, S.E., Woods, R. and Porporato, A. (2019) Quantifying asynchronicity of precipitation and potential evapotranspiration in Mediterranean climates. *Geophysical Research Letters* 46, 14692–14701.

Foster, L.M., Bearup, L.A., Molotch, N.P., Brooks, P.D. and Maxwell, R.M. (2016) Energy budget increases reduce mean streamflow more than snow–rain transitions: Using integrated modeling to isolate climate change impacts on rocky mountain hydrology. *Environmental Research Letters* 11, 044015.

Gebrehiwot, S.G., Ellison, D., Bewket, W., Seleshi, Y., Inogwabini, B.I. *et al.* (2019) The Nile basin waters and the West African rainforest: Rethinking the boundaries. *Water* 6, e1317.

Graham, L.P., Andréasson, J. and Carlsson, B. (2007) Assessing climate change impacts on hydrology from an ensemble of regional climate models, model scales and linking methods – A case study on the Lule River basin. *Climatic Change* 81, 293–307.

Grossiord, C., Forner, A., Gessler, A., Granier, A., Pollastrini, M. *et al.* (2015) Influence of species interactions on transpiration of Mediterranean tree species during a summer drought. *European Journal of Forest Research* 134, 365–376.

Grossiord, C., Buckley, T.N., Cernusak, L.A., Novick, K.A., Poulter, B. *et al.* (2020) Plant responses to rising vapor pressure deficit. *New Phytologist* 226, 1550–1566.

Hallema, D.W., Sun, G., Caldwell, P.V., Norman, S.P., Cohen, E.C. *et al.* (2018) Burned forests impact water supplies. *Nature Communications* 9, 1307.

Harrison, J.L., Sanders-DeMott, R., Reinmann, A.B., Sorensen, P.O., Phillips, N.G. *et al.* (2020) Growing-season warming and winter soil freeze/thaw cycles increase transpiration in a northern hardwood forest. *Ecology* 101, e03173.

Hewlett, J.D. (1982) *Principles of Forest Hydrology*. University of Georgia Press, Athens, Georgia.

Holmes, P.M., Richardson, D.M., Esler, K.J., Witkowski, E.T.F. and Fourie, S. (2005) A decision-making framework for restoring riparian zones degraded by invasive alien plants in South Africa. *South African Journal of Science* 101, 553–564.

Hu, X., Fu, Z., Sun, G., Wang, B., Liu, K. *et al.* (2024) Importance of forest stand structures for gross rainfall partitioning on China's Loess Plateau. *Journal of Hydrology* 631, 130671.

Hua, F., Bruijnzeel, L.A., Meli, P., Martin, P.A., Zhang, J. *et al.* (2022) The biodiversity and ecosystem service contributions and trade-offs of forest restoration approaches. *Science* 376, 839–844.

IPCC (2021) Contribution of Working Group I to the Sixth Assessment Report of the Intergovernmental Panel on Climate Change. Geneva, Switzerland: IPCC.

Iroumé, A., Jones, J. and Bathurst, J.C. (2021) Forest operations, tree species composition and decline in rainfall explain runoff changes in the nacimiento experimental catchments, south central chile. *Hydrological Processes* 35, e14257.

Jasechko, S., Sharp, Z.D., Gibson, J.J., Birks, S.J., Yi, Y. *et al.* (2013) Terrestrial water fluxes dominated by transpiration. *Nature* 496, 347–350.

Jones, J., Almeida, A., Cisneros, F., Iroumé, A., Jobbágy, E. *et al.* (2017) Forests and water in South America. *Hydrological Processes* 31, 972–980.

Jones, J.A., Creed, I.F., Hatcher, K.L., Warren, R.J., Adams, M.B. *et al.* (2012) Ecosystem processes and human influences regulate streamflow response to climate change at long-term ecological research sites. *BioScience* 62, 390–404.

Jung, M., Reichstein, M., Ciais, P., Seneviratne, S.I., Sheffield, J. *et al.* (2010) Recent decline in the global land evapotranspiration trend due to limited moisture supply. *Nature* 467, 951–954.

Kayitesi, N., Guzha, A. and Mariethoz, G. (2022) Impacts of land use land cover change and climate change on river hydro-morphology – A review of research studies in tropical regions. *Journal of Hydrology* 615, 128702.

Klos, R.J., Wang, G.G., Bauerle, W.L. and Rieck, J.R. (2009) Drought impact on forest growth and mortality in the southeast USA: An analysis using forest health and monitoring data. *Ecological Applications* 19, 699–708.

Köplin, N., Schädler, B., Viviroli, D. and Weingartner, R. (2013) The importance of glacier and forest change in hydrological climate-impact studies. *Hydrology and Earth System Sciences* 14, 619–635.

Lara, A., Jones, J., Little, C. and Vergara, N. (2021) Streamflow response to native forest restoration in former eucalyptus plantations in south Central Chile. *Hydrological Processes* 35, e14270.

Le Maitre, D.C., Wilgen, B.W., Gelderblom, C.M., Bailey, C., Chapman, R.A. *et al.* (2002) Invasive alien trees and water resources in South Africa: Case studies of the costs and benefits of management. *Forest Ecology and Management* 160, 143–159.

Lee, R. (1980) *Forest Hydrology*. Columbia University Press, New York.

Li, Q., Zhang, M., Wei, X., Winkler, R., Spencer, S. *et al.* (2021) Roles of forest disturbance and climate variability on streamflow components in snow-dominated paired watersheds at multiple temporal scales. *Hydrological Processes* 35, e14414.

Liu, N., Harper, R.J., Smettem, K.R.J., Dell, B. and Liu, S. (2019) Responses of streamflow to vegetation and climate change in Southwestern Australia. *Journal of Hydrology* 572, 761–770.

Lu, N., Sun, G., Feng, X.M. and Fu, B.J. (2013) Water yield responses to climate change and variability across the North-South Transect of Eastern China (NSTEC). *Journal of Hydrology* 481, 96–105.

Mack, E., Lilja, R., Claggett, S., Sun, G. and Caldwell, P. (2022) Forests to Faucets 2.0: Connecting Forests, Water, and Communities. General Technical Report WO-99. Forest Service, USDA Washington, DC.

Marsooli, R., Lin, N., Emanuel, K. and Feng, K.R. (2019) Climate change exacerbates hurricane flood hazards along US Atlantic and Gulf coasts in spatially varying patterns. *Nature Communications* 10, 3785.

McCulloch, J.S. and Robinson, M. (1993) History of forest hydrology. *Journal of Hydrology* 150, 189–216.

McNulty, S.G., Archer, E.R.M. and Gush, M.B. (2024) Unknown tipping-points: A method for anticipating future forest disturbance risk. In: McNulty, S.G. (ed.) *Future Forests: Mitigation and Adaptation to Climate Change*. Elsevier, Amsterdam, Netherlands, pp. 279–293.

Meier, R., Schwaab, J., Seneviratne, S.I., Sprenger, M., Lewis, E. *et al.* (2021) Empirical estimate of forestation-induced precipitation changes in Europe. *Nature Geoscience* 14, 473–478.

Meinzer, F.C., Woodruff, D.R., Domec, J.C., Goldstein, G., Campanello, P.I. *et al.* (2008) Coordination of leaf and stem water transport properties in tropical forest trees. *Oecologia* 156, 31–41.

Meli, P., Ellison, D., Ferraz, S.F.D.B., Filoso, S. and Brancalion, P.H. (2024) On the unique value of forests for water: Hydrologic impacts of forest disturbances, conversion, and restoration. *Global Change Biology* 30, e17162.

Mukherjee, S., Amatya, D.M., Campbell, J.L., Gryczkowski, L., Panda, S. *et al.* (2024a) A watershed-scale multi-approach assessment of design flood discharge estimates used in hydrologic risk analyses for forest road stream crossings and culverts. *Journal of Hydrology* 632, 130698.

Mukherjee, S., Panda, S., Amatya, D.M., Dobre, M., Campbell, J.L. *et al.* (2024b) Hydro-geomorphological assessment of culvert vulnerability to flood-induced soil erosion using an ensemble modeling approach. *Environmental Modelling and Software* 183, 106243.

Oishi, A.C., Miniat, C.F., Novick, K.A., Brantle, S.T., Vose, J.M. *et al.* (2018) Warmer temperatures reduce net carbon uptake, but do not affect water use, in a mature southern Appalachian forest. *Agricultural and Forest Meteorology* 252, 269–282.

Perry, T.D. and Jones, J.A. (2017) Summer streamflow deficits from regenerating Douglas-fir forest in the Pacific Northwest, USA. *Ecohydrology* 10, e1790.

Piao, S.L., Liu, Z., Wang, T., Peng, S.S., Ciais, P. *et al.* (2017) Weakening temperature control on the inter-annual variations of spring carbon uptake across northern lands. *Nature Climate Change* 7, 359–363.

Ren, J.N., Adam, J.C., Hicke, J.A., Hanan, E.J., Tague, C.L. *et al.* (2021) How does water yield respond to mountain pine beetle infestation in a semiarid forest? *Hydrology and Earth System Sciences* 25, 4681–4699.

Roudier, P., Andersson, J.C.M., Donnelly, C., Feyen, L., Greuell, W. *et al.* (2016) Projections of future floods and hydrological droughts in Europe under a +2°C global warming. *Climatic Change* 135, 341–355.

Rumman, R., Atkin, O.K., Bloomfield, K.J. and Eamus, D. (2018) Variation in bulk-leaf ^{13}C discrimination, leaf traits and water-use efficiency-trait relationships along a continental-scale climate gradient in Australia. *Global Change Biology* 24, 1186–1200.

Senent-Aparicio, J., Liu, S.T., Pérez-Sánchez, J., López-Ballesteros, A. and Jimeno-Sáez, P. (2018) Assessing impacts of climate variability and reforestation activities on water resources in the head-waters of the Segura River basin (SE spain). *Sustainability* 10, 3277.

Skelton, R.P., West, A.G. and Dawson, T.E. (2015) Predicting plant vulnerability to drought in biodiverse regions using functional traits. *Proceedings of the National Academy of Sciences USA* 112, 5744–5749.

Smith-Tripp, S., Griffith, A., Pasquarella, V.J. and Matthes, J.H. (2021) Impacts of a regional multiyear insect defoliation event on growing-season runoff ratios and instantaneous streamflow characteristics. *Ecohydrology* 14, e2332.

Snyder, R.L., Moratiel, R., Song, Z.W., Swelam, A., Jomaa, I. *et al.* (2011) Evapotranspiration response to climate change. *Acta Horticulturae* 922, 91–98.

Springgay, E. (2019) Forests as nature-based solutions for water. *Unasylva* 251, 3–13.

Stehle, T. (2018) Land of smoke and fire: Reflections on the Southern Cape and Tsitsikamma wildfire disaster. *SA Forestry Online*. Available at: https://saforestryonline.co.za/articles/land-of-smoke-and-fire/ (accessed 16 April 2025).

Sun, G. and Vose, J.M. (2016) Forest management challenges for sustaining water resources in the Anthropocene. *Forests* 7, 68.

Sun, G., Wei, X., Hao, L., Sanchis, M.G., Hou, Y. *et al.* (2023) Forest hydrology modeling tools for watershed management: A review. *Forest Ecology and Management* 530, 120755.

Swann, A.L.S., Hoffman, F.M., Koven, C.D. and Randerson, J.T. (2016) Plant responses to increasing CO_2 reduce estimates of climate impacts on drought severity. *Proceedings of the National Academy of Sciences USA* 113, 10019–10024.

Szejner, P., Wright, W.E., Belmecheri, S., Meko, D., Leavitt, S.W. *et al.* (2018) Disentangling seasonal and interannual legacies from inferred patterns of forest water and carbon cycling using tree-ring stable isotopes. *Global Change Biology* 24, 5332–5347.

Teutschbein, C., Grabs, T., Karlsen, R.H., Laudon, H. and Bishop, K. (2015) Hydrological response to changing climate conditions: Spatial streamflow variability in the boreal region. *Water Resources Research* 51, 9425–9446.

Thurman, L.L., Stein, B.A., Beever, E.A., Foden, W., Geange, S.R. *et al.* (2020) Persist in place or shift in space? Evaluating the adaptive capacity of species to climate change. *Frontiers in Ecology and the Environment* 18, 520–528.

Urrutia-Jalabert, R., Malhi, Y., Barichivivh, J., Lara, A. *et al.* (2015) Increased water use efficiency but contrasting tree growth patterns in *Fitzroya cupressoides* forests of southern Chile during recent decades. *Journal of Geophysical Research: Biogeosciences* 120, 2505–2524.

Vadeboncoeur, M.A., Green, M.B., Asbjornsen, H., Campbell, J.L., Adams, M.B. *et al.* (2018) Systematic variation in evapotranspiration trends and drivers across the northeastern United States. *Hydrological Processes* 32, 3547–3560.

Vose, J.M., Miniat, C.F., Luce, C.H., Asbjornsen, H., Caldwell, P.V. *et al.* (2016) Ecohydrological implications of drought for forests in the United States. *Forest Ecology and Management* 380, 335–345.

Ward, E.J., Ram, O., Seok, K.H., Dohyoung, K., Pantana, T.N. *et al.* (2018) Evapotranspiration and water yield of a pine-broadleaf forest are not altered by long-term atmospheric [CO2] enrichment under native or enhanced soil fertility. *Global Change Biology* 24, 4841–4856.

Wattenbach, M., Zebisch, M., Hattermann, F., Gottschalk, P., Goemann, H. *et al.* (2007) Hydrological impact assessment of afforestation and change in tree-species composition – A regional case study for the Federal State of Brandenburg (Germany). *Journal of Hydrology* 346, 1–17.

Wei, H., Zhang, Y., Huang, Q., Chiew, F.H., Luan, J. *et al.* (2024) Direct vegetation response to recent CO_2 rise shows limited effect on global streamflow. *Nature Communications* 15, 9423.

Wei, X.H., Li, Q., Zhang, M.F., Giles-Hansen, K., Liu, W.F. *et al.* (2017) Vegetation cover – another dominant factor in determining global water resources in forested regions. *Global Change Biology* 24, 786–795.

Whitley, R., Taylor, D., Macinnis-Ng, C., Zeppel, M., Yunusa, I. *et al.* (2013) Developing an empirical model of canopy water flux describing the common response of transpiration to solar radiation and VPD across five contrasting woodlands and forests. *Hydrological Processes* 27, 1133–1146.

Wieser, G., Grams, T.E.E., Matyssek, R., Oberhuber, W. and Gruber, A. (2015) Soil warming increased whole-tree water use of pinus cembra at the treeline in the central tyrolean alps. *Tree Physiology* 35, 279–288.

Winkler, R., Boon, S., Zimonick, B. and Baleshta, K. (2010) Assessing the effects of post-pine beetle forest litter on snow albedo. *Hydrological Processes* 24, 803–812.

Wu, Y., Liu, S. and Abdul-Aziz, O.I. (2012) Hydrological effects of the increased CO_2 and climate change in the upper Mississippi River basin using a modified SWAT. *Climatic Change* 110, 977–1003.

Xu, G.B., Liu, X.H., Belmecheri, S., Chen, T., Wu, G.J. *et al.* (2018) Disentangling contributions of CO_2 concentration and climate to changes in intrinsic water-use efficiency in the arid boreal forest in China's altay mountains. *Forests* 9, 642.

Xu, H., Xiao, J.F. and Zhang, Z.Q. (2020) Heatwave effects on gross primary production of northern mid-latitude ecosystems. *Environmental Research Letters* 15, 074027.

Yang, Q.C., Tian, H.Q., Li, X., Tao, B., Ren, W. *et al.* (2015) Spatiotemporal patterns of evapotranspiration along the North American east coast as influenced by multiple environmental changes. *Ecohydrology* 8, 714–725.

Yu, Z.X., Sun, G., Cai, T.J., Hallema, D.W. and Duan, L.L. (2019) Water yield responses to gradual changes in forest structure and species composition in a subboreal watershed in Northeastern China. *Forests* 10, 211.

Zapater, M., Bréda, N., Bonal, D., Pardonnet, S. and Granier, A. (2013) Differential response to soil drought among co-occurring broad-leaved tree species growing in a 15- to 25-year-old mixed stand. *Annals of Forest Science* 70, 31–39.

Zhang, C.Q., Zhang, B.A., Li, W.H. and Liu, M.C. (2014) Response of streamflow to climate change and human activity in Xitiaoxi River basin in China. *Hydrological Processes* 28, 43–50.

Zhang, M.F. and Wei, X.H. (2021) Deforestation, forestation and water supply. *Science* 371, 990–991.

Zhang, Y., Peña-Arancibia, J., McVicar, T., Chiew, F.H.S., Vaze, J. *et al.* (2016) Multi-decadal trends in global terrestrial evapotranspiration and its components. *Scientific Reports* 6, 19124.

Zhang, M.F., Liu, N., Harper, R., Li, Q., Liu, K. *et al.* (2017) A global review on hydrological responses to forest change across multiple spatial scales: Importance of scale, climate, forest type and hydrological regime. *Journal of Hydrology* 546, 44–59.

Zhang, Z., Zhang, L., Xu, H., Creed, I.F., Blanco, J.A. *et al.* (2023) Forest water-use efficiency: Effects of climate change and management on the coupling of carbon and water processes. *Forest Ecology and Management* 534, 120853.

Zhou, G.Y., Wei, X.H., Chen, X.Z., Zhou, P., Liu, X.D. *et al.* (2015) Global pattern on the effects of climate and land cover on water yield. *Nature Communications* 6, 5918.

5 Trends and Characteristics of Hydroclimatic Extremes on Reference Watersheds in Experimental Forests in the USA

D.M. Amatya[1]*, S. Mukherjee[2], E. Keppeler[3], P. Caldwell[4], S.L. Johnson[5], S. Sebestyen[6], J. Campbell[7], B. Rau[8], K. Elder[9], D. Misra[10] and P. Wohlgemuth[11]

[1]Santee Experimental Forest, Southern Research Station, USDA Forest Service, Cordesville, South Carolina, USA; [2]Oak Ridge Institute of Science and Education, Santee Experimental Forest, Southern Research Station, USDA Forest Service, Cordesville, South Carolina, USA; [3]Caspar Creek Experimental Watershed, Pacific Southwest Research Station, USDA Forest Service, Fort Bragg, California, USA; [4]Center for Integrated Forest Science, Southern Research Station, USDA Forest Service, Otto, North Carolina, USA; [5]Formerly HJ Andrews Experimental Forest, Pacific Northwest Research Station, USDA Forest Service, Corvallis, Oregon USA; [6]Marcell Experimental Forest, Northern Research Station, USDA Forest Service, Grand Rapids, Minnesota, USA; [7]Hubbard Brook Experimental Forest, Northern Research Station, USDA Forest Service, Durham, New Hampshire, USA; [8]Fernow Experimental Forest, Northern Research Station, USDA Forest Service, Parsons, West Virginia, USA; [9]Fraser Experimental Forest, Rocky Mountain Research Station, USDA Forest Service, Ft. Collins, Colorado, USA; [10]University of Alaska Fairbanks, Caribou-Poker Creek Research Watershed, Fairbanks, Alaska, USA; [11]Formerly with San Dimas Experimental Forest, Pacific Southwest Research Station, USDA Forest Service, Riverside, California, USA

Abstract

This chapter provides an overview and comparisons of the precipitation intensity–duration–frequency (PIDF) and flood (FFRQ) and low-flow (LFRQ) frequencies for return intervals of 25 years or more at ten relatively undisturbed reference watersheds in the US Forest Service Experimental Forest (EF) network. We demonstrate potential effects of recent climate change on the PIDFs, FFRQ and LFRQ developed with high-resolution temporal data at these ten sites with widely contrasting hydrogeological, topographical, climatic and ecological characteristics. Similarly, we evaluate the on-site-based FFRQ and LFRQ with those published by the US Geological Survey for the regions including our EF sites. This evaluation enables us to better predict PIDFs and FFRQs, frequently used by forest managers/engineers but relatively less studied in forest hydrology, and to prepare for future forest and water management in response to further environmental change.

*Corresponding author: devendra.m.amatya@usda.gov

DOI: 10.1079/9781800625310.0005

5.1 Introduction

There is growing evidence globally that climate change and associated extreme conditions will impact the intensity and frequency of precipitation and hydrological responses (Mukherjee *et al.*, 2023). This response is particularly true in regions with high moisture availability and in wet months causing more frequent and severe flooding (Gimeno *et al.*, 2022). For example, Gu *et al.* (2017) found that alterations in precipitation seasonality based on distributions across 728 stations in China were likely being driven by changes in the pathways of seasonal vapor flux and tropical cyclones.

There is an undisputed relationship between precipitation and flooding, with flooding events following extreme precipitation being reported all over the world (Papalexiou and Montanari, 2019). For example, Jalowska *et al.* (2021) documented an increasing trend in the frequency and intensity of extreme precipitation events and associated flooding within the south-eastern USA using historical climate records. Studies also describe dramatic ecosystem responses to extreme precipitation events with plausible regime shifts in the intensity and quantity of runoff within some ecosystems (Jayakaran *et al.*, 2014; Amatya *et al.*, 2016a; Jalowska *et al.*, 2021; Campbell *et al.*, 2022; Sun *et al.*, 2023). Similarly, streamflow response to increasing precipitation was shown to be altered by forest management (Kelly *et al.*, 2016). Extreme precipitation and floods have attracted a great deal of scientific interest globally due to the particular threat to human activities (Gimeno *et al.*, 2022).

Although trends in precipitation extremes have not yet translated into observable increases in flood risks, except for cyclones as noted by Jalowska *et al.* (2021), a recent study by Wright *et al.* (2019) nevertheless highlights the need for prompt updating of hydrological design standards, taking into consideration recent changes in extreme precipitation properties. Increased extreme rainfall alone does not necessarily lead to increased flooding (Blöschl *et al.*, 2019). There are many factors that affect flood response in addition to precipitation intensity (PI), including the duration and extent of precipitation events, antecedent soil moisture conditions, catchment size, vegetation cover, catchment imperviousness and roughness (Sharma *et al.*, 2018; Yochum *et al.*, 2019) and channel morphology (Wondzell and Swanson, 1999). The drivers of streamflow timing (e.g. response time of streamflow) depend on the magnitude of the event. Lesser extreme flood event timings tend to correspond with antecedent soil moisture, while more extreme flood timings depend more on rainfall timing (Wasko *et al.*, 2020).

Precipitation intensity–duration–frequency (PIDF) analysis based on long-term historical data is frequently used to describe the extreme PI, temporal distribution and frequency of such intensities for decision making by water managers (Srivastava *et al.*, 2019; Amatya *et al.*, 2021; Nerantzaki and Papalexiou, 2022; Mukherjee *et al.*, 2023, 2024), and to estimate design flood magnitudes in ungauged watersheds (Eisenbies *et al.*, 2007). Mukherjee *et al.* (2023) noted that the need for updated PIDF estimations using the most recent data has grown significantly due to recent increases in intense precipitation and the associated impacts on transportation and infrastructure.

The assumption of stationarity has long served as the basis for the statistical analysis of hazards and the design of engineering structures, by defining the magnitude of events with a given frequency of occurrence, such as the stationary 100-year design flood (Salas *et al.*, 2018). However, with climate change, the validity of stationarity in water-resource planning is being questioned (Milly *et al.*, 2008); thus, it is critical to evaluate the stationarity of climate variables, especially precipitation (Wang and Sun, 2020). Using long-term high-resolution historical data, Amatya *et al.* (2021) tested for stationarity in annual maximum PIs at multiple rain gauges within three US Department of Agriculture (USDA) Forest Service experimental forest (USFS-EF) sites (Fig. 5.1), finding increasing trends in peak rainfall intensities at the Santee (SAN) and Coweeta (CHL) sites, but not Alum Creek in Arkansas (not shown), comparable to previous findings of increased frequency of intense precipitation, especially for fall months (Laseter *et al.*, 2012; Burt *et al.*, 2018). Accordingly, Amatya *et al.* (2021) used the assumption of stationarity in their estimates of the PIDF, consistent with de Luca and Galasso (2018), who argued that it is not essential to adopt non-stationary models. The results of de

Fig. 5.1. USDA Forest Service experimental forests evaluated in this and previous studies.

Luca and Galasso (2018) emphasized the importance of the observations of the past for improving the knowledge of hydrological processes under future climate change. Mukherjee *et al.* (2023) tested for non-stationarity of extreme PIs of multiple durations at three additional USFS-EFs (Hubbard Brook, Fraser and HJ Andrews (HJA)) and the three sites used by Amatya *et al.* (2021), but using data extended through 2021.

National Oceanic and Atmospheric Administration (NOAA)'s Atlas 14-based gridded PIDF values based on regional frequency analysis of precipitation recorded at various locations (Bonnin *et al.*, 2006; Perica *et al.*, 2013) are commonly used for design applications to determine the PIDF and associated confidence limits. However, for some durations and frequencies, significant differences between the NOAA and on-site-derived PIs were reported at USFS-EF gauge sites resulting in recommendations for use of the NOAA PIDF values for 1 hour duration and the on-site-derived values for longer durations (Amatya *et al.*, 2021) for conservative design applications. Mukherjee *et al.* (2023) applied this comparative approach to further improve PIDF estimates used for culvert sizing and other engineering and ecological applications in six small, ungauged forested watersheds. The results showed considerable differences between the on-site and NOAA Atlas 14 PIDFs at these six EFs relating to storm durations and gauge elevations, particularly at the steeper HJA and CHL sites. Expanding further on these results, Mukherjee *et al.* (2023) evaluated normalized peak design discharges (Q_p) for 1387 hydrological unit code 16–20 watersheds in the White Mountain National Forest (New Hampshire) and in the six EFs. Consistent with earlier findings by Amatya *et al.* (2021), the authors concluded that the rational method outperformed the US Geological Survey (USGS) regional regression equations (USGS, 1982) in predicting Q_p in three small, high-relief forest headwater watersheds (both gauged and ungauged), and the USGS regional regression equations performed better than the rational method for larger watersheds. These results have important implications for road crossings and culvert design and maintenance, particularly in high-gradient, high-sediment transport systems characteristic of steep forested landscapes, where enhanced resiliency to extreme precipitation and flood risk induced by spatial heterogeneity (Preece *et al.*, 2021) and climate change is needed. Development of more accurate duration–frequency analyses for EF headwater watersheds is crucial for informed management of roads – approximately 600,000 km with at least 40,000 stream crossings – in our national forests (Heredia *et al.*, 2016).

This chapter provides an overview and comparisons of the PIDF and flood (FFRQ) and low-flow (LFRQ) frequencies for 25-, 50- and 100-year return intervals at ten relatively undisturbed reference watersheds in the USFS-EF network (Fig. 5.1, Table 5.1). We demonstrate potential effects of recent climate change on the PIDFs, FFRQ and LFRQ in these ten reference watersheds with widely contrasting hydrogeological, topographical, climatic and ecological characteristics. Similarly, we evaluate the on-site-based FFRQ and LFRQ with those published by the USGS for the regions including our EF sites. This evaluation will enable us to better predict PIDFs and FFRQs and prepare for future management in response to further environmental change (Wright *et al.*, 2019; Amatya *et al.*, 2021; Mukherjee *et al.*, 2023).

While Amatya *et al.* (2016b) evaluated the response of streamflow to variation in annual precipitation magnitude, form and seasonality, and evapotranspiration at multiple EFs, this cross-site comparison study used high-resolution data from headwater forested catchments for statistical analysis and risk assessment of climate and land-use change on ecosystem functions.

5.2 Site Description

Full descriptions of the ten reference watersheds have been given previously in the first edition of this book (Amatya *et al.*, 2016b). The key characteristics and acronyms are shown in Table 5.1.

5.3 Data and Methods

Data record periods, gauge IDs, and the temporal scales of precipitation and streamflow measurements are presented in Table 5.1.

First, trends in annual maximum PI and streamflow were assessed for significance ($\alpha = 0.05$) using Sen's slope estimator (Sen, 1968) and the modified Mann–Kendall (M-K) trend test (Mann, 1945; Kendall, 1975). Secondly, extreme value analysis was performed as follows.

5.3.1 PIDF estimation

The generalized extreme value (GEV) distribution (Coles *et al.*, 2001) was used for frequency analysis of extreme PI. The GEV distribution is a flexible statistical framework to effectively model upper tail behavior of extreme events, making it well suited for characterizing the frequency and magnitude of rare rainfall events.

The block maxima method (Coles *et al.*, 2001) was utilized to extract the yearly maximum values of 1 and 24 hour PI estimates. This method is widely used in extreme value analysis to capture characteristics of the upper tail of climate data.

The choice of the GEV distribution over other extreme value modeling approaches is consistent with the methodology adopted in NOAA's Atlas 14 (Perica *et al.*, 2018). By employing the same GEV framework, our results are comparable with the standards of the NOAA.

The GEV is a three-parameter distribution comprising location (μ), scale (σ) and shape (ξ) parameters (Coles *et al.*, 2001). These parameters specify the center of the distribution, the deviation around μ and the tail behavior of the distribution, influencing the frequency and magnitude of extreme events. The theoretical cumulative distribution function of a real-valued

Table 5.1. Comparative characteristics of the ten selected reference watersheds (with their acronyms) from the USFS Long-term Experimental Forests network.

Watershed characteristics	Caribou-Poker (CPCR), Alaska	Caspar Creek (CCEW), California	Coweeta (CHL), North Carolina	Fernow (FERN), West Virginia	Fraser (FRS), Colorado	HJ Andrews (HJA), Oregon	Hubbard Brook (HBR), New Hampshire	Marcell (MARC), Minnesota	San Dimas (SDEF), California	Santee (SAN), South Carolina
Physiographical region as per classification by Fenneman	Yukon–Tanana Northern Plateaus Province, 12	Pacific Mountain System, 23f, Pacific Boarder Province	Appalachian Highlands, 5b, Blue Ridge Province	Appalachian Highlands, 8c, Appalachian Plateau	Rocky Mountain Systems, 15, Southern Rocky Mountain	Pacific Mountain System,22b, Sierra-Cascade Mountain	Appalachian Highlands, 9b, New England Province	12b, Interior Plain, Central Lowland	Pacific Mountain System, 23g, Pacific Boarder Province	Atlantic Plain Coastal Plain
Climatic region as per classification by Köppen (Peel *et al.*, 2007)	Dfc, continental subarctic or boreal taiga	Csb, temperate/ mesothermal, Mediterranean	Cfb, marine temperate	Dfb, continental warm summer	Dsc, continental cold winter and cool, short, dry summer	Csb, temperate/ mesothermal dry summer	Dfb, continental warm summer	Dfb, continental warm summer	Csa, Mediterranean hot summer	Cfa, temperate humid subtropical
Watershed #/ year gauging started	CPCRW –C2, 1969	North Fork, 1962	WS18, 1936	WS4, 1951	East St. Louis, 1943	WS2, 1952	WS 3, 1957	S2, 1960	Bell 3, 1938	(WS80) 1968
Latitude/ longitude	65.17°N, 147.50°W	39.36°N, 123.74°W	35.05°N, 83.43°W	39.03°N, 79.67°W	39.89°N, 105.88°W	44.21°N, 122.23°W,	44.0°N, 71.7°W	47.514°N, 93.473°W	34.20°N, 117.78°W	33.17°N, 79.77°W
Elevation (m a.m.s.l.)a	210–826	40–330	726–993	670–866	2907–3719	572–1079	527–732	420–430	755–1080	3.7–10
Average slope (%)	31	49	52	20	16	41	21	3	34	<3
Drainage area (ha)	520	479	12.5	38.7	803	60	42.4	9.7	25	160
Vegetation type/average leaf area index (LAI) (m^2/m^2)	Boreal Forest/ LAI = 4.1	Second-growth coast redwood/ Douglas fir forest/ LAI = 11.7	Mixed deciduous forest/LAI = 6.2	Mixed deciduous hardwoods/ LAI = 4.5	Mixed spruce/ fire and pine/ LAI = 3.44	Conifer primarily Douglas fir and western hemlock/ LAI = 12	Northern hardwood/ LAI = 6.3	Deciduous uplands; black spruce-*Sphagnum* bog/LAI = N/A	Mixed chaparral/ LAI = 2.2	Pine mixed hardwood/ LAI = 2.8
Dominant geology/aquifer	Yukon–Tanana metamorphic complex/ discontinuous permafrost	Marine shales and sandstones, Coastal Belt of the Franciscan Complex	Quartz dioritic gneiss predominant, Coweeta Group	Sedimentary; Hampshire formation sandstone and shales	Gneiss and schist, glacial till	Volcanic tuffs and breccias covered with andesite colluvium	Metasedimentary/ mica schist, calc-silicate granulite, Silurian Rangeley formation	Glacial till overlying deep outwash sands above pre-Cambrian bedrock	Pre-Cambrian metamorphics and Mesozoic granitics	Sedimentary/Santee limestone

Dominant soil type/depth (m)	Olnes Silt Loam – *Typic Cryorthents*; Fairplay Silt Loam – *Fluvaquentic Endoaquolls*; Ester Silt Loam – *Typic Histoturbels*0.2–0.5m	Vandame Series Ultisols (*Typic Haplohumults*) 1–1.5 m	Coweeta–Evard–Saunook complex (fine loamy, mixed, *Mesic* and *Humic Hapludults*); >1.5 m	Loamy-skeletal, mixed mesic *Typic Dystrudepts*1 m	Leighcan Series, loamy-skeletal, *Typic Dystrocyepts* <1.5 m	50% andesite colluvium (unnamed soil series), 20% Limberlost series, loam to green brecciaUp to 1.2 m	Lyman-Tunbridge–Becket series, *Typic Haplorthods*C horizon depth 0–9 m	Warba Series fine loamy, mixed, superactive, frigid haplic Glossudalfs (0.5 m); Loxely peat Dysic, frigid Typic Haplosaprists (≤7 m)	Trigo–Exchequer series loamy, mixed, thermic, shallow, *Typic Xerorthents* 0.1–0.5 m	Wahee Series clayey, mixed, thermic *Aeric Ochraquults* 1.5 m
Long-term mean precipitation (mm)	262	1146	2010[2]	1458	750	2300	1350	780	715	1388
Precipitation period (gauge – timescale) used in this study	1993–2021 (CPEAK – hourly)	1985–2021 (NFC620 – hourly)	1976–2021 (RG41 –hourly)	1951–2022 (S4 – daily)	2004–2021 (HQTRS – hourly)	1979–2018 (PRIMET – hourly)	1977–2021 (RG1 – hourly)	1962–2022 (MEF South – daily)	1977–2015 (Bell 3 – hourly)	1977–2021 (Met25 – hourly)
Long-term mean temperature (°C)	−3.0	10.7	12.9b	9.3	1.0	8.4	5.9	3.4	14.4	18.5
Long-term mean potential evapotranspiration (PET) (mm)	466c	660c	1013[b,d]	560d	383c	546c	550c	552	753c	1149 (Hargreaves-Samani)c
Dryness index	1.13	0.50	0.50	0.38	0.51	0.24	0.41	0.71	1.05	0.71
Long-term mean streamflow (mm)	80	621	997	640	337	1321	860	171	84	290
Period of streamflow record	1969–present	1962–present	1936–present	1951–present	1943–present	November 1952–present	1957–present	1961–2017	1938–1960; 1964–2001; 2013–present	1969–1981, 1989–1999, 2003–present

Continued

Table 5.1. Continued

Watershed characteristics	Caribou-Poker (CPCR), Alaska	Caspar Creek (CCEW), California	Coweeta (CHL), North Carolina	Fernow (FERN), West Virginia	Fraser (FRS), Colorado	HJ Andrews (HJA), Oregon	Hubbard Brook (HBR), New Hampshire	Marcell (MARC), Minnesota	San Dimas (SDEF), California	Santee (SAN), South Carolina
Streamflow period (gauge – timescale) used in this study	1969–2021 (C2 – instant)	1985–2017 (NF – instant)	1938–2015 (WS18 – instant)	1980–2015 (WS4 – instant)	1943–2021 (ELOUI – instant)	1952–2018 (WS2 – instant)	1957–2021 (WS3 – instant)	1962–2017 (S2 WS – instant)	1939–2001 (Bell3 —instant)	1965–2020 (WS80 –instant)
Surface runoff/ flow generation	Saturation excess flow	Infiltration excess overland flow limited to compacted surfaces	Rare surface runoff, direct channel and fast shallow subsurface flow from variable source area	Minimal surface runoff	Rare, only during snowmelt	Minimal surface runoff – high porosity	Minimal surface runoff	Infiltration excess over frozen and saturation excess flow over unfrozen soils	Rare hillslope flow except after fire when infiltration excess flow	Saturation excess flow
Subsurface flow/drainage	Shallow subsurface flow	Transient subsurface stormflow and soil pipe preferential flow	Shallow lateral flow via soils with high conductivity	Lateral subsurface flow to channel	Shallow subsurface (macropores, coarse soils) and groundwater	Shallow subsurface lateral flow	Lateral subsurface flow	Shallow subsurface with some seepage to an underlying groundwater aquifer	Groundwater flow unknown but presumably high rate of shallow lateral flow	Shallow lateral subsurface flow with negligible deep seepage
Average water table dynamics/ depth (m)	Unknown	1–8 m	>1.5m except near stream	Unknown	Unknown	Unknown	Variable water table depth	~0.3m in the bog; 0.5m in uplands	Unknown; potentially very deep	Shallow, ~1.0m
Riparian areas for hydrology	None	Limited due to steep topography	Limited due to steep topography	Limited due to steep topography	Limited to valleys, fens and bogs	Limited	Limited due to steep topography	33% of area is a peatland	~2%	15%
Major or extreme natural disturbance	1967 Fairbanks Flood	Windstorm (1995); landslide (2006); drought (2020–2021)	Chestnut blight (1920s-1930s), drought, hurricanes, hemlock wooly adelgid (2003–present)	Chestnut blight; hurricanes (windstorms; SuperStorm Sandy (2012)	Pine bark beetle epidemic	None in reference	Hurricane (1938); ice storm (1998)	Peatland wildfire (1864); potential for derecho, tornados and wildfires	Wildfire (Meixner and Wohlgemuth, 2003)	Hurricane Hugo (1989); indirect effects of Hurricane Joaquin (2015)
Other specific hydrological features	3% permafrost underlain	Fog input	–	–	Snowmelt-dominated hydrological regime	–	Discontinuous dense pan C horizon	Drainage from bog dome and uplands; some deep seepage to the aquifer	Extremely high levels of nitrate from chronic air pollution	Compared with pre-Hugo, flow reversal in paired watersheds after Hugo extreme discharge topping the rating curve and site flooding (Joaquin)

Key publication(s) on extreme climatic events or related to extreme events	Bolton *et al.* (2004); Jones and Rinehart (2010); van Cleve *et al.* (2023)	Cafferata and Reid (2013); Richardson *et al.* (2021)	Laseter *et al.* (2012); Burt *et al.* (2013); Mukherjee *et al.* (2023)	Adams *et al.* (1994)	Mukherjee *et al.* (2023)	Rothacher *et al.* (1967); Daly *et al.* (2019); Johnson *et al.* (2021, 2023); Mukherjee *et al.* (2023)	Campbell *et al.* (2011); Mukherjee *et al.* (2023)	Sebestyen *et al.* (2011, 2021); Verry *et al.* (2011)	Dunn *et al.* (1988); Wohlgemuth (2016)	Hook *et al.* (1991); Jayakaran *et al.* (2014); Amatya *et al.* (2016b, 2021); Mukherjee *et al.* (2023)
Forest experimental watershed contact	Jamie Hollingsworth: jhollingsworth@alaska.edu	Joe Wagenbrenner: joseph.wagenbrenner@usda.gov	Peter Caldwell: peter.v.caldwell@usda.gov	Benjamin Rau: benjamin.rau@usda.gov	Kelly Elder: kevin.elder@usda.gov	Sherri Johnson: sherri.johnson@oregonstate.edu	Amey Bailey: amey.bailey@usda.gov	FS-MarcellExpForest SM.FS.MarcellEF@usda.gov	Pete Wohlgemuth: peter.wohlgemuth@usda.gov (retired)	Devendra Amatya: devendra.m.amatya@usda.gov
Experimental forest website	www.lter.uaf.edu/bnz_cpcrw.cfm	www.fs.usda.gov/research/psw/forestsandranges/locations/casparcreek	www.srs.fs.usda.gov/coweeta/	www.nrs.fs.fed.us/ef/locatios/wv/fernow	www.fs.usda.gov/efr/fraser	http://andrewsforest.oregonstate.edu	www.hubbardbrook.org	www.nrs.fs.fed.us/ef/marcell/	www.fs.fed.us/psw/ef/san_dimas	www.fs.usda.gov/research/srs/forestsandranges/locations/santee

N/A, not applicable.
[a]a.m.s.l., above mean sea level.
[b]Water years 1938–2013. Water year taken as starting in May and ending in April.
[c]PET estimated from Thornthwaite (1948) method for all, except for the Santee (SAN) site from the Hargreaves–Samani method (Hargreaves and Samani, 1982).
[d]PET estimated from an evaporation pan (Patric and Goswami, 1968).
[e]PET estimated from Hamon (1963) method with corrections.

random variable, x, following the GEV distribution can be expressed as:

$$F_{GEV}(x \mid \mu,\sigma,\varepsilon)=\left[-\left(1+\frac{\varepsilon}{\sigma}(x-\mu)\right)^{-1/\varepsilon}\right],\mu \in R, \sigma > 0, \varepsilon \neq 0 \quad (5.1)$$

Parameter estimation for the GEV distribution was carried out using the maximum likelihood estimation (MLE) method with both stationary and non-stationary assumptions for the location and scale parameters (Martins and Stedinger, 2000; Coles *et al.*, 2001). MLE is a widely used statistical technique that is used to find parameter values that maximize the likelihood of the observed data. In the context of the GEV distribution, MLE involves finding the values of XII, σ and XIV that maximize the likelihood function.

To account for potential non-stationarities due to climate change in PIs, we employed a time-varying approach, allowing the location parameter to change over time (Perica *et al.*, 2018), an approach particularly relevant for extreme value analysis where characteristics of extreme events vary with time (Cheng and AghaKouchak, 2014).

Specifically, we adopted a time-varying model for the location parameter:

$$\mu(t) = \mu_0 + \mu_1 t \quad (5.2)$$

where, μ_0 is the initial location parameter at time t = 0, and μ_1 is the rate of change of the location parameter.

By incorporating time-varying models, we can capture potential changes in the frequency and magnitude of extreme rainfall events over time. The p-quantile of the GEV distribution was then estimated as,

$$q_p = \left[\left(-\frac{1}{\ln(1-p)}\right)^{\varepsilon} - 1\right] \times \frac{\sigma}{\varepsilon} + \mu, (\varepsilon \neq 0) \quad (5.3)$$

where (1 − p) is the non-exceedance probability.

The non-stationary assumptions were employed to identify the best GEV model following the Akaike information criteria, Bayesian information criterion and the likelihood ratio test (Ansa *et al.*, 2021; Kim *et al.*, 2017). Non-stationarity was confirmed only if all these criteria were fulfilled. The uncertainties associated with the estimation of PI quantiles were quantified based on the 90% confidence intervals (CI)s using the delta method (Cox, 1990).

5.3.2 High- and low-flow frequency analysis

Availability of long-term streamflow data including for both extreme high- and low-flow periods is crucial for the efficient management of water resources and infrastructure. The USGS requires 30 years of streamflow data to designate long-term stream gauges (Jian *et al.*, 2015). We conducted the high- and low-flow frequency analysis only for the sites that have at least 30 years of annual maximum and annual minimum flows after eliminating the outliers and zero flows. Additionally, our analysis only included the sites at which the number of outliers and zero flows do not exceed 25% of the total number of data years.

Observations that deviate significantly from the overall data trend are potentially influential 'outliers.' In the context of annual peak flows, low outliers may represent floods generated by processes distinct from those of larger floods (National Research Council, 1988, 1995). Due to their atypical nature, the inclusion of these zero flows values (common in dry areas) and outliers can significantly impact the statistical parameters derived from the data, particularly for small samples. The low and high outliers were identified based on the threshold estimation guidelines recommended by the Interagency Advisory Committee on Water Data (1982).

High-flow frequency analysis was conducted on long-term streamflow data sets by fitting annual maxima to the log-Pearson type III (LPIII) distribution, a widely established statistical method for predicting flood frequency. The LPIII distribution is endorsed by US federal agencies and many flood frequency analysts (England *et al.*, 2019), and is used in USGS PEAKFQW version 5.2.0 (Feaster *et al.*, 2009).

A key advantage of this method is effective extrapolation of event data for return periods exceeding the range of observed flood events.

Low-flow statistics quantify the magnitude and frequency of low-flow conditions as the minimum average streamflow over a specified time period. Low-flow frequencies are calculated by fitting an annual minima series of N days average streamflow to a known statistical distribution, where N can range from 1 to 365 days (Feaster and Guimaraes, 2014). The 10 year recurrence interval of the annual minimum 7 day average (7Q10) is commonly used (Feaster and Guimaraes, 2014). In probabilistic terms, the estimated 7Q10 value represents the 10% probability that the annual minimum 7 day average streamflow of any given year will be equal to or lower than this value (Riggs, 1985). We estimated 7Q10 by fitting the logarithms (base 10) of the annual minimum 7 day average streamflow (ANMIN7Q) to the LPIII distribution.

The goodness of LPIII fit was tested using the Anderson–Darling test (Laio, 2004). To fit LPIII distribution, the mean, standard deviation and skew coefficient of the logarithms of the streamflow were calculated. Estimates of the non-exceedance flows for a specified recurrence interval T were calculated using the following equation:

$$\log(Q_T) = X + K \times S \tag{5.4}$$

where Q_T is the annual maximum flow (ANMAXQ) for high-flow analysis or ANMIN7Q for low-flow analysis (l/s), T is the return interval (years), X is the mean of the logarithms of the annual flow values (ANMAXQ or ANMIN7Q), K is a frequency factor that is a function of the return interval and the weighted coefficient of skew, and S is the standard deviation of the logarithms of the annual flow values (ANMAXQ or ANMIN7Q). The weighted skew coefficient and frequency factor were estimated (Haan *et al.*, 1994). The uncertainty associated with the estimation of the non-exceedance flows was quantified through the calculation of the standard error via a resampling approach. This involved generating 10,000 random samples from the LPIII distribution using the estimated model parameters.

5.4 Results and Discussion

5.4.1 Characteristics of precipitation extremes

The mean and variability of annual maximum PIs (ANMAXPI) for all sites are shown in Fig. 5.2 for the 1 and 24 h-duration storms. We focused on the ANMAXPI values for each site because of the high probability of extreme precipitation events translating into floods (Amatya *et al.*, 2021). The rain gauge at SAN and CHL recorded the highest mean 1 hour ANMAXPI magnitude of 4.8 cm/h and 3.4 cm/h, respectively, over a long-term period of more than 40 years. The annual variability of 1 hour ANMAXPI was found to be relatively higher at CHL, CPCR and SAN. The mean 24 hour ANMAXPI magnitude was relatively similar across all the sites investigated, except for the very low mean 24 hour ANMAXPI magnitude of 0.1 cm/h at FRS. The annual variability of 24 hour ANMAXPI was found to be relatively higher in CPCR and SAN. Overall, for both 1 hour and 24 hour storms, the mean ANMAXPI magnitude was found to be similar at SAN and CHL, the two south-eastern sites.

The results of trend analysis for the 1 and 24 hour ANMAXPI are shown in Supplementary Figs S5.1 and S5.2 and Tables 5.2 and 5.3. A long term decline in 1 hour PI was most pronounced and significant at the CPCR site. Only the CHL site exhibited a pronounced statistically significant increasing long-term trend in 24 hour PI (Fig.S5.2, Table 5.3). These findings highlight the complex temporal dynamics underlying the 1 and 24 hour PI. Further investigation into the specific factors driving these diverse long-term and recent trends is warranted to elucidate the underlying mechanisms and inform future management strategies.

5.4.2 High-flow characteristics

The mean and variability of annual maximum streamflow per unit drainage area (ANMAXQ) for all sites are shown in Fig. 5.3 and Table 5.4.

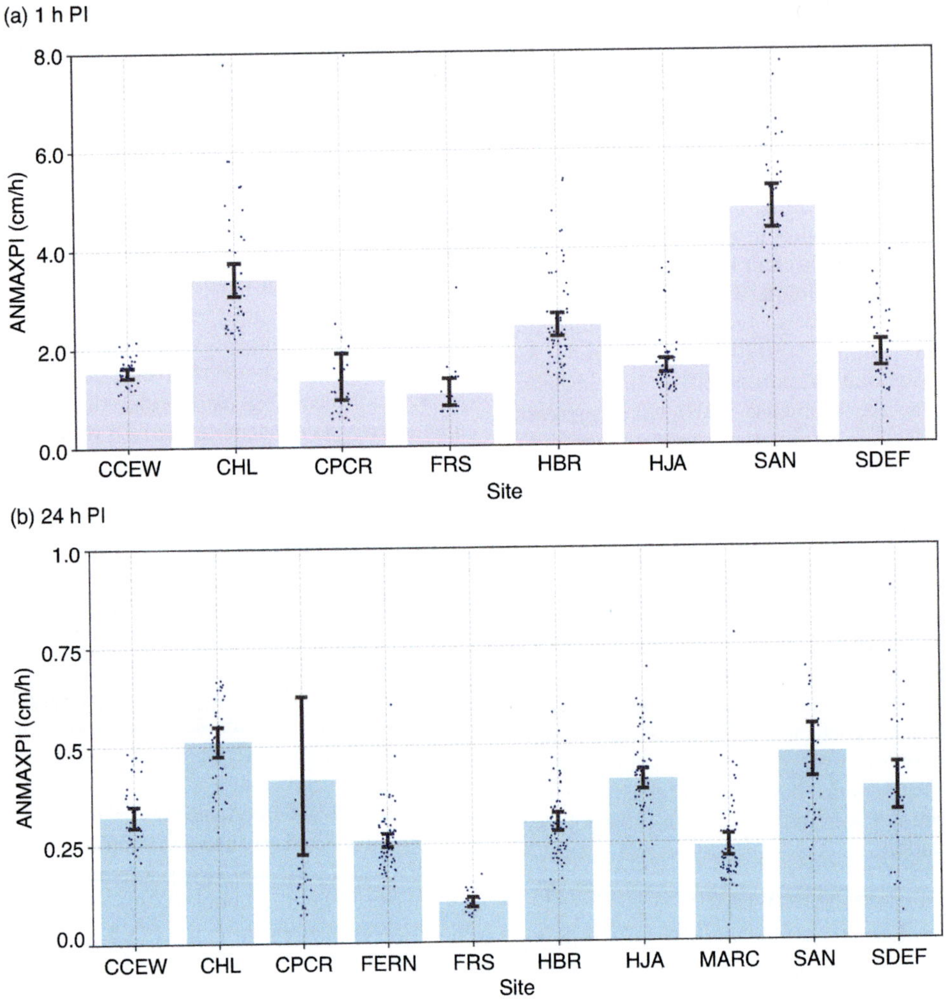

Fig. 5.2. Mean (box plots) and variability (1 SD, shown by whiskers) of annual maximum precipitation intensities (ANMAXPI, cm/h) for all sites for the 1 hour (a) and 24 hour (b) duration storms. PI, precipitation intensity. See Table 5.1 for site abbreviations.

The SAN and HBR EF gauging stations recorded extremely high mean ANMAXQ magnitudes of 11.8 and 15.5 l/s/ha, respectively. The annual variability of ANMAXQ was large for the HBR EF (9 l/s/ha) and even larger for the SAN EF (20 l/s/ha). The CCEW, CHL, FERN and HJA sites showed moderately high mean ANMAXQ values of 7, 5, 5 and 7 l/s/ha, respectively, while the CPCR, FRS, MARC and SDEF sites showed very low values for mean ANMAXQ of about 0.65, 1.1, 2.4 and 3.5 l/s/ha, respectively.

The long-term trend analysis of ANMAXQ revealed a decline across most of the sites, except for the CHL and HBR sites, which exhibited increasing trends (Fig. S5.3). Notably, for the HBR site, ANMAXQ showed a statistically significant ($P<0.05$) upward trajectory. None of the sites showed a statistically significant trend in ANMAXQ for the recent 20 year period, despite a suggested downward trend for most sites and an upward trend for CHL, FRS, HBR, and SAN (Table 5.4).

Table 5.2. Results of trend analysis of annual maximum 1 h precipitation intensity including τ and P values.

Site	Start year	End year	No. of years	M-K trend (τ)	M-K trend (P value)	M-K trend (τ for recent 20 years)	M-K trend (P value for recent 20 years)
CCEW	1985	2022	38	–0.02	0.85	–0.12	0.48
CHL	1976	2021	46	0.13	0.21	0.21	0.20
CPCR	1993	2021	29	–0.27	0.04	0.05	0.80
FRS	2004	2021	19	–0.19	0.28	–0.19	0.28
HBR	1956	2021	66	0.04	0.62	0.05	0.77
HJA	1957	2018	62	0.08	0.36	0.02	0.92
SAN	1977	2021	45	0.03	0.80	–0.09	0.58
SDEF	1975	2015	36	0.18	0.12	–0.20	0.23

Table 5.3. Results of trend analysis of annual maximum 24 h precipitation intensity including τ and P values.

Site	Start year	End year	No. of years	M-K trend (τ)	M-K trend (P value)	M-K trend (τ for recent 20 years)	M-K trend (P value for recent 20 years)
CCEW	1985	2022	38	0.03	0.81	0.11	0.54
CHL	1976	2021	46	0.26	0.01	0.16	0.35
CPCR	1993	2021	29	–0.24	0.07	0.22	0.18
FERN	1951	2022	72	0.03	0.67	–0.03	0.90
FRS	2004	2021	18	–0.06	0.76	–0.06	0.76
HBR	1956	2021	66	0.12	0.14	–0.02	0.92
HJA	1957	2018	62	–0.064	0.46	–0.28	0.08
MARC	1961	2022	62	–0.01	0.90	–0.20	0.23
SAN	1977	2021	45	0.13	0.22	0.08	0.63
SDEF	1975	2015	36	0.06	0.62	–0.11	0.52

5.4.3 Low-flow characteristics

The mean and variability of annual minimum 7-day average flow per unit drainage area (ANMIN7Q) for all sites are shown in Fig. 5.4. The SDEF, HBR and CCEW gauging stations recorded the lowest mean ANMIN7Q magnitudes of 0.0016, 0.002 and 0.004 l/s/ha. The annual variability of ANMIN7Q was relatively larger for the CHL and CPCR EFs. Trend analysis of long-term ANMIN7Q revealed a significant increase in the HBR and SDEF sites, and a significant decline at the FRS site (Fig. S5.4, and Table 5.5). During the last two decades, a significant positive trend was evident at the CHL and CPCR sites. The CCEW and HJA sites showed a non-significant negative trend and positive trend, respectively, in the most recent decades, as well as over the long-term period.

Overall, the results of the trend analysis of climatic extremes for both the peak and low-flow extremes revealed mixed patterns across the sites in varying climatic zones (Table 5.1). These findings suggest that the response of peak

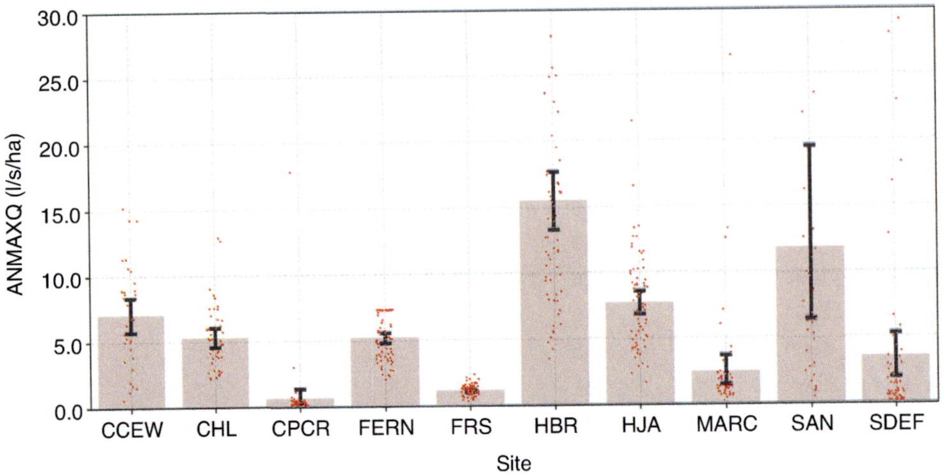

Fig. 5.3. Mean (box plots) and variability (1 SD, shown by whiskers) of annual maximum specific discharge (ANMAXQ, l/s/ha) for the selected study sites.

Table 5.4. Results of trend analysis of ANMAXQ including *τ* and *P* values.

Site	Start year	End year	No. of years	M-K trend (*τ*)	M-K trend (*P* value)	M-K trend (*τ* for recent 20 years)	M-K trend (*P* value for recent 20 years)
CCEW	1985	2017	33	–0.02	0.90	–0.18	0.28
CHL	1976	2020	45	0.06	0.58	0.00	1.00
CPCR	1969	2022	52	–0.05	0.61	–0.13	0.46
FERN	1951	2023	73	–0.13	0.10	–0.13	0.45
FRS	1943	2021	78	–0.03	0.73	0.34	0.04
HBR	1958	2021	64	0.28	0.00	0.04	0.82
HJA	1952	2019	68	–0.08	0.34	–0.03	0.87
MARC	1962	2017	56	–0.07	0.45	–0.29	0.08
SAN	1969	2016	31	–0.03	0.81	0.15	0.38
SDEF	1938	2001	57	–0.05	0.62	–0.17	0.30

streamflow to PIs induced by climate change are complex and are influenced by site-specific factors, such as climate, land use and land cover changes, as well as watershed characteristics including storage dynamics (Berghuijs *et al.*, 2016; Wasko *et al.*, 2020). The trend analysis highlights the importance of considering precipitation and streamflow when assessing water resources in the context of climate change. While increasing precipitation may lead to increased water availability in some areas, it may also lead to more frequent and intense flooding events. Understanding the trends in both precipitation and streamflow is crucial for developing effective water-management strategies that can adapt to changing climate, conditions. Understanding of the response of streams to precipitation inputs is still limited especially in northern headwater catchments (Tetzlaff *et al.*, 2013). Ali *et al.* (2015) published runoff initiation thresholds

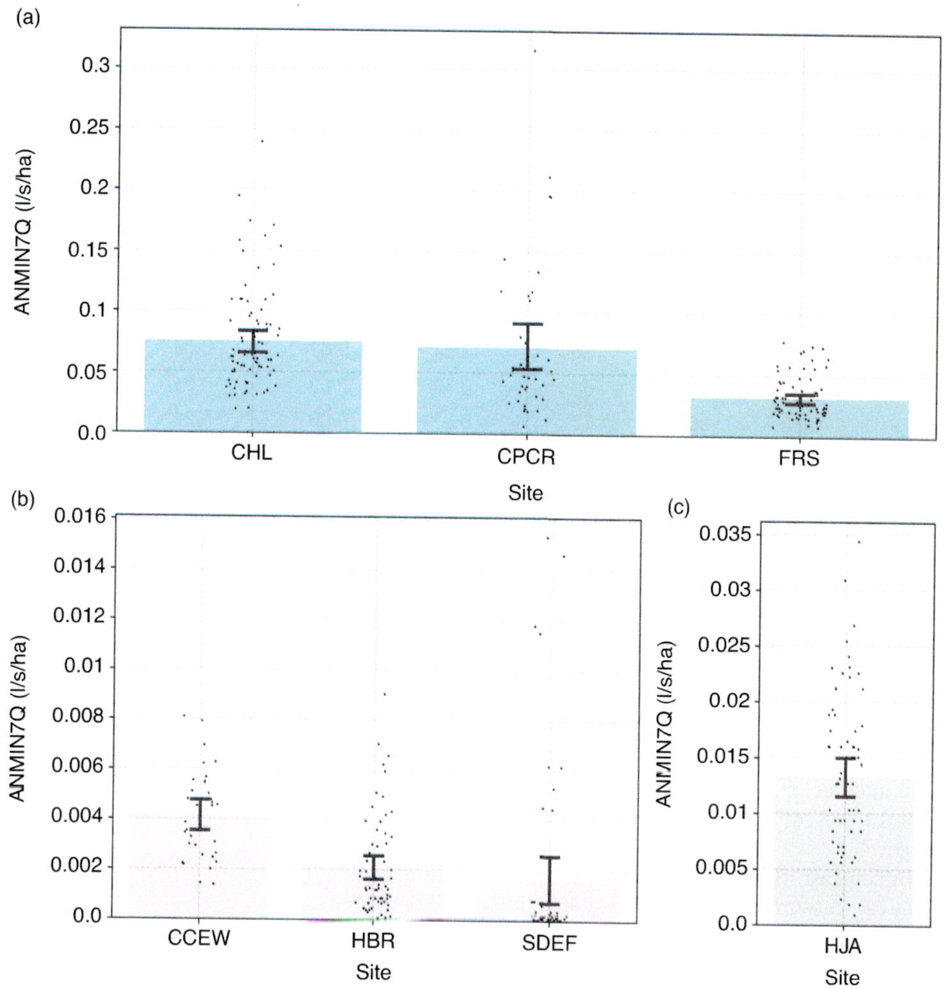

Fig. 5.4. (a–c) Mean (box plots) and variability (1 SD, shown by whiskers) of annual minimum 7-day average specific discharge (ANMIN7Q, l/s/ha) for the selected study sites.

and effective precipitation input thresholds for rainfall- and snowmelt-driven events for nine northern forested catchments that included HBR and HJA.

5.4.4 PIDF

The results of the frequency analysis of 1 and 24 hour PI for the selected sites are shown in Fig. 5.5 and Table S5.1. The results indicated that there was considerable variation in the magnitude and uncertainty of 1 and 24 hour PI values across different locations and return intervals. The mean 24 hour PI estimates for the 25-, 50- and 100 year return intervals ranged from 0.164 to 0.972 cm/h, 0.182 to 1.181 cm/h and 0.201 to 1.425 cm/h, respectively. SAN and CHL had the highest PI estimates for both durations and all return intervals, indicating that they experienced the most intense rainfall events. For example, the mean PI estimate for the 100 year

Table 5.5. Results of trend analysis of ANMIN7Q including τ and P values.

Site	Start year	End year	No. of years	M-K trend (τ)	M-K trend (P value)	M-K trend (τ for recent 20 years)	M-K trend (P value for recent 20 years)
CCEW	1985	2017	33	−0.16	0.18	−0.30	0.06
CHL	1936	2021	86	−0.04	0.60	0.41	0.01
CPCR	2012	2022	43	0.17	0.12	0.45	0.01
FRS	1943	2021	78	−0.25	0.00	−0.13	0.44
HBR	1957	2022	66	0.21	0.01	−0.02	0.95
HJA	1959	2019	68	0.14	0.09	−0.01	0.97
SDEF	1938	2001	57	0.31	0.00	−0.21	0.21

return interval for SAN was 0.972 cm/h for the 24 hour PI and 9.492 cm/h for the 1 hour PI, which were more than four times higher than the FRS mean PI estimates of 0.201 cm/h and 2.939 cm/h, respectively. However, they also had the widest CIs, suggesting that there was a high degree of uncertainty in the PI estimates for these sites. The 95% CIs for the 100 year return interval for SAN were 0.680–2.169 and 7.115–11.868 cm/h for the 24 and 1 hour PI, respectively.

In contrast, the FRS and CCEW sites had the lowest PI estimates for both durations and all return intervals, indicating that they had the least intense rainfall events. They also had the narrowest CIs, suggesting that there was a low degree of uncertainty in the PI estimates for these sites. The 95% CIs for the 100 year return interval for FRS were 0.080–0.323 and 1.705–4.173 cm/h for the 24 and 1 hour PI, respectively. This low degree of uncertainty could be due to the availability of data, the homogeneity of the terrain, or the stability of the climatic and hydrological conditions. The other sites (CPCR, HBR, HJA and SDEF) had intermediate PI estimates and CIs.

These results are consistent with previous findings (Amatya *et al.*, 2021; Mukherjee *et al.*, 2023) and provide important information for the design and management of water resources and infrastructure, as well as for the assessment and mitigation of flood risks. The PI estimates and CIs can be used to estimate the design storm for each site, which is the rainfall event with a specified return interval that is used as the basis for hydrological and hydraulic calculations. The design storm can help determine the capacity and performance of drainage systems, culverts and other road–stream crossing structures (Rosenzweig *et al.*, 2019). The results can also help identify the areas that are more prone to flooding and the factors that contribute to the uncertainty of the PI estimates. These results can inform the development of adaptation and resilience strategies to cope with the potential impacts of extreme rainfall events.

5.4.5 High-flow quantiles

The mean estimates of 25-, 50- and 100-year return intervals of annual maximum peak flow (ANMAXQ) per unit area along with their 95% upper and lower bounds, as a measure of the uncertainties for each of the sites, are shown in Fig. 5.6 and Table S5.2. The results suggested that the sites have different values of high-flow-specific discharges and different degrees of uncertainty.

The SAN site had the highest mean estimate of ANMAXQ for all the return intervals, indicating that it is prone to more severe floods than the other sites. For the 100-year return interval, the mean estimate was 107.2 l/s/ha, with a lower bound of 101.3 l/s/ha and an

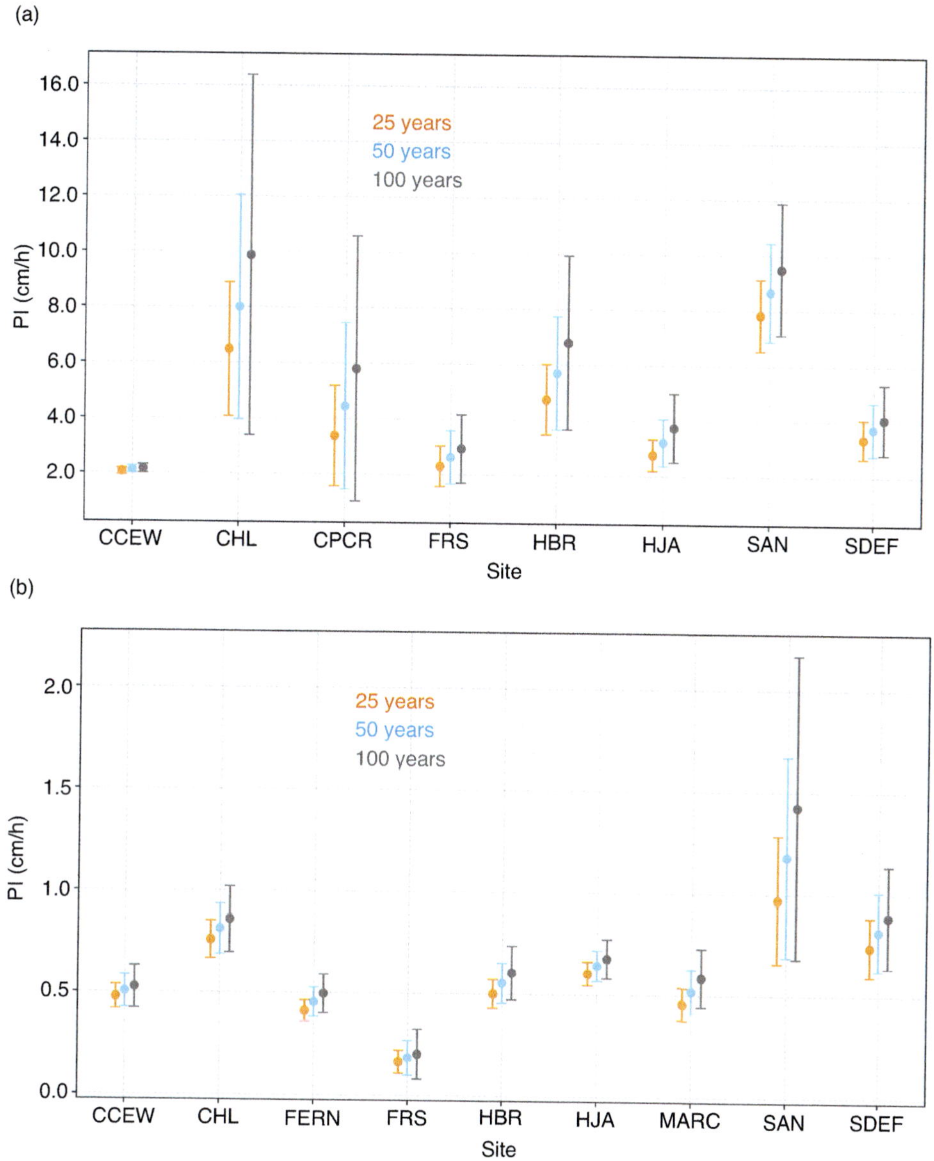

Fig. 5.5. (a) 1 hour and (b) 24 hour PI with 25-, 50- and 100-year events derived from the on-site rain gauge data for all sites. The error-bar plots show the estimates (filled circle) and 95% confidence bounds (error bars). Note that 1 hour PI results were not available for the FERN and MARC sites and therefore are not shown.

upper bound of 113.0 l/s/ha. SDEF had the widest CI for all the return intervals, indicating that it has the most uncertainty in the estimation of ANMAXQ. For the 100-year return interval, the mean estimate was 115.8 l/s/ha, with a lower bound of 97.7 l/s/ha and an upper bound of 133.8 l/s/ha. This means that the true value of ANMAXQ could be anywhere between 97.7 and 133.8 l/s/ha, which is a large range of variation. The FRS site had the lowest mean

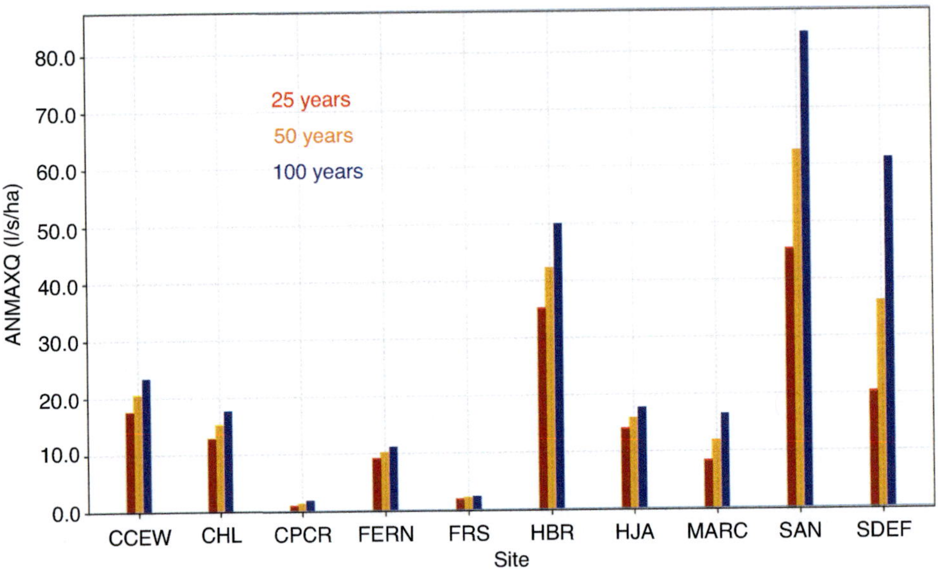

Fig. 5.6. Results for 25-year, 50-year and 100-year peak flow discharge per unit area (ANMAXQ in l/s/ha) for the selected stream gauges.

estimate of ANMAXQ and the narrowest CIs, indicating that it has the least risk of flooding and the most reliable estimation of ANMAXQ. For the 100-year return interval, the mean estimate was 2.2 l/s/ha, with a lower bound of 2.0 l/s/ha and an upper bound of 2.3 l/s/ha. The HBR site had a mean estimate of ANMAXQ for all the return intervals, indicating that it also has a high risk of flooding. For the 100-year return interval, the mean estimate was 66.5 l/s/ha, with a lower bound of 65.3 l/s/ha and an upper bound of 67.7 l/s/ha.

The HJA site had a relatively lower mean estimate of ANMAXQ compared with the HBR, SAN and SDEF sites, indicating that it also has a moderate risk of flooding and a reliable estimation of ANMAXQ. For the 100-year return interval, the mean estimate was 36.2 l/s/ha, with a lower bound of 36.0 l/s/ha and an upper bound of 36.4 l/s/ha. CCEW had a similar mean estimate of ANMAXQ for all the return intervals, indicating that it has a moderate risk of flooding. For the 100-year return interval, the mean estimate was 22.3 l/s/ha, with a lower bound of 21.9 l/s/ha and an upper bound of 22.8 l/s/ha. At the CHL site, the mean estimate of the 100-year return interval was 13.9 l/s/ha, with a lower bound of 13.6 l/s/ha and an upper bound of 14.2 l/s/ha.

Both the MARC and FERN sites had a low to moderate risk of flooding and fairly reliable estimations of ANMAXQ with quite similar mean estimates of 9.0 l/s/ha and 9.6 l/s/ha, respectively, for the 100-year return interval, with a slightly wider range of bounds for MARC than for FERN. The CPCR site, with a mean estimate is 3.6 l/s/ha, a lower bound of 3.5 l/s/ha and an upper bound of 3.8 l/s/ha for the 100-year return interval has a low risk of flooding but a very uncertain estimation of ANMAXQ.

The lower and upper bounds of the 95% CIs provide valuable information about the uncertainty associated with the estimated peak flow discharges. These intervals indicate a range within which the true discharge values are likely to fall, with 95% confidence. Sites like FERN, MARC and CHL exhibited relatively narrow CIs across all return periods, suggesting higher confidence in their discharge estimations. However, sites like SAN, CCEW and HBR had wider CIs, indicating greater uncertainty in their discharge estimations.

Overall, the results are consistent with previous studies (Amatya *et al.*, 2021; Mukherjee *et al.*, 2024) and can be used in forest road-culvert design to prevent failures caused by flooding. The high-flow quantiles can be used by engineers and land managers to reduce both economic and societal burdens through decreased failure rates, minimized maintenance costs and preserved ecological values within forested watersheds. However, meticulous selection of the most suitable design strategy remains paramount for both new and renovated road-crossing structures, including site inspections associated with identifying the geomorphological and post-wildfire flood vulnerabilities. Equally crucial is meticulous installation procedures, including alignment checks, stream dimension verification and thorough substrate composition assessment (Hansen *et al.*, 2009). Finally, implementation of effective monitoring protocols and strategies for the drainage culverts will ensure sustained flood resilience throughout their lifespan. In addition, the high flood quantiles can also have impacts on the carrying capacities of long-term gauging stations. For example, Amatya *et al.* (2016a) found some of the SAN gauging stations inundated with flows exceeding the established rating curve after the extreme rainfall of Hurricane Joaquin (3–4 October 2015) when over 590 mm of precipitation fell in 48 hour (Figs 5.2, 5.3 and 5.6) causing sustained ponding.

5.4.6 Low-flow quantiles

The results in Table 5.6 and Fig. 5.7 show the low-flow quantiles in L/s, with 95% confidence bounds, estimated using the annual minimum of 7-day average streamflow data for the six sites.

7Q10 varied across the sites, highlighting the heterogeneity of low-flow conditions in forested watersheds due to climate and precipitation patterns. The low-flow estimates at the SDEF site across all return intervals ranged from 0.0032 l/s/ha (5th percentile) to 0.0036 l/s/ha (95th percentile) for a 25-year return interval and from 0.1044 l/s/ha (5th percentile) to 0.1308 l/s/ha (95th percentile) for a 100-year return interval, reflecting the relatively stable precipitation pattern of this site. Similar to SDEF, the HJA site exhibited a stable low-flow uncertainty.

Among our ten sites, CHL and CPCR demonstrated the highest 10-year low flow of 0.127 and 0.124 l/s/ha, respectively (Table 5.6). This low 7Q10 suggests a low susceptibility to drought and minimal ecological stress during low-flow periods. CCEW exhibited moderately low 7Q10 magnitudes (Table 5.6) with a 10-year low flow of 0.18 l/s/ha. This value indicates a less-permanent flow regime, where the stream may become discontinuous during drought periods, as was observed in 2021 (Keppeler *et al.*, 2024). Typical of California, the CCEW aquatic ecosystem experiences seasonal drought, and stream organisms are adapted to more limited summer habitats. Water-resource managers regulate withdrawal and consumptive use during the low-flow season. Overall, this analysis highlights the diverse flow characteristics across the ten sites. While some sites exhibited consistent and predictable flow regimes, others presented more complex patterns requiring further investigation, particularly SDEF's potential transition from intermittent to perennial flow. By analyzing low-flow statistics, ecologists can quantify the potential impacts of low-flow events on forest health and productivity. For example, declining fall precipitation has reduced fall flows, thereby impeding salmonid spawning migration in northern California (Keppeler *et al.*, 2024). This information enables the development of data-driven management strategies to mitigate low-flow stress on ecosystems and prioritize conservation efforts for vulnerable watersheds. Ultimately, this approach ensures the sustainable management and conservation of forested landscapes in the face of changing climatic conditions.

5.5 Conclusions

This study was conducted using long-term measured high-resolution annual maximum PIs and extreme high and low streamflow rates from ten USDA Forest Service EFs. We first evaluated the long-term temporal characteristics and trends of hydrometeorological variables, assuming their stationarity, followed by a cross-site comparison

Table 5.6. Minimum 7-day flow (ANMIN7Q) for 2-, 5-, 10-, 25-, 50- and 100-year return intervals (RIs) with lower and upper 95% confidence intervals (bounds) (l/s/ha).

Site	RI (years)	Non-exceedance probability	Estimates	Lower bound	Upper bound
CCEW	2	0.5	0.10	0.10	0.10
	5	0.8	0.15	0.15	0.15
	10	0.9	0.18	0.18	0.18
	25	0.96	0.23	0.23	0.23
	50	0.98	0.26	0.26	0.267
	100	0.99	0.30	0.29	0.31
CHL	2	0.5	0.066	0.066	0.066
	5	0.8	0.102	0.102	0.102
	10	0.9	0.128	0.127	0.129
	25	0.96	0.164	0.162	0.166
	50	0.98	0.193	0.191	0.194
	100	0.99	0.223	0.221	0.226
CPCR	2	0.5	0.053	0.052	0.054
	5	0.8	0.093	0.092	0.094
	10	0.9	0.126	0.124	0.128
	25	0.96	0.175	0.171	0.178
	50	0.98	0.217	0.212	0.221
	100	0.99	0.264	0.257	0.271
FRS	2	0.5	0.028	0.028	0.028
	5	0.8	0.045	0.044	0.045
	10	0.9	0.057	0.057	0.057
	25	0.96	0.075	0.075	0.076
	50	0.98	0.090	0.089	0.091
	100	0.99	0.106	0.105	0.108
HBR	2	0.5	0.0012	0.0012	0.0014
	5	0.8	0.0028	0.0028	0.0028
	10	0.9	0.0042	0.0042	0.0042
	25	0.96	0.0064	0.0064	0.0064
	50	0.98	0.0083	0.0080	0.0083
	100	0.99	0.0101	0.0099	0.0104
HJA	2	0.5	0.0010	0.0010	0.0010
	5	0.8	0.0023	0.0022	0.0023
	10	0.9	0.0033	0.0033	0.0033
	25	0.96	0.0048	0.0047	0.0048
	50	0.98	0.0060	0.0060	0.0062
	100	0.99	0.0073	0.0072	0.0075
SDEF	2	0.5	0.0000	0.0000	0.0000
	5	0.8	0.0012	0.0008	0.0012

Continued

Table 5.6. Continued

Site	RI (years)	Non-exceedance probability	Estimates	Lower bound	Upper bound
	10	0.9	0.0032	0.0032	0.0036
	25	0.96	0.0140	0.0128	0.0152
	50	0.98	0.0408	0.0368	0.0452
	100	0.99	0.1176	0.1044	0.1308

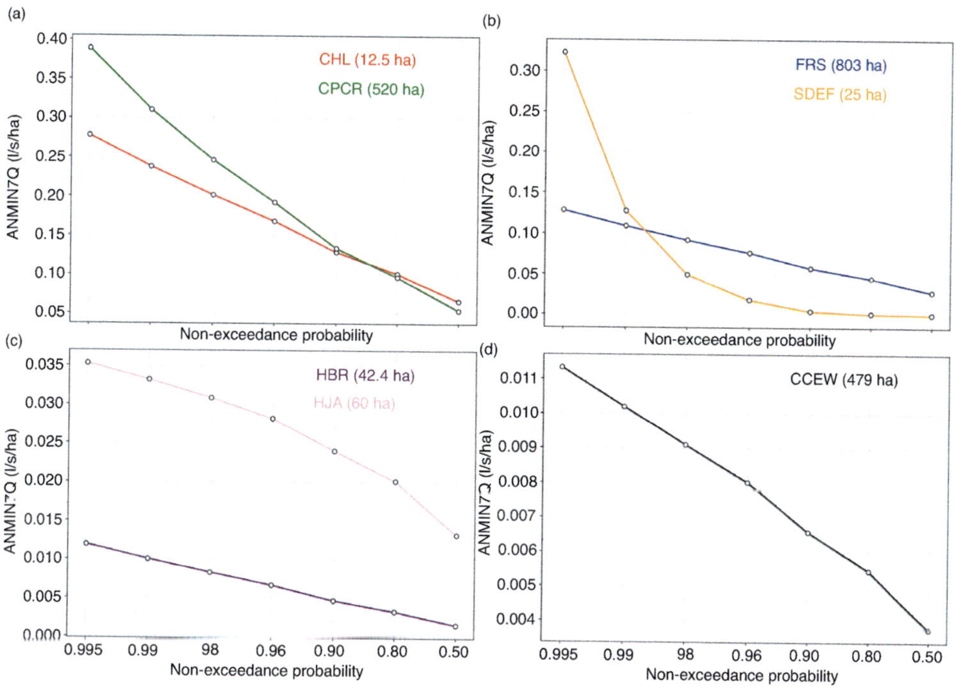

Fig. 5.7. Low-flow frequency curves for the annual minimum 7-day average streamflow per unit drainage area (ANMIN7Q, l/s/ha) for the selected stream gauges.

of the extreme (≥25 year return interval) PIDF and high (≥25 years) and low (≥10 years) flow frequencies of discharges. Frequencies of 25 years or more are commonly used in design of road cross-drainage and stormwater management structures and other similar ecological applications. Similarly, 10 year frequency low-flow quantiles are generally used in assessing environmental flows for aquatic ecosystems.

The results suggested that the SAN, CHL and HBR sites had the highest mean values and variability of 1 and 24 hour annual maximum PIs (ANMAXPI), while the FRS and CPCR sites had the lowest. Trend analysis suggested a significant ($\alpha = 0.05$) long-term decrease in 1 hour PI at the CPCR site and an increase in 24 hour PI at the CHL site. These results are consistent with the results of the PIDF analysis yielding the highest and most uncertain PIs for the 25-, 50- and 100 year return intervals at the CHL (for 1 hour) and SAN (for 24 hour) sites. The extreme high-flow characteristics and trend analysis indicated a similar complex spatiotemporal pattern across the sites, with the highest mean annual maximum specific discharge (ANMAXQ) at the low-gradient SAN site and the high-gradient HBR site, both on the east coast. A significant positive long-term trend of ANMAX was also found at the HBR site. A relatively high interannual variability of ANMAXQ at the SAN,

HBR and SDEF sites is also consistent with their higher 25-, 50- and 100 year ANMAXQ values than all other sites. In contrast, FRS and CPCR yielded the lowest risk of flooding based on their low 100 year ANMAXQ values. The other sites had intermediate values and uncertainties. Similarly, the assessment of extreme low-flow characteristics revealed that SDEF, HBR and CCEW had the lowest mean and variability of ANMIN7Q. Trend analysis revealed diverse temporal patterns of ANMIN7Q, with some sites showing long-term and recent temporal increases, some showing the opposite, and others showing divergent trends in the recent 20 year period. The extreme low-flow frequency analysis indicated the highest ANMIN7Q values of 0.127 and 0.124 l/s/ha at the CHL and CPCR sites, respectively, suggesting a low susceptibility to drought and minimal ecological stress during low-flow periods at these sites. In contrast, SDEF, situated in the dry chaparral of southern California, exhibited the lowest flow magnitude among all sites, with a 10 year low flow of 0.0016 l/s/ha.

Overall, the findings showed sites with different impacts of climate change and variability on streamflow regimes, and that site-specific factors, such as climate, land use and land cover changes, and watershed characteristics, play a role in determining the response of streams to precipitation inputs. Consistent with Wright *et al.* (2019), we believe that, although trends in rainfall extremes may not have necessarily translated into observable increases in flood risks, these results on extreme precipitation and flood frequencies none the less would be of use in hydrological design applications, taking into consideration recent changes in extreme rainfall properties at the USFS FE watersheds (Amatya *et al.*, 2016a; Amatya *et al.*, 2021; Mukherjee *et al.*, 2023; Mukherjee *et al.*, 2024). In addition, the derived flood frequencies from these reference watersheds may serve the purpose of comparing them with those from the paired treatment watersheds at these EFs to correctly evaluate the treatment effects as opposed to a chronologically paired approach, as argued by Alila *et al.* (2009).

Although this study has provided valuable insights into the trends and variability of hydrometeorological variables across ten USDA Forest Service EFs, and their implications for flood risk, water availability and ecological health, there are still some limitations and challenges that need to be addressed in future research. First, the assumption of stationarity may not be valid for some sites, especially under the influence of climate change and human activities. Therefore, alternative methods that account for non-stationarity (Cheng and AghaKouchak, 2014) may be more suitable for detecting and quantifying changes in hydrometeorological variables. Second, the data record availability and quality may result in high uncertainty, warranting a need for more data collection and quality control to improve the robustness of the results and reduce uncertainties. Third, the spatial and temporal scales of the analysis may influence the interpretation and application of the findings. For instance, the annual and seasonal trends may not capture the subseasonal or daily variations that are important for hydrological design and management, particularly the timing of occurrence of annual peak discharge relative to the timing of the precipitation intensities in these small headwater forest watersheds. Similarly, the site-specific factors may not reflect the regional or global patterns of climate change and variability. Even where data are available for relatively long periods (20–50 years) and the distribution of discharge is fairly well known, there is little confidence in estimates of high-flow discharges for return periods of over 50 years (Eisenbies *et al.*, 2007). Therefore, more comprehensive and multi-site and multi-scale analyses globally are needed to better understand the hydrometeorological processes and their interactive effects, as a result of climate change, on forest management of streamflow, particularly during extreme events, as was shown for the Coweeta watersheds (Kelly *et al.*, 2016), and the effects of extreme precipitation on hydrological behavior transformation in watersheds, as shown by Jayakaran *et al.* (2014).

References

Adams, M.B., Kochenderfer, J.N., Wood, F., Angradi, T.R. and Edwards, P.J. (1994) Forty Years of Hydrometeorological Data from the Fernow Experimental Forest. General Technical Report NE-184. Forest Service: Northeastern Forest Experiment Station.

Ali, G., Tetzlaff, D., McDonnell, J.J., Soulsby, C., Carey, S. *et al.* (2015) Comparison of threshold hydrologic response across northern catchments. *Hydrological Processes* 29, 3575–3591.

Alila, Y.P., Kuras, K., Schnorbus, M. and Hudson, R. (2009) Forests and floods: A new paradigm sheds light on age-old controversies. *Water Resources Research* 45, W08416.

Amatya, D.M., Harrison, C.A. and Trettin, C.C. (2016a) Hydro-meteorologic assessment of October 2015 extreme precipitation event on Santee Experimental Forest watersheds, South Carolina. *Journal of South Carolina Water Resources* 3, 19–30.

Amatya, D.M., Campbell, J., Wohlgemuth, P., Elder, K.E., Sebestyen, S. *et al.* (2016b) Hydrological processes of reference watersheds in experimental forests, USA. In: Amatya, D.M., Williams, T.M., Bren, L. and Jong, C. (eds) *Forest Hydrology: Processes, Management, and Applications*. CAB International, Wallingford, UK, pp. 219–239.

Amatya, D.M., Tian, S., Marion, D.A., Caldwell, P., Laseter, S. *et al.* (2021) Estimates of precipitation IDF curves and design discharges for road-crossing drainage structures: Case study in four small forested watersheds in the southeastern US. *Journal of Hydrologic Engineering* 26, 05021004.

Ansa Thasneem, S., Chithra, N.R. and Thampi, S.G. (2021) Assessment of nonstationarity and uncertainty in precipitation extremes of a river basin under climate change. *Environmental Modeling and Assessment* 26, 295–312.

Berghuijs, W.R., Woods, R.A., Hutton, C.J. and Sivapalan, M. (2016) Dominant flood generating mechanisms across the United States. *Geophysical Research Letters* 43, 4382–4390.

Blöschl, G., Hall, J., Viglione, A., Perdigão, R.A., Parajka, J. *et al.* (2019) Changing climate both increases and decreases European river floods. *Nature* 573, 108–111.

Bolton, W.R., Hinzman, L.D. and Yoshikawa, K. (2004) Water balance dynamics of three small catchments in a sub-Arctic boreal forest. In: Kane, D.L. and Yang, D. (eds) *Proceedings of a Northern Research Basins Water Balance Workshop, 15–19 March 2004, Victoria, BC*, International Association of Hydrological Sciences, Victoria, Canada, pp. 213–223.

Bonnin, G.M., Martin, D., Lin, B., Parzybok, T., Yekta, M. *et al.* (2006) *Precipitation-frequency atlas of the United States Vol. 1, Version 5.0: Semiarid Southwest (Arizona, Southeast California, Nevada, New Mexico, Utah)*, NOAA Atlas 14. US Department of Commerce, Silver Spring, Maryland.

Burt, T.P., Miniat, C., Laseter, S.H. and Swank, W.T. (2018) Changing patterns of daily precipitation totals at the Coweeta Hydrologic Laboratory, North Carolina, USA. *International Journal of Climatology* 38, 94–104.

Cafferata, P.H. and Reid, L.M. (2013) *Application of long-term watershed research to forest management in California: 50 years of learning from Caspar Creek Experimental watersheds*. California Forestry Report No. 5. California: Natural Resources Agency, Department of Forestry and Fire Protection.

Campbell, J., Driscoll, C.T., Jones, J.A., Boose, E.R., Dugan, H.A. *et al.* (2022) Forest and freshwater ecosystem responses to climate change and variability at US LTER sites. *BioScience* 72, 851–870.

Campbell, J.L., Driscoll, C.T., Pourmokhtarian, A. and Hayhoe, K. (2011) Streamflow responses to past and projected future changes in climate at the Hubbard Brook Experimental forest, New Hampshire, USA. *Water Resources Research* 47, W02514.

Cheng, L. and AghaKouchak, A. (2014) Nonstationary precipitation intensity-duration-frequency curves for infrastructure design in a changing climate. *Scientific Reports* 4, 7093.

Coles, S., Bawa, J., Trenner, L. and Dorazio, P. (2001) *An Introduction to Statistical Modeling of Extreme Values*. Springer, London.

Cox, C. (1990) Fieller's theorem, the likelihood and the delta method. *Biometrics* 46, 709–718.

Daly, C., Schulze, M. and McKee, W. (2019) Meteorological data from benchmark stations at the HJ Andrews Experimental Forest, 1957 to present. Available at: https://andlter.forestry.oregonstate.edu/data/abstractdetail.aspx?dbcode=ms001 (accessed 17 April 2025).

de Luca, D. and Galasso, L. (2018) Stationary and non-stationary frameworks for extreme rainfall time series in Southern Italy. *Water* 10, 1477.

Dunn, P.H., Barro, S.C., Wells, W.G.I., Poth, M.A., Wohlegemuth, P.M. *et al.* (1988) *The San Dimas Experimental Forest: 50 Years of Research*. General Technical Report PSW-105. Berkeley, California: US Forest Service.

Eisenbies, M., Aust, W., Burger, J. and Adams, M.B. (2007) Forest operations, extreme flooding events, and considerations for hydrologic modeling in the Appalachians – A review. *Forest Ecology and Management* 242, 77–98.

England, J.F.J., Cohn, T.A., Faber, B.A., Stedinger, J.R., Thomas, W.O.J. *et al.* (2019) *Guidelines for Determining Flood Flow Frequency*, Bulletin 17C (No. 4-B5). US Geological Survey. Available at: www.usgs.gov/publications/guidelines-determining-flood-flow-frequency-bulletin-17c (accessed 17 April 2025).

Feaster, T.D. and Guimaraes, W.B. (2014) Low-flow Frequency and Flow Duration of Selected South Carolina Streams in the Catawba-Wateree and Santee River Basins through March 2012. US Geological Survey Open-File Report 2014-1113. Available at: https://pubs.usgs.gov/publication/ofr20141113 (accessed 17 April 2025).

Feaster, T.D., Gotvald, A.J. and Weaver, J. (2009) *Magnitude and Frequency of Rural Floods in the Southeastern United States, 2006: Volume 3, South Carolina*. US Geological Survey. Available at: https://pubs.usgs.gov/publication/sir20095156 (accessed 17 April 2025).

Gimeno, L., Sorí, R., Vazquez, M., Stojanovic, M., Algarra, I. *et al.* (2022) Extreme precipitation events. *WIREs Water* 9, e1611.

Gu, X., Zhang, Q., Singh, V.P. and Shi, P. (2017) Nonstationarity in timing of extreme precipitation across China and impact of tropical cyclones. *Global Planet Change* 149, 153–165.

Haan, C.T., Barfield, B.J. and Hayes, J.C. (1994) *Design Hydrology and Sedimentology for Small Catchments*. Elsevier, London.

Hamon, W.R. (1963) Estimating potential evapotranspiration. *Journal of the Hydraulics Division* 87, 107–120.

Hansen, B., Nieber, J.L. and Lenhart, C. (2009) *Cost analysis of alternative culvert installation practices in Minnesota*. University of Minnesota Digital Conservancy, Minnesota.

Hargreaves, G.H. and Samani, Z.A. (1982) Estimating potential *evapotranspiration*. *Journal of Irrigation and Drainage Division* 108, 223–230.

Heredia, N., Roper, B., Gillespie, N. and Roghair, C. (2016) *Technical Guide for Field Practitioners: Understanding and Monitoring Aquatic Organism Passage at Road Stream Crossings*. Technical Report no.TR-101. Center, USDA. Washington, DC: Forest Service, National Stream and Aquatic Ecology.

Hook, D.D., Buford, M.A. and Williams, T.M. (1991) Impact of hurricane hugo on the South Carolina coastal pine forest. *Journal of Coastal Research SI* 291–300.

Interagency Advisory Committee on Water Data (1982) *Guidelines for Determining Flood Flow Frequency – Bulletin 17B of the Hydrology Subcommittee*. US Geological Survey, Office of Water Data Coordination, Reston, Virginia.

Jalowska, A.M., Spero, T.L. and Bowden, J.H. (2021) Projecting changes in extreme rainfall from three tropical cyclones using the design rainfall approach. *Npj Climate and Atmospheric Science* 4, 23.

Jayakaran, A., Williams, T.M., Ssegane, H.S., Amatya, D.M., Song, B. *et al.* (2014) Hurricane impacts on a pair of coastal forested watersheds: Implications of selective hurricane damage to forest structure and streamflow dynamics. *Hydrology and Earth System Science* 18, 1151–1164.

Jian, X., Wolock, D.M., Jenter, H.L. and Brady, S. (2015) *Streamflow of 2014: Water Year Summary*, Fact Sheet No. 2015-3026. US Geological Survey. Available at: https://pubs.usgs.gov/fs//2015/3026/ (accessed 17 April 2025).

Johnson, S., Henshaw, D., Downing, G., Wondzell, S., Schulze, M. *et al.* (2021) Cascade Mountains, Oregon. *Hydrological Processes* 35, e14187.

Johnson, S., Wondzell, S. and Rothacher, J. (2023) Stream discharge in gaged watersheds at the HJ Andrews Experimental Forest, 1949 to present. Available at: https://andlter.forestry.oregonstate.edu/data/abstractdetail.aspx?dbcode=hf004 (accessed 17 April 2025).

Jones, J.B. and Rinehart, A.J. (2010) The long-term response of stream flow to climatic warming in headwater streams of interior Alaska. *Canadian Journal of Forest Research* 40, 1210–1218.

Kelly, C.N., McGuire, K.J., Miniat, C.F. and Vose, J.M. (2016) Streamflow response to increasing precipitation extremes altered by forest management. *Geophysical Research Letters* 43, 3727–3736.

Kendall, M.G. (1975) *Rank Correlation Methods*, 4th edn. Charles Griffin, London.

Keppeler, E., Wagenbrenner, J., Dymond, S. and Dralle, D. (2024) Streamflow response to drought in a managed coast redwood catchment. *Journal of the American Water Resources Association* 60, 928–952.

Kim, H., Kim, S., Shin, H. and Heo, J.H. (2017) Appropriate model selection methods for nonstationary generalized extreme value models. *Journal of Hydrology* 547, 557–574.

Laio, F. (2004) Cramer–von Mises and Anderson–Darling goodness of fit tests for extreme value distributions with unknown parameters. *Water Resources Research* 40, W09308.

Laseter, S.H., Ford, C.R., Vose, J.M. and Swift, L.W. (2012) Long-term temperature and precipitation trends at the Coweeta Hydrologic Laboratory, Otto, North Carolina, USA. *Hydrology Research* 43, 890–901.

Mann, H.B. (1945) Non-parametric tests against trend. *Econometrica* 13, 163–171.

Martins, E.S. and Stedinger, J.R. (2000) Generalized maximum-likelihood generalized extreme-value quantile estimators for hydrologic data. *Water Resources Research* 36, 737–744.

Meixner, T. and Wohlgemuth, P.M. (*et al*2003) Climate variability, fire, vegetation recovery, and watershed hydrology. In: Renard, K.G., McElroy, S.A., Gburek, W.J., Canfield, H.E., Russell, L. et al. (eds) *The First Interagency Conference on Research in the Watersheds*, Benson, Arizona, October 27–30, 2003, pp. 651–656.

Milly, P.C., Betancourt, J., Falkenmark, M., Hirsch, R.M., Kundzewicz, Z.W. *et al.* (2008) Stationarity is dead: Whither water management. *Science* 319, 573–574.

Mukherjee, S., Amatya, D.M., Jalowska, A., Campbell, J.L., Johnson, S.L. *et al.* (2023) Comparison of on-site versus NOAA's extreme precipitation intensity-duration-frequency estimates for six forest headwater catchments across the continental United States. *Stochastic Environmental Research and Risk Assessment* 37, 4051–4070.

Mukherjee, S., Amatya, D.M., Campbell, J.L., Gryczkowski, L., Panda, S. *et al.* (2024) A watershed-scale multi-approach assessment of design flood discharge estimates used in hydrologic risk analyses for forest road stream crossings and culverts. *Journal of Hydrology* 632L, 130698.

National Research Council (1988) *Estimating Probabilities of Extreme Floods – Methods and Recommended Research*. National Academy Press, Washington, DC.

National Research Council (1995) *Flood Risk Management and The American River Basin; an Evaluation*. National Academy Press, Washington, DC.

Nerantzaki, S.D. and Papalexiou, S.M. (2022) Assessing extremes in hydroclimatology: A review on probabilistic methods. *Journal of Hydrology* 605, 127302.

Papalexiou, S.M. and Montanari, A. (2019) Global and regional increase of precipitation extremes under global warming. *Water Resources Research* 55, 4901–4914.

Patric, J.H. and Goswami, N. (1968) Evaporation pan studies – forest research at Parsons. *West Virginia Agriculture and Forestry* 1, 6–10.

Peel, M.C., Finlayson, B.L. and McMahon, T.A. (2007) Updated world map of the Köppen-Geiger climate classification. *Hydrology and Earth System Sciences* 11, 1633–1644.

Perica, S., Martin, D., Pavlovic, S., Roy, I., Laurent, M. *et al.* (2013) *NOAA Atlas 14 Precipitation-Frequency Atlas of the United States*. National Oceanic and Atmospheric Administration, Silver Spring, Maryland.

Perica, S., Pavlovic, S., Laurent, M., Trypaluk, C., Unruh, D. *et al*. (2018) Precipitation-Frequency Atlas of the United States. Version 2.0. Texas (Vol. 11). Texas: National Oceanic Atmospheric Administration.

Preece, J.R., Shinker, J.J., Riebe, C.S. and Minckley, T.A. (2021) Elevation-dependent precipitation response to El Niño-Southern Oscillation revealed in headwater basins of the US Central Rocky Mountains. *International Journal of Climatology* 41, 1199–1210.

Richardson, P.W., Seehafer, J.E., Keppeler, E.T., Sutherland, D.G. and Wagenbrenner, J.W. (2021) Fifty-eight years and counting of watershed science at the Caspar Creek experimental watersheds in Northern California. *Hydrological Processes* 35, e14207.

Riggs, H.C. (1985) *Streamflow Characteristics*. Elsevier, Amsterdam, Netherlands.

Rosenzweig, B., Ruddell, B.L., McPhillips, L., Hobbins, R., McPhearson, T. *et al.* (2019) Developing knowledge systems for urban resilience to cloudburst rain events. *Environmental Science and Policy* 99, 150–159.

Rothacher, J., Dyrness, C.T. and Fredriksen, R.L. (1967) *Hydrologic and Related Characteristics of Three Small Watersheds in the Oregon Cascades*. Pacific Northwest Forest and Range Experiment Station, Forest Service, US Department of Agriculture, Portland, Oregon.

Salas, J., Obeysekera, J. and Vogel, R. (2018) Techniques for assessing water infrastructure for nonstationary extreme events: A review. *Hydrologic Sciences Journal* 63, 325–352.

Sebestyen, S.D., Dorrance, C., Olson, D.M., Verry, E.S., Kolka, R.K. *et al.* (2011) Long-term monitoring sites and trends at the Marcell Experimental Forest. In: Kolka, R.K., Sebestyen, S.D., Verry, E.S. and Brooks, K.N. (eds) *Peatland Biogeochemistry and Watershed Hydrology at The Marcell Experimental Forest*. CRC Press, Boca Raton, Florida, pp. 15–71.

Sebestyen, S.D., Lany, N.K., Roman, D.T., Burdick, J.M., Kyllander, R.L. *et al.* (2021) Hydrological and meteorological data from research catchments at the Marcell Experimental Forest, Minnesota, USA. *Hydrologic Processes* 35, e14092.

Sen, P.K. (1968) Estimates of the regression coefficient based on Kendall's tau. *Journal of the American Statistical Association* 63, 1379–1389.

Sharma, A., Wasko, C. and Lettenmaier, D.P. (2018) If precipitation extremes are increasing, why aren't floods? *Water Resources Research* 54, 8545–8551.

Srivastava, A., Grotjahn, R., Ullrich, P.A. and Risser, M. (2019) A unified approach to evaluating precipitation frequency estimates with uncertainty quantification: Application to Florida and California watersheds. *Journal of Hydrology* 578, 124095.

Sun, G., Tiwari, K.R., Hao, L., Amatya, D., Liu, N. *et al.* (2023) Climate change and forest hydrology in future forests. In: McNulty, S.G. (ed.) *Future Forests: Mitigation and Adaptation to Climate Change*. Elsevier, Amsterdam, Netherlands, pp. 95–124.

Tetzlaff, D., Carey, S. and Soulsby, C. (2013) Catchments in the future North: Interdisciplinary science for sustainable management in the 21st century. North-Watch Workshop, Potsdam, Germany, 21–25 May 2012, *Hydrological Processes* 27, 635–774.

Thornthwaite, C.W. (1948) An approach toward a rational classification of climate. *Geographical Review* 38, 55–94.

USGS (1982) *Guidelines for Determining Flood Frequency*. Bulletin 17B of the Hydrology Subcommittee. Revised September 1981, Editorial Corrections, March 1982. US Department of Interior Geological Survey, Reston, Virginia.

van Cleve, K., Chapin, F.S., Ruess, R., Mack, M.C., Jones, J.B. *et al.* (2023) *Caribou-Poker Creeks Research Watershed: Daily Flow Rates for C2, C3, C4*. Version 24. Environmental Data Initiative.

Verry, E.S., Brooks, K.N., Nichols, D.S., Ferris, D.R. and Sebestyen, S.D. (2011) Watershed hydrology. In: Kolka, R.K., Sebestyen, S.D., Verry, E.S. and Brooks, K.N. (eds) *Peatland Biogeochemistry and Watershed Hydrology at the Marcell Experimental Forest*. CRC Press, Boca Raton, Florida, pp. 193–212.

Wang, H. and Sun, F. (2020) On the stationarity of annual precipitation over China (1959–2018). *Journal of Hydrometeorology* 21, 881–890.

Wasko, C., Nathan, R. and Peel, M.C. (2020) Trends in global flood and streamflow timing based on local water year. *Water Resources Research* 56, e2020WR027233.

Wohlgemuth, P.M. (2016) Long-term hydrologic research on the San Dimas Experimental Forest: Lessons learned and future directions. In: *Proceedings of the 5th Interagency Conference on Research in the Watersheds*, General Technical Report GTR-SRS-xxx. Southern Research Station, US Department of Agriculture, Forest Service, Ashville, North Carolina.

Wondzell, S.M. and Swanson, F.J. (1999) Floods, channel change, and the hyporheic zone. *Water Resources Research* 35, 555–567.

Wright, D.B., Bosma, C.D. and Lopez-Cantu, T. (2019) US hydrologic design standards insufficient due to large increases in frequency of rainfall extremes. *Geophysical Research Letters* 46, 8144–8153.

Yochum, S.E., Scott, J.A. and Levinson, D.H. (2019) Methods for assessing expected flood potential and variability: Southern Rocky Mountains region. *Water Resources Research* 55, 6392–6416.

6 Forest Cover Changes and Hydrology in Large Watersheds

X. Wei[1]*, Q. Li[2], M. Zhang[3], W. Liu[4], H. Fan[5] and Y. Hou[1,3]

[1]Department of Earth and Environmental Sciences, University of British Columbia (Okanagan campus), Kelowna, Canada; [2]College of Forestry, Northwest A&F University, Yangling, China; [3]School of Resources and Environment, University of Electronic Science and Technology of China, Chengdu, China; [4]Institute of Ecology and Environmental Science, Nanchang Institute of Technology, Nanchang, China; [5]Jiangxi Agriculture University, Nanchang, China

Abstract

Forest and watershed management strategies are expanding to large landscapes or watersheds. Understanding of hydrological impacts of forest cover changes in large watersheds is essential for protecting water supply and many other aquatic functions. However, hydrological responses to forest cover changes in large watersheds are complex and inconsistent due to diverse climates, vegetation types, topographies and their interactions. The lack of unified research methodology also contributes to such complexity. This chapter reviews existing studies and research methods on the hydrological impacts of forest cover changes in large watersheds. It also proposes future priorities for advancing research on the forest–water nexus in large watersheds within the context of a changing environment.

6.1 Introduction

Forests play an important role in the water cycle by influencing rainfall interception, evapotranspiration, soil infiltration and storage, and streamflow. The impacts of forest changes caused by either natural or human forces (e.g. wildfires, deforestation, reforestation, urbanization) on hydrology have been studied for over a century by various methods including the experimental paired-watershed approach and hydrological modeling (see also Chapter 17, this volume). A general understanding is that deforestation can substantially increase annual streamflow, magnify peak flows and alter base flows, while reforestation can decrease annual streamflow and reduce peak flows (Wei *et al.*, 2017; Zhang *et al.*, 2017; Creed and van Noordwijk, 2018). However, these results have been drawn mainly from the experimental watershed studies conducted at small spatial scales (<100 km^2, most of which are less than 10 km^2), and they cannot simply be extrapolated to large watersheds (>1000 km^2) because of more complexities of landforms (e.g. managed and natural forests, wetlands, lakes, open lands) and their interactions (Zhang *et al.*, 2017). In addition, forest management strategies and planning have been shifting from forest stands to large landscapes or watersheds in order to balance different

*Corresponding author: adam.wei@ubc.ca

 Forest Hydrology: Processes, Management and Assessment (eds Leon J. Bren *et al.*)
DOI: 10.1079/9781800625310.0006

management objectives or trade-offs. These clearly highlight a critical need for conducting separate research on the impacts of forest cover changes in water in large watersheds. The objectives of this chapter are: (i) to briefly summarize the impacts of forest cover changes in hydrology in large watersheds; (ii) to describe various existing research methods in evaluating forest cover change effects on hydrology in large watersheds; and (iii) to identify research challenges and future research priorities.

Studying the impacts of forest cover changes on water in large watersheds is challenging. The first challenge is the lack of an efficient, commonly accepted methodology. The greatest difficulty in a large watershed study lies in separating the effects of forest changes (e.g. disturbances) and climate variability on hydrology (Wei and Zhang, 2010). Forest cover changes and climatic variability are generally viewed as two major drivers interactively influencing streamflow in large forested watersheds (Wei and Zhang, 2010; de Niel and Willems, 2019). It is commonly accepted that the effects of climate variability on hydrology must be excluded in order to quantify the hydrological impacts of forest cover changes in large watersheds. Paired-watershed experiments or physical-based hydrological models are commonly used to study the hydrological effects of forest cover changes in small watersheds but have limitations when applied to large watersheds (Winkler *et al.*, 2017; Sun *et al.*, 2023). Paired-watershed experiments are generally not feasible for large watersheds, given the great difficulty in locating suitable control watersheds (Fohrer *et al.*, 2005). Similarly, physical-based hydrological models, such as the distributed hydrology soils and vegetation model (DHSVM), MIKE-SHE, variable infiltration capacity (VIC) and other models, are only applicable to watersheds that are well monitored with extensive, long-term available data on vegetation, soil, topography, land use, climate and hydrology (Kirchner, 2009; Kirchner and Allen, 2020). Moreover, the empirical relationships between different watershed processes and components used in hydrological models are drawn mainly from small watershed studies and may be problematic when transferred to large watersheds (McMillan *et al.*, 2022). Therefore, the most commonly used methods in small watersheds have limited utility in forest hydrological studies on large watersheds.

Second, the lack of a suitable indicator for representing and integrating various types of forest cover changes or disturbances and post-disturbance vegetation recovery is another challenge in large watershed studies (Wei and Zhang, 2010). For example, in a large watershed, different types of forest disturbances (both natural and anthropogenic) are accumulated over space and time. To quantitatively represent cumulative forest disturbances over time at a watershed scale, an integrated indicator other than a simple indicator such as total disturbed area or forest cover rate is needed. A suitable forest disturbance indicator for a large watershed should represent not only all types of disturbances and intensity ranges but also their cumulative forest disturbance histories and subsequent recovery processes following disturbances over space and time (Wei and Zhang, 2010). Equivalent clear-cut area (ECA) is defined as the area that has been harvested, insect infested or burned, with a reduction factor to account for hydrological recovery due to forest regeneration after disturbances (Winkler and Boon, 2017). The indicator of cumulative equivalent clear-cut area (CECA; the sum of annual equivalent clear-cut area) has been used successfully in the Pacific Northwest to test watershed-scale forest logging or wildfire and their effects on various watershed processes including aquatic habitat, hydrology and aquatic biology (Wei and Zhang, 2010). For example, the annual ECA and CECA of all forest disturbances including logging, wildfires and mountain pine beetle infestation accounted for 13.4% and 31.2% of the watershed area in 2004, respectively, in the Baker Creek watershed (1570 km^2) located in British Columbia, Canada (Zhang and Wei, 2012). Other indicators such as remote sensing-based normalized difference vegetation index (Yu *et al.*, 2023), leaf area index (Parker, 2020) and total watershed sapwood area (Jaskierniak *et al.*, 2015) have also been applied. However, no full comparisons have yet been made to determine which indicators or indices are more suitable than others.

Finally, the lack of suitable study watersheds can also constrain forest hydrological studies in large watersheds. In order to detect the effects of cumulative forest changes on hydrology, a large watershed must experience significant forest changes or disturbances (e.g. >10% CECA) and must also include a sufficiently long period without forest disturbances (or with limited forest disturbances) as a comparable reference or control period. Long-term data on forest cover change history, climate and hydrology must also be available. Moreover, large watersheds are more prone to anthropogenic activities (e.g. channelization, reservoir or dam and/or road constructions including legacy water management structures such as levees and impoundments) and their hydrological impacts (Palmer and Ruhi, 2019). Given the fact that the majority of large watersheds are poorly monitored, it is rather challenging to find suitable study watersheds to assess forest changes and their effects on hydrology.

Despite these limitations, the topic of forest cover changes and hydrology in large watersheds has received growing attention, mainly because many practices and policies of natural resource management are planned and operated on large landscapes, watersheds or even regional scales. The scientific information on large watersheds is critically needed to support the design of natural resource management strategies, especially given the fact that climate change (e.g. global warming) and anthropogenic activities (e.g. logging, urbanization and land conversion) are dramatically and extensively altering watershed processes and ecosystem functions, and are leading to more frequent and catastrophic forest disturbances (e.g. insect infestation and wildfire) (Adams *et al.*, 2020). A comprehensive understanding of the impacts of forest cover changes on water in large watersheds is essential for the sustainability of a long-term water supply and the protection of watershed ecosystem functions under a changing environment.

6.2 Forest Cover Changes and Water in Large Watersheds

Forest cover changes and water in large watersheds have received growing attention in the past few decades mainly because of the increasing demand for scientific information on large-scale watersheds or landscapes to support sustainable natural resources management. In spite of limited studies, significant progress employing different methods such as statistical analyses (Wei *et al.*, 2013; Biederman *et al.*, 2022) and modeling (Restaino *et al.*, 2016; Barnhart *et al.*, 2021) has been made on this subject.

A global synthesis based on 162 published case studies on forest cover changes (deforestation and reforestation) and annual water yield (AWY) in large watersheds (>1000 km^2) showed that deforestation increases AWY, while reforestation decreases it, which is generally consistent with the results from small paired-watershed experiments (Li *et al.*, 2017). The synthesis also showed that greater changes in forest cover caused larger AWY responses regardless of change directions (deforestation or reforestation impacts) (Fig. 6.1). However, it is important to highlight that there are some contradictory results. Hou *et al.* (2022) detected significant reductions in annual streamflow (31 and 37 mm) caused by cumulative forest disturbance in two boreal forested watersheds in northern British Columbia, Canada. These reductions might be due to increased evapotranspiration as a result of the cumulative replacement of fast-growing pioneering species following mountain pine beetle infestation. Goeking and Tarboton (2020) also found more inconsistent and even contradictory conclusions in snow-dominated watersheds.

Forest cover changes not only substantially alter AWY or annual mean flow but also change peak flows. However, few studies have been conducted on assessing forest cover changes and peak flows in large watersheds. In addition, the results on peak flow responses to forest cover changes are inconsistent, with large variations. Many studies have shown that hydrological responses of peak flows to the alteration of forest covers are not significant in large-scale basins. For instance, a study in north-eastern Ontario, Canada, by Buttle and Metcalfe (2000) found limited streamflow responses in some large watersheds (ranging from 401 to 11,900 km^2) to land cover changes (5–25%) and no definitive changes in annual peak flows. Wilk *et al.* (2001) did not find any significant hydrological change in peak flows in the Nam Pong River basin (12,100 km^2) in north-east Thailand after a reduction of forest cover from 80% in 1957 to 27% in 1995, which may have been due to shaded trees left in the agriculture areas and secondary growth in the abandoned plots. In

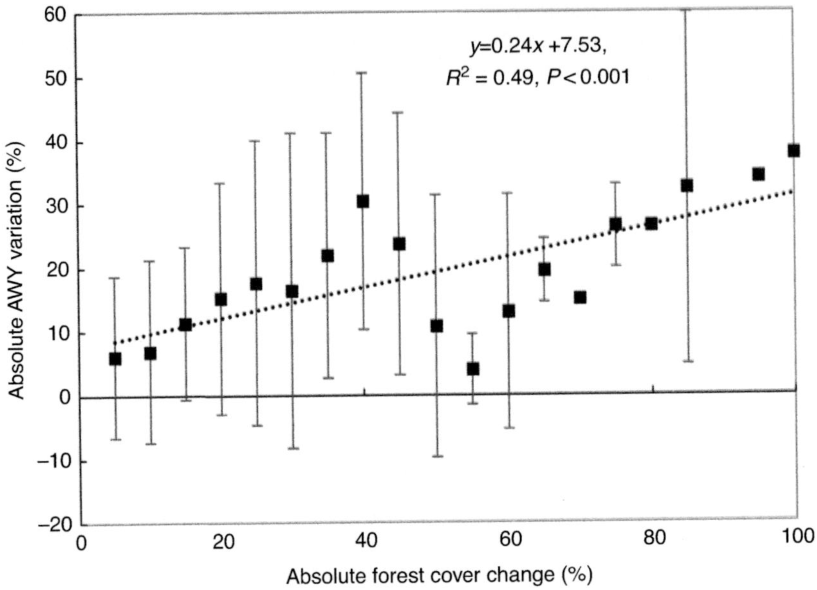

Fig. 6.1. Absolute changes (%) in annual water yield (AWY) responding to the absolute changes in forest cover (for every 5% forest cover change) (Li *et al.*, 2017). Bars show standard deviations.

contrast, some studies have shown that peak or high flows are significantly increased by deforestation in large watersheds. For example, peak flows or high flow regimes were dramatically increased in several large watersheds located in the interior of British Columbia, Canada, including the Willow River watershed (Lin and Wei, 2008), Deadman River watershed (Giles-Hansen *et al.*, 2019) and Baker River watershed (Zhang *et al.*, 2016).

An interesting case study on the comparison of peak flow responses to forest disturbance between two neighboring large watersheds (Bowron River and Willow River watersheds, located in the interior of British Columbia) is worth mentioning here (Zhang and Wei, 2014). Both watersheds experienced similar forest disturbance levels (25–30% ECA). The results showed that forest harvesting in the Willow watershed dramatically increased annual and spring mean flows as well as annual and spring peak flows, whereas forest harvesting caused insignificant changes in these hydrological variables in the Bowron watershed. The contrasted differences in hydrological responses are due to the differences in topography, spatial heterogeneity, forest harvesting characteristics and climate between the two watersheds. The relative uniform topography and climate in the Willow watershed may promote hydrological synchronization effects, whereas larger variations in elevations, together with forest harvesting that occurred at lower elevations, may cause hydrological desynchronization effects in the Bowron watershed. The contrasted results demonstrate that the effects of forest disturbance on hydrology in large watersheds are likely watershed specific, and any attempt to generalize hydrological responses to forest changes must be carried out with caution.

Studies on low or base flow responses to forest cover changes in large watersheds are even rarer. The results from small watershed studies showed that the responses of low flows to logging could be positive, negative or negligible (Coble *et al.*, 2020; Moore *et al.*, 2020), while reforestation generally decreased low flow (Filoso *et al.*, 2017). Due to more complexities in landforms, channel morphology and topographies in large watersheds, it is generally expected that the responses of low flows to forest changes in large watersheds are more varied. For example, in a large, severely disturbed watershed, Baker River watershed (equivalent clear-cut percentage ~60%), low flows were significantly increased (Zhang *et al.*, 2016). Interestingly, Zhou *et al.* (2010) also found that large-scale reforestation (forest recovery) plays a positive role in redistributing water from the wet season to the dry season and, consequently, in increasing water yield in the dry season. Nevertheless, more case studies are needed before any meaningful conclusions on forest cover changes and low flows in large watersheds can be provided.

6.3 Research Methods

Current approaches to evaluating hydrological responses associated with forest changes in large watersheds can be classified into two general categories: hydrological modeling and non-modeling (Wei *et al.*, 2013). The selection of a suitable research approach depends mainly on the purpose of research, data availability and the number of available watersheds. Here we describe six methods commonly applied in this subject.

6.3.1 Hydrological modeling

Hydrological models are frequently used in large watershed hydrological research. Hydrological models can be divided into lumped, semi-distributed and distributed models in light of their spatial representations. Lumped hydrological models treat a watershed as a whole system or entity, and do not consider the detailed spatial representations of watershed elements and processes. Semi-distributed models divide a watershed into several sub-basins. However, the spatial heterogeneity is expressed only to some extent, and not in great detail. Unlike lumped or semi-distributed models, distributed models can well represent a watershed by assigning input data and physical characteristics to grids or elements within the delineated sub-basins. Physical-based distributed models are able to provide distributed approximations or predictions of hydrological variables across watersheds, and thus have a better representation of reality. However, a physical-based distributed model requires a large data set and input parameters on various processes, components and their interactions. For large-scale watershed research, a semi-distributed model is commonly used because of a general absence of detailed data and input parameters at large scales.

The one-factor-at-a-time approach commonly used in sensitivity analysis (Abou-Taleb and Galal, 2018) is also used in association with hydrological models to distinguish the impact of climate factors and land cover change on watershed hydrology (Zhang *et al.*, 2018; Ceppi *et al.*, 2023). In a hypothetical example, the impacts of climatic variability and land-use change on streamflow are assessed with the available data in the period of 1960–2000. First, we keep the land cover in 1960 unchanged over the simulation period while climate change is allowed from 1960 to 2000. Then, we simulate the streamflow change (ΔQC), which can be treated as the impact of climatic variability on hydrology. Second, we keep the climate of 1960 unchanged while land cover is changed, and then calculate the streamflow change (ΔQL) as the impact of land cover change. Finally, we assume the changes of both climate and land cover, and then calculate the streamflow change ($\Delta QL + C$). In this way, the relative contributions of forest and land cover changes and climatic variability to hydrology can be computed.

Various hydrological models have been used successfully to quantitatively study the effects of climate change and forest change/land cover change on hydrology (Sun *et al.*, 2023); some examples are the soil and water assessment tool (SWAT) (Hu *et al.*, 2021; Senbeta and Romanowicz, 2021), DHSVM (Du *et al.*, 2016) and MIKE-SHE (Wang *et al.*, 2013; Lagos *et al.*, 2024). In spite of increased applications, hydrological models are still based on our current theories, which are deeply rooted in the physics of small-scale processes. This gives rise to difficulties in representing non-linear hydrological processes and their interactions at all scales across heterogeneous landscapes. In addition, calibrating and testing a model may not always assure its validity, as there are some inherent drawbacks in the approaches of parameter calibration and validation (Kirchner, 2006). We often overparameterize our models to meet high accuracy levels, ignoring the equifinality problem where different parameter sets for a model might yield the same results during calibration but distinctly different predictions when conditions are altered (Kirchner, 2006).

6.3.2 Breakpoints and double mass curves

Double mass curves (DMCs) and their modifications are a non-modeling, simple and intuitive method, widely used in paired-watershed experiments and long-term trend analysis of hydrometeorological elements. A DMC draws a curve between river flows from the treatment watershed against those from the control watershed in paired-watershed experiments, or two cumulative hydrometeorological variables in individual watersheds to test the consistency of two variables or analyze the trend change and its strength (Biederman *et al.*, 2015; Hallema *et al.*, 2017). The DMC method can be also used to separate the relative influences of forest change and climatic effects on hydrology (Wei and Zhang, 2010; Aryal and Zhu, 2020). For example, a modified DMC (MDMC; Wei and Zhang, 2010) between accumulated annual streamflow and accumulated effective precipitation (the difference between total precipitation and estimated evapotranspiration) can be constructed for a large forested watershed (Fig. 6.2). In this way, the climatic effect on annual streamflow can be eliminated. In a period of limited forest disturbance, the curve should produce a straight line, i.e. a baseline that describes the linear relation between annual streamflow and annual effective precipitation. With continuous forest disturbance, a breakpoint in this curve (e.g. year 1986 in Fig. 6.2) would suggest a change of annual streamflow caused by forest disturbance. In other words, a step change or regime shift occurs in the slope of the MDMC, and the slope before the breakpoint is different from that afterward. However, this visually detected breakpoint needs confirmation of its statistical significance, such as non-parametric tests, and the ARIMA (autoregressive integrated moving average) intervention model (Box and Pierce, 1970). The difference between actual observations and predictions after the breakpoint can be calculated, and is regarded as the cumulative impact of forest changes.

6.3.3 Sensitivity-based approach

The sensitivity-based approach (a non-modeling method) is similar to the elasticity method, and is used to calculate the effect of climate variability on streamflow. Perturbations in both

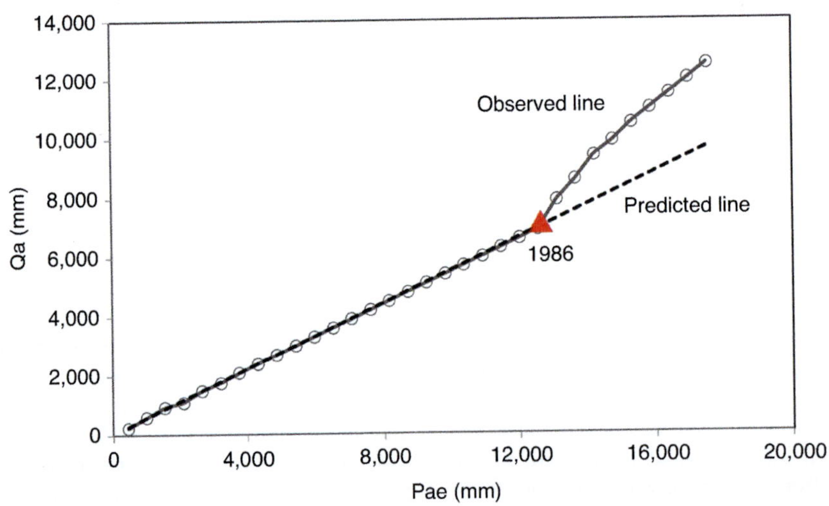

Fig. 6.2. A hypothetical example of the application of a modified double mass curve between accumulated annual streamflow (Q_a) and accumulated annual effective precipitation (P_{ae}) for quantifying the effects of forest changes on annual streamflow.

precipitation and potential evapotranspiration (PET) can lead to changes in water balance. It can be assumed that a change in mean annual streamflow can be determined using the following expression (Wei *et al.*, 2013; Lv *et al.*, 2019):

$$\Delta Q_{\text{clim}} = \beta \times \Delta P + \gamma \times \Delta PET \tag{6.1}$$

where ΔQ_{clim}, ΔP and ΔPET are changes in streamflow due to climate variability, precipitation and potential evapotranspiration, respectively, and β and γ are the sensitivity coefficients of streamflow to precipitation and potential evapotranspiration, respectively, expressed as:

$$\beta = \frac{\left(1+2x+3wx^2\right)^2}{\left(1+w+wx^2\right)^2} \tag{6.2}$$

and

$$\gamma = \frac{\left(1+2wx\right)}{\left(1+x+wx^2\right)^2} \tag{6.3}$$

where x is the mean annual dryness index (equal to PET/P), and w is the vegetation factor with values for forest, grassland and shrubland of 2, 0.5 and 1, respectively (Zhang *et al.*, 2001).

This method is suitable for the analysis of a single basin, and for quantitative calculation of the impact of climate variables on streamflow. Once the effects of climatic variability on flow are estimated, the effects of forest disturbance or land use changes can be deducted from total streamflow variations. The method has been used successfully for several case studies (Lv *et al.*, 2019; Hou *et al.*, 2022). There may be two challenges in this method. First, it is not easy to determine w values for specific forest types. Where there are always different types of forests in a large watershed, how to select a specific w value remains challenging. Second, the effect of forest disturbance or land-use changes on hydrology is indirectly estimated from total hydrological variations and the effects of climatic variability. Thus, its reliability is dependent on the accuracy of the other two terms.

6.3.4 Simple water balance

The water balance methods provide a framework to determine changes in the water balance components. A simple water balance model can be used to determine the influence of climate and vegetation on streamflow at a watershed scale, using the equation:

$$P = ET + Q + \Delta S \tag{6.4}$$

where P is precipitation, ET is actual evapotranspiration, Q is streamflow and ΔS is the change in catchment water storage. When averaged over a long period, deep percolation (recharge) and changes in soil moisture storage are often only 5–10% of the annual water balance, and therefore the change in catchment water storage (ΔS) can be neglected (Han *et al.*, 2020).

Precipitation and actual evapotranspiration constitute the most important variables to influence streamflow change at the watershed scale. Precipitation, which varies both in temporal trend and spatial distribution, is regarded as independent of vegetation types (Zhang *et al.*, 2001), which mainly reflect changes of climate. However, actual evapotranspiration is a complex process. There are various ways to estimate watershed-scale evapotranspiration. For example, following the Budyko hypothesis (Budyko, 1974), the simple two-parameter model for estimating actual evapotranspiration was developed (Eqn. 6.5; Zhang *et al.*, 2001). The model is consistent with the previous theoretical work and shows good agreement with more than 250 catchment-scale measurements from the world (Zhang *et al.*, 2001):

$$ET = \left[\frac{1+w\times\left(\frac{PET}{P}\right)}{\left(1+w\times\left(\frac{PET}{P}\right)+\left(\frac{PET}{P}\right)^{-1}\right)^2}\right] \times P \tag{6.5}$$

where ET is actual evapotranspiration, P is precipitation, PET is potential evapotranspiration and w is the plant-available water coefficient estimated in the same way as in the sensitivity-based approach (Zhang *et al.*, 2001).

The following steps have been used to describe how a simple water balance method is implemented to estimate the effects of forest cover change on hydrology. First, according to Eqn. 6.5, calculate the actual evapotranspiration using the original data including precipitation, and required data for estimating PET (e.g. air temperature, relative humidity, wind speeds and sunshine hours). Second, estimate annual streamflow using Eqn. 6.4. In this step, the change of annual streamflow is influenced by both climate variability and vegetation changes, and thus the calculated annual streamflow can be defined as Q_{v+c} (i.e. $Q_{v+c} = P - ET$). Third, remove the decreasing or increasing trends of precipitation and other climate variables (e.g. air temperature, relative humidity, wind speeds and sunshine hours) in the data series to make them a stationary time series (Xu *et al.*, 2006), recalculate PET_{new} and ET_{new} using the detrended climate variables, and estimate annual streamflow according to Eqns. 6.4 and 6.5 (i.e. $Q_v = P - ET_{new}$). In this step, the change in annual streamflow mainly reflects the influences of vegetation changes, and thus the recalculated annual streamflow can be defined as Q_v. Finally, calculate the difference between Q_{v+c} and Q_v; thus, the change of annual streamflow caused by the climate variability ($Q_c = Q_{v+c} - Q_v$) can be estimated.

A simple water balance method provides a new way to distinguish the impact of climatic variables and vegetation factors on hydrological change. However, the choice of w values and the difficulty associated with the removal of the decreasing or increasing trends in climate data may introduce some errors in this method.

6.3.5 Time trend method

In the time trend method, a relationship is established between streamflow and climatic variables before the occurrence of the basin's vegetation perturbations and is then used to predict the streamflow response of post-perturbation assuming undisturbed basin conditions. The typical time trend approach is to divide the whole study period into the calibration period and prediction period. The model accuracy depends on the length of the calibration or pre-perturbation periods (Zhao *et al.*, 2010).

During the calibration period, the streamflow is calculated as:

$$Q_1 = a \times P_1 + b \quad (6.6)$$

During the prediction period, the expected streamflow and the change in streamflow attributed to forest/vegetation changes are calculated as:

$$Q_2 = a \times P_2 + b \quad (6.7)$$

and

$$\Delta Q_V = \bar{Q_2} - \bar{Q'_2} \quad (6.8)$$

where P is precipitation, Q is streamflow, Q' is the predicted streamflow for the catchment after treatment (using Eqn. 6.6 developed during the calibration period), ΔQ_v represents the change in mean annual streamflow because of vegetation change, subscripts 1 and 2 represent the calibration period and prediction period, a and b are the fitted regression coefficients, Q_2 is the average observed streamflow in the prediction period, and Q'_2 is the average predicted streamflow calculated by Eqn. 6.6 using the regression coefficients from the calibration period.

This method uses a simple regression to express the relationship between precipitation and streamflow both before and after forest cover changes. The method only requires data for precipitation, streamflow and other meteorological variables, and the requirement for detailed forest cover changes can be ignored to some extent. The method may accept discontinuous data. Depending on different hydrometeorological characteristics in a study basin, time trend method performance at yearly hydrological variables is better than on the variables at monthly or daily intervals. This is because the rainfall–runoff relationship in a watershed at an annual interval is much stronger than those at daily and monthly ones. Guardiola-Claramonte *et al.* (2011) proposed considering the impact of temperature on the relationship between precipitation and streamflow.

6.3.6 Tomer–Schilling framework

Tomer and Schilling (2009) developed a coupled water-energy balance framework that requires the long-term annual time series of P, ET ($P - Q$) and PET to assess whether unused available energy ($PET - ET$) and water ($P - ET$) are related to climate and/or to land management in agricultural and woodland catchments (Tomer and Schilling, 2009; Peña-Arancibia *et al.*, 2012). The conceptual framework qualitatively discriminates whether the dominant drivers of observed changes are related to land cover change and/or climate. The framework relating changes in land cover change and/or climate to the observed changes in the excess amounts of water (P_{ex}) and excess amounts of energy (E_{ex}) as fractions is illustrated in Peña-Arancibia *et al.* (2012).

The Tomer–Schilling framework assumes that land cover change will affect *ET* but not *P* or *PET*, acknowledging that the effects of land cover change on *P* and *PET* can be considered indirectly at this scale and would possibly be of second order when compared with changes in *ET* in the woodland environment. Thus, land cover change will cause ecohydrological shifts toward increased P_{ex} and E_{ex}, or toward decreased P_{ex} and E_{ex}. Changes in climate are required to cause increased P_{ex} and decreased E_{ex}, due to the temporal increase in the *P*/*PET* ratio and vice versa:

$$P_{ex} = \frac{(P-ET)}{P} \tag{6.9}$$

$$E_{ex} = \frac{(PET-ET)}{PET} \tag{6.10}$$

The Tomer–Schilling framework is an effective and qualitative analysis tool for hydrological processes. The method not only analyzes the long-term impacts of climate and forest cover on hydrology but also explains the main factors of hydrological responses in different time periods. However, as with all the qualitative methods, the impact of each variable of climate or vegetation cannot be evaluated quantitatively, and this indicates the same difficulties in studying the common law of hydrological responses under different topographical and vegetative conditions.

6.4 Future Research Priorities

Here, we propose the following future research priorities on this subject of large-scale studies of forest cover changes and water. First, more case studies are needed. Assessing the relative contributions of forest or land cover changes and climatic variability to hydrology is limited in large watersheds, and the results are variable. Zhang and Wei (2014) compared two adjacent large watersheds located in the interior of British Columbia, Canada, and found contrasting conclusions under similar forest disturbance levels. They further concluded that the effects of forest change on hydrology in large watersheds are likely watershed specific. This clearly demonstrates that more case studies are needed before general conclusions can be derived.

Second, a research priority should be given to further development and improvement of existing research methods. Although a number of research methods are currently available for studying the impacts of forest cover change and climate variability on hydrology, there is no commonly accepted method. The lack of commonly accepted methods may limit our ability to compare the results of different studies. With the availability of long-term data, machine learning can be used to quantify the relative role of forest cover change in hydrology (Cheng *et al.*, 2022; Slater *et al.*, 2024), particularly on these short-interval variables (peak or low flows) where water balance-based methods have limited utilities (Floriancic *et al.*, 2022; Slater *et al.*, 2024).

Third, in large watershed studies, analytical results are based largely on data quality and spatial coverage. Inherent spatial variabilities in, for example, precipitation, temperature, solar radiation, humidity, wind speed, surface albedo and canopy characteristics in large watersheds and uncertainties in parameter estimates constrain our ability to derive robust conclusions. For instance, only a few climatic stations are located in large watersheds, particularly at high elevations. The existing climatic stations are often located in easily accessible places. The spatial variability of precipitation thus cannot easily be addressed in large watersheds. Future studies should be designed to specifically address such uncertainties.

Finally, more research should be focused on mechanisms, processes and their interactions. It is difficult to study the mechanisms and processes in large watersheds, mainly due to the lack of suitable methodology and data for assessing the complicated interactions and cumulative behaviors across various spatial scales. However, such research is critical for explaining and verifying the findings obtained through statistical and modeling approaches. To address this, techniques such as stable isotopes and geochemical tracing could provide useful insights into the processes or mechanisms (e.g. flow paths, residence time, surface water and groundwater interactions).

References

Abou-Taleb, K.A. and Galal, G.F. (2018) A comparative study between one-factor-at-a-time and minimum runs resolution-IV methods for enhancing the production of polysaccharide by *Stenotrophomonas daejeonensis* and *Pseudomonas geniculate*. *Annals of Agricultural Sciences* 63, 173–180.

Adams, M.A., Shadmanroodposhti, M. and Neumann, M. (2020) Causes and consequences of Eastern Australia's 2019–20 season of mega-fires: A broader perspective. *Global Change Biology* 26, 3756–3758.

Aryal, Y. and Zhu, J. (2020) Effect of watershed disturbance on seasonal hydrological drought: An improved double mass curve (IDMC) technique. *Journal of Hydrology* 585, 124746.

Barnhart, T., Vukomanovic, J., Bourgeron, P. and Molotch, N. (2021) Future land cover and climate may drive decreases in snow wind-scour and transpiration, increasing streamflow at a Colorado, USA headwater catchment. *Hydrological Processes* 35, e14416.

Biederman, J.A., Somor, A.J., Harpold, A.A., Gutmann, E.D., Breshears, D.D. *et al*. (2015) Recent tree die-off has little effect on streamflow in contrast to expected increases from historical studies. *Water Resources Research* 51, 9775–9789.

Biederman, J.A., Robles, M.D., Scott, R.L. and Knowles, J.F. (2022) Streamflow response to wildfire differs with season and elevation in adjacent headwaters of the Lower Colorado River Basin. *Water Resources Research* 58, e2021WR030687.

Box, G.E.P. and Pierce, D.A. (1970) Distribution of residual autocorrelations in autoregressive-integrated moving average time series models. *Journal of the American Statistical Association* 65, 1509–1526.

Budyko, M.I. (1974) *Climate and Life*. Academic Press, New York.

Buttle, J.M. and Metcalfe, R.A. (2000) Boreal forest disturbance and streamflow response, northeastern Ontario. *Canadian Journal of Fisheries and Aquatic Sciences* 57, 5–18.

Ceppi, A., Chaves González, N.A., Davolio, S. and Ravazzani, G. (2023) Can meteorological model forecasts initialize hydrological simulations rather than observed data in ungauged basins? *Meteorological Applications* 30, e2165.

Cheng, S., Cheng, L., Qin, S., Zhang, L., Liu, P. *et al*. (2022) Improved understanding of how catchment properties control hydrological partitioning through machine learning. *Water Resources Research* 58, e2021WR031412.

Coble, A.A., Barnard, H., Du, E., Johnson, S., Jones, J. *et al*. (2020) Long-term hydrological response to forest harvest during seasonal low flow: Potential implications for current forest practices. *Science of The Total Environment* 730, 138926.

Creed, I.F. and van Noordwijk, M. (2018) *Forest and Water on a Changing Planet: Vulnerability, Adaptation and Governance Opportunities*. IUFRO World Series Vol. 38. Vienna: IUFRO.

de Niel, J. and Willems, P. (2019) Climate or land cover variations: What is driving observed changes in river peak flows? A data-based attribution study. *Hydrology and Earth System Sciences* 23, 871–882.

Du, E., Link, T.E., Wei, L. and Marshall, J.D. (2016) Evaluating hydrologic effects of spatial and temporal patterns of forest canopy change using numerical modelling. *Hydrological Processes* 30, 217–231.

Filoso, S., Bezerra, M.O., Weiss, K.C.B. and Palmer, M.A. (2017) Impacts of forest restoration on water yield: A systematic review. *PLOS ONE* 12, e0183210.

Floriancic, M.G., Spies, D., Meerveld, I.H.J. and Molnar, P. (2022) A multi-scale study of the dominant catchment characteristics impacting low-flow metrics. *Hydrological Processes* 36, e14462.

Fohrer, N., Haverkamp, S. and Frede, H.G. (eds) (2005) Assessment of the effects of land use patterns on hydrologic landscape functions: Development of sustainable land use concepts for low mountain range areas. *Hydrological Processes* 19, 659–672.

Giles-Hansen, K., Li, Q. and Wei, X. (2019) The cumulative effects of forest disturbance and climate variability on streamflow in the Deadman River Watershed. *Forests* 10, 196.

Goeking, S.A. and Tarboton, D.G. (2020) Forests and water yield: A synthesis of disturbance effects on streamflow and snowpack in western coniferous forests. *Journal of Forestry* 118, 172–192.

Guardiola-Claramonte, M., Troch, P.A., Breshears, D.D., Huxman, T.E., Switanek, M.B. *et al*. (2011) Decreased streamflow in semi-arid basins following drought-induced tree die-off: A counter-intuitive and indirect climate impact on hydrology. *Journal of Hydrology* 406, 225–233.

Hallema, D.W., Sun, G., Caldwell, P.V., Norman, S.P., Cohen, E.C. *et al*. (2017) Assessment of wildland fire impacts on watershed annual water yield: Analytical framework and case studies in the United States. *Ecohydrology* 10, e1794.

Han, J., Yang, Y., Roderick, M.L., McVicar, T.R., Yang, D. *et al.* (2020) Assessing the steady-state assumption in water balance calculation across global catchments. *Water Resources Research* 56, e2020WR027392.

Hou, Y., Wei, X., Vore, M., Déry, S.J., Pypker, T. *et al.* (2022) Cumulative forest disturbances decrease runoff in two boreal forested watersheds of the northern interior of British Columbia, Canada. *Journal of Hydrology* 605, 127362.

Hu, J., Wu, Y., Wang, L., Sun, P., Zhao, F. *et al.* (2021) Impacts of land-use conversions on the water cycle in a typical watershed in the southern Chinese Loess Plateau. *Journal of Hydrology* 593, 125741.

Jaskierniak, D., Kuczera, G., Benyon, R. and Wallace, L. (2015) Using tree detection algorithms to predict stand sapwood area, basal area and stocking density in *Eucalyptus regnans* forest. *Remote Sensing* 7, 7298–7323.

Kirchner, J.W. (2006) Getting the right answers for the right reasons: Linking measurements, analyses and models to advance the science of hydrology. *Water Resources Research* 42, W03S04.

Kirchner, J.W. (2009) Catchments as simple dynamical systems: Catchment characterization, rainfall-runoff modeling, and doing hydrology backward. *Water Resources Research* 45, W02429.

Kirchner, J.W. and Allen, S.T. (2020) Seasonal partitioning of precipitation between streamflow and evapotranspiration, inferred from end-member splitting analysis. *Hydrology and Earth System Sciences* 24, 17–39.

Lagos, M.S., Muñoz, J.F., Suárez, F.I., Fuenzalida, M.J., Yáñez-Morroni, G. *et al.* (2024) Investigating the effects of channelization in the Silala River: A review of the implementation of a coupled MIKE-11 and MIKE-SHE modeling system. *WIREs Water* 11, e1673.

Li, Q., Wei, X., Zhang, M., Liu, W., Fan, H. *et al.* (2017) Forest cover change and water yield in large forested watersheds: A global synthetic assessment. *Ecohydrology* 10, e1838.

Lin, Y. and Wei, X. (2008) The impact of large-scale forest harvesting on hydrology in the willow watershed of Central British Columbia. *Journal of Hydrology* 359, 141–149.

Lv, X., Zuo, Z., Ni, Y., Sun, J. and Wang, H. (2019) The effects of climate and catchment characteristic change on streamflow in a typical tributary of the Yellow River. *Scientific Reports* 9, 14535.

McMillan, H.K., Gnann, S.J. and Araki, R. (2022) Large scale evaluation of relationships between hydrologic signatures and processes. *Water Resources Research* 58, e2021WR031751.

Moore, R.D., Gronsdahl, S. and McCleary, R. (2020) Effects of forest harvesting on warm-season low flows in the Pacific Northwest: A review. *Confluence: Journal of Watershed Science and Management* 4, 29.

Palmer, M. and Ruhi, A. (2019) Linkages between flow regime, biota, and ecosystem processes: Implications for river restoration. *Science* 365, eaaw2087.

Parker, G.G. (2020) Tamm review: Leaf area index (LAI) is both a determinant and a consequence of important processes in vegetation canopies. *Forest Ecology and Management* 477, 118496.

Peña-Arancibia, J.L., van Dijk, A.I., Guerschman, J.P., Mulligan, M., Bruijnzeel, L.A. *et al.* (2012) Detecting changes in streamflow after partial woodland clearing in two large catchments in the seasonal tropics. *Journal of Hydrology* 416–417, 60–71.

Restaino, C.M., Peterson, D.L. and Littell, J. (2016) Increased water deficit decreases Douglas fir growth throughout western US forests. *Proceedings of the National Academy of Sciences USA* 113, 9557–9562.

Senbeta, T. and Romanowicz, R. (2021) The role of climate change and human interventions in affecting watershed runoff responses. *Hydrological Processes* 35, e14448.

Slater, L., Coxon, G., Brunner, M., McMillan, H., Yu, L. *et al.* (2024) Spatial sensitivity of river flooding to changes in climate and land cover through explainable AI. *Earth's Future* 12, e2023EF004035.

Sun, G., Wei, X., Hao, L., Sanchis, M.G., Hou, Y. *et al.* (2023) Forest hydrology modeling tools for watershed management: A review. *Forest Ecology and Management* 530, 120755.

Tomer, M.D. and Schilling, K.E. (2009) A simple approach to distinguish land-use and climate-change effects on watershed hydrology. *Journal of Hydrology* 376, 24–33.

Wang, S., Zhang, Z., McVicar, T., Guo, J., Tang, Y. *et al.* (2013) Isolating the impacts of climate change and land use change on decadal streamflow variation: Assessing three complementary approaches. *Journal of Hydrology* 507, 63–74.

Wei, X. and Zhang, M. (2010) Quantifying streamflow change caused by forest disturbance at a large spatial scale: A single watershed study. *Water Resources Research* 46, W12525.

Wei, X., Liu, W. and Zhou, P. (2013) Quantifying the relative contributions of forest change and climatic variability to hydrology in large watersheds: A critical review of research methods. *Water* 5, 728–746.

Wei, X., Li, Q., Zhang, M., Giles-Hansen, K., Liu, W. *et al.* (2017) Vegetation cover – another dominant factor in determining global water resources in forested regions. *Global Change Biology* 24, 786–795.

Wilk, J., Andersson, L. and Plermkamon, V. (2001) Hydrological impacts of forest conversion to agriculture in a large river basin in Northeast Thailand. *Hydrological Processes* 15, 2729–2748.

Winkler, R. and Boon, S. (2017) *Equivalent Clearcut Area as an Indicator of Hydrologic Change in Snow-dominated Watersheds of Southern British Columbia*. Extension Note 118. Government of British Columbia, Canada. Available at: www.for.gov.bc.ca/hfd/pubs/Docs/En/En118.htm (accessed 17 April 2025).

Winkler, R., Spittlehouse, D. and Boon, S. (2017) Streamflow response to clear-cut logging on british Columbia's Okanagan Plateau. *Ecohydrology* 10, e1836.

Xu, C., Gong, L., Jiang, T., Chen, D. and Singh, V.P. (2006) Analysis of spatial distribution and temporal trend of reference evapotranspiration and pan evaporation in Changjiang (Yangtze River) catchment. *Journal of Hydrology* 327, 81–93.

Yu, Y., Yu, P., Wang, Y., Wan, Y., Wang, B. *et al.* (2023) Natural revegetation has dominated annual runoff reduction since the Grain for Green Program began in the Jing River basin, Northwest China. *Journal of Hydrology* 625, 129978.

Zhang, M. and Wei, X. (2012) The effects of cumulative forest disturbance on streamflow in a large watershed in the central interior of British Columbia, Canada. *Hydrology and Earth System Sciences* 16, 2021–2034.

Zhang, M. and Wei, X. (2014) Contrasted hydrological responses to forest harvesting in two large neighbouring watersheds in snow hydrology dominant environment: Implications for forest management and future forest hydrology studies. *Hydrological Processes* 28, 6183–6195.

Zhang, L., Dawes, W.R. and Walker, G.R. (2001) Response of mean annual evapotranspiration to vegetation changes at catchment scale. *Water Resources Research* 37, 701–708.

Zhang, L., Nan, Z., Yu, W., Zhao, Y. and Xu, Y. (2018) Comparison of baseline period choices for separating climate and land use/land cover change impacts on watershed hydrology using distributed hydrological models. *Science of the Total Environment* 622–623, 1016–1028.

Zhang, M., Wei, X. and Li, Q. (2016) A quantitative assessment on the response of flow regimes to cumulative forest disturbances in large snow-dominated watersheds in the interior of British Columbia, Canada. *Ecohydrology* 9, 843–859.

Zhang, M., Liu, N., Harper, R., Li, Q., Liu, K. *et al.* (2017) A global review on hydrological responses to forest change across multiple spatial scales: Importance of scale, climate, forest type and hydrological regime. *Journal of Hydrology* 546, 44–59.

Zhao, F., Zhang, L., Xu, Z. and Scott, D.F. (2010) Evaluation of methods for estimating the effects of vegetation change and climate variability on streamflow. *Water Resources Research* 46, W03505.

Zhou, G., Wei, X., Luo, Y., Zhang, M., Li, Y. *et al.* (2010) Forest recovery and river discharge at the regional scale of Guangdong Province, China. *Water Resources Research* 46, W09503.

7 Hydrology of Cold-region Montane Forested Watersheds

M. Dobre[1]*, A. Srivastava[1], W.J. Elliot[2], K.J. Elder[3], T.E. Link[1] and E.S. Brooks[1]

[1]University of Idaho, Moscow, Idaho, USA; [2]Forestry Sciences Laboratory, Rocky Mountains Research Station, USDA Forest Service, Moscow, Idaho, USA; [3]Fraser Experimental Forest, Rocky Mountain Research Station, USDA Forest Service, Fort Collins, Colorado, USA

Abstract

This chapter explores the snow and hydrological processes characteristic of steep, cold-region montane forested watersheds, particularly in western North America, which rely heavily on snowmelt as a critical water source for ecosystems and human use. While many hydrological principles discussed in other chapters apply to these high-elevation regions, the unique interactions among surface runoff, shallow lateral flow and baseflow merit special attention, particularly in the context of snow processes and steep terrain. The principles outlined here are broadly relevant to forested ecosystems that transition from chaparral or agricultural landscapes at lower elevations to subalpine forested ecosystems at higher elevations.

7.1 Introduction

High-elevation forests are often considered the 'water towers' for much of the world. From the Americas to Europe, Africa, Asia and Australia, large urban areas and irrigation projects look to runoff from higher-elevation forests to meet their water supply needs. These areas not only provide surface water to keep streams flowing and lakes and reservoirs filled but also are often important areas for recharging groundwater reservoirs that are later tapped by wells to supply critical water needs. Worldwide, it is estimated that more than a billion people rely on snow-covered areas and glaciers for their water supply (Immerzeel *et al.*, 2020).

The accumulation of snow in high-elevation forests is especially important in arid, semi-arid and Mediterranean climates, as melting snow recharges groundwater and provides low flows throughout the dry season. In continental climates, convective storms can also contribute to snow accumulation in the winter; however, high elevations typically receive more summer precipitation due to orographic effects. This chapter describes how precipitation is routed through forested watersheds as lateral flow and baseflow, rather than surface runoff, which is more common in non-forested or disturbed watersheds.

In most high-elevation forests, the key hydrological processes include winter snow accumulation and melt, higher summer

*Corresponding author: mdobre@uidaho.edu

 Forest Hydrology: Processes, Management and Assessment (eds Leon J. Bren *et al.*)
DOI: 10.1079/9781800625310.0007

precipitation compared with lower elevations, shallow subsurface flow from gradually melting snow in the spring, and continuous baseflow from deep seepage that persists through the dry season. Some of these processes were introduced in Chapters 1 and 2 (this volume). This chapter focuses mainly on hydrological processes in the mountainous and subalpine temperate coniferous forests of western North America, including the states of Idaho, Oregon, Washington, northern California and Montana (west of the Continental Divide) in the USA, and southern British Columbia in Canada. This region is characterized mainly by a maritime climate on the west coast and a continental climate inland. The transition zone between these climates can experience either maritime or continental conditions, depending on prevailing air masses (Hubbart *et al.*, 2007). The main difference between watersheds in these climates is the amount and form of winter precipitation, which not only influences hydrological processes but also has cascading effects on the development of soil and vegetation. For example, higher winter rainfall and peak snow accumulation in maritime climates contribute to deeper soil moisture retention and support more extensive root systems, while drier conditions in continental climates can lead to shallower soils and drought-tolerant vegetation (Zhang *et al.*, 2024).

The maritime climate, found along the Pacific coast, is dominated by warm, moist air masses from the Pacific Ocean. Similar climates are present in countries bordering the Mediterranean Sea and other regions influenced by maritime air masses in winter. Further east, continental convective storms dominate the hydrology, characterized by significant temperature variations, with colder winters and hotter summers. While this chapter primarily addresses hydrological processes in western North America, the principles apply to other steep, snow-affected forested areas worldwide that experience similar climates.

7.2 Snow Processes

Snow processes in forested watersheds occur primarily in the atmosphere and canopy and on the ground. In the atmosphere, snow crystals form in the clouds and, as they fall, they may either melt into rain, remain as snow or become a mix of both, depending on upper and lower atmospheric temperatures and vapor pressures. Upon reaching the forest canopy, the snow can be intercepted, where it may eventually be released to the ground, melt or sublimate. On the ground, snow can accumulate or melt, with melt rates influenced by energy inputs such as shortwave radiation from the sun, longwave radiation from the atmosphere and/or the canopy, convective heat from warm air or soil, and latent heat from humid air. Sublimation is influenced by wind speed and vapor pressure deficit, and can also occur from the snow under certain conditions, especially when energy inputs are high, the air is dry and/or blowing snow occurs.

Snow accumulation is influenced by the amount of precipitation falling as snow, wind speed and direction, and interception by the forest canopy, which can reduce snowpack by up to 40% due to the combined effects of interception and subsequent sublimation (Varhola *et al.*, 2010). Snowmelt is driven primarily by shortwave radiation; however, forest canopy structure, including tree height, canopy density and tree spacing, affects the degree of shading and thus the amount of incoming shortwave radiation that reaches the forest floor. In dense forests, reduced shortwave radiation may be offset by increases in the longwave radiation from the canopy (Pomeroy *et al.*, 2009). In less dense forests, gaps in the canopy result in highly variable snowmelt rates, creating 'hotspots' where total incoming radiation exceeds open areas, or 'cold holes' where radiation is less than under dense canopies (Lawler and Link, 2011). Additionally, shading reduces shortwave radiation, and lower albedo from leaves and dust increases the absorption of solar energy by the snowpack. This dual effect, combined with enhanced longwave radiation from trees, contributes to the 'radiative paradox,' where net incoming radiation increases with forest canopy cover (Seyednasrollah *et al.*, 2013). Sensible and latent energy fluxes are also affected, with forests reducing wind speed and thereby diminishing the convective transfer of heat and moisture.

Altering the forest canopy can increase streamflow discharge, but the response of snow accumulation and melt to management activities varies among high-elevation forests. These responses depend on interactions between the canopy, snow interception, snowmelt rates and runoff, which are influenced by elevation, slope, aspect and the dominant air mass (maritime or continental), as discussed by Dickerson-Lange *et al.* (2021). This categorization is further complicated by year-to-year variability, as a location may exhibit characteristics of different regimes or a combination of responses within a single year. Reducing the canopy can also decrease evapotranspiration, potentially increasing water yield under certain climatic conditions. Additionally, snow redistribution by wind plays a critical role in forested landscapes, shaping snowpack distribution and water availability. Disturbances such as wildfire or tree die-offs can amplify wind-driven snow redistribution, further altering snow retention and potentially reducing water yield from affected watersheds.

Building on the understanding of how forest canopy structure influences snow accumulation and melt, recent modeling work by Sun *et al.* (2022) further explored the effects of canopy density on snowpack duration across different climate types, offering insights for forest-management strategies aimed at enhancing snow retention. The authors found that in wet/warm (winter precipitation >300 mm and winter temperatures >−1°C) and dry/cold (winter precipitation <300 mm and winter temperatures <−1°C) climates, reducing canopy density generally increased snowpack duration, with low-density forests maintaining snow longer than open areas. From a forest-management perspective, thinning the canopy in these climates is more effective than clearing for enhancing snowpack duration and storage. In wet/cold climates (winter precipitation >300 mm and winter temperatures <−1°C), snowpack duration is less sensitive to changes in canopy density.

7.2.1 Maritime climates

Maritime climates, with their proximity to large bodies of water, experience milder winters characterized by frequent fluctuations above and below freezing. Abundant precipitation, driven by atmospheric rivers carrying moisture from nearby large bodies of water, often falls as rain or a mix of rain and snow. These atmospheric rivers deliver substantial snowfall over short durations, especially as moisture-laden air encounters mountain ranges and undergoes orographic lifting, resulting in rapid snowpack buildup. The warm, humid air masses also contain sensible and latent heat, which can quickly melt the snowpack through convective heat transfer and condensation (Marks *et al.*, 1998).

The snowpack in these regions is dynamic, with frequent accumulation and melting throughout winter. The relatively warm temperatures and high ablation rates lead to an early onset of snowmelt, typically between March and April, as higher atmospheric temperatures and increased solar irradiance as the year progresses accelerate the melting process (Trujillo and Molotch, 2014). Additionally, frequent temperature swings near the freezing point make maritime climates particularly susceptible to rain-on-snow (ROS) events. Many ROS events are driven by atmospheric rivers, where warm rains fall on shallow snowpacks, triggering sudden and intense snowmelt. This can lead to significant runoff from saturated soils and increase the risk of flooding, which requires careful monitoring and management.

At the smaller scale of forest canopies, these warmer, moisture-rich conditions further affect how snow is intercepted, retained and released from the canopy. Warmer temperatures during snowfall events can lead to the formation of cohesive snow crystals, potentially enhancing interception efficiency (Lundquist *et al.*, 2021). These conditions, however, also increase the likelihood of snow unloading from the canopy through melting or ROS events. While initially promoting snow interception, warmer temperatures accelerate unloading through melt and drip, reducing the total snow retained in the canopy. Additionally, the removal of canopy cover, such as through forest clearing, increases wind speed within forest clearings, enhancing turbulent energy exchanges at the snow surface and further accelerating snowmelt (Marks *et al.*, 1998). This can result in either shorter or longer durations of snow cover in forested areas compared with open areas. Finally, high humidity, mild temperatures and frequent cloud cover contribute to lower sublimation rates, as the moist air reduces the vapor gradient and cloud cover limits shortwave radiation.

7.2.2 Continental climates

Continental climates, typically located farther inland and away from large bodies of water, experience more extreme temperature variations between winter and summer. Winters in these regions are colder, with extended periods of subfreezing temperatures, leading to prolonged snow accumulation seasons. Snowfall is often steady but less frequent compared with maritime climates, with less influence from atmospheric rivers. Snow processes in continental climates are driven primarily by cold, dry air masses, resulting in lighter, less dense snowpacks. Snow accumulation can persist for months due to the consistently low winter temperatures and reduced ablation rates. In these regions, snowmelt typically occurs later in the season, often starting in late spring or early summer, as temperatures rise gradually. Snowmelt is influenced primarily by shortwave radiation, which increases as the days lengthen, and warmer daytime temperatures lead to a slower, more controlled melt compared with maritime regions. ROS events are less common in continental climates due to the colder temperatures, but when they do occur, they are often associated with unusual warm spells or transitional periods in late winter or early spring. The slower, steady snowmelt and reduced frequency of ROS events generally lower the risk of rapid runoff and flooding, although snowmelt floods can still pose challenges in low-lying areas.

Snow–vegetation interactions in continental forests differ significantly from maritime forests. For example, during snowfall events, lower temperatures produce less cohesive snow crystals, such as needles and columns, which reduce initial interception efficiency. However, colder temperatures also increase branch stiffness, enabling canopies to support larger snow loads (Klamerus-Iwan *et al.*, 2020). This structural support may offset the initial lower interception efficiency, allowing significant snow accumulation in the canopy over time. The dry conditions typical of continental climates promote snow sublimation, reducing the snowpack in both the canopy and on the ground. Wind further enhances sublimation losses, especially in open areas, by increasing snow unloading and redistributing intercepted snow, often leading to uneven snow distribution.

While topography and the size and amount of forest canopy gaps are important for snow accumulation, the snowmelt is driven primarily by the net available radiant energy, which accounts for a large proportion of the total snowmelt energy. The radiative fluxes for snowmelt are incoming shortwave radiation (direct and diffuse) and longwave radiation (diffuse). In addition, longwave radiation from the forest canopy and trunks is also significant (Lundquist *et al.*, 2013). The proportion of each of these components that reaches the snow surface is dependent on many factors such as elevation, aspect, latitude, day length, cloudiness and solar angle.

7.3 Forests and Avalanches

Landscapes in snowy regions share complex relationships with avalanches that are driven by interconnections of terrain, weather, climate, ecology and land use. Forests may ameliorate or exacerbate avalanche issues, and avalanches may influence forest dynamics.

7.3.1 Snow avalanche anatomy and characteristics

Avalanche paths have a starting zone, track and runout zone (Fig. 7.1). The starting zone is the area of snow accumulation and is where failure usually occurs. In this zone, an internal failure plane with less stable snow on the top and frozen snow, rock or soil beneath the failure plane forms. An increase in downhill forces exceeding resistive forces on this plane and/or lubrication from meltwater at the failure plane will release the top layer, creating a slab of snow cascading downhill (Fig. 7.1, zone 1). The track is a portion of the path where snow is transported from the starting zone, often gaining additional mass (Fig. 7.1, zone 2). The runout zone is where the avalanche decelerates, stops and deposits debris (Fig. 7.1, zone 3). In large, well-defined paths, especially in forested areas, these zones are obvious (Fig. 7.1, path B). In smaller paths or those above the treeline, the delineation of different zones may be subtle (Fig. 7.1, path A).

Fig. 7.1. Typical avalanche paths near Loveland Pass, Colorado. Path A is located entirely above the treeline. Path B intersects the treeline and deposits debris in the valley bottom. Both paths have starting zones (1), tracks (2) and runout zones (3), delineated approximately by black lines in this figure. The transition between the zones and the paths' margins and extents vary with the magnitude and season of individual events. In extreme events, all of path A may become part of the starting zone for path B. In minor events, the starting zone alone of path B may contain a starting zone, track and runout zone, like path A.

Starting zones are often located above the treeline in alpine areas where wind loading on lee slopes enhances snow deposition. These starting zones feed tracks that intersect the treeline and flow through a well-defined path and runout zone without significant tree cover. Paths entirely below the treeline often have clear starting zones, tracks and runout zones defined by forest margins or avalanche gullies developed over time. Snowpack character, terrain and vegetation can affect path morphology and extent, with variations in avalanche paths occurring across different snow climates from maritime to continental and high arctic.

In maritime climates, avalanches tend to be a two-step process. The first step is the accumulation of a snowpack of sufficient depth to initiate an avalanche. This is generally followed by a rainfall event on top of melting snow, until the snowpack is sufficiently saturated such that, as with the slab avalanche, the internal stresses from the heavy wet snow exceed the resistance to stress from the rainfall-weakened snowpack (Conway and Raymond, 1989). Such avalanches more typically are confined to historical chutes free of large trees, maintained by frequent landslides.

7.3.2 Forest effects on avalanches

Forests affect avalanches through both passive and active processes. Passive processes include forest influences on snowpack accumulation, distribution and snowpack energy balance. Trees may act as anchors for snow on slopes but may also create weak points where avalanches can start. Forests also significantly affect the local energy balance of the snowpack by diminishing incoming shortwave radiation, increasing longwave radiation and reducing turbulent energy exchanges due to suppressed wind. This energy balance drives snow metamorphism, ultimately controlling snowpack structure and stability, which varies greatly between forested and unforested slopes.

Active processes affected by trees include avalanche track delineation, roughness or friction on moving snow, entrainment of woody material, mass in the avalanche flow, and runout zone dynamics that resist or arrest flow. Trees provide significant friction or flow impediment for avalanches, dissipating kinetic energy and altering avalanche velocity. Small trees may bend or break with little impact, while large trees can completely stop an avalanche flow. Large trees on the margins of runout zones may experience frequent events that are incapable of causing significant damage, but longer-return events may periodically devastate the established boundaries (Schläppy *et al.*, 2014).

7.3.3 Avalanches and hydrology

Snow avalanches effectively move large volumes of snow from one location to another, causing two notable hydrological changes. First, snow is moved from one energy balance regime to another, such as from zone B1 to B3 in Fig. 7.1. Even small elevation changes can result in late-lying snow deposits if the accumulation is significant, as in zone A3 in Fig. 7.1. Snow in high-alpine zones experiences high incoming shortwave radiation and large outgoing longwave radiation losses, with cold temperatures and high winds. In valley bottoms, reduced shortwave radiation and longwave losses occur due to atmospheric transmittance and shading from valley walls, with warmer temperatures and lower winds, causing faster snowmelt in these areas. Second, the structure and distribution of the snowpack are altered. Snowpacks in the starting zone are typically shallow, low-density and spread over a large area, except in wind deposits. Avalanching transforms this into a high-density, deep snowpack with a smaller surface area.

The hydrological effect of avalanching depends on changes in snowpack properties and the energy balance between the starting zone and deposition area. Snow typically melts faster and earlier at lower elevations, but deep avalanche deposits in mountain valley bottoms can last well into the melt season, often outlasting the surrounding snowpack by weeks or months. Martinec and de Quervain (1971) observed that an avalanche in the Swiss Alps increased early-season streamflow and decreased late-season streamflow compared with modeled non-avalanche results. However, other researchers, such as Sosedov and Seversky (1965), found different outcomes. The effects of avalanching on runoff regimes vary and require local path- or event-specific investigations.

Avalanches change the hydrological pathways of snow that would otherwise have melted in the starting zone. Snowmelt in upper basin areas undergoes hillslope processes like overland flow, infiltration, subsurface flow and possible evapotranspiration (see Section 7.4). In contrast, snowmelt from avalanche deposits in valley bottoms has a short, direct path to the stream channel. The snowmelt from these locations may have different biogeochemical signatures due to variations in pathways, residence time, and exposure to soils, vegetation and geology (Sánchez-Murillo *et al.*, 2015).

Avalanches can also alter runoff by damming streams, leading to impounded flow that is later released catastrophically, potentially causing downstream flooding. This can change downstream vegetation and channel morphology, and impact life and infrastructure. Additionally, avalanches may affect ice-covered lake surfaces, causing a plunger effect that rapidly expels stored water, leading to downstream flooding (Williams *et al.*, 1992).

7.4 Hydrology in Steep Watersheds

High-elevation steep watersheds, including montane and subalpine regions with or without forests, have similar soil water and baseflow

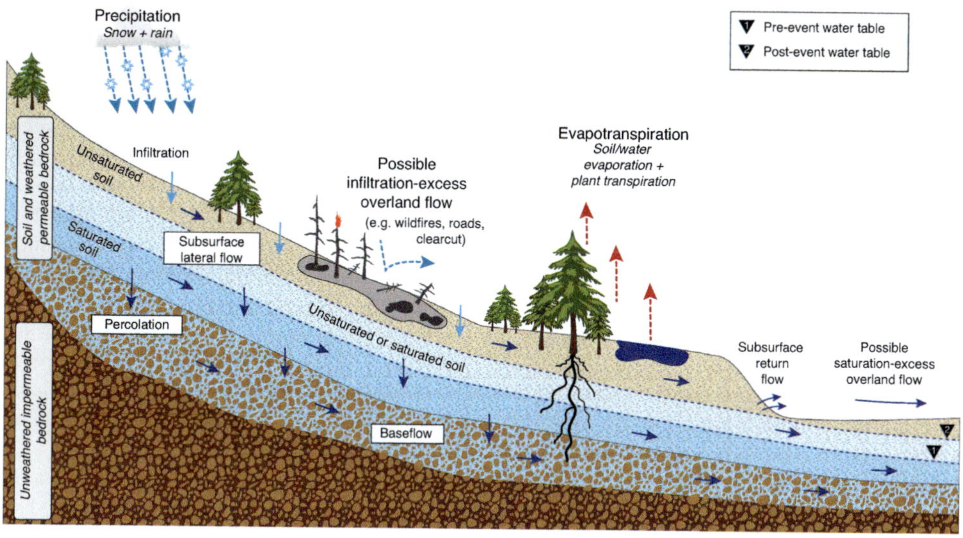

Fig. 7.2. Schematic of hydrological pathways on a steep, concave, forested hillslope.

processes but differ in vegetation impacts on hydrology. Chapter 2 provided a good introduction to forest hydrological processes, and this chapter expands on those principles, focusing on watersheds where snow accumulation and melt dominate the hydrological response. Fig. 7.2 illustrates the dominant flow processes described in this section.

Upland forested watersheds are characterized by steep slopes, shallow soils, the absence of floodplains, high precipitation and low evapotranspiration, unlike lower-elevation watersheds. Consequently, streams respond quickly to storms. Water from rainfall or snowmelt moves through hillslopes into streams via overland flow, subsurface lateral flow and baseflow (Fig. 7.2), although in real-world applications or hydrological modeling, it is difficult to accurately quantify the contributions of each of these components to the streamflow hydrograph (Beven, 2021). As slope steepness increases, so does flow velocity. These pathways respond differently to snowmelt versus rainfall in terms of runoff amount, peak flows and timing of contributions to streamflow. Each pathway is influenced by the interactions among climate, vegetation, topography, soil characteristics and geology. In forest soils, macropores, soil pipes from decayed tree roots, lateral and vertical tree root systems, animal burrows and large cracks provide permeable conduits for rapid water movement, enhancing subsurface lateral flow processes (Bronstert *et al.*, 2023).

7.4.1 Rain- versus snow-dominated hydrological systems

Both rain and snow are forms of precipitation; however, they affect the hydrological cycle in different ways. Rain increases soil moisture, promotes evaporation and can lead to runoff, impacting river flows and water availability. Snow, in contrast, accumulates on the ground, increasing surface albedo, which reflects more sunlight and can delay melting until warmer seasons. This delayed melting affects streamflow timing and water supply, particularly in regions dependent on snowmelt. The temperature threshold at which precipitation is partitioned between rain and snow is influenced by atmospheric conditions, although it generally varies between −0.4°C and +2.4°C. Continental climates tend to have warmer thresholds, while maritime climates have cooler ones (Jennings *et al.*, 2018).

In steep mountainous regions, precipitation increases and temperatures decrease with elevation, affecting the phases of precipitation (rain, sleet or snow). Higher-elevation watershed processes are driven by snowfall and snowmelt, while moderate elevations are dominated by rain or ROS events. Lower-elevation runoff is mostly constrained by low precipitation. Both snow- and rain-dominated processes show large seasonal variability and interact with vegetation. In snowmelt-dominated regions, runoff usually occurs in late spring or early summer, with significant water percolating into the bedrock (Wilson and Guan, 2004). In rainfall-dominated watersheds, peak flows are linked to seasonal weather patterns and can occur at any time, including summer. In watersheds dominated by more maritime climates, peak flows are typically associated with ROS events driven by atmospheric rivers or seasonal increases in shortwave radiation.

7.4.2 Infiltration and runoff processes

The two primary mechanisms that generate surface runoff are infiltration excess and saturation excess (Fig. 7.2). In steep, undisturbed forested watersheds, infiltration-excess or Hortonian overland flow is rare due to the high infiltration capacity of organic litter and permeable soils, which typically exceed rainfall intensity. Water preferentially moves vertically into subsurface layers instead of over the surface. However, infiltration-excess overland flow can occur when rainfall or snowmelt rates exceed infiltration capacity, particularly on forest roads, logging trails (due to soil compaction), after wildfires (due to reduced infiltration from litter loss and water repellency) and on frozen soils. Infiltration-excess runoff is also common in areas with continental climates experiencing high-intensity thunderstorms or monsoonal rainfall events.

Saturation-excess runoff occurs whenever the soil water-storage capacity is exceeded. This occurs most often in topographically convergent landscape positions in steep, humid ecosystems. The saturation-excess source areas vary in size depending on overall wetness of the watershed, which has led some to refer to these landscapes as dominated by 'variable source area' hydrology. The soil becomes saturated due to prolonged wet conditions, snowmelt, high infiltration rates in shallow soils or its location in convergent areas of the landscape, such as at the base of a hill. At the base of steep hills, the slope often levels out, and the convergence of subsurface flow leads to increased soil moisture that results in saturated conditions and exfiltration of water (Fig. 7.2). Under these saturated conditions, soil cannot absorb additional rainfall or snowmelt, resulting in surface runoff, even if the melt rate or rainfall intensity is below the soil's saturated hydraulic conductivity. In some cases, the subsurface lateral flow can clearly be seen seeping out of road cuts or at steep breaks in that landscape, a phenomenon known as 'subsurface return flow' (Fig. 7.2). These seepage locations are often referred to as spring lines.

Baseflow is generally the return of water that has percolated below the root zone into fissured bedrock beneath forest soils and/or water accumulated on top of the soil–bedrock interface forming a perched water table that drains slowly to the nearest channel or seeps into the groundwater. This water can take from days to years to find its way to forest streams. The rate of return depends on the amount of water that has percolated and the geology (Sánchez-Murillo *et al.*, 2014; Brooks *et al.*, 2016).

In steep forested areas, infiltration-excess overland flow is rare due to the high infiltration capacity of forest soils. Instead, saturation-excess runoff and subsurface return flow are the primary mechanisms of runoff generation, while baseflow contributes to sustained streamflow during non-storm periods. (Brooks *et al.*, 2004; Srivastava *et al.*, 2013). The contribution of each runoff pathway depends on site-specific conditions, such as local climate, soil properties, bedrock permeability, topography and soil water content.

7.4.3 Soil water

In steep forest hydrology, important soil properties include thickness, bedrock permeability, drainable porosity (the difference between soil water content at saturation and field capacity), topography, antecedent soil water conditions and plant rooting depth. The spatial variability of these properties affects soil storage capacity, water availability for evapotranspiration, and vertical and lateral water movement, thereby defining runoff generation processes. Soil texture interacts with soil thickness to influence hydrological responses on steep forested hillslopes. Coarse-textured soils have higher infiltration rates and hydraulic conductivities but retain less water than finer-textured soils. In steep watersheds, rocky soils retain less water than those with fewer rocks (van Wesemael *et al.*, 2000).

Soil thickness and properties affecting soil water availability are important influences on the relative proportion of subsurface lateral flow, bedrock percolation and vegetation water uptake. Topography influences the variability of soil thickness and texture, with shallower soils common on upland slopes and deeper soils in lowlands (van Wesemael *et al.*, 2000). Shallower soils, having less water-holding capacity, saturate faster, leading to more subsurface lateral flow, bedrock infiltration and lower evapotranspiration due to limited water for shallow-rooted vegetation. A shallow soil is more likely to become saturated during a wet spell, leading to saturation-excess runoff. In contrast, deeper soils, with higher water-storage capacity, saturate less often and retain more water, resulting in less subsurface lateral flow and greater evapotranspiration for deep-rooted vegetation.

Figure 7.3 shows the comparison of observed and simulated hillslope hydrological responses from two watersheds with different soil thickness located in the Lake Tahoe Basin, Nevada (Brooks *et al.*, 2016). The Blackwood Creek watershed, with shallow soils, showed

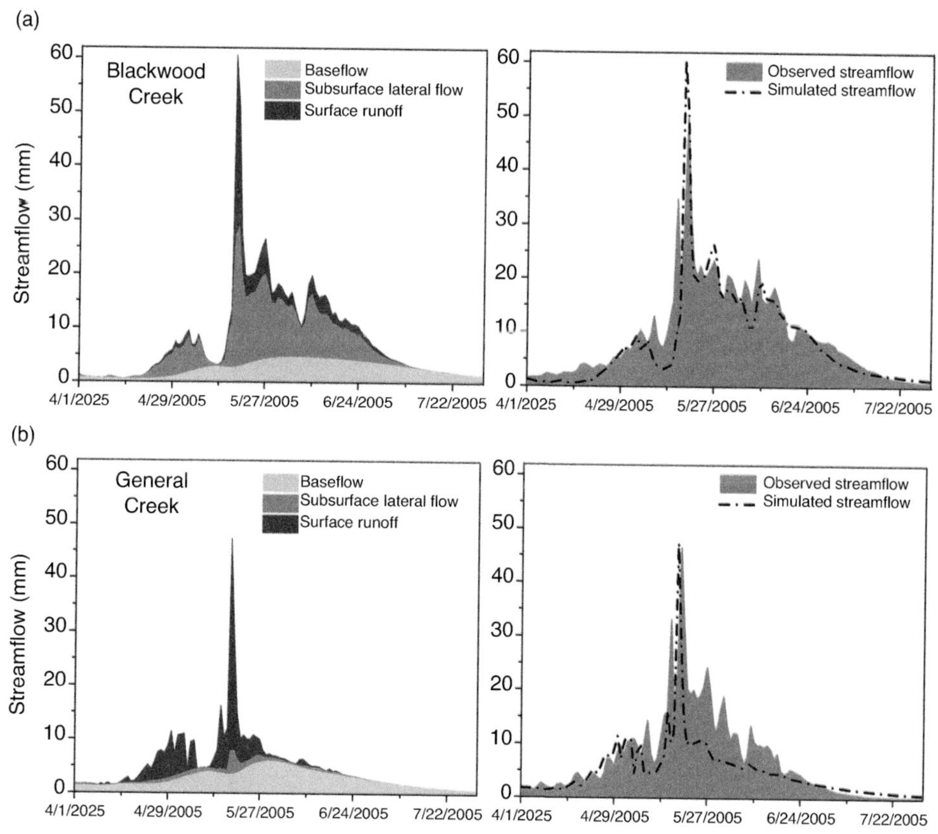

Fig. 7.3. Simulated and observed hydrological responses from two watersheds located in the Lake Tahoe Basin: (a) Blackwood Creek watershed (shallow soils); (b) General Creek watershed (deep soils). Reprinted from Brooks *et al.* (2016), with permission from Elsevier.

major contributions to streamflow from subsurface lateral flow in May and June (Fig. 7.3a), while the General Creek watershed, with deep soils, showed minimal subsurface lateral flow and primarily baseflow contributions (Fig. 7.3b).

Runoff generation mechanisms depend on soil water content at the time of rainfall or snowmelt, known as 'antecedent soil water.' In general, higher soil water content leads to a faster and more intense runoff response. Continuous processes such as soil evaporation, plant transpiration, subsurface lateral flow and bedrock percolation gradually reduce soil water content between events. Thus, runoff mechanisms depend on current soil water conditions and the site's hydrological history. Soil water conditions are influenced by climatic factors (e.g. precipitation, temperature, shortwave radiation) and seasonal weather variability.

Soil water content during winter is often low because the previous growing season's evapotranspiration depleted it, and temperatures dropped below freezing before the onset of winter precipitation. In spring, rising temperatures cause snowmelt, increasing soil water content and activating runoff pathways (Fig. 7.2), often aided by early-spring ROS or later-spring rain on wet soil. In the summer or dry season, evapotranspiration reduces soil water content, especially in regions with dry summers like the western USA and southern Europe. Soil water content is dynamic, influenced by climatic conditions, soil properties (drainable porosity, thickness, texture), vegetation (species, age, rooting depth), underlying geology (bedrock permeability, soil depth) and topography (van Wesemael *et al.*, 2000).

The above-mentioned complex interactions make it challenging to fully understand high-elevation steep hydrological processes and explain different hydrological responses in similar watersheds. In recent decades, researchers have focused on studying small watersheds in detail, leading to new theories on variability in hydrological responses. These theories emphasize the importance of spatial variability in soil and vegetative properties, microclimates, snowmelt dynamics, and surface and groundwater interactions. Current research in snow-dominated high-elevation mountainous relies heavily on complex computer models (see Chapter 16, this volume), stable isotopes or machine learning techniques to understand these interactions.

7.4.4 Bedrock permeability

The role of bedrock permeability in controlling water movement can be understood, in a simplistic view, by comparing watersheds with weathered bedrock with those with unweathered bedrock. In areas with weathered bedrock, which is more permeable due to fractures and porosity, water is more likely to move from the soil into the bedrock. As a result, less water flows laterally through the soil because gravity quickly pulls the water down into the permeable weathered bedrock. In this condition, seasonal evapotranspiration may be lower because the soil retains less water for plants to access later in the growing season. In contrast, areas with unweathered bedrock, which is less permeable and more solid, will have less water percolating into the bedrock. This leads to more water remaining in the soil layer, resulting in greater lateral flow through the soil and more water available for plants during the growing season.

However, realistically, as shown in Fig. 7.2, a systematically varying weathering zone that develops within bedrock beneath soils and above unweathered bedrock transforms the poorly conductive bedrock into a dynamic water reservoir. During the wet season, infiltrating precipitation accumulates in the weathered bedrock, where excess water either moves laterally through fractures or percolates deeper to recharge groundwater. The weathered bedrock, often referred to as a source of 'rock moisture,' stores water in the unsaturated zone, distinct from soil moisture (Dralle *et al.*, 2018; Rempe and Dietrich, 2018). In the dry season, vegetation gradually depletes rock moisture, sustaining transpiration and evapotranspiration later into the season than would be possible with soil moisture alone.

7.4.5 Topography

The hillslope topography plays a major role in mountainous hydrological processes. Based on

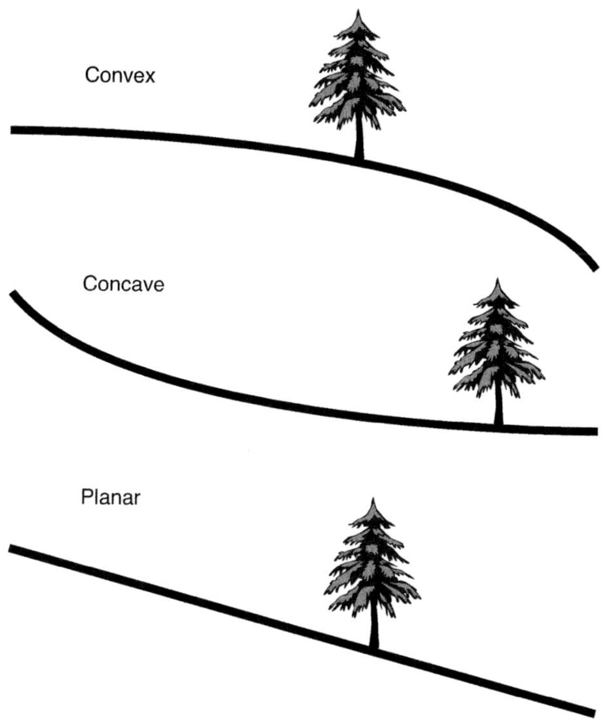

Fig. 7.4. Convex, concave and planar slope shapes.

slope length and the variability in steepness, most of the hillslopes can be characterized as convex, concave or planar (Fig. 7.4). Different sections within a hillslope profile significantly affect soil saturation and flow behavior. Fig. 7.2 depicts an idealized concave hillslope with well-drained soils of uniform thickness and low-permeability bedrock. The slope is steep at the top and gentler at the bottom, illustrating subsurface lateral flow and saturation-excess overland flow or subsurface return flow to the stream. Before a storm, baseflow defines a water table near the stream. During or after a storm, direct precipitation on the valley bottom and subsurface lateral flow elevate the water table and soil water content. For small to moderate events, soil is unlikely to become saturated, and therefore saturation-excess runoff will be rare and subsurface lateral flow will dominate the volume of runoff to streamflow. However, during large events, more of the soil profile saturates, leading to saturation-excess runoff and subsurface return flow as dominant pathways. Soil thickness influences soil saturation near the valley bottom and runoff pathways. Convex and planar slopes are less likely to experience saturation-excess or subsurface return flows unless geology diverts subsurface flow to the surface.

At a landscape scale, soil depth is not uniform but rather tends to be deeper in upland swales and less deep on upland ridges. Soils tend to be deeper on the lower parts of concave slopes and less deep on shoulder areas of convex slopes. Underlying geology can further affect soil depth, leading to a diversity of potential seepage points as lateral flows are forced to the surface or as surface runoff seeps back into the soil profile (McDonnell *et al.*, 2007).

7.4.6 Impacts of forest access on hydrology

Forest access is associated with nearly all forests to support fuel management, timber harvest, fire suppression and for recreation (Gucinski *et al.*,

2001; Elliot, 2013). Access includes 'temporary' skid trails, low-volume forest roads, paved roads, off-road vehicle trails and other trails. All access corridors reduce vegetative cover, increase soil compaction and decrease infiltration rates. The increased runoff may simply infiltrate into nearby forested areas, or may increase watershed runoff, particularly from 2 to 5 year return period storms (Gucinski *et al.*, 2001; DeArmond *et al.*, 2021). The impact of roads on runoff is less noticeable during larger or smaller storms.

Access corridors may become vegetated, but compaction and low hydraulic conductivity can persist for years on lightly compacted skid trails and for decades on more compacted trails and roads (DeArmond *et al.*, 2021). Off-road vehicle trails are particularly problematic due to soil pulverization from traffic followed by high erosion rates, which can expose bedrock and significantly increase runoff potential (Meadows *et al.*, 2008).

In addition to road compaction generating runoff, roads excavated into steep hillsides can intercept subsurface lateral flow, increasing surface runoff and peak flows. This interception may extend stream networks further into steep forested hillsides, further increasing peak flows.

Mitigation practices such as revegetation and mulching may decrease surface erosion but are unlikely to reduce other hydrological impacts. Mechanical ripping of closed roads can increase infiltration from lower-intensity storms. Cross-drains can reduce runoff concentration, downslope channel erosion and sediment delivery, especially if they direct runoff into wetlands or forests instead of streams (Copstead *et al.*, 1998). Paving roads decreases sediment delivery from the road surface, but the increased runoff can cause downslope or downstream erosion. Removing unneeded roads is a common watershed improvement practice.

7.5 Projected Impacts of Climate Change

The climates that are the basis of the knowledge presented in this chapter are not constant. Predictions of future climates around the world suggest a continued increase in temperatures and an increase or decrease in annual precipitation (Stocker *et al.*, 2013). During the 20th century, air temperatures have increased, decreasing the mountain snowpack, especially at elevations below 1800 m, and threatening water resources (Milly *et al.*, 2008). Higher temperatures are expected to decrease snow accumulation, affect the timing and rate of spring snowmelt (Musselman *et al.*, 2017), and may reduce annual runoff (Berghuijs *et al.*, 2014).

Snowmelt during the snow accumulation season is increasing and is projected to continue rising under warmer conditions, leading to a loss of seasonal snow water storage in North America's mountainous regions and elsewhere (Musselman *et al.*, 2021; Scaff *et al.*, 2024). However, greater snowfall amounts can mediate these effects. Higher snowfall is associated with reduced melt during the accumulation season, particularly in the continental interior of the western USA, which may mitigate some impacts of warming on snowmelt. Conversely, decreased snowfall in maritime regions will likely amplify the effects of warming on winter melt (Marshall *et al.*, 2020). These findings underscore the importance of accurately capturing changes in precipitation amounts and distributions in climate change projections.

Major changes are expected in the extent and distribution of snow-covered areas and the timing of peak snow water equivalent (Klos *et al.*, 2014; Marshall *et al.*, 2019). While high-elevation mountains will experience temperature-related changes, mid- and low-elevation mountains in the rain–snow transition zone are most sensitive to climate change. In the western USA, a shift from snow to rain by the mid-21st century is estimated to decrease winter snow-dominated areas by 30% and shorten snow cover duration from 5 months to 3 months (Klos *et al.*, 2014). This will likely increase early-spring flows and decrease summer flows, impacting overall hydrology in maritime climates. Watershed managers may need to increase storage within watersheds using reservoirs to mitigate the effects of a shorter snowfall season (Marshall *et al.*, 2019). Additionally, more streams may dry up before the end of the dry season (Rauscher *et al.*, 2008). A decrease in snowpack will also reduce albedo, complicating energy balance analyses in climate change models and potentially leading to greater temperature increases (Manninen and Stenberg, 2009).

Increasing temperatures and more variable snowmelt will cause vegetation to transpire earlier, leading to earlier soil water depletion (Wieder *et al.*, 2022). In drier forests, this could shift plant communities toward drought-tolerant species (Rehfeldt *et al.*, 2006). Earlier soil water loss will also increase the risk of insect damage, wildfire and changes in plant communities, impacting hydrology (Westerling *et al.*, 2006). This trend is already evident with more frequent and intense fires in both hemispheres. Wildfire and tree mortality often increase peak flows and sediment delivery from degraded and burned watersheds (see Chapter 10, this volume; Elliot, 2013).

7.6 Conclusions

Mountainous and snow-dominated forests are major sources of water on every continent. Snow accumulation and melt rates are big drivers of hydrological processes. Temperature, radiation, and the interaction between rain and snow are major factors in generating runoff. Surface and subsurface flow processes route water from saturated hillsides to deeper groundwater storage or directly to streams, resulting in hydrographs with long durations and low peak flows. Avalanches can redistribute large amounts of snow on steep landscapes, but the effects on watershed hydrology are site specific and small on a large or regional watershed scale. Flooding is associated with high rainfall and snowmelt rates, and on ROS events occurring when soils are saturated. Future climates will likely result in more flooding from rainfall events and less snow accumulation, especially at lower elevations. The impact that future climate scenarios have on streamflows cannot be generalized but will depend on elevation, source of air masses and complex interactions among the climate, topography, soil and vegetation. Hydrological models will continue to play an important role in better understanding these complex interactions.

References

Berghuijs, W., Woods, R. and Hrachowitz, M. (2014) A precipitation shift from snow towards rain leads to a decrease in streamflow. *Nature Climate Change* 4, 583–586.

Beven, K. (2021) The era of infiltration. *Hydrology and Earth System Science* 25, 851–866.

Bronstert, A., Niehoff, D. and Schiffler, G. (2023) Modelling infiltration and infiltration excess: The importance of fast and local processes. *Hydrological Process* 37, e14875.

Brooks, E.S., Boll, J. and McDaniel, P.A. (2004) A hillslope-scale experiment to measure lateral saturated hydraulic conductivity. *Water Resources Research* 40, W04208.

Brooks, E.S., Dobre, M., Elliot, W.J., Wu, J.Q. and Boll, J. (2016) Watershed-scale evaluation of the Water Erosion Prediction Project (WEPP) model in the Lake Tahoe basin. *Journal of Hydrology* 533, 389–402.

Conway, H. and Raymond, C.F. (1989) *Prediction of Snow Avalanches in Maritime Climates*. WA-RD 203.2 Technical Report. Olympia, Washington: Washington Department of Transport.

Copstead, R.L., Johansen, D.K. and Moll, J. (1998) *Water/Road Interaction: Introduction to Surface Cross Drains*. Report 9877 1806 – SDTDC. Technology and Development Program, USDA Forest Service, San Dimas, California.

DeArmond, D., Ferrax, J.B.S. and Higuchi, N. (2021) Natural recovery of skid trails: A review. *Canadian Journal of Forest Research* 51, 948–961.

Dickerson-Lange, S.E., Vano, J.A., Gersonde, R. and Lundquist, J.D. (2021) Ranking forest effects on snow storage: A decision tool for forest management. *Water Resources Research* 57, e2020WR027926.

Dralle, D.N., Hahm, W.J., Rempe, D.M., Karst, N.J., Thompson, S.E. *et al.* (2018) Quantification of the seasonal hillslope water storage that does not drive streamflow. *Hydrological Processes* 32, 1978–1992.

Elliot, W.J. (2013) Erosion processes and prediction with WEPP technology in forests in the Northwestern US. *Transactions of the ASABE* 52, 563–579.

Gucinski, H., Furniss, M.J., Ziemer, R.R. and Brookes, M.H. (2001) *Forest Roads: A Synthesis of Scientific Information*. General Technical Report PNW-GTR-509. Pacific Northwest Research Station, Forest Service, USDA, Portland, Oregon.

Hubbart, J.A., Link, T.E., Gravelle, J.A. and Elliot, W.J. (2007) Timber harvest impacts on water yield in the continental/maritime hydroclimate region of the US. *Journal of Forestry* 53, 169–180.

Immerzeel, W.W., Lutz, A.F., Andrade, M., Bahl, A., Biemans, H. *et al.* (2020) Importance and vulnerability of the world's water towers. *Nature* 577, 364–369.

Jennings, K.S., Winchell, T.S., Livneh, B. and Molotch, N.P. (2018) Spatial variation of the rain–snow temperature threshold across the Northern Hemisphere. *Nature Communications* 9, 1148.

Klamerus-Iwan, A., Link, T.E., Keim, R.F. and Van Stan, J.T.II. (2020) Storage and routing of precipitation through canopies. In: Vab Stan, J.T.II., Gutmann, E.D. and Friesen, J. (eds) *Precipitation Partitioning by Vegetation: A Global Synthesis*. Springer Nature, Cham, Switzerland, pp. 17–34.

Klos, P.Z., Link, T.E. and Abatzoglou, T.J. (2014) Extent of the rain-on-snow transition zone in the western US under historic and projected climate. *Geophysical Research Letters* 41, 4560–4568.

Lawler, R.R. and Link, T.E. (2011) Quantification of incoming all-wave radiation in discontinuous forest canopies with application to snowmelt prediction. *Hydrological Processes* 25, 3322–3331.

Lundquist, J.D., Dickerson-Lange, S.E., Lutz, J.A. and Cristea, N.C. (2013) Lower forest density enhances snow retention in regions with warmer winters: A global framework developed from plot-scale observations and modeling. *Water Resources Research* 49, 6356–6370.

Lundquist, J.D., Dickerson-Lange, S., Gutmann, E., Jonas, T., Lumbrazo, C. *et al.* (2021) Snow interception modelling: Isolated observations have led to many land surface models lacking appropriate temperature sensitivities. *Hydrological Process* 35, e14274.

Manninen, T. and Stenberg, P. (2009) Simulation of the effect of snow covered forest floor on the total forest Albedo. *Agricultural and Forest Meteorology* 149, 303–319.

Marks, D., Kimball, J., Tingey, D. and Link, T.E. (1998) The sensitivity of snowmelt processes to climate conditions and forest cover during rain-on-snow: A case study of the 1996 Pacific Northwest flood. *Hydrological Processes* 12, 1569–1587.

Marshall, A.M., Abatzoglou, J.T., Link, T.E. and Tennant, C.J. (2019) Projected changes in interannual variability of peak snowpack amount and timing in the Western United States. *Geophysical Research Letters* 46, 8882–8892.

Marshall, A.M., Link, T.E., Robinson, A.P. and Abatzoglou, J.T. (2020) Higher snowfall intensity is associated with reduced impacts of warming upon winter snow ablation. *Geophysical Research Letters* 47, e2019GL086409.

Martinec, J. and de Quervain, M. (1971) The effect of snow displacement by avalanches on snow melt and runoff. In: *Proceedings of the Snow and Ice Symposium, August 1971, Moscow.* IAHS-AISH Publication No. 104, IAHS, Wallingford, UK, pp. 364–377.

McDonnell, J.J., Sivapalan, M., Vaché, K., Dunn, S., Grant, G. *et al.* (2007) Moving beyond heterogeneity and process complexity: A new vision for watershed hydrology. *Water Resources Research* 43, W07301.

Meadows, D., Foltz, R. and Geehan, N. (2008) *Effects of All-terrain Vehicles on Forested Lands and Grasslands*. Publication no.0823-1811-SDTDC. San Dimas Technology and Development Center, San Dimas, California.

Milly, P.C.D., Betancourt, J., Falkenmark, M., Hirsch, R.M., Kundzewicz, Z.W. *et al.* (2008) Stationarity is dead: Whither water management? *Science* 319, 573–574.

Musselman, K.N., Clark, M.P., Liu, C., Ikeda, K. and Rasmussen, R. (2017) Slower snowmelt in a warmer world. *Nature Climate Change* 7, 214–219.

Musselman, K.N., Addor, N., Vano, J.A. and Molotch, N.P. (2021) Winter melt trends portend widespread declines in snow water resources. *Nature Climate Change* 11, 418–424.

Pomeroy, J.W., Marks, D., Link, T., Ellis, C., Hardy, J. *et al.* (2009) The impact of coniferous forest temperatures on incoming longwave radiation to melting snow. *Hydrological Processes* 23, 2513–2525.

Rauscher, S.A., Pal, J.S., Diffenbaugh, N.S. and Benedetti, M.M. (2008) Future changes in snowmelt-driven runoff timing over the Western US. *Geophysical Research Letters* 35, L16703.

Rehfeldt, G.E., Crookston, N.L., Warwell, M.V. and Evans, J.S. (2006) Empirical analyses of plant–climate relationships for the Western United States. *International Journal of Plant Sciences* 167, 1123–1150.

Rempe, D.M. and Dietrich, W.E. (2018) Direct observations of rock moisture, a hidden component of the hydrologic cycle. *Proceedings of the National Academy of Sciences USA* 115, 2664–2669.

Sánchez-Murillo, R., Brooks, E.S., Elliot, W.J., Gazel, E. and Boll, J. (2014) Baseflow recession analysis in the inland Pacific Northwest of the United States. *Hydrogeology Journal* 23, 287–303.

Sánchez-Murillo, R., Brooks, E.S., Elliot, W.J. and Boll, J. (2015) Isotope hydrology and baseflow geochemistry in natural and human-altered watersheds in the Inland Pacific Northwest, USA. *Isotopes in Environmental and Health Studies* 51, 231–254.

Scaff, L., Krogh, S.A., Musselman, K., Harpold, A., Li, Y. *et al.* (2024) The impacts of changing winter warm spells on snow ablation over Western North America. *Water Resources Research* 60, e2023WR034492.

Schläppy, R., Eckert, N., Jomelli, V., Stoffel, M., Grancher, D. *et al.* (2014) Validation of extreme snow avalanches and related return periods derived from a statistical-dynamical model using tree-ring techniques. *Cold Regions Science and Technology* 99, 12–26.

Seyednasrollah, B., Kumar, M. and Link, T.E. (2013) On the role of vegetation density on net snowcover radiation at the forest floor. *Journal of Geophysical Research: Atmospheres* 118, 8359–8374.

Sosedov, I. and Seversky, I. (1965) On the hydrological role of snow avalanches in the northern slope of the Zailiysky Alatua. In: *Proceedings of the International Symposium on the Scientific Aspects of Snow and Ice Avalanches, 5–10 April 1965, Davos, Switzerland*, IAHS-AISH Publication No. 69. IAHS, Wallingford, UK, pp. 78–85.

Srivastava, A., Dobre, M., Wu, J.Q., Elliot, W.J., Bruner, E.A. *et al.* (2013) Modifying WEPP to improve streamflow simulation in a Pacific Northwest watershed. *Transactions of the ASABE* 56, 603–611.

Stocker, T.F., Qin, D., Plattner, G.-K., Tignor, M.M.B., Allen, S.K. *et al.* (2013) *Climate Change 2013: The Physical Science Basis. Contribution of Working Group I to the Fifth Assessment Report of the Intergovernmental Panel on Climate Change*. Cambridge University Press, Cambridge, UK/New York.

Sun, N., Yan, H., Wigmosta, M.S., Lundquist, J., Dickerson-Lange, S. *et al.* (2022) Forest canopy density effects on snowpack across the climate gradients of the Western United States mountain ranges. *Water Resources Research* 58, e2020WR029194.

Trujillo, E. and Molotch, N.P. (2014) Snowpack regimes of the Western United States. *Water Resources Research* 50, 5611–5623.

van Wesemael, B., Mulligan, M. and Poesen, J. (2000) Spatial patterns of soil water balance on intensively cultivated hillslopes in a semi-arid environment: The impact of rock fragments and soil thickness. *Hydrological Processes* 14, 1811–1828.

Varhola, A., Coops, N.C., Weiler, M. and Moore, R.D. (2010) Forest canopy effects on snow accumulation and ablation: An integrative review of empirical results. *Journal of Hydrology* 392, 219–233.

Westerling, A.L., Hidalgo, H.G., Cayan, D.R. and Swetnam, T.W. (2006) Warming and earlier spring increase western US forest wildfire activity. *Science* 313, 940–943.

Wieder, W.R., Kennedy, D., Lehner, F., Musselman, K.N., Rodgers, K.B. *et al.* (2022) Pervasive alterations to snow-dominated ecosystem functions under climate change. *Proceedings of the National Academy of Sciences USA* 119, e2202393119.

Williams, M., Elder, K., Soiseth, C. and Kattelmann, R. (1992) Aquatic ecology as a function of avalanche runout into an alpine lake. In: *Proceedings of the 1992 International Snow Science Workshop, Breckenridge, Colorado, USA*, pp. 47–56.

Wilson, J.L. and Guan, H. (2004) Mountain-block hydrology and mountain-front recharge. In: Hogan, J.F., Philips, F.M. and Scanlon, B.R. (eds) *Water Science and Applications Series*, Vol. 9. American Geophysical Union, Washington, DC, pp. 113–137.

Zhang, Y.K., Hartemink, A.E., Vanwalleghem, T., Bonfatti, B.R. and Moen, S. (2024) Climate and land use changes explain variation in the a horizon and soil thickness in the United States. *Communications Earth and Environment* 5, 129.

8 Hydrology of Flooded and Wetland Forests

L.J. Bren[1]*, T.M. Williams[2] and D.M. Amatya[3]

[1]The University of Melbourne, Creswick, Victoria, Australia; [2]Clemson University, Clemson, South Carolina, USA; [3]Santee Experimental Forest, Southern Research Station, USDA Forest Service, Cordesville, South Carolina, USA

Abstract

Flooded or wetland forests are commonly continuously inundated for many months. Trees that can survive this have specialized mechanisms for allowing roots to respire. The forests can be classified by the source of the water causing inundation – rainfall, groundwater, flow from an adjacent watercourse or tidal inundation. The latter may include specialized vegetation at the ocean salt/freshwater interface known as 'mangroves'. These seem to require some input of freshwater, although the mechanism of how this assists their survival is unclear. The hydroperiod is the usual annual period for which a forest is inundated. Flooded forests usually contain a range of vegetation aligned to the hydroperiod of specific sites. Flooding of the forests gives them an unusual, distinctive and highly valued character, both in the vegetation and the associated fauna. Often, these forests have large accumulations of carbon in anoxic soils. Many flooded forests are under some sort of threat relating to changes in their inundation due to the value of water for other uses. The resulting problems may be difficult to resolve and require compromise to achieve resolution. The advent of climate change and the impacts of this on rainfall in large catchments pose a particularly serious threat to the long-term survival of this group of forests around the world. Oxidation of dried-out 'wetland soils' can be a major source of greenhouse gases.

8.1 Introduction

In this chapter, we examine the hydrology of forested areas that are subject to prolonged soil saturation by precipitation, groundwater or surface flooding. These include flood-plain forests, estuarine tidal forests, mangrove forests, the forested portions of peatlands and the tree-dominated wetlands defined by the Ramsar Convention on Wetlands (www.ramsar.org, accessed 18 April 2025). A broad outline of the ecology of all wetlands has been described by Mitsch and Gosselink (2015). Forested wetlands are a subset of this all-embracing term and generally lack the flooding extremes of wetlands. Trees found in these regions commonly share with other wetland plants a metabolism that allows translocation of oxygen through stem tissues to the roots, thereby facilitating root respiration (Kozlowski, 2002). This may involve specialized tissue development. Fig. 8.1 gives a view of one such area in an Australian flooded river red gum (*Eucalyptus camaldulensis*) forest.

As most forest species (except certain mangroves) cannot regenerate under continuous flooding (Lugo, 1988), the wet limit of forested wetlands requires at least periodic drying of the

*Corresponding author: lbren@iinet.net.au

DOI: 10.1079/9781800625310.0008

Fig. 8.1. View of an Australian flooding forest. This shows young river red gum invading a 'grass plain' (although the ground species is mainly water milfoil, *Myriophyllum* spp.).

soil surface. How wet a soil has to be to be classed as a 'wetland' is ambiguous but is generally associated with soil saturation that is of sufficient duration to produce visible soil features (creating a 'hydric soil') and to limit the growth of plants not specially adapted to saturated soils. This is of importance in conservation because it restricts weed invasion, as many weeds cannot tolerate prolonged inundation. Around the world, the local definition of 'wetlands' commonly involves an interplay of science, economics, and local and national politics. The terminology can be context dependent, ambiguous and often unsatisfactory to apply to areas outside of where the definition was formulated. We have used the term 'flooded forests' or 'forested wetlands' interchangeably. Terms such as mires, fens, swamps, bogs or marshes are often ambiguous as to whether the presence of trees or forests is implied.

The most common model of a flooding (freshwater) forest is one of forest on a flood plain. The streams carry silt and, over time, this causes the river bed to rise above the adjoining plain. The river may also form natural 'levees' – walls that constrain flow in the channel; sometimes these are 'improved' by human activity to bring them to a constant river level. Periodically, an excess of water may flow down the channel and cause the water to overtop these levees or to pass through gaps ('crevasses') in them. This allows flooding from the river into the forest. Often, this leads to deep groundwater recharge on the flood plain. The origin of this water is from streamflow derived by storms upstream. Usually, once the water has left the river channel, it flows through the forest (sometimes designated a 'flowing wetland'). The water quality tends to reflect the water quality of the source river. Because forests are hydraulically rough, the water slows down compared with its velocity in open channels. This causes deposition of sediment, and many flooded forests are 'rising' topographically because of this. Sooner or later, the source of water diminishes, and the water 'dries up' – this can be due to a reduction in river levels, water flowing offsite, perhaps infiltration to deeper aquifers and evapotranspiration. In some cases, the soil surface may be below saturation but at a matric potential close to that of free water. In such cases, it is common to find a 'water table' (phreatic surface) not far below the land surface. The term 'saturation' implies no air in the soil matrix, implying that the matric potential, Ψ, is below the 'air entry' value for that soil. Rawls *et al.* (1982) listed geometric values of Ψ from 7.3 cm (sand) to 36.5 cm (clay). The Rawls *et al.* (1982) criterion may be a reasonable methodology for defining 'saturated'.

The period of time for which the forests may reasonably be expected to be flooded (on a mean annual basis) is called the 'hydroperiod', and for most flooded forests might reasonably be expected to be in the 3–9 month range, given the usual annual seasonal behaviour. Forests at the low end of the range are sometimes called 'riparian forests' in which the trees can tolerate flooding but do not acquire special adaptations to cope with prolonged inundation (Meixner *et al.*, 2006). Flooded forests may also have a high 'frequency of flooding'. This term is vague but implies the number of times the forest might be expected to have separate flooding events (with a drying out period interposed between flooding) in each year. Such terminology is useful but can also be ambiguous (i.e. a forest may have a substantial number of flooding events in a given year, but if each is of a short duration, then the hydroperiod may not be long). A useful first task in any quantification of a flooded forest is to define the hydroperiod and the flooding frequency as best one can.

Flooding has a number of effects. First, it provides the forests with an inflow of water and nutrients (both dissolved and attached to sediment); in some cases (e.g. Australian red gum forests; see below), the water complements the inadequate local rainfall. Without this, the trees survive but have little growth; with flooding, there is an immediate flush of growth. Second, it gives plants with the ability of surviving long periods of inundation an advantage on such sites by excluding those less able. Third, it creates a diverse range of environments that foster 'life' – in terms of both flora and fauna. Flooded forests commonly have specialized flora and fauna occupying a diverse range of habitats. These are greatly valued on ecological and economic grounds.

8.2 the Origins of Flood Water

A useful (but not always easy) classification is based on the origin of the water causing inundation.

8.2.1 Precipitation-excess wetlands

Precipitation-excess wetlands are found in moist climatic regions and occur where precipitation is greater than transpiration. Annual precipitation is likely to equal or exceed annual evapotranspiration and precipitation must exceed evapotranspiration for a significant portion of the year for wetland conditions to occur. Restricted surface and/or subsurface drainage are commonly present. These wetlands generally have shallow surface drainage (if any). In many cases, these wetlands are found in relatively young geological formations where mature drainage patterns have not yet developed. Groundwater flows are restricted vertically by a confining layer or, in the horizontal plane, by a lack of slope and/or low hydraulic conductivity of the media. Brinson (1993) classed these wetlands as 'mineral' or 'organic' flats, while Semeniuk and Semeniuk (1995) called them 'damplands', and Tiner (2017) called them 'seasonally variable wetlands'. Emphasis should be put on their 'seasonal' nature because they may have (very) long periods of dry land between inundation. Because of their dependence on recent rainfall, they can be very ephemeral.

These are the only possible wetlands on topographic convex surfaces at watershed divides. In the bog–fen terminology, all bogs are rainwater wetlands. These wetlands must be convex enough and high enough in the landscape such that groundwater and surface water leave the site. Rainwater wetlands are most sensitive to climate and management. Forest species, density and age all alter the transpiration component. Ditching draining, bedding, burning, road construction or any other soil alteration will alter the runoff component. This means that their hydrology can be inadvertently changed but also creates opportunities for hydrological modification to improve suitability for selected species (Amatya *et al.*, 2024).

Although examples of these can be found all over the world, the greatest extent is found in the northern hemisphere, including north-eastern and north-central USA, eastern and most of northern Canada, north-eastern Europe, Fennoscandia, and Russia in Eurasia. Common tree species are pines (*Pinus* spp.) or birch (*Betula* spp.) in Eurasia, and spruce (*Picea mariana*), larch (*Larix laricina*) and birch in North America. Sometimes, these are called by the generic term 'mires', and a distinction is made between bogs (precipitation fed) and fens (groundwater fed). Among others, Payette and Delwaide (2003) showed that boreal forests at high latitudes are climate-sensitive ecosystems that respond directly to environmental forcing by changing their boundaries (retreating/advancing) and/or altering their abundance at local and regional scales. Fires and excessive inundation have caused events such as mass mortality of the forests and post-fire deforestation. They noted that 'future moisture changes ... will result in important rearrangements of wetland ecosystems'.

In some glaciated landscapes, a distinction can be made in undisturbed peat bogs between the 'acrotelm' and the 'catotelm'; the former overlies the latter. The boundary between the two layers is defined by the transition from peat containing living plants (acrotelm) to peat consisting of dead plant material (catotelm). This typically coincides with the lowest level of the water table (i.e. the catotelm is the zone of permanent inundation). Trees may be a component of the acrotelm, with roots penetrating into the catotelm in forested peatlands – in such cases, the trees may have physiological adaptations to allow respiration of the roots.

The southern hemisphere lacks large areas of land at high (southern) latitudes, so there is very limited development of such forest on glaciated landscapes. It does, however, have large areas of precipitation-excess wetland forest developed on tropical peatland, particularly in Borneo and New Guinea. Wösten *et al.* (2008) found such an area with a 10 year average annual rainfall of 2570 mm and an estimated evapotranspiration of 1500 mm. They also found large variations in the rainfall range (1848–3749 mm); a practical consequence is that such forests may suffer substantially from annual drought and become fire prone during such times. Such tropical swamp forests are commonly sought after for competitive uses, particularly oil palm plantation development. This has been controversial.

8.2.2 Groundwater-fed forested wetlands

Occasionally, a substantial groundwater aquifer will pass water into an adjacent wetland area; commonly this would be where a permeable hillside adjoins a flood plain or a large old lake has filled with sediment and developed forest. In such a case, the contribution of the hillslope can be substantial and prolonged, and can keep the forest supplied with water in excess of its transpiration capacity for substantial periods. Waddington *et al.* (1993) gave a detailed description of one such area in Toronto, Canada. The 'swamp' consisted of fluvio-glacial wash overlaid by peat soils. The forest consisted of cedars (*Thuja occidentalis*) and hemlock (*Tsuga* spp). Groundwater entered the wetland through a series of springs at the margin of the wetland, creating many diffuse, saturated zones that joined together to form small streamlets. Chemical separation of groundwater contribution from the rainwater composition suggested that 70% of the runoff was 'pre-event' water (i.e. groundwater).

Groundwater outflows are stable for long periods, but such wetlands can be subject to offsite influences relating to groundwater drawdown (pumping), drainage or a change in recharge of the aquifer. Such change can occur many kilometres away from the wetland and can be hard to recognize. Simon *et al.* (2023) discussed such a case in Hungary in which a poplar wetland deteriorated. Similar issues were cited by van Deventer *et al.* (2021) in South Africa and by Durigan *et al.* (2022) in Brazil; in both of these cases, the authors felt that existing environmental legislation was inadequate to recognize the threats to the forested wetlands and to counter the observed changes.

8.2.3 Forested wetlands due to surplus surface water flow

In these, the combination of rainfall plus water from adjacent sources (particularly rivers or lakes) exceeds (or in Australia, helps to meet) the evapotranspiration needs of the trees. These wetlands are sometimes called 'fringing forests' because they fringe a lake or river, although this 'fringe' may extend many tens of kilometres from the water source. Riverine forested wetlands are the 'prototype' forest described in Section 8.1. Often such forests are of great economic benefit and have large, ecosystem-dependent communities living in or near them. A common hydrological variation is that a major river receives flow from high rainfall mountains and passes through an incised channel to an alluvial plain of lower rainfall. On this plain, the incised channel may give way to a braided channel with complex branching and interconnections. Sometimes, the channel braiding is the outcome of a sediment-filled lake bed forming. This develops a forest cover with an intricate mosaic of waterways, giving a wide diversity of hydroperiods and vegetation types. Such areas can be greatly valued for their biological diversity (e.g. the Okavango Delta in north-western Botswana, and the Achafalyala Delta in the USA).

8.3 Three Continents and Three Examples of Flooded Forests

Flooded forests are found around the world. Three brief examples illustrating this are given here. In each case, management of the forests involves an interplay of state and federal politics involving considerations other than the hydrological well-being of the resource. Although differing in location and tree characteristics, the areas share the feeling of being in a flooded forest – with the smell of water-saturated landscapes and unusual species (of flora and fauna) never being far away. In each case, there can be long periods without flooding, in which case the forest becomes the more common 'dryland' variant – and this can go on for many months. Readers are encouraged to extend the examples here – there are many that have achieved international fame because of their uniqueness and/or distinctiveness in form or landscape. Other areas are hardly recognized as flooded forests because they are 'taken for granted'.

In each of the cases below, the forests have their own benefits in terms of forest products and biodiversity. They also serve as large, transient water storages. These reduce peak flows in the river and protect towns along the river from flooding. Readers are encouraged to view videos extolling how the biodiversity of these forests 'comes alive' on media such as YouTube. Although this is miraculous in itself, the really important factor is that of adding water. All else is a consequence of that.

8.3.1 The red gum forests of the River Murray in Australia

In economic terms, the River Murray is Australia's most important river. It starts in the well-watered hills and mountains of eastern Australia and flows west and then south to enter the sea near Adelaide (South Australia). A ubiquitous tree found close to the river is river red gum (*E. camaldulensis*). At many locations, it forms extensive forests, of which the Barmah-Millewa Forest (straddling the New South Wales/Victoria state border) is the best known.

In most of these flooded forests, river red gum is the only woody species; other vegetation comprises grasses or reeds. These forests have a hydroperiod of 4–6 months. Typically, annual rainfall in these areas is around 400 mm. The tree is thought to transpire about 1100 mm/year if the water is available. The annual flooding provides the balance of water between the annual rainfall and the trees' needs. If this does not occur, then the tree is reduced to a dryland state. When the forest floods, it 'comes alive' – the crowns dramatically expand and change colour from a greenish-yellow to a dark green, and the trees actively grow. This change is accompanied by major bird-breeding events and a vast increase in faunal activity (tortoises, kangaroos and echidnas are commonly encountered).

The forests have a long history of sustaining indigenous people. More latterly, the forests sustained a substantial timber industry supplying heavy timber used in construction. The higher flood frequency areas have traditionally been too wet to sustain river red gum. However, changes in the last six decades associated with irrigation water use have made these areas flood less frequently. The result is that river red gum is 'invading' these grass plains, changing the 'character' of the area, as discussed by Bren (1992) and Colloff (2014), and is viewed by many as a major and undesirable floral change.

River red gum is Australia's most widespread eucalypt and is found in both dry and wet sites. It is also widely planted around the world in industrial plantations. These take advantage of the ability of the tree to survive both prolonged flooding and prolonged drought. Extensive genetic work using Australian red gum as a 'base' (e.g. Butcher *et al.*, 2009) have shown that there is a wide genetic variation in the species, reflecting its provenance.

8.3.2 The swamp cypress forests of the Lower Mississippi River, USA

Bald cypress (*Taxodium distichum*) forests once dominated the swamps of the Lower Mississippi River, but today only small areas remain, of which the Atchafalaya Delta is a major area. In general, the tree occurs on seasonally inundated wetlands with flat topography and humid climates. The tree is deciduous (in winter) and has 'knees' – roots that protrude out of the water. These were once thought to intake oxygen for the roots under anaerobic conditions, but this has been shown not to be the case. They are often found in association with 'tupelo' (*Nyssa* spp.). Even a brief Internet perusal will show the value of the forests for a diversity of food stuffs and the distinctive Cajun cooking style based on these. The misty forests with towering, shapely trees festooned with moss also attract artists and photographers.

The Atchafalaya Delta is an alternative course of the Mississippi River and includes large areas of flooded forest (see Piazza, 2014, for a full account). The area is a maze of forested wetlands (and other vegetation) growing on ever-changing deposition and erosion features. The main course of this river is of huge importance for navigation and water supply, and the high energy of this system combined with engineering projects to maintain non-forest values have created a huge diversity of habitats (and many management conundrums). Much effort has been put into stopping main river flows from passing down the Atchafalaya River to the sea; this makes a definition of the 'natural flow regime' a tenuous task. Threats include salinity from channel construction to the ocean, channel instability (and the possibility of the main river spontaneously 'jumping' to an alternative course to the ocean), diversions of river flow to ameliorate flooding, and penetration of sea water into freshwater environments associated with oil drilling.

The flooding serves to exclude invasive species and create a 'unique environment'. Unlike the red gum forests above, the rainfall is high enough for the trees to thrive without flooding. It is almost impossible to overstate the importance of these 'forests'.

8.3.3 The willow forests of the Danube River, Europe

The Danube River has a length of 800 km and crosses 15 countries from its origins in southern Germany to its confluence with the Black Sea (Hein *et al.*, 2016). The forests have a rich diversity, including willows, poplars, oak, ash, hornbeam, lime, elms and maple. The forests are well suited to the Danube's frequent flooding, with flexible branches and roots that help stabilize riverbanks and withstand strong winds and strong current. The flood plain of the Danube is dynamic, with sediment deposition establishing new areas. The stages of this development series lead from herbaceous vegetation on freshly created pioneer areas to a phase with willow bushes to the establishment of the first wetland forest trees and the so-called 'softwood riparian forest'.

Mölder and Schneider (2011) examined the composition of Danubian hardwood flood-plain forests and found that the distribution of forest types was quite stable, but there was much greater diversity in the understorey herb-layer vegetation along the river. They concluded that there was a 'remarkable congruence between the results of their flood-plain vegetation analysis and the longitudinal river eutrophication patterns'. Thus, with increasing distance downstream, the river had higher nutrient concentrations, and this promoted the growth of taller forbs, which outcompeted other plant species. They concluded that anthropogenic eutrophication is a factor in the growth of the Danubian hardwood flood-plain forests.

As in the two previous examples, the ability of the channel to carry navigation is of national importance. Practically, this means many series of locks and dams, and the concept of a 'natural river' has not existed for centuries on this river. The meshing of forest-management needs with river-management needs across international boundaries is a fine balancing act with a strong emphasis on the ability to compromise.

8.4 Common Management Issues of Flooded Forests

8.4.1 Are flooded forests homogeneous ecosystems?

'Flooded forests' may be large areas (>1000 km^2) encompassing a wide range of flooding frequencies and durations. Many studies (e.g. Correa-Araneda *et al.*, 2012) have shown that, given time, the vegetation will order itself to adapt to the diverse environments offered. Thus, some areas may be permanently flooded wetlands; it is unlikely that these will be forested; rather, the flora will be specialized wetland plants (particularly reeds, rushes and some grasses). As the flooding frequency or duration decreases, there will be a range of habitats, which will be occupied by trees or forests. The higher end will be occupied by specialist flood-tolerant species. The lower flood frequencies (often higher in the topography) commonly have trees that can survive a few months of flooding. These may grade into species that can tolerate occasional flooding. Using the term 'flooded forest' usually glosses over many types of vegetation communities (not always forests) found within the forest. There will be a similar gradation of fauna within the forests. Shallow flooding allows animals to splash through the flood waters (e.g. kangaroos, deer), but deeper flooding favours specialized swimming animals (e.g. beavers, otters) or animals that can live in and move through the tree crowns.

8.4.2 River development and flood modification

Forested wetlands along rivers exist in an intimate relationship with the flow of the river. High flows have the energy to push freshwater to distant parts of the forest. In contrast, low flows can usually only inundate areas in direct contact with the river. Rivers are valuable state and national resources. Modifications include the construction of large dams and reservoirs for water supply (domestic and irrigation) and to maintain river flows to allow irrigation and river navigation, and the removal of high flow peaks to avoid flooding of towns and farms. In general, these reduce the peak flows of the river and increase the low flows (in particular, this is often a major aim of the structures). As well as this, the dams will often allow settlement of sediment entering them so that forests are no longer the recipient of the sediment and nutrients that they once were. In some such cases, the water entering the forests may pick up sediment rather than deposit it. The seasonality of river flows may be changed dramatically to meet irrigation requirements.

Such changes have been well documented in the scientific literature (e.g. Day *et al.*, 2012; King and Keim, 2019). Quantification of the changes is an excellent first step, but a more difficult issue is stopping or reversing such changes, as large economic and developmental forces may be at work. Often, large dams and water resource developments can be symbols of national pride and development, and the possibility of a future ecological change in a remote flood-plain forest will not stop the surge of national pride invoked by the developments. Usually, the change involves collection of inter-related issues. Bargaining points may include the magnitude and timing of flows (within the limits set by the structures), the provision of pumping to get water to distant parts of the forest (expensive but sometimes used), occasional provision of additional water to 'peak' smaller flood freshets, and occasional lowering of river water levels to allow topographically lower forest areas to 'dry out'. In the last few decades, forest managers have become acquainted with the quantitative hydrology of their flooded forests and the economics of river management; this makes them much more potent forces in negotiating better outcomes.

The use of rivers as state or national boundaries gives rise to an unusual set of issues. In the Barmah–Millewa river red gum forests of Australia, the south bank of the River Murray is defined as the boundary between the states of Victoria and New South Wales, with the river flowing more or less through the middle of this large, semi-natural area. A practical consequence is that any 'diversion' of the river water to one side must be matched by a diversion to the other side to maintain parity between the states. The river flows through a (currently but not historically) stable channel in a braided stream network; a river avulsion in a large flow would presumably make a sudden, large change to the boundaries of the two states, with many issues (ecological, economic and social) possibly entailed.

8.4.3 Coming to grips with the hydrology of flooded forests

To deal properly with management needs and/or threats to a flooded forest ideally requires some years of quantifying the hydrology. This is a mixture of gaining experience, assembling available hydrographic data, collecting your own hydrographic and forest data, and somehow assembling the information to give a coherent 'model' of the forest. Each of these can be a difficult task.

The first task is quantifying the source of the water. An excellent first step is mapping the extent of flooding in the forest at a given date. In the past, this was done by observers and, over a period of years, a catalogue of flooding information can be built up (see Bren *et al.*, 1988, for an example of an analysis of 34 maps containing annual and intermediate flood levels). The forest maps should be tied to discrete annual hydrographs if they are to be used in statistical modelling.

Fig. 8.2 is derived from Bren *et al.* (1988) for the Barmah river red gum forests along the River Murray in Australia and shows the percentage of the forest flooded as a function of mean monthly flow in the river (expressed in daily units) during annual periods of active flooding. Among other things, this shows that the forest can be viewed as a 'finite container' – flows of about 90,000 MegaLitres/day effectively 'wet' the forest completely. Higher flows do not increase the areal extent of flooding but do increase the depth. This is effectively a 'static model' but was useful in quantifying what changes in river flow levels did mean in annual forest flooding. These showed that river regulation associated with irrigation had decreased the size of forest flooding overall but had also increased the month-to-month variability of flooding. The maintenance of summer-time irrigation flows at a time when the River Murray would have very little flow meant that some areas close to the river were not getting a 'drying period', thereby changing their ecology.

Multi-temporal satellite data and use of personal aerial observation devices has allowed improved mapping of flood dynamics and special distribution of vegetation more effectively (e.g. Martinez and Le Toan, 2007). By combining water level reports with maps of different flood stages, the flooding patterns can be discerned, along with the vegetation succession processes. Martinez and Le Toan (2007) concluded that 'the spatial distribution of vegetation communities is governed by flood stress and can be modelled as a function of the mean annual exposure to floods'.

Ideally, a dynamic model of forest flooding would involve quantification of the following:

1. The relationship between stage height in the water source and flow in the source.
2. The hydraulic properties of points of entry and exit of water into the forests. Most flooded forests have many of these. In some cases, there are 'water gates', levees or other

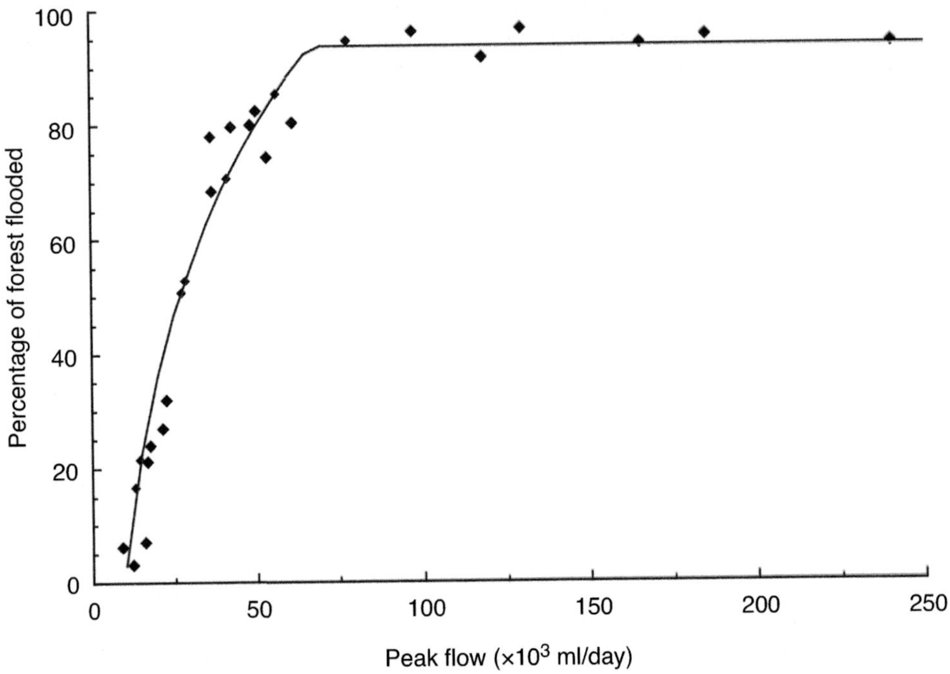

Fig. 8.2. Percentage of the Barmah Forest flooded as a function of the peak daily flow during the flooding period. Modified from Bren *et al.* (1988). The unit ml refers to a million litres.

forms of flow control on water entry and exit. Having such water gates set correctly and controlling them (often against locals who 'know better') can be a formidable task.

3. The height of the forest floor relative to the source of water and its variation across the extent of the forest. Modern LiDAR (light detection and ranging) techniques (e.g. Chen *et al.*, 2020) have made this feasible.
4. Hydraulic roughness of flow paths and its variation with water levels at different points of the forest. This is particularly important as flooded forests are not homogeneous entities. For many wetland communities, the hydraulic roughness decreases with increasing depth. The flow paths can be a chaotic maze in a natural flooded forest (Fig. 8.3).
5. Agreement between flooding data and the model. Assessment of this is easier said than done – not the least is the difficulty of movement in a flooded forest.

Such data can provide parameters for a large, spatially based hydraulic model (commonly based on cellular divisions of the flooded area). Once such a model is developed, this can become a trusted tool, allowing examinations of 'what-ifs' and interpretation of the impact on the forest caused by changes in river flow over long periods, in terms of forest growth and biodiversity. Such modelling is good practice in dealing with urban flooding, so there is no reason why it should not be feasible in forest situations. It is, however, a long and expensive exercise in quantification, requires high levels of analytical and modelling skills (and excellent equipment), and dedicated (and long-serving) staff to bring to fruition. The first task is to quantify the impacts of management actions on the hydrological behaviour of the forest. The next step – assessing the impacts on the biological resource – is another demanding area.

At the time of writing, such an approach is feasible in that commercial software capable of modelling is available, LiDAR measurements of floodplains and vegetation is often a routine operation, and there is growing experience in complex flood modelling (particularly in dealing

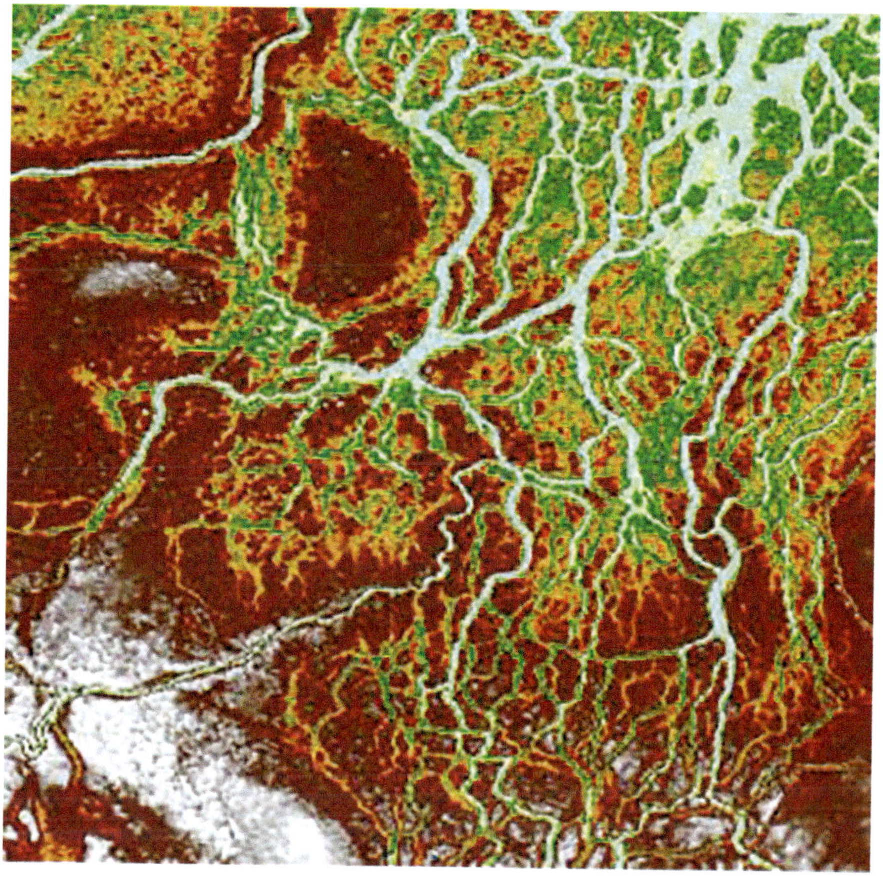

Fig. 8.3. A chaotic maze of flow paths is shown by a LiDAR image in a river red gum forest in Australia. The point of inflow to the forest is at the lower left-hand corner. Image courtesy of Ben Tate of Water Technology Pty Ltd and the Department of Environment and Primary Industry (Victoria, Australia).

with urban flooding). Putting these together could be done by a large, well-resourced forest management operation. This has been viewed as a desirable aim for many years.

The model being espoused here is one in which the forest manager can define the impact of given levels of river water in the associated flow system on their forest, in both the long and short term. As forest control and water control are not usually vested in one organization, the forest manager can approach the water manager with quantitative suggestions on water issues. Ideally, the two organizations would work together to overcome problems for the good of the public. Both will, of course, have many other constraints in achieving some sort of compromise.

8.4.4 Blackwater events

Blackwater is water with a high concentration of dissolved organic carbon (Kerr *et al.*, 2013); characteristically, this gives the water a black appearance, hence the name. A blackwater event is when this water is discharged from a flooded forest into the river mainstream. These commonly occur in lowland flood plains, during flood events. This water may become hypoxic, particularly when combined with high temperatures. This leads to lethal and sublethal effects on aquatic biota. A major tributary of the Amazon River flowing out of the Amazon forest is named the Rio Negro because of the blackness of its water.

The source of the organic carbon is leachate from surface plant litter. This represents an abundant food source for aquatic bacteria. When carbon loads are high, when flooding occurs during periods of high temperature or when floodwaters are pooled for extended periods, the metabolic activity of these bacteria may consume dissolved oxygen at a faster rate than it can be replenished. This results in hypoxia, which may be detrimental to aquatic species. Hypoxic blackwater events occur periodically in streams that are normally well oxygenated and have low to moderate concentrations of dissolved organic carbon (typically <10 mg/L). River regulation may limit forest flooding and lead to an accumulation of organic matter so that when blackwater events occur, they may be more serious than in a less regulated system.

Blackwater events have been well reported in the literature, including in the Atchafalaya River in Louisiana (Fontenot *et al.*, 2001), the Rio Grande catchment of New Mexico (Valett *et al.*, 2005) and Australia (Howitt *et al.*, 2007; Hladyz *et al.*, 2011; Baldwin, 2021). Each of these discharge events was associated with inundation of the forested floodplains. Between flood events, forest litter accumulated on the flood plain and in dry channels. When inundated, carbon compounds were leached from this decomposing litter, contributing dissolved organic carbon to the water column; Kerr *et al.* (2013) has given a fuller description of the various processes involved.

Ultimately, at least some of this water returns to the main channel, particularly in the initial flush of the flooding event. This can lead to death of species less able to tolerate low oxygen levels. In the Australian case, it is quite common for River Murray crayfish (*Euastacus armatus*) to emerge (typically by climbing trees) to avoid the water for substantial periods; this makes them vulnerable to desiccation, exploitation and predation (King *et al.*, 2012). In the the Atchafalaya Delta in the USA, Piazza (2014) showed red swamp crawfish using the same strategy (this time on water hyacinth) to escape anoxic water. A 2023 survey of River Murray crayfish populations (Raymond *et al.*, 2023) suggested a slow recovery from a blackwater episode in 2010. Such events often include difficult judgements as to whether it is a 'natural' event or, somehow, exacerbated by river management. Neville *et al.* (2021) looked at the dynamics of a blackwater stream in North Carolina and concluded that their results highlighted the potential for extreme flooding events to impact biogeochemical processes in large rivers long after flood waters subside. The association of blackwater events and higher temperatures does suggest some management pathways for avoiding these in well-regulated systems. In some forests with a prolonged 'dry period' (e.g. river red gum), this may include fuel-reduction burning to reduce the organic litter load.

8.5 Intertidal Forests

Tidal variation makes ocean levels fluctuate on a bi-diurnal basis, with the amplitude varying due to other variables. Low-lying forests are exposed to the flow of water, either directly by the tidal inflow and outflow, or by pressure changes in groundwater. Sometimes, in sandy areas such as islands, an increase in tidal levels will cause a rise in fresh groundwater levels due to its floating as a Ghyben–Herzberg lens on the underlying more saline (and denser) seawater (e.g. Tribble, 2008).

8.5.1 Intertidal freshwater forests

Figure 8.4 shows an example of such a forest from a low-lying area near Charleston, South Carolina (where such forests are common). Typically, they are low-lying areas close to the coast with a maze of waterways allowing flow in and out between the forest and coastal areas. A range of hardwood and softwood species more usually found in upland forests occur. These have some ability to tolerate salinity (for short periods at least). Williams *et al.* (2019) noted that these forests have been understudied. In general, these forests occur on areas where freshwater is passing to the sea. Commonly, high tides will restrict the outflow of freshwater to the sea (due to high sea levels), causing increased inundation by the freshwater outflow. However, occasional high tidal surges (particularly when accompanied by low freshwater outflow) allow

Fig. 8.4. Occasional flooding in the Francis Marion National Forest near Charleston, South Carolina. Usually, these areas are not flooded, but this was associated with cyclonic heavy rainfall in a coastal area. Photograph courtesy of Rick Wrenn (retired, USDA Forest Service).

the penetration of ocean water upstream. Thus, the forests must have some ability to tolerate saline ocean water if they are to survive. Odum (1988) showed that these forests occur at salinities below 0.5 parts per thousand but suggested that salinities of 2 parts per thousand may lead to vegetation change. Hydrology within these forests can be difficult to interpret because of river discharge, tidal stage and bidirectional flow, local precipitation, evapotranspiration, groundwater and prevailing winds (Anderson and Lockaby, 2011).

8.5.2 Freshwater hydrology of mangrove forests: a mystery?

Intertidal mangrove forests occur along tropical, subtropical and some temperate coasts, and often overlap with high and increasing densities of human populations. Friess (2025) noted that mangrove forests show 'amazing diversity in structure and process' across scales ranging from local to global. This diversity includes the definition of 'What is a mangrove?' They provide important ecosystem services including fish, timber, fuelwood, coastal protection, pollution control and cultural values. In some cases, they are associated with peat formation, but more commonly they lead to land being reclaimed from the sea. Drivers such as aquaculture, agriculture and urban development have sometimes led to mangrove deforestation. Mangroves are further threatened by resource overextraction. Mangrove forests are particularly impacted by changes such as relative sea-level rise and fluctuations in sea level linked to climate oscillations (Friess *et al.*, 2019). Friess (2025) argued that research into mangrove forests has a geographical bias that avoids lower-income countries where these forests commonly occur, and argues that this distorts our 'science'. He also noted that high-ranking scientific journals sometimes reject papers on mangroves as 'too peculiar' (or other reasons unrelated to their scientific merit).

The term 'mangrove' encompasses about 80 different species of which *Avicennia* and *Rhizophora* spp. are the best known. There is a wide variation in tree size, salinity tolerance and requirements within this group (Gopal and Chauhan, 2018; Beniwal *et al.*, 2024). Mangroves grow in areas with low-oxygen soil where slow-moving waters allow fine sediment to accumulate. The trees develop dense tangles of prop roots that sometimes make the trees appear to be standing on stilts above the water. The tangle of roots (sometimes called a 'mangal') allows the trees to handle the bi-diurnal rise and fall of tides. The roots slow the movement of tidal waters, causing sediment to settle out of the water and build up the muddy bottom, and providing habitat for fish and other organisms.

Radabaugh *et al.* (2021) showed that mangroves experience stress or mortality when gas exchange by aerial root structures is hindered by sediment burial or constant inundation. The resulting loss of below-ground roots and increased decomposition can lead to a loss of surface elevation through peat collapse, furthering the degree of inundation. Signs of forest stress include pools of stagnant water (often with their own colour, distinctive biota and very low oxygen levels) and abundant adventitious root growth.

The role of freshwater hydrology in the survival of mangroves is not well understood but is known to be important in both mitigating salinity extremes and in providing nutrients to the growing plants. Particularly in estuarine environments, a reduction in freshwater inflow to the estuary may result in decline or death of 'freshwater' mangroves species less adopted to salinity. Wang *et al.* (2011) suggested that mangroves are obligate halophytes, basing this conclusion on the observations that most mangroves cannot survive indefinitely in freshwater. Studies such as that of Peng *et al.* (2023) have examined the role of both groundwater and surface mixing of fresh runoff with tidal water. This study did not support the model of the mangrove forest 'hanging on' by being dependent on a flow of freshwater in an otherwise saline environment, although there was some mixing in both the freshwater and the groundwater. A common hypothesis is that mangal roots are often located in (relatively) fresh, upwelling groundwater from the adjacent land, and that this mitigates extremes of salinity. This hypothesis appears hard to either prove or disprove.

Gopal (2014) noted the sensitivity of many mangroves to reductions in freshwater flowing from land into estuaries, but also noted the lack of detailed studies on the matter. Such results suggest that a chapter such as this can only make generalizations on the role of fresh water in the survival of mangroves. It may well be possible to manage onshore land uses in such a way that, for instance, groundwater flows from the land into the root zones of mangroves are optimized, but we do not have the ability to make refined predictions or to do the arithmetic of production to look for economic trade-offs. The area is ripe for research.

In the tropics a variety of mangrove species are able to grow along the gradient that is described for temperate tidal wetlands. Mangroves may be able to compensate for sea-level rises by capturing fine sediments by production of extensive aerial root systems (Stringer *et al.*, 2016). These root systems can also contribute to carbon assimilation.

Hydrology is probably the single most important determinant of ecological processes in wetlands (Gosselink and Turner, 1978). There are few quantitative water budgets for mangroves, which limits our understanding of how hydrology controls ecological processes of mangrove ecosystems (Twilley and Chen, 1998). In their study in Florida, these authors used a finite-difference hydrology model (HYMAN). They reported that the relative significance of tides and rainfall deficits (rainfall < evapotranspiration) on the seasonal patterns of soil saturation (water levels) and soil salinity, which at higher elevations in the intertidal zone, where tidal inundation frequency is reduced, is more sensitive to changes in rainfall deficit.

Similarly, Dai *et al.* (2018) highlighted the value of a tool to assess carbon dynamics in mangroves, especially for regional or large mangrove forests, and estimated their role in mitigating climate change because of their high carbon density, the threats to their integrity from land-use change and sea-level rises, and functional linkages of the many goods and services. Based on their modelling analysis, the authors reported that the responses of gross and net primary productivity and above-ground biomass to alterations of mean daily temperature were quadratic, or increasing or decreasing non-linearly with an increment or decrement in mean daily temperature, but leaf production was linear. Similarly, other mangrove carbon components, such as dissolved organic carbon, particulate organic carbon, dissolved inorganic carbon and black carbon, respond substantially to variations of the ecological drivers.

8.6 Contribution of Forest Wetlands to Climate Change

8.6.1 Wetlands and stored carbon

Wetlands are a major repository of stored carbon representing one of world's largest carbon sinks (Poulter *et al.*, 2022). Wetlands are also the largest and most uncertain biological source of atmospheric methane, with hydrological fluctuations exacerbating this uncertainty (Cui *et al.*, 2024). Water-table level controls of biogeochemical cycles influence emissions of greenhouse gases such as methane (CH_4), carbon dioxide (CO_2) and nitrogen oxides (NOx), which can have an influence on regional and global climate (Paschalis *et al.*, 2017). Wetland hydrology (which reflects the local water balance) is directly affected by climatic change and variability (Zhu *et al.*, 2017).

Wetland forest hydrology can be an important driver of climate impacts. Zhang *et al.* (2019) aimed to improve the approach for quantifying the hydrological resilience of wetland ecosystems to climate variability and climate change by developing metrics including the variations of the groundwater table, overland flow and saltwater table.

Wetland soils below the water table have little aerobic decomposition so that even small productivity gains result in carbon sequestration for long periods. However, unless the soil is frozen, prolonged anaerobic decomposition depletes the soil of alternative electron acceptors and carbon is reduced to CH_4, a potent greenhouse gas. As drainage increases, the hydroperiod decreases, and the rates of productivity and decomposition increase (Fig. 8.5). However, as drainage increases, aerobic decomposition will spread to the older peat deposits, and decomposition can increase beyond productivity. Wetland management in the past has not considered climatic effects but has concentrated only on increased plant production. This focus is changing in the 21st century with recognition of the ecological values of wetlands and the challenges of climate change. In the future, the goal will be to maintain wetland hydroperiods in the zone

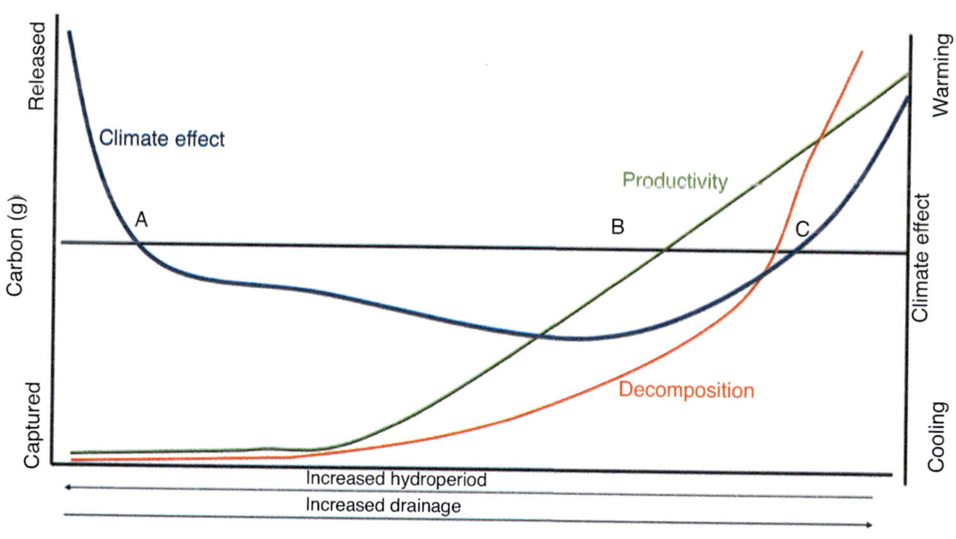

Fig. 8.5. Idealized reaction of forested wetland carbon sequestration and the climate warming effect of varying degrees of soil saturation. The horizontal line is carbon neutral. Point A is saturation to the extent that methanogenesis occurs. Point C is soils drained to the point to promote peat oxidation. Point B is an ideal wetland hydrology that promotes maximum growth but does not promote peat oxidation.

between A and C of Fig. 8.5. Points A, B and C are dependent on, at least, temperature, species, concentrations of alternative electron acceptors (e.g. oxygen, nitrogen, iron and manganese) and the many factors that determine the rate and quantity of these receptors, and the rate and quality of water flowing through the system. Northern-hemisphere peatlands and coastal wetlands both represent the systems facing the most critical climatic and land-use change challenges. Sea-level rises due to climatic change have added new stress to coastal wetlands (Aguilos *et al.*, 2021).

8.6.2 Northern bogs and fens

In the far northern regions of the planet, the interaction of climate warming, permafrost melt and methane production pose the threat of rapid global warming. In northern Canada, melting permafrost is creating a new landscape, by creating meltwater fens that can coalesce into drainage channels (Haynes *et al.*, 2023). The interplay of forest, snow accumulations, depth of the active layer, heat transfers in meltwater and degree of refreezing is an area of active and important research. In Canadian discontinuous permafrost areas, permafrost thaw and forest cover loss were found to be related to forest plateau edge area. As the perimeter:area ratio increased (i.e. areas became less circular and more irregular), the thaw of permafrost forest plateaus increased (Baltzer *et al.*, 2014). The decrease in bogs and increase in fen has resulted in enhanced CH_4 release (AminiTabrizi *et al.*, 2020).

South of the permafrost regions, vast areas of peatland forest exist across most of the northern hemisphere. Northern peatland forests are characterized by low evapotranspiration and extended periods of soil saturation with very wet conditions in the absence of adequate natural drainage (Skaggs *et al.*, 2016). During the 20th century, drainage of peatland for forestry was commonly practised. Although the practice was widespread, it was most extensive in Finland and Russia. The objectives of drainage were for accessibility and enhanced growth of merchantable timber. Paavilainen and Päivänen (1995) presented a detailed review of the history, methods and results of forest drainage of peatlands. Although preservation of soil organic matter was a goal of Nordic silviculture, excessive drainage has resulted in carbon release and soil subsidence, including decreased pH (Minkkinen *et al.*, 2008), decomposition rates of soil organic matter (Domisch *et al.*, 2000) and soil carbon stock (Minkkinen and Laine, 1998; Laiho, 2006). Peat oxidation and carbon release are more likely in peatlands cleared for agriculture due to a greater need for access to the fields with annual crops. Peatlands are also responsible for the majority of CH_4 emissions from wetlands globally (Wang *et al.*, 2021). Hydrological changes induced by climatic and anthropogenic disturbance may substantially alter CH_4 emission in peatlands. Wang *et al.* (2021) showed that restoration through a non-inundated rewetting would not stimulate CH_4 emission in drained/degraded low-latitude shrub bogs such as the pocosin wetlands (mainly *Pinus serotina*) found in the Carolinas area of the USA.

In the 21st century, new drainage initiatives are limited and there are efforts to restore peatlands by plugging drainage ditches. Climate-smart restoration could include controlled drainage to achieve conditions between point A and C of Fig. 8.5, with permanent sequestration favoured near point A and maximum carbon capture approaching point C.

8.6.3 Climate change and tidal freshwater forested wetlands

Alluvial rivers eventually lead to the coast where the river is subject to ocean tides. The tidal influence results in a progression of wetlands as hydroperiods and salinity vary along the gradient from river flood plain to estuary fringe. As sea-level rises due to changing climate, these wetlands will be subject to stress and migration (Morris and Sundberg, 2024).

The hydraulics of the tidal river creates various zones where river flow and tidal momentum interact. The vegetation and ecology of these zones are determined by river flows and the tide. These vegetation zones will change as climate change alters river flows and enhances sea-level rises (Williams *et al.*, 2019). Currently, the zones reflect the slow rise of sea level since the early Holocene. Rising sea level will cause the zones of tidal influence in the river to move inland and shift species by mortality of the least flood-intolerant species, and later the least salt-tolerant species. In the upper reaches of this zone, the topography will become hummock and hammock with root mortality resulting in subsidence of the hammocks (Krauss *et al.*, 2023). As this zone advances, the hummocks will be filled as tidal sedimentation covers the alluvial river topography. While the forest species die from inundation and salt stress, the system will be a source of carbon to the atmosphere as both CO_2 and CH_4. Methane production will increase as the hydroperiod of the decomposing forest is prolonged by tidal subsidy. The site will cease to be a source of CO_2 and CH_4 as marsh grass productivity increases and seawater sulfate becomes available as an electron receptor for anaerobic decomposers (Helton *et al.*, 2019).

8.7 Conclusions

Forested wetlands are found throughout the world and in a variety of landscape positions. They are valued for their unique biology and their distinctive environments. As well as directly providing habitation, they provide various ecosystem services including water quality improvement, groundwater recharge to wildlife habitat and carbon sequestration.

Forested wetlands can be classified by the source of the water that causes a surplus of inflow over loss. Precipitation may be the only cause of water surplus for sites on nearly level slopes. These sites are characterized by poor nutrition and slow tree growth. Organic matter may accumulate on the surface due to slow decomposition. Drainage of these sites to improve plantation tree growth has been common around the world.

Sites where the inflow surplus is due to groundwater additions to precipitation are generally located on concave landscape positions. The length of flooding generally depends on the storage characteristics of the aquifer(s) feeding the site and ranges from small, short-lived contributions to aquifers with small storage to nearly constant saturation when fed by aquifers with large storage. The most common location occurs at the abrupt change in slope from hillside to valley bottom.

The third common source of water is from contributions ('subsidy') from adjacent rivers. In the common case, water overflows from the river into the flood-plain forest. This water than infiltrates, evaporates or makes its way back to the river at a lower topographical position. Such areas can be large and important. Because of the direct linkage between these areas and the adjacent river, they are heavily influenced by river-management projects, which commonly reduce inflows (decreasing the hydroperiod) but in some cases increase the hydroperiod.

An important but not fully understood class of forested wetlands is associated with forests found in direct association with the ocean – sometimes along coastlines or in the lower portions of estuaries. These are usually subjected to tidal forcing, as well as other hydrological inputs. It is well known that they depend to some extent on freshwater contributions from the land to ameliorate the ocean salinity, but the exact mechanism by which this is achieved is not fully understood.

References

Aguilos, M., Sun, G., Noormets, A., Domec, J.C., McNulty, S. *et al.* (2021) Effects of land-use change and drought on decadal evapotranspiration and water balance of natural and managed forested wetlands along the southeastern US lower coastal plain. *Agricultural and Forest Meteorology* 303, 108381.

Amatya, D.M., Williams, T.M., Skaggs, RW., Nettles, J.E. and Mehan, S. (2024) Silvicultural practices and water table dynamics of coastal forested wetlands. *Journal of Natural Resources and Agricultural Ecosystems* 2, 191–211.

AminiTabrizi, R., Wilson, R.M., Fudyma, J.D., Hodgkins, S.B., Heyman, H.M. *et al*. (2020) Controls on soil organic matter degradation and subsequent greenhouse gas emissions across a permafrost thaw gradient in Northern Sweden. *Frontiers of Earth Science* 8, 557961.

Anderson, C.J. and Lockaby, B.G. (2011) Forested wetland communities as indicators of tidal influence along the Apalachicola River. *Wetlands* 31, 895–906.

Baldwin, D.S. (2021) Water quality in the Murray–Darling Basin: The potential impacts of climate change. In: Hart, B.T., Bond, N.R., Byron, N., Pollino, C.A. and Stewardson, M.J. (eds) *Murray-Darling Basin, Australia: Its Future Management*. Elsevier, Amsterdam, pp. 137–159.

Baltzer, J.L., Veness, T., Chasmer, L.E., Sniderhan, A.E. and Quinton, W.L. (2014) Forests on thawing permafrost: Fragmentation, edge effects, and net forest loss. *Global Change Biology* 20, 824–834.

Beniwal, D., Dhull, S.S., Gulia, V. and Rani, J. (2024) Avicennia: A mangrove genus unveiled through its phytochemistry, pharmacological, and ecological importance. *Rendiconti Lincei. Scienze Fisiche e Naturali* 35, 907–929.

Bren, L.J., O'Neill, I.C. and Gibbs, N.L. (1988) Use of map analysis to elucidate flooding in an Australian Riparian River red gum forest. *Water Resources Research* 24, 1152–1162.

Bren, L.J. (1992) Tree invasion of an intermittent wetland in relation to changes in the flooding frequency of the River Murray, Australia. *Australian Journal of Ecology* 17, 395–408.

Brinson, M.M. (1993) *A Hydrogeomorphic Classification for Wetlands*. US Army Engineer Waterways Experiment Station, Vicksburg, Mississippi.

Butcher, P.A., McDonald, M.W. and Bell, J.C. (2009) Congruence between environmental parameters, morphology and genetic structure in Australia's most widely distributed eucalypt, *Eucalyptus camaldulensis*. *Tree Genetics & Genomes* 5, 189–210.

Chen, Y.H., Mossa, J. and Singh, K.K. (2020) Floodplain response to varied flows in a large coastal plain river. *Geomorphology* 354, 107035.

Colloff, M. (2014) *Flooded Forest and Desert Creek: Ecology and History of the River Red Gum*. CSIRO Publishing, Canberra.

Correa-Araneda, F.J., Urrutia, J., Soto-Mora, Y., Figueroa, R. and Hauenstein, E. (2012) Effects of the hydroperiod on the vegetative and community structure of freshwater forested wetlands, Chile. *Journal of Freshwater Ecology* 27, 459–470.

Cui, S., Liu, P., Guo, H., Kalla Nielsen, C., Pullens, J.W. *et al*. (2024) Wetland hydrological dynamics and methane emissions. *Communications Earth and Environment* 5, 470.

Dai, Z., Trettin, C.C., Frolking, C. and Birdsey, R.A. (2018) Mangrove carbon assessment tool: Model development and sensitivity analysis. *Estuarine, Coastal, and Shelf Science* 208, 23–25.

Day, J., Hunter, R., Keim, R.F., DeLaune, R., Shaffer, G. *et al*. (2012) Ecological response of forested wetlands with and without large-scale Mississippi River input: Implications for management. *Ecological Engineering* 46, 57–67.

Domisch, T., Finer, L., Laiho, R., Karsisto, M. and Laine, J. (2000) Decomposition of scots pine litter and the fate of released carbon in pristine and drained pine mires. *Soil Biology and Biochemistry* 32, 1571–1580.

Durigan, G., Munhoz, C.B., Zakia, M.J.B., Oliveira, R.S., Pilon, N. *et al*. (2022) Cerrado wetlands: Multiple ecosystems deserving legal protection as a unique and irreplaceable treasure. *Perspectives in Ecology and Conservation* 20, 185–196.

Fontenot, Q.C., Rutherford, D.A. and Kelso, W.E. (2001) Effects of environmental hypoxia associated with the annual flood pulse on the distribution of larval sunfish and shad in the Atchafalaya River Basin, Louisiana. *Transactions of the American Fisheries Society* 130, 107–116.

Friess, D.A. (2025) The diversity of mangrove forests and geographical biases in their research. *Bulletin of Marine Science* 101. DOI: 10.5343/bms.2024.0005.

Friess, D.A., Rogers, K., Lovelock, C.E., Krauss, K.W., Hamilton, S.E. *et al*. (2019) The state of the world's mangrove forests: Past, present, and future. *Annual Review of Environment and Resources* 44, 89–115.

Gopal, B. (2014) Mangroves are wetlands, not forests: Some implications for their management. In: Hanum-Faridah, I., Latiff, A., Hakeem, K.R. and Ozturl, M. (eds) *Mangrove Ecosystems of Asia: Status, Challenges and Management Strategies*. Springe, New York, pp. 439–453.

Gopal, B. and Malavika Chauhan, M. (2018) The transboundary Sundarbans mangroves (India and Bangladesh). In: Finlayson, C.M., Milton, G.R., Prentice, R.C. and Davidson, N.C. (eds) *The Wetland Book: II: Distribution, Description, and Conservation*. Springer, Dordrecht, Netherlands, pp. 1733–1742.

Gosselink, J.G. and Turner, R.E. (1978) The role of hydrology in freshwater wetland ecosystems. In: *Proceedings of the Symposium Freshwater Marshes: Present Status, Future Needs, February 1977, Rutgers University, New Brunswick, New Jersey*, pp. 63–78.

Haynes, K.M., Frederick, I., Disher, B., Carpino, O. and Quinton, W.L. (2023) Long-term trends in wetland event response with permafrost thaw-induced landscape transition and hummock development. *Ecohydrology* 16, e2515.

Hein, T., Schwarz, U., Habersack, H., Nichersu, I. and Preiner, S. (2016) Current status and restoration options for floodplains along the Danube River. *Science of the Total Environment* 543, 778–790.

Helton, A.M., Ardòn, M. and Bernhardt, E.S. (2019) Hydrologic context alters greenhouse gas feedbacks of coastal wetland salinization. *Ecosystems* 22, 1108–1125.

Hladyz, S., Watkins, S.C., Whitworth, K.L. and Baldwin, D.S. (2011) Flows and hypoxic blackwater events in managed ephemeral river channels. *Journal of Hydrology* 401, 117–125.

Howitt, J.A., Baldwin, D.S., Rees, G.N. and Williams, J.L. (2007) Modelling blackwater: Predicting water quality during flooding of lowland river forests. *Ecological Modelling* 203, 229–242.

Kerr, J.L., Baldwin, D.S. and Whitworth, K.L. (2013) Options for managing hypoxic blackwater events in river systems: A review. *Journal of Environmental Management* 114, 139–147.

King, A.J., Tonkin, Z. and Lieshcke, J. (2012) Short-term effects of a prolonged blackwater event on aquatic fauna in the Murray River, Australia: Considerations for future events. *Marine and Freshwater Research* 63, 576–586.

King, S.L. and Keim, R.F. (2019) Hydrologic modifications challenge bottomland hardwood forest management. *Journal of Forestry* 117, 504–514.

Kozlowski, T.T. (2002) Physiological–ecological impacts of flooding on riparian forest ecosystems. *Wetlands* 22, 550–561.

Krauss, K.W., Noe, G.B., Duberstein, J.A., Cormier, N., From, A.S. *et al.* (2023) Presence of hummock and hollow microtopography reflects shifting balances of shallow subsidence and root zone expansion along forested wetland river gradients. *Estuaries and Coasts* 47, 1750–1763.

Laiho, R. (2006) Decomposition in peatlands: Reconciling seemingly contrasting results on the impacts of lowered water levels. *Soil Biology and Biochemistry* 38, 2011–2024.

Lugo, A.E. (1988) The future of the forest: Ecosystem rehabilitation in the tropics. *Environment: Science and Policy for Sustainable Development* 30, 16–45.

Martinez, J. and Le Toan, T. (2007) Mapping of flood dynamics and spatial distribution of vegetation in the Amazon floodplain using multitemporal SAR data. *Remote Sensing of Environment* 108, 209–223.

Meixner, H., Schnauder, I., Bölscher, J. and Iordache, V. (2006) *Hydraulic, Sedimentological and Ecological Problems of Multifunctional Riparian Forest Management – RIPFOR. Guideline for Endusers*. Berliner Geographische Abhandlungen Heft 66, Berlin.

Minkkinen, K. and Laine, J. (1998) Long-term effect of forest drainage on the peat carbon stores of pine mires in Finland. *Canadian Journal of Forest Research* 28, 1267–1275.

Minkkinen, K., Byrne, K.A. and Trettin, C. (2008) Climate impacts of peatland forestry. In: Strack, M. (ed.) *Peatlands and Climate Change*. International Peatland Society, Jyväskylä, Finland, pp. 98–122.

Mitsch, W.J. and Gosselink, J.G. (2015) *Wetlands*, 5th edition. Wiley, Hoboken, New Jersey.

Mölder, A. and Schneider, E. (2011) On the beautiful diverse Danubian floodplain forest vegetation and flora under the influence of river eutrophication. *River Research and Applications* 27, 881–894.

Morris, J.T. and Sundberg, K. (2024) Responses of coastal wetlands to rising sea-level revisited: The importance of organic production. *Estuaries and Coasts* 47, 1735–1749.

Neville, J.A., Emanuel, R.R., Nichols, E.G. and Vose, J. (2021) Extreme flooding and nitrogen dynamics of a blackwater river. *Water Resources Research* 57, e2020WR029106.

Odum, W.E. (1988) Comparative ecology of tidal freshwater and salt marshes. *Annual Review of Ecological Systems* 19, 147–176.

Paavilainen, E. and Päivänen, J. (1995) *Peatland Forestry – Ecology and Principles*. Springer, New York.

Paschalis, A., Katul, G.G., Fatichi, S., Palmroth, S. and Way, D. (2017) On the variability of the ecosystem response to elevated atmospheric CO_2 across spatial and temporal scales at the duke forest FACE experiment. *Agricultural and Forest Meteorology* 232, 367–383.

Payette, S. and Delwaide, A. (2003) Shift of conifer boreal forest to lichen–heath parkland caused by successive stand disturbances. *Ecosystems* 6, 540–550.

Peng, K., Heiss, J.W., Xie, X., Yan, L., Deng, Y. *et al.* (2023) Groundwater discharge and saltwater-freshwater mixing in a mangrove wetland over tidal cycles: A field and modelling study. *Journal of Hydrology* 620, 129472.

Piazza, B.P. (2014) *The Atchafalaya River Basin: History and Ecology of an American Wetland*. Texas A&M University Press, College Station, Texas.

Poulter, B., Fluet-Chouinard, E., Hugelius, G., Koven, C., Fatoyinbo, L. *et al.* (2022) A review of global wetland carbon stocks and management challenges. In: Kraus, K.R., Zhu, Z. and Stagg, C.L. (eds) *Wetland Carbon and Environmental Management*. Wiley Hoboken, New Jersey, pp. 3–21.

Radabaugh, K.R., Dontis, E.E., Chappel, A.R., Russo, C.E. and Moyer, R.P. (2021) Early indicators of stress in mangrove forests with altered hydrology in Tampa Bay, Florida, USA. *Estuarine, Coastal and Shelf Science* 254, 107324.

Rawls, W.J., Brakensiek, D.L. and Saxtonn, K.E. (1982) Estimation of soil water properties. *Transactions of the ASAE* 25, 1316–1320.

Raymond, S., Duncan, M., Tonkin, Z. and Robinson, W. (2023) *Barmah-Millewa Fish and Crayfish Condition Monitoring (2007–2023)*. Report for the Murray–Darling Basin Authority. Arthur Rylah Institute for Environmental Research, Department of Energy, Environment, and Climate Action, Heidelberg, Australia.

Semeniuk, C.A. and Semeniuk, V. (1995) A geomorphic approach to global classification for inland wetlands. *Classification and Inventory of the World's Wetlands* 118, 103–124.

Simon, S., Déri-Takács, J., Szijártó, M., Szél, L. and Mádl-Szőnyi, J. (2023) Wetland management in recharge regions of regional groundwater flow systems with water shortage, Nyírség region, Hungary. *Water* 15, 3589.

Skaggs, R.W., Tian, S., Chescheir, G.M., Amatya, D.M. and Youssef, M.A. (2016) Forest drainage. In: Amatya, D.M., Williams, T.M., Bren, L. and Jong, C. (eds) *Forest Hydrology: Processes, Management and Assessment*. CAB International, Wallingford, UK, pp. 124–140.

Stringer, C.E., Trettin, C.C. and Zarnoch, S.J. (2016) Soil properties of mangroves in contrasting geomorphic settings within the Zambezi River Delta, Mozambique. *Wetlands Ecological Management* 24, 139–152.

Tiner, R.W. (2017) *Wetland Indicators: A Guide to Wetland Formation, Identification, Delineation, Classification, and Mapping*. CRC Press, Boca Raton, Florida.

Tribble, G. (2008) *Ground Water on Tropical Pacific Islands – Understanding a Vital Resource*. Series Number 1312. US Geological Survey, Reston, Virginia.

Twilley, R.R. and Chen, R. (1998) A water budget and hydrology model of a basin mangrove forest in Rookery Bay, Florida. *Marine and Freshwater Research* 49, 309–323.

Valett, H.M., Baker, M.A., Morrice, J.A., Crawford, C.S., Molles, M.C. *et al.* (2005) Biogeochemical and metabolic responses to the flood pulse in a semiarid floodplain. *Ecology* 86, 220–234.

van Deventer, H., Adams, J.B., Durand, J.F., Grobler, R., Grundling, P.L. *et al.* (2021) Conservation conundrum – red listing of subtropical-temperate coastal forested wetlands of South Africa. *Ecological Indicators* 130, 108077.

Waddington, J.M., Roulet, N.T. and Hill, A.R. (1993) Runoff mechanisms in a forested groundwater discharge wetland. *Journal of Hydrology* 147, 37–60.

Wang, H., Ho, M., Flanagan, N. and Richardson, C. (2021) The effects of hydrological management on methane emissions from southeastern shrub bogs of the USA. *Wetlands* 41, 87.

Wang, W., Yan, Z., You, S., Zhang, Y. and Chen, L. (2011) Mangroves: Obligate or facultative halophytes? A review. *Trees* 25, 953–963.

Williams, T.M., Amatya, D.M., Conner, W., Panda, S., Dong, C. *et al.* (2019) Tidal forested wetlands: Mechanisms, threats, and management tools. In: An, S. and Verhoeven, J.T.A. (eds) *Wetlands: Ecosystem Services, Restoration and Wise Use,* Ecological Studies 238. Springer, Cham, Switzerland, pp. 129–158.

Wösten, J.H.M., Clymans, E., Page, S.E., Rieley, J.O. and Limin, S.H. (2008) Peat–water interrelationships in a tropical peatland ecosystem in Southeast Asia. *Catena* 73, 212–224.

Zhang, Y., Li, W., Sun, G. and King, J.S. (2019) Coastal wetland resilience to climate variability: A hydrologic perspective. *Journal of Hydrology* 568, 275–284.

Zhu, J., Sun, G., Li, W., Zhang, Y., Miao, G. *et al.* (2017). Modeling the potential impacts of climate change on the water table level of selected forested wetlands in the southeastern United States. *Hydrology and Earth System Sciences* 21, 6289–6305.

9 Tropical Forest Hydrology

T. Kumagai[1,2*], H. Kanamori[3] and N.A. Chappell[4]
[1]University of Tokyo, Bunkyo-ku, Tokyo, Japan; [2]University of Hawai'i, Honolulu, Hawaii, USA; [3]Kobe Gakuin University, Kobe, Japan; [4]Lancaster Environment Centre, Lancaster University, Lancaster, UK

Abstract
The consequences of land-use and climate change in tropical forest regions are considered among the greatest environmental concerns for the survival of the human population today and include increases in stream flow and flood frequency, loss of soils, changes in nutrient fluxes, warmer and drier climates and concomitant hydrological ecosystem services at the local scale. To tackle all aspects of these hydrological problems, we have broad recognition and tacit acceptance that progress on these complex issues benefits from fundamental pieces of knowledge on all of the hydrological components in various types of tropical forests and synthesizing them in terms of the global implications. Therefore, this chapter's main objective is to provide a pan-tropical perspective on tropical forest hydrology.

9.1 Introduction

The objective of this chapter is to provide the state of knowledge on the characteristics of hydrological components of precipitation, evapotranspiration and streamflow generation in tropical forest regions by a review of tropical literature. This uses a pan-tropical perspective partly via the use of pan-tropical maps of the hydrological components. We will suggest research needs and strategies for studying tropical forest hydrology in a dramatically changing world.

The tropical domain has the largest proportion of the world's forests (45%; FAO, 2020). One may imagine that the tropical forest regions are generally characterized by higher incident radiative energy, a constantly higher temperature and a large amount of ecosystem water resources, enough to support their higher primary productivity and active water cycling. In reality, tropical forests play a significant role in the global carbon budget (e.g. Beer *et al.*, 2010, Pan *et al.*, 2011) and are a significant source of global hydrological fluxes, profoundly influencing both global and regional climates (Avissar and Werth, 2005; Spracklen *et al.*, 2012; Poveda *et al.*, 2014; Kanamori *et al.*, 2018).

The tropical forest regions can be roughly classified into two types based on seasonal variations in precipitation: tropical evergreen rainforest with a 'rainforest climate' and tropical seasonal forest with a 'monsoon climate' (Kumagai *et al.*, 2009). Each region can be classified into more climate and forest types (Walsh, 1996; Tanaka *et al.*, 2008). Tropical evergreen rainforests exist only in regions of ample water resources (Kumagai *et al.*, 2005; Kumagai and Porporato, 2012b). Tropical seasonal forests have to cope with seasonal

*Corresponding author: tomoomikumagai@gmail.com

 Forest Hydrology: Processes, Management and Assessment (eds Leon J. Bren *et al.*)
DOI: 10.1079/9781800625310.0009

droughts (Vourlitis *et al.*, 2008; Miyazawa *et al.*, 2014; Kumagai *et al.*, 2015). As the existence of these tropical forests achieves a delicate balance in terms of their ecosystem water resources, hydrological change in either region could result in significant impacts on ecological patterns and processes (Malhi *et al.*, 2009; Phillips *et al.*, 2009; Kumagai and Porporato, 2012a), in turn affecting feedbacks to the atmosphere (Meir *et al.*, 2006; Bonan, 2008; Kumagai *et al.*, 2013; van der Ent *et al.*, 2014; Takahashi *et al.*, 2017). Some studies have shown that a marked change in regional climate will occur first in the tropics, making tropical forest ecosystems particularly vulnerable in the future (Mora *et al.*, 2013).

Humans have been modifying tropical forest land cover for food and energy production and for developing tropical countries. Such modifications (i.e. land-use change) are being combined with climate change and should be anticipated to impact the regional hydro-climate as well as the local freshwater resources (Bruijnzeel, 1990; Giambelluca, 2002; van der Ent *et al.*, 2010; Wohl *et al.*, 2012; Kumagai *et al.*, 2013; Takahashi *et al.*, 2017). What distinguishes the current modifications is that the entire hydro-climate is now subject to these modifications. The consequences of land-use and climate change in tropical forest regions are considered among the greatest environmental concerns for the survival of the human population today and include changes in streamflow and flood frequency, loss of productive soils, changes in nutrient fluxes, a warmer and drier climate, and concomitant changes in hydrological ecosystem services at the local scale (Bruijnzeel, 2004; Chappell *et al.*, 2004; Krusche *et al.*, 2011; Lawrence and Vandecar, 2015; Gómez *et al.*, 2023). Progress on these complex issues benefits from fundamental knowledge of all of the hydrological components in various types of tropical forests and synthesizing them globally (Bruijnzeel, 2004).

9.2 Pan-Tropical Climatic Regime and Forest Type

Pan-tropic maps of precipitation characteristics are derived from the NOAA's PRECipitation REConstruction over Land (PREC/L) long-term gridded precipitation data set (Fig. 9.1) (Chen *et al.*, 2002). High radiative energy received around the equator generates the updraft of hot and humid air. Convergence of this hot and humid air (i.e. the low atmospheric pressure system) results in the belt of precipitation (Fig. 9.1a) associated with the Intertropical Convergence Zone (ITCZ) around the equator. Updraft in the ITCZ moves both northward and southward in the tropopause and descends around the Tropic of Cancer and the Tropic of Capricorn. Around both tropics (the so-called 'horse latitudes'), hot and dry air divergence typically creates arid areas and deserts (Fig. 9.1a). This circulation of air masses ascending around the equator and descending around the tropics is called the Hadley circulation.

According to the Köppen climate classification, tropical climates are defined as conditions in which the air temperatures do not fall below 18°C in the coolest month, where the annual mean precipitation (mm) is more than the Köppen arid boundary ($P_K = 20 \times (T_Y + x)$ measured in mm, where T_Y is annual mean air temperature (°C) and x is the index representing the precipitation patterns such as 0, 7 and 14 for dry summer and wet winter, wet throughout the year, and dry winter and wet summer, respectively). These climates are further subdivided into rainforest (Af), monsoon (Am) and savannah (Aw) classes. The rainforest climate occurs in the zone where all 12 months have a mean precipitation of at least 60 mm. The monsoon and savannah climates have a mean precipitation of less than 60 mm in the least precipitation month but have the least monthly precipitation, more and less than 100–0.04 × (annual mean precipitation in mm), respectively.

As Feng *et al.* (2013) suggested, the areas with the largest amount of precipitation always exist in the regions with the most aseasonal precipitation patterns but not vice versa (Fig. 9.1). The rainforest climate is characterized by both a large annual amount of precipitation (Fig. 9.1a) and little or negligible seasonal variation in precipitation (Fig. 9.1b), which are caused by a stationary low atmospheric pressure system around the equator. Tropical evergreen forests can exist in this region because of their constant water requirements throughout the year. High rainfall intensity of short duration is distinctive in the rainforest climate, and in both diurnal

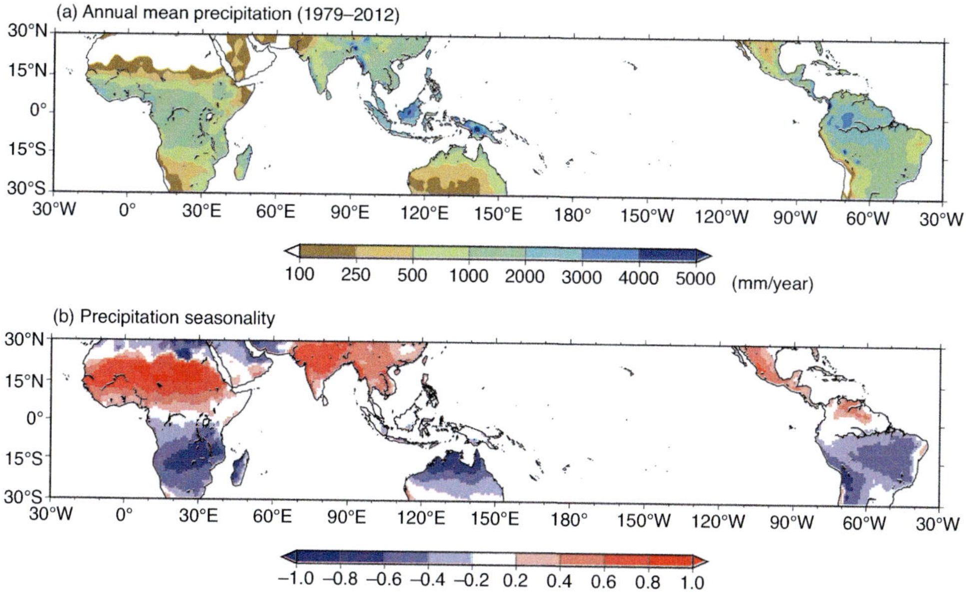

Fig. 9.1. Pan-tropic maps of (a) annual mean precipitation and (b) precipitation seasonality represented by the ratio of difference in precipitation between June–August and December–February to annual precipitation (constructed using NOAA's PRECipitation REConstruction over Land (PREC/L) with values averaged over 1979–2012).

and seasonal variations, there is strongly spatial heterogeneous rainfall distribution over a scale of a single to thousands of square kilometers of tropical rainforests (Krusche *et al.*, 2011; Kumagai and Kume, 2012; Kanamori *et al.*, 2013).

By contrast, the monsoon and the savannah climates are characterized by their degree of seasonality in precipitation caused primarily by seasonal fluctuations of the ITCZ. It should be noted that annual precipitation decreases with the increased seasonality of precipitation but not vice versa (see Feng *et al.*, 2013; Fig. 9.1). There is a less intense dry season and larger precipitation in the monsoon climate than in the savannah climate. Tropical seasonal forests can exist in monsoon areas with climate adaptations that cope with water stress during the dry season (e.g. full- or semi-deciduous traits, seasonal variations in stomatal behavior). This, in turn, means that tropical seasonal forests can vary from evergreen to deciduous forests corresponding to the local precipitation regimes, for example influenced by altitude and topography (Tanaka *et al.*, 2008).

The tropics are alternatively divided into three regions based on biotic and geological history (US Department of Energy, 2012): (i) the neo-tropical ecozone (NEO) of South America, Central America and the Caribbean; (ii) the Afro-tropical ecozone (AFR) of sub-Saharan Africa; and (iii) the Indo–Malay–Australasia tropical ecozone (IMA) including regions of India, South-east Asia, southern China, New Guinea and northern Australia. The forests of the Amazon Basin in the neotropics represent the single largest block of intact tropical forests, occupying around 40% of forest biomass in the tropics (Saatchi *et al.*, 2011). South and Central American and Caribbean forests often experience hurricane-induced disturbances (Boose *et al.*, 1994, 2004; Negrón-Juárez *et al.*, 2010). Among the Afro-tropical forests, those of the Congo Basin represent the second-largest intact block of tropical forests (Pan *et al.*, 2011). Central African forests do not experience tropical cyclones but are still subject to large storm events and suffer from severe drought and special wet periods, particularly during the El Niño and the La Niña phenomena (US Department of Energy,

2012). While the dominant plant family in neo-tropical and Afro-tropical forests is Leguminosae, Dipterocarpaceae dominate within South-east Asian forests. Dipterocarps are among the most valuable timber species globally, leading to intense logging pressure within the IMA tropical forests. Indeed, over the late half of the 20th century, more timber was exported from Borneo Island in this region than from neo-tropical and Afro-tropical regions combined (Curran *et al.*, 2004). Although South-east Asian tropical forests represent only about 11% of the world's tropical forests in terms of area, they had the highest relative deforestation rate around 2000 (e.g. Canadell *et al.*, 2007; Pan *et al.*, 2011). An analysis of climatic trends in global tropical rainforest regions over the period 1960–1998 showed that precipitation declined more significantly in South-east Asia than in Amazonia and that the El Niño–Southern Oscillation (ENSO) is a particularly important driver of drought in South-east Asia (Malhi and Wright, 2004).

9.3 Pan-Tropical Evapotranspiration

Evapotranspiration from the forest canopy can be simply partitioned into wet canopy evaporation or canopy interception, and dry canopy evaporation or transpiration. Kume *et al.* (2011) reviewed tropical forest canopy interception ratios (i.e. wet canopy evaporation divided by gross precipitation) from the data of 40 tropical forests, including both rainforests and seasonal forests, and reported that most ratios measured in tropical forests were in the range of 10–20% with a mean of 17%. In contrast, the reported transpiration rates in tropical forests had large temporal and spatial variation in the value of 2.3–4.6 mm/day (e.g. Shuttleworth *et al.*, 1984; Roberts *et al.*, 1993; Cienciala *et al.*, 2000; Kumagai *et al.*, 2004). Bruijnzeel (1990) suggested that in humid tropical forests, the average annual transpiration was 1045 mm (range 885–1285 mm). The resultant variation in the annual tropical forest evapotranspiration range was from around 1000 to 1500 mm and the ratios of annual evapotranspiration to annual precipitation ranged mostly between 30% and 90% (Kume *et al.*, 2011). Kume *et al.* (2011) found that, apparently, in tropical forests, when the annual precipitation was less than 2000 mm, evapotranspiration increased with increases in precipitation, and when precipitation exceeded 2000 mm, evapotranspiration reached a plateau of about 1500 mm.

There are more complex situations controlling the evapotranspiration of tropical forests. Trees in the tropical monsoon climatic regions tend to cope with seasonal drought by stomatal control, leaf fall and root water uptake at deeper soil depths (Igarashi *et al.*, 2015). Many pristine tropical forests have been cleared for conversion to rubber trees and oil palm plantations (Carlson *et al.*, 2012; Fox *et al.*, 2012). The water use of such plantations is comparable to or exceeds that of the original vegetation (Guardiola-Claramonte *et al.*, 2010; Kumagai *et al.*, 2015; Giambelluca *et al.*, 2016). Significant portions of the cleared tropical land quickly revert to secondary vegetation and, in terms of hydrological characteristics, increasingly resemble the original forest over time (Giambelluca, 2002). Giambelluca *et al.* (2003) pointed out that evapotranspiration from fragmented forests tends to be enhanced by conditions in surrounding clearings. Bruijnzeel *et al.* (2011) claimed the need to pay special attention to the hydrometeorology in tropical montane cloud forests, which are fog- and cloud-affected, and thus small and large amounts of the transpiration and canopy interception are caused due to the small water vapor deficit and the cloudwater, respectively.

Although it is complex to generalize on evapotranspiration, we can compare evapotranspiration from tropical forests with that of other regions. Annual evapotranspiration (*ET*) from ground observations (e.g. watershed water balance, micrometeorological measurements and soil–water balance) can be classified by annual mean temperature (*T*) and represented as a function of latitude (Fig. 9.2, adapted from Komatsu *et al.*, 2012). Despite a large variation in *ET* at each latitude (~500 mm), which might be caused by altitude, forest type, local climate of the forest sites and differences in hydrological observation methods, there is a strong linear relationship between latitude and *ET*. In humid climates of the ITCZ, the high precipitation (*P*) is expected to cause higher *ET* (Fig. 9.2). Note that there are large variations in *ET* in the tropical forest regions, but many can be explained by the

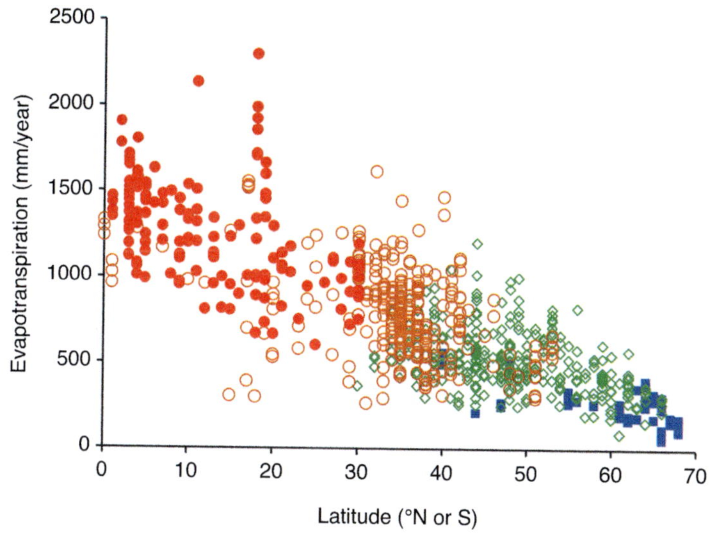

Fig. 9.2. Annual evapotranspiration as a function of latitude, classified by the annual mean temperatures (T, °C): closed circles, $T>20$; open circles, $10<T<20$; open diamonds, $0<T<10$; closed squares, $T<0$. Data on evapotranspiration, latitude and temperature taken from Komatsu *et al.* (2012).

difference in T, probably because of the altitude effects.

From this relationship between annual P and T, Komatsu *et al.* (2012) modified the Zhang *et al.* (2001) evapotranspiration model where:

$$ET = P\frac{1+w\frac{ET_0}{P}}{1+w\frac{ET_0}{P}+\frac{P}{ET_0}} \quad (9.1)$$

in which

$$ET_0 = 0.488T^2 + 27.5T + 412 \quad (9.2)$$

where w is a coefficient representing plant water availability (= 2.0). ET_0 is the potential evaporation, which was defined as a constant value calculated by the Priestley and Taylor (1972) equation in Zhang *et al.* (2001) but was modified by Komatsu *et al.* (2012) so that it can appropriately describe the temperature effect. Equations 9.1 and 9.2 successfully reproduce ET as a function of latitude (Fig. 9.2). In the tropical rainforest and monsoon climates, it suggests that higher ET eventuates through the conditions of a plentiful P and higher T.

Theoretical (Eqns. 9.1 and 9.2) and observed relationships between annual P and ET are investigated in Fig. 9.3a. The shallower slope in the relation between ET and P in the neotropics (Fig. 9.3a, top) and IMA tropics (Fig. 9.3a, bottom) compared with the steeper relation in the Afro-tropics (Fig. 9.3a, middle) can be explained by the saturation curve, Eqn. 9.1. It is also surprising that such a simple formulation as Eqn. 9.1 describes the pan-tropical ET characteristics because Fig. 9.3 contains the data not only from natural seasonal and rain forests but also secondary forests and plantations. From an interpretation of Budyko's vegetation categorization using the relationships among the radiation dryness index, net radiation and vegetation types (Budyko and Miller, 1974), tropical forests can exist within the ranges of annual mean net radiation of 110–133 W/m^2 and annual mean precipitation of 1370–5000 mm. The lowest precipitation limit for the presence of tropical forest (i.e. 1370 mm/year) is found to be broadly comparable with the lowest value of P in each ecozone (Fig. 9.3a).

The local water-use ratio (LEUR) is defined as ET divided by P and represents the ratio of P water recycling from ET. Fig. 9.3b shows these data for each tropical ecozone. The LEUR data show a significant decrease with P decreasing for all ecozones. This implies that in areas with smaller P (mainly in AFR), P is effectively recycled from ET so that the vegetation regulates its own regional water resources. In areas with

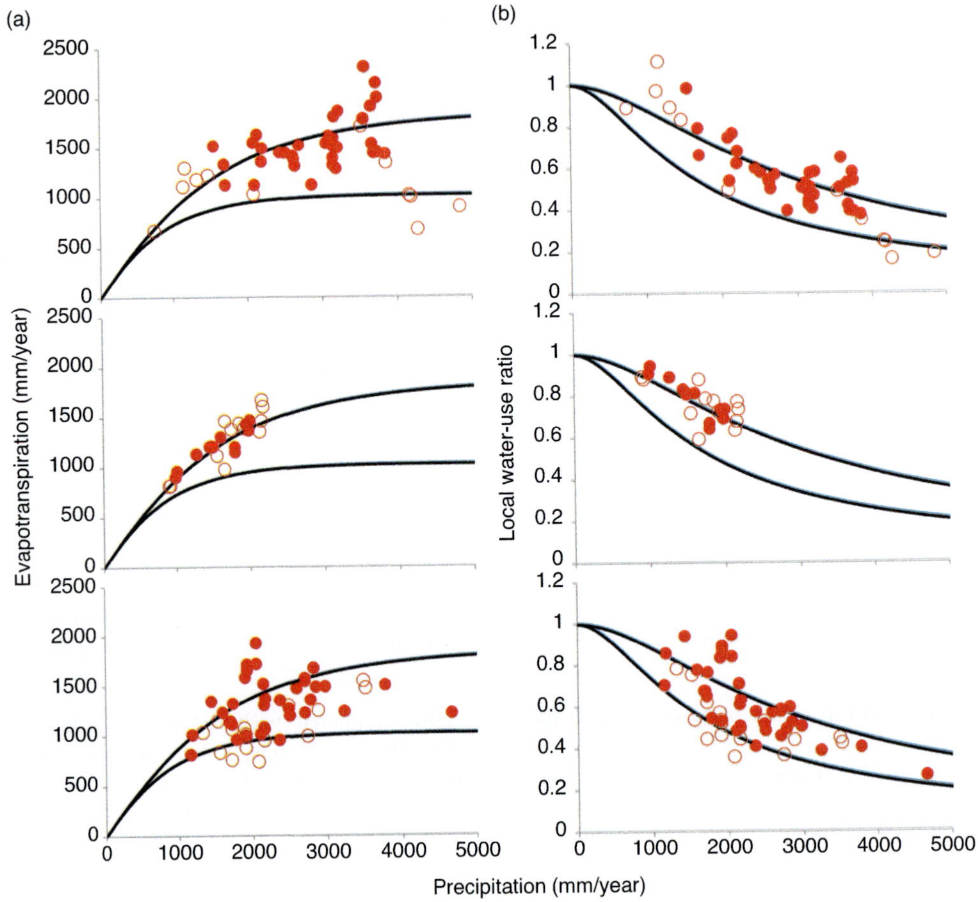

Fig. 9.3. Relationships between precipitation and (a) annual evapotranspiration and (b) local water-use ratio defined as evapotranspiration divided by precipitation, classified by the annual mean temperatures (T, °C): closed circles, $T>25$; open circles, $20<T<25$. Solid lines denote the theoretical relationships: upper and bottom lines for each panel assume $T=30$ and 15, respectively. Top, neotropical ecozone; middle, Afrotropical ecozone; bottom, Indo–Malay–Australasia tropical ecozone. Data on evapotranspiration and precipitation taken from Komatsu *et al.* (2012).

larger P (such as NEO and IMA), P is also supplied by moist air from the surrounding terrain and oceans.

The atmosphere–land surface water balance method (Oki *et al.*, 1995) with global data on the P (Fig. 9.1) and atmospheric water vapor divergence derived from ERA-Interim gridded four-dimensional meteorology data set (Dee *et al.*, 2011) produced a pan-tropical map of ET and the LEUR. Although ET is well reproduced in areas with a relatively small amount of P, it is likely that large P in the method has induced an overestimation of ET in such areas. Nevertheless, characteristics of spatial variation in the ET are well represented. Comparatively small ET can be seen in AFR and a much larger ET in Borneo and New Guinea Islands in the IMA tropics and Amazonian headwaters in the neotropics. Areas with lower LEUR values tend to overlap areas with larger P (compared with Fig. 9.1a), and this is consistent with the patterns shown in Fig. 9.3b. The exception is Borneo Island (IMA ecozone), where, despite a plentiful P, higher values of LEUR are estimated due to near-zero annual atmospheric water vapor divergence/convergence (Kumagai *et al.*,

2013; Kanamori *et al.*, 2018). Here, we should note that, for considering the extent to which P relies on terrestrial ET (i.e. moisture recycling), the role of global wind patterns, topography and land cover should be interpreted more in the context of continental moisture recycling (van der Ent *et al.*, 2010). Notably, Poveda *et al.* (2014) examined how 'aerial river' pathways modified by the effects of topography, orography and land-cover types contribute to precipitation patterns in tropical South America.

9.4 Pan-tropical Streamflow Generation

Fig. 9.4 shows the annual streamflow per unit drainage area (the strict definition of the term 'runoff') against latitude (determined at the confluence with the ocean; adapted from Wohl *et al.*, 2012). The streamflow for December–February (winter in the northern hemisphere) and June–August (winter in the southern hemisphere) averaged per latitude was also estimated using the atmosphere–land surface water balance method with assumptions that deep groundwater flow contributes to streamflow at the scale of large watersheds and that changes in calculated domain water storage can be neglected (Oki *et al.*, 1995) and added to Fig. 9.4. The negative values of computed 3 month streamflow are probably attributable to the role of dynamic subsurface storage. It is apparent that the annual streamflow from tropical basins is typically much greater than that from temperate basins due primarily to the greater precipitation amounts in the humid tropics. Furthermore, the range in annual observed streamflow is much larger within the latitudes 0 to 23.4° due to the presence of extensive areas with dry climates (arid (BW) and semi-arid (BS) areas) in the tropics.

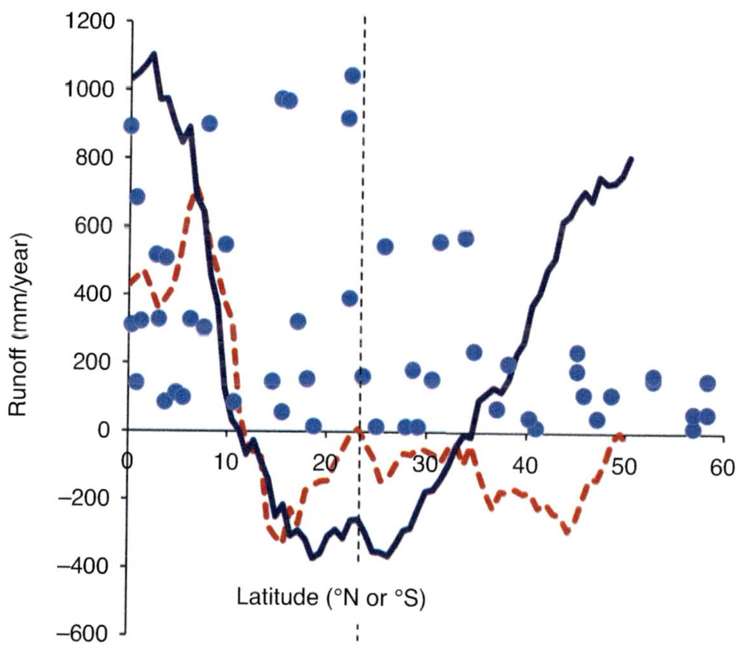

Fig. 9.4. Annually scaled streamflow per unit drainage area as a function of latitude at the river mouth (circles), adapted from Wohl *et al.* (2012). Relationships between latitude and annually scaled and latitude-averaged streamflow computed from the atmosphere–land surface water balance method for December–February in the northern hemisphere (solid line) and June–August in the southern hemisphere (dashed line) are also shown. The vertical dotted line denotes the latitude of the tropics north and south of the equator (23.4°N and S).

The greater streamflow present within the parts of the tropics with a rainforest climate and monsoon climate indicates that considerably more precipitation travels through watersheds toward streams than at other latitudes. The greater water flows within such tropical basins mean that the magnitudes of chemical and particle transport are likely to be greater (Lepcha *et al.*, 2024), and the watershed systems are more sensitive to disturbance (Wohl *et al.*, 2012). Consequently, tropical forest hydrology has implications for the other scientific disciplines of biogeochemistry and geomorphology.

Some 89% of the channel network of the globe's streams/rivers comprises first-, second- and third-order channels (see Table 2 in Downing *et al.*, 2012). This means that most streamflow is generated in the network of such low-order channels. Experimental watersheds typically comprise channels of first- to third-order size and so are ideal locations for study of the pathways of rainwater to the channel network via surface and/or subsurface pathways. These routes of water migration to channels are known as the pathways of 'streamflow generation' or simply 'runoff pathways' (Bonell, 2004; Burt and McDonnell, 2015). Most experimental watersheds used for the study of runoff pathways are in temperate regions, with few in tropical regions (Bonell, 2004; Burt and McDonnell, 2015). Fig. 9.5a shows the locations of some key experimental watersheds with a long history of research on streamflow generation pathways in the tropics. As these are commonly located beneath tropical forests, the findings are pertinent to tropical forest environments.

Streamflow generation pathways that have been observed within these tropical forest environments include the following:

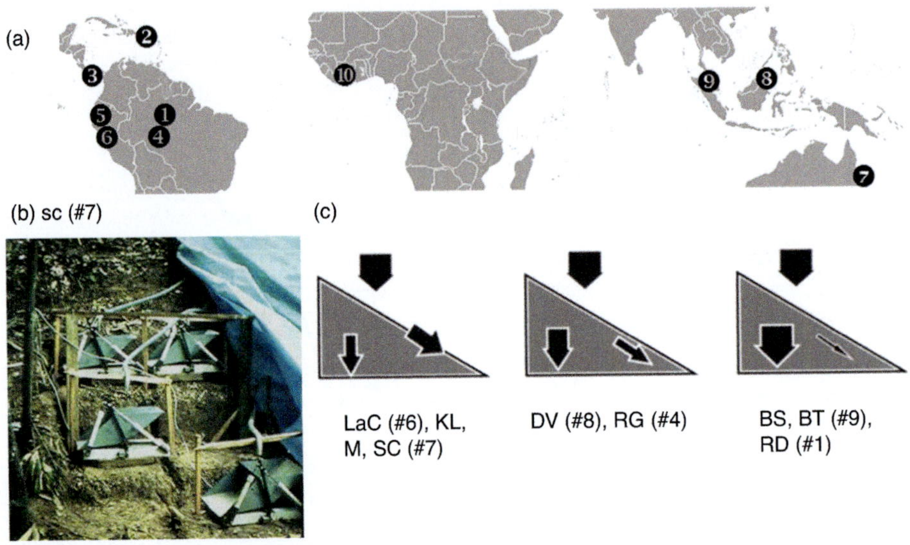

Fig. 9.5. (a) Location of ten experimental basins in the tropics with a long history of research into streamflow generation pathways: 1, Reserva Ducke (Brazil); 2, Bisley II (Puerto Rico); 3, Barro Colorado and Lutzito (Panama); 4, Rancho Grande and Jurena Ultisol (Brazil); 5, Rio San Francisco (Ecuador); 6, La Cuenca (Peru); 7, South Creek (Australia); 8, Danum Valley (Malaysia); 9, Bukit Tarek and Bukit Berembun (Malaysia); 10, Tai Forest (Ivory Coast). See Elsenbeer (2001), Bonell (2004), Hugenschmidt *et al.* (2014), and Barthold and Woods (2015) for an extensive list of publications for these and other experimental sites in the tropics. (b) The "throughflow trough! system used by Bonell and Gilmour (1978) to demonstrate the presence of shallow lateral flows in the organic and mineral soils at the South Creek Experimental Watershed (no. 7), Queensland, Australia. (c) Inferred dominant streamflow generation pathways presented in Elsenbeer (2001). Experimental watersheds: LaC: La Cuenca (#6); KL: Kiani Lestari; M: Mendolong; SC: South Creek (#7); DV: Danum Valley (#8); RG: Rancho Grande (#4); BS: Bukit Soeharto; BT: Bukit Tarek (#9); RD: Reserva Ducke (#1) (see Elsenbeer, 2001, for details).

1. *Infiltration-excess overland flow.* This water flow on slopes outside of channels is caused by precipitation falling at a rate faster than the local coefficient of permeability at the ground surface (equivalent to the 'infiltration capacity' or surface 'saturated hydraulic conductivity'; e.g. Cheng *et al.*, 2023).
2. *Saturation overland flow by direct precipitation.* Where rainfall falls on to the ground at a rate less than that of the infiltration capacity but where pores are already saturated, then no further infiltration can take place, and new water travels over the surface.
3. *Subsurface flow.* Where rainfall infiltrates, some will evaporate from the soil or support transpiration from vegetation, and the remainder will travel toward streams below the surface. Most of this water will enter the streams via the channel bed and banks, but some will return to the surface prior to reaching a channel (so-called return flow) and flow over the surface as saturation overland flow. Flow beneath the surface may be very shallow where lithomorphic soils overlie impermeable geology but may be over 100m deep where permeable soils overlie permeable geology (whether unconsolidated materials or rock).

Some ambiguity in the definition of water pathways arises from the definition of the ground surface. Some scientists define overland flow as water moving over the surface of a mineral A (or E) soil horizon, while others use the definition of water moving laterally above the typically overlying organic horizons (i.e. L, H and/or O horizons).

Streamflow generation studies in the tropics that have directly observed the presence of overland flow where the rainfall intensity exceeds the infiltration capacity are very limited, but the Tai Forest basin (Bonell, 2004) and Basper basin (Cheng *et al.*, 2023) are good examples. Studies that have observed overland flow on permeable but saturated soils include the South Creek (Bonell and Gilmour, 1978) and La Cuenca (Elsenbeer and Vertessy, 2000) basins. Several studies have direct evidence of lateral (downslope) flows within the subsurface generated during rainstorms. Bonell and Gilmour (1978) observed the presence of these flows using so-called 'throughflow troughs' (Fig. 9.5b), while Chappell and Sherlock (2005) observed the presence of these flows by monitoring tracer migration.

The more pertinent question is not whether a particular pathway is present but whether it is the dominant pathway producing more than 50% of the observed total streamflow, particularly during storm events (i.e. without any manipulation of hydrographs by separation methods). Too often, researchers infer the dominance of a particular pathway simply from the observed presence of the pathway, rather than relating measured flows per unit basin area from that pathway to those observed in the stream per unit basin area. For example, the measured lateral flows per unit basin area of Gilmour *et al.* (1980) are only a tiny fraction of the observed streamflow per unit basin area. Consequently, the importance of measured near-surface flows at this site (and many other sites) has been overemphasized in comparison with the unmeasured flows within deeper soil and unconsolidated rock strata. The resultant ambiguity and misinterpretation of the dominant pathway is amply illustrated by inconsistencies in the inferred dominant pathways presented in the reviews of Elsenbeer (2001), Bonell (2004) and Barthold and Woods (2015), as shown in Fig. 9.5c and Table 9.1.

A further area of concern is the focus of most studies on pathways in the solum (i.e. A and B soil horizons) alone, as highlighted by Bonell and Balek (1993) and Bonell (2004). There is an increasing awareness that at some tropical sites, soils may be developed on unconsolidated geological materials that are permeable and have deeper pathways within these strata. Examples of experimental watersheds developed on these deeper porous media in the tropics include the Lake Calado microbasin near the Reserva Ducke basin in Brazil (Lesack, 1993), the Jungle Falls basin in Singapore (Chappell and Sherlock, 2005; Rahardjo *et al.*, 2010), the Bukit Timah (Noguchi *et al.*, 2005) and Bukit Berembun (Chappell *et al.*, 2004) basins in Malaysia, and the O Thom II basin in Cambodia (Shimizu *et al.*, 2007). Equally, other low-order basins in the tropics may be located on rock aquifers or rocks with fracture systems that produce very deep water pathways. The Arboleda basin near La Selva, Costa Rica (Genereux *et al.*, 2005) and the Jatunhuaycu hillslope in the tropical Andes

Table 9.1. Inferred dominant streamflow generation pathways presented in Table 14.1 of Bonell (2004). See Bonell (2004) for further details of the experimental watersheds and pathway definitions.

Dominant pathway	Experimental watershed
Predominantly vertical pathways	Reserva Ducke (#1), Bukit Tarek (#9), Fazenda Dimona, Mgera
Predominantly lateral pathways (infiltration-excess overland flow)	Tai Forest (#10), Basper (Cheng *et al.*, 2023)
Predominantly lateral pathways (saturation-excess overland flow)	South Creek (#7), Barro Colorado (#3), La Cuenca (#6), Maburae
Predominantly lateral pathways (subsurface stormflow within the soil)	Danum Valley (Sungai Baru Barat) (#8), Rancho Grande (#4), Bisley II (#2), Dodmane, Kannike, Ife, ECEREX, Sekancing (Sayama *et al.*, 2021), Agua Salud (Bush *et al.*, 2020)
Predominantly lateral pathways (subsurface stormflow at the soil-bedrock Interface)	Kuala Belalong

each # represents the location in Fig. 9.5a.

(Lahuatte *et al.*, 2022) have such pathways. To alert researchers to the potential presence and role of these deeper streamflow generation pathways, Chappell *et al.* (2007) developed a perceptual model of runoff pathways where a type III system has a dominance of pathways via unconsolidated geological materials, and a type IV system has a dominance of pathways via rock aquifers or fracture systems (Fig. 9.6). Experimental basins lacking any evidence of major pathways through geological strata (solid rock or unconsolidated materials) are classified as either type II systems (where a B soil horizon is present, e.g. Chappell *et al.*, 1998) or type I systems (where a lithomorphic soil is developed on steep, impermeable mountain slopes). The potential presence of deeper pathways within existing experimental watersheds needs to be a key focus for new research. With complete observational evidence, including information gained from both hydrometric and tracer studies (as recommended by Barthold and Woods, 2015), researchers may be closer to developing a unified, numerical model of the dominant pathways of streamflow generation applicable across all tropical forest environments.

9.5 Research Needs

The projected growth in atmospheric greenhouse gases within the coming century, as predicted by the Intergovernmental Panel on Climate Change's (IPCC) A1B (balanced across all sources) scenario, will significantly increase tropical surface temperatures ranging from ~3°C to 5°C for South-east Asia, the Amazon and West Africa (IPCC, 2007). In addition, the A1B scenario predicts concomitant modifications to precipitation patterns: a general increase in precipitation for West Africa, intensification of the seasonality of precipitation for South-east Asia (i.e. more and less precipitation in the wet and dry seasons, respectively) and a general decrease in precipitation for the southern and eastern Amazon. More dramatically, as an example, the 1997–1998 El Niño was the strongest in the 20th century, and its associated drought in Borneo Island in South-east Asia was the most severe (statistically, a drought such as in 1998 may occur once in ~360 years). Tree mortality rates during that drought were 6.37%/year, as compared with 0.89%/year during the pre-drought period (1993–1997) in the studied forest site in western Borneo (Nakagawa *et al.*, 2000). Climate models project an increase in the frequency and intensity of ENSO activity (El Niño and La Niña events) in response to global warming (Capotondi, 2015). This will likely also induce severe droughts and devastating floods in the South-east Asian tropics (Capotondi, 2015). Again, tropical forest regions are crucial for climate mitigation through carbon sequestration and evapotranspiration. In this context, Takamura *et al.* (2023) characterized the relationship between the ENSO-induced

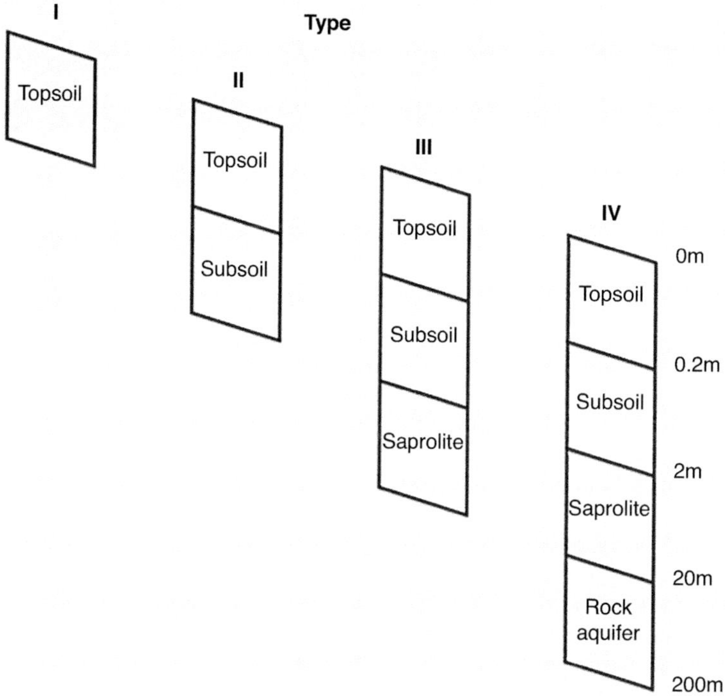

Fig. 9.6. Schematic representation of water pathway systems within tropical forests classified according to the presence of as many as four strata (A soil horizon or topsoil, B soil horizon or subsoil, unconsolidated geological materials and solid rock) named types I to IV. Example depths and the logarithmic depth scale are given. Modified from Chappell *et al.* (2007) and US Department of Energy (2012).

climatic conditions and atmosphere–forest canopy exchange in a South-east Asian rainforest, suggesting the long-term climatic influence on the carbon and water cycles and ecosystem regulation mechanisms in tropical rainforests.

In addition, the tropics are known to be a very active domain in terms of changes in land cover. According to an investigation by Hansen *et al.* (2013) using Earth observation satellite data from 2000 to 2012, the tropics were the only domain to exhibit a statistically significant increasing trend in annual forest loss. This was 2101 km^2/year and tropical rainforests loss was 32% of the global forest loss. South American tropical dry forests have been lost at the highest rate in the lost tropical forests. Furthermore, Hansen *et al.* (2013) confirmed that the tropical forest loss rate in Brazil has decreased from 40,000 km^2 in 2003–2004 to 20,000 km^2 in 2010–2011, while forest loss in Indonesia drastically increased from 8000 km^2 in 2000–2001 to 20,000 km^2 in 2010–2011, indicating that an increase in Asian tropical forest loss 'compensated' for a decrease in Amazonian tropical forest loss. Despite a drastic reduction in the deforestation rate in the whole world in 2015–2020 compared with the rate in 1990–2010, deforestation and land-use change in the tropical domain are still ongoing, especially in the African tropics. The tropical deforestation intensity is in the order Africa >> South America > South east Asia (FAO, 2020).

In the humid tropics, there is plenty of precipitation, and a combination of climate change and drastic changes in land covers must induce pressure for freshwater resources and loss of soil via changes in local hydrological processes. Tropical forest evapotranspiration is generally greater than evapotranspiration from grasslands. Hence changing land covers, such as from forests to pastures, would reduce evapotranspiration and increase streamflow, resulting in, in some cases, increasing flood frequencies (Bruijnzeel, 2001, 2004). The generation of localized infiltration-excess overland flow on compacted soil surfaces caused by forestry operations accelerates the geomorphological process of soil erosion (Bruijnzeel, 2004; Sidle *et al.*, 2006; Sidle and Ziegler, 2012; Douglas, 2022). Such land degradations would become increasingly worse by altered precipitation regimes through climate change (Peña-Arancibia *et al.*, 2010; Wohl *et al.*, 2012). However, the hydrological properties of secondary vegetation, such as evapotranspiration and surface infiltration, may quickly resemble those of the original forest again (Giambelluca, 2002; Bruijnzeel, 2004).

From a global perspective, it should be emphasized that tropical forests are a major source of global hydrological fluxes. Changes in evapotranspiration rates could significantly impact both global and regional climates, in turn potentially affecting feedback to the atmosphere (Bonan, 2008). Many previous research contributions to the knowledge of hydro-climate in the tropics have suggested that: (i) forest cutting has more effect on local and global climate under maritime conditions than under continental conditions (van der Molen *et al.*, 2006; Takahashi *et al.*, 2017; Kanamori *et al.*, 2018); (ii) rainwater is recycled earlier by wet canopy evaporation than via transpiration (van der Ent *et al.*, 2014); and (iii) in the maritime continent, changes in sea surface temperature influence the precipitation regimes more than the land-cover changes (Bruijnzeel, 2004). Furthermore, large-scale tropical deforestation and selective logging could result in warmer and drier conditions at the local scale, and also the teleconnections from the converted tropical lands could pose a considerable risk to agriculture in other regions due to impacts on precipitation against a background of warmer temperatures (Lawrence and Vandecar, 2015).

As the first step for considering the future hydrological impact in the tropics, we need an understanding of the current and basic tropical hydrological cycling, as well as the hydrological interactions among the earth's surface and subsurface, vegetation and atmosphere, based on long-term and networked data acquisition and organization. According to Cui *et al.* (2022), who investigated water resource availability altered by global vegetation changes, we can determine how much the local terrestrial water availability has decreased with the reduction of vegetation covers in the tropical forest regions. As Wohl *et al.* (2012) pointed out, field-based hydrological measurements in many tropical countries have been less explored than those in temperate regions, and to make matters worse, are waning. We should note that the lack of long-term observations throughout the tropics leads to a failure to validate hydro-climate models and, thus, an impossibility for extrapolating future tropical forest hydrology using the output of the models, which must be built with reference to reliable observations and should use these observations as inputs (e.g. Duvert *et al.*, 2022). Besides making more effort to organize a field-based hydrological observation network over the tropics, as shown in this chapter, remote sensing technologies and global-scale climate and geoinformation databases can serve among the most promising tools to complement the lack of tropical field-observation sites.

Acknowledgements

The work was supported in part by Grant-in-Aid for Scientific Researches (#21H05316, #15H02645 and #25281005) and the granted project 'Program for Risk Information on Climate Change' from the Ministry of Education, Science and Culture, Japan. We are deeply grateful to L.A. Bruijnzeel and the editors for their constructive comments on the manuscript.

References

Avissar, R. and Werth, D. (2005) Global teleconnections resulting from tropical deforestation. *Journal of Hydrometeorology* 6, 134–145.

Barthold, F.K. and Woods, R.A. (2015) Stormflow generation: A meta-analysis of field evidence from small, forested catchments. *Water Resources Research* 51, 3730–3753.

Beer, C., Reichstein, M., Tomelleri, E., Ciais, P., Jung, M. *et al.* (2010) Terrestrial gross carbon dioxide uptake: Global distribution and covariation with climate. *Science* 329, 834–838.

Bonan, G.B. (2008) Forests and climate change: Forcing, feedbacks, and the climate benefits of forests. *Science* 320, 1444–1449.

Bonell, M. and Gilmour, D.A. (1978) The development of overland flow in a tropical rainforest catchment. *Journal of Hydrology* 39, 365–382.

Bonell, M. and Balek, J. (1993) Recent scientific developments and research needs in hydrological processes of the humid tropics. In: Bonell, M., Hufschmidt, M.M. and Gladwell, J.S. (eds) *Hydrology and Water Management in the Humid Tropics*. Cambridge University Press, Cambridge, UK, pp. 167–260.

Bonell, M. (2004) Runoff generation in tropical forests. In: Bonell, M. and Bruijnzeel, L.A. (eds) *Forests, Water and People in the Humid Tropics: Past, Present and Future Hydrological Research for Integrated Land and Water Management*. UNESCO International Hydrology Series. Cambridge University Press, Cambridge, UK, pp. 314–406.

Boose, E.R., Foster, D.R. and Fluet, M. (1994) Hurricane impacts to tropical and temperate forest landscapes. *Ecological Monographs* 64, 369–400.

Boose, E.R., Serrano, M.I. and Foster, D.R. (2004) Landscape and regional impacts of hurricanes in Puerto Rico. *Ecological Monographs* 74, 335–352.

Bruijnzeel, L.A. (1990) *Hydrology of Moist Tropical Forests and Effects of Conversion: A State of Knowledge Review*. International Hydrological Programme, UNESCO, Paris.

Bruijnzeel, L.A. (2004) Hydrological functions of tropical forests: Not seeing the soil for the trees? *Agriculture, Ecosystems & Environment* 104, 185–228.

Bruijnzeel, L.A. (2001) Forest hydrology. In: Evans, J. (ed.) *The Forests Handbook, Vol. 1: An Overview of Forest Science*. Blackwell, Oxford, UK, pp. 301–343.

Bruijnzeel, L.A., Mulligan, M. and Scatena, F.N. (2011) Hydrometeorology of tropical montane cloud forests: Emerging patterns. *Hydrological Processes* 25, 465–498.

Budyko, M.I. and Miller, D.H. (1974) *Climate and Life*. Academic Press, New York.

Burt, T.P. and McDonnell, J.J. (2015) Whither field hydrology? The need for discovery science and outrageous hydrological hypotheses. *Water Resources Research* 51, 5919–5928.

Bush, S.A., Stallard, R.F., Ebel, B.A. and Barnard, H.R. (2020) Assessing plot-scale impacts of land use on overland flow generation in Central Panama. *Hydrological Processes* 34, 5043–5069.

Canadell, J.G., Le Quere, C., Raupach, M.R., Field, C.B., Buitenhuis, E.T. *et al.* (2007) Contributions to accelerating atmospheric CO_2 growth from economic activity, carbon intensity, and efficiency of natural sinks. *Proceedings of the National Academy of Sciences USA* 104, 18866–18870.

Capotondi, A. (2015) Extreme La Niña events to increase. *Nature Climate Change* 5, 100–101.

Carlson, K.M., Curran, L.M., Ratnasari, D., Pittman, A.M., Soares-Filho, B.S. *et al.* (2012) Committed carbon emissions, deforestation, and community land conversion from oil palm plantation expansion in West Kalimantan, Indonesia. *Proceedings of the National Academy of Sciences USA* 109, 7559–7564.

Chappell, N.A. and Sherlock, M.D. (2005) Contrasting flow pathways within tropical forest slopes of Ultisol soil. *Earth Surface Processes and Landforms* 30, 735–753.

Chappell, N.A., Franks, S.W. and Larenus, J. (1998) Multi-scale permeability estimation in a tropical catchment. *Hydrological Processes* 12, 1507–1523.

Chappell, N.A., Tych, W., Yusop, Z., Rahim, N.A. and Kasran, B. (2004) Spatially-significant effects of selective tropical forestry on water, nutrient and sediment flows: A modelling-supported review. In: Bonell, M. and Bruijnzeel, L.A. (eds) *Forests, Water and People in the Humid Tropics: Past, Present and Future Hydrological Research for Integrated Land and Water Management*. UNESCO International Hydrology Series, Cambridge University Press, Cambridge, UK, pp. 513–532.

Chappell, N.A., Sherlock, M., Bidin, K., Macdonald, R., Najman, Y. *et al.* (2007) Runoff processes in Southeast Asia: Role of soil, regolith and rock type. In: Sawada, H., Araki, M., Chappell, N.A.,

LaFrankie, J.V. and Shimizu, A. (eds) *Forest Environments in the Mekong River Basin*. Springer, Tokyo, Japan, pp. 3–23.

Chen, M., Xie, P., Janowiak, J.E. and Arkin, P.A. (2002) Global land precipitation: A 50-yr monthly analysis based on gauge observations. *Journal of Hydrometeorology* 3, 249–266.

Cheng, Z., Zhang, J., Yu, B., Chappell, N.A., Meerveld, H.J. *et al.* (2023) Stormflow response and "effective" hydraulic conductivity of a degraded tropical *imperata* grassland catchment as evaluated with two infiltration models. *Water Resources Research* 59, e2022WR033625.

Cienciala, E., Kucera, J. and Malmer, A. (2000) Tree sap flow and stand transpiration of two *Acacia mangium* plantations in Sabah, Borneo. *Journal of Hydrology* 236, 109–120.

Cui, J., Lian, X., Huntingford, C., Gimeno, L., Wang, T. *et al.* (2022) Global water availability boosted by vegetation-driven changes in atmospheric moisture transport. *Nature Geoscience* 15, 982–988.

Curran, L.M., Trigg, S.N., McDonald, A.K., Astiani, D., Hardiono, Y.M. *et al.* (2004) Lowland forest loss in protected areas of Indonesian Borneo. *Science* 303, 1000–1003.

Dee, D.P., Uppala, S.M., Simmons, A.J., Berrisford, P., Poli, P. *et al.* (2011) The ERA-interim reanalysis: Configuration and performance of the data assimilation system. *Quarterly Journal of the Royal Meteorological Society* 137, 553–597.

Douglas, I. (2022) *Water and the rainforest in Malaysian Borneo: Hydrological research at the Danum Valley Field Centre*. Springer, Cham, Switzerland.

Downing, J.A., Cole, J.J., Duarte, C.M., Middelburg, J.J., Melack, J.M. *et al.* (2012) Global abundance and size distribution of streams and rivers. *Inland Waters* 2, 229–236.

Duvert, C., Lim, H.S., Irvine, D.J., Bird, M.I., Bass, A.M. *et al.* (2022) Hydrological processes in tropical Australia: Historical perspective and the need for a catchment observatory network to address future development. *Journal of Hydrology: Regional Studies* 43, 101194.

Elsenbeer, H. (2001) Hydrologic flowpaths in tropical soilscapes – A review. *Hydrological Processes* 15, 1751–1759.

Elsenbeer, H. and Vertessy, R.A. (2000) Stormflow generation and flowpath characteristics in an Amazonian rainforest catchment. *Hydrological Processes* 14, 2367–2381.

FAO (2020) *FAO (2020) Global Forest Resource Assessment 2020: Main Report*. Food and Agriculture Organization of the United Nations, Rome.

Feng, X., Porporato, A. and Rodriguez-Iturbe, I. (2013) Changes in rainfall seasonality in the tropics. *Nature Climate Change* 3, 811–815.

Fox, J., Vogler, J.B., Sen, O.L., Giambelluca, T.W. and Ziegler, A.D. (2012) Simulating land-cover change in montane mainland Southeast Asia. *Environmental Management* 49, 968–979.

Genereux, D.P., Jordan, M.T. and Carbonell, D. (2005) A paired watershed budget study to quantify interbasin groundwater flow in a lowland rainforest, Costa Rica. *Water Resources Research* 41, W04011.

Giambelluca, T.W. (2002) Hydrology of altered tropical forest. *Hydrological Processes* 16, 1665–1669.

Giambelluca, T.W., Ziegler, A.D., Nullet, M.A., Truong, D.M. and Tran, L.T. (2003) Transpiration in a small tropical forest patch. *Agricultural and Forest Meteorology* 117, 1–22.

Giambelluca, T.W., Mudd, R.G., Liu, W., Ziegler, A.D., Kobayashi, N. *et al.* (2016) Evapotranspiration of rubber (*Hevea brasiliensis*) cultivated at two sites in Southeast Asia. *Water Resources Research* 52, 660–679.

Gilmour, D.A., Bonell, M. and Sinclair, D.F. (1980) *An investigation of storm drainage processes in a tropical rainforest catchment*. Australian Water Resources Council Technology, Canberra.

Gómez, A.M., Parra, A., Pavelsky, T.M., Wise, E.K., Villegas, J.C. *et al.* (2023) Ecohydrological impacts of oil palm expansion: A systematic review. *Environmental Research Letters* 18, 033005.

Guardiola-Claramonte, M., Troch, P.A., Ziegler, A.D., Giambelluca, T.W., Durcik, M. *et al.* (2010) Hydrologic effects of the expansion of rubber (*Hevea brasiliensis*) in a tropical catchment. *Ecohydrology* 3, 306–314.

Hansen, M.C., Potapov, P.V., Moore, R., Hancher, M., Turubanova, S.A. *et al.* (2013) High-resolution global maps of 21st-century forest cover change. *Science* 342, 850–853.

Hugenschmidt, C., Ingwersen, J., Sangchan, W., Sukvanachaikul, Y., Duffner, A. *et al.* (2014) A three-component hydrograph separation based on geochemical tracers in a tropical mountainous headwater catchment in Northern Thailand. *Hydrology and Earth System Science* 18, 525–537.

Igarashi, Y., Kumagai, T., Yoshifuji, N., Sato, T., Tanaka, N. *et al.* (2015) Environmental control of canopy stomatal conductance in a tropical deciduous forest in Northern Thailand. *Agricultural and Forest Meteorology* 202, 1–10.

IPCC (Intergovernmental Panel on Climate Change) (2007) *Climate Change 2007: Impacts, Adaptation and Vulnerability. Contribution of Working Group II to the Fourth Assessment Report of the Intergovernmental Panel on Climate Change*. Cambridge University Press, New York.

Kanamori, H., Yasunari, T. and Kuraji, K. (2013) Modulation of the diurnal cycle of rainfall associated with the MJO observed by a dense hourly rain gauge network at Sarawak, Borneo. *Journal of Climate* 26, 4858–4875.

Kanamori, H., Kumagai, T., Fujinami, H., Hiyama, T. and Yasunari, T. (2018) Effects of long- and short-term atmospheric water cycles on the water balance over the maritime continent. *Journal of Hydrometeorology* 19, 1413–1427.

Komatsu, H., Cho, J., Matsumoto, K. and Otsuki, K. (2012) Simple modeling of the global variation in annual forest evapotranspiration. *Journal of Hydrology* 420, 380–390.

Krusche, A.V., Ballester, M.V.R. and Leite, N.K. (2011) Hydrology and biogeochemistry of terra firme lowland tropical forests. In: Levia, D.F., Carlyle-Moses, D. and Tanaka, T. (eds) *Forest Hydrology and Biogeochemistry: Synthesis of Past Research and Future Directions*. Ecological Studies 216. Springer, New York, pp. 187–201.

Kumagai, T. and Porporato, A. (2012a) Drought-induced mortality of a Bornean tropical rain forest amplified by climate change. *Journal of Geophysical Research-Biogeosciences* 117, G02032.

Kumagai, T. and Kume, T. (2012) Influences of diurnal rainfall cycle on CO_2 exchange over Bornean tropical rainforests. *Ecological Modelling* 246, 91–98.

Kumagai, T. and Porporato, A. (2012b) Strategies of a Bornean tropical rainforest water use as a function of rainfall regime: Isohydric or anisohydric? *Plant, Cell and Environment* 35, 61–71.

Kumagai, T., Saitoh, T.M., Sato, Y., Morooka, T., Manfroi, O.J. *et al*. (2004) Transpiration, canopy conductance and the decoupling coefficient of a lowland mixed dipterocarp forest in Sarawak, Borneo: Dry spell effects. *Journal of Hydrology* 287, 237–251.

Kumagai, T., Saitoh, T.M., Sato, Y., Takahashi, H., Manfroi, O.J. *et al*. (2005) Annual water balance and seasonality of evapotranspiration in a Bornean tropical rainforest. *Agricultural and Forest Meteorology* 128, 81–92.

Kumagai, T., Yoshifuji, N., Tanaka, N., Suzuki, M. and Kume, T. (2009) Comparison of soil moisture dynamics between a tropical rainforest and a tropical seasonal forest in Southeast Asia: Impact of seasonal and year-to-year variations in rainfall. *Water Resources Research* 45, W04413.

Kumagai, T., Kanamori, H. and Yasunari, T. (2013) Deforestation-induced reduction in rainfall. *Hydrological Processes* 27, 3811–3814.

Kumagai, T., Mudd, R.G., Giambelluca, T.W., Kobayashi, N., Miyazawa, Y. *et al*. (2015) How do rubber (*Hevea brasiliensis*) plantations behave under seasonal water stress in Northeastern Thailand and Central Cambodia? *Agricultural and Forest Meteorology* 213, 10–22.

Kume, T., Tanaka, N., Kuraji, K., Komatsu, H., Yoshifuji, N. *et al*. (2011) Ten-year evapotranspiration estimates in a Bornean tropical rainforest. *Agricultural and Forest Meteorology* 151, 1183–1192.

Lahuatte, B., Mosquera, G.M., Páez Bimos, S., Calispa, M., Vanacker, V. *et al*. (2022) Delineation of water flow paths in a tropical Andean headwater catchment with deep soils and permeable bedrock. *Hydrological Processes* 36, e14725.

Lawrence, D. and Vandecar, K. (2015) Effects of tropical deforestation on climate and agriculture. *Nature Climate Change* 5, 27–36.

Lepcha, P.T., Pandey, P.K. and Pandey, V. (2024) Quantification of the impact of land cover and climate change on water and sediment yield in sub-tropical Himalayas in upstream Teesta River basin. Sikkim. *Remote Sensing Applications: Society and Environment* 34, 101146.

Lesack, L.F.W. (1993) Water balance and hydrologic characteristics of a rain forest catchment in the Central Amazon Basin. *Water Resources Research* 29, 759–773.

Malhi, Y. and Wright, J. (2004) Spatial patterns and recent trends in the climate of tropical rainforest regions. *Philosophical Transactions of the Royal Society B: Biological Sciences* 359, 311–329.

Malhi, Y., Aragao, L.E.O.C., Galbraith, D., Huntingford, C., Fisher, R. *et al*. (2009) Exploring the likelihood and mechanism of a climate-change-induced dieback of the Amazon rainforest. *Proceedings of the National Academy of Sciences USA* 106, 20610–20615.

Meir, P., Cox, P.M. and Grace, J. (2006) The influence of terrestrial ecosystems on climate. *Trends in Ecology and Evolution* 21, 254–260.

Miyazawa, Y., Tateishi, M., Komatsu, H., Ma, V., Kajisa, T. *et al*. (2014) Tropical tree water use under seasonal waterlogging and drought in Central Cambodia. *Journal of Hydrology* 515, 81–89.

Mora, C., Frazier, A.G., Longman, R.J., Dacks, R.S., Walton, M.M. *et al.* (2013) The projected timing of climate departure from recent variability. *Nature* 502, 183–187.

Nakagawa, M., Tanaka, K., Nakashizuka, T., Ohkubo, T., Kato, T. *et al.* (2000) Impact of severe drought associated with the 1997–1998 El Niño in a tropical forest in Sarawak. *Journal of Tropical Ecology* 16, 355–367.

Negrón-Juárez, R.I., Chambers, J.Q., Guimaraes, G., Zeng, H., Raupp, C.F. *et al.* (2010) Widespread Amazon forest tree mortality from a single cross-basin squall line event. *Geophysical Research Letters* 37, L16701.

Noguchi, S., Abdul Rahim, N. and Tani, M. (2005) Runoff characteristics in a tropical rain forest catchment. *Japan Agricultural Research Quarterly* 39, 215–219.

Oki, T., Musiake, K., Matsuyama, H. and Masuda, K. (1995) Global atmospheric water balance and runoff from large river basins. *Hydrological Processes* 9, 655–678.

Pan, Y., Birdsey, R.A., Fang, J., Houghton, R., Kauppi, P.E. *et al.* (2011) A large and persistent carbon sink in the world's forests. *Science* 333, 988–993.

Peña-Arancibia, J.L., Dijk, A.I.J.M., Mulligan, M. and Bruijnzeel, L.A. (2010) The role of climatic and terrain attributes in estimating baseflow recession in tropical catchments. *Hydrology and Earth System Sciences* 14, 2193–2205.

Phillips, O.L., Aragao, L.E.O.C., Lewis, S.L., Fisher, J.B., Lloyd, J. *et al.* (2009) Drought sensitivity of the Amazon rainforest. *Science* 323, 1344–1347.

Poveda, G., Jaramillo, L. and Vallejo, L.F. (2014) Seasonal precipitation patterns along pathways of South American low-level jets and aerial rivers. *Water Resources Research* 50, 98–118.

Priestley, C.H.B. and Taylor, R.J. (1972) On the assessment of surface heat flux and evaporation using large-scale parameters. *Monthly Weather Review* 100, 81–92.

Rahardjo, H., Nio, A., Leong, E. and Song, N. (2010) Effects of groundwater table position and soil properties on stability of slope during rainfall. *Journal of Geotechnical and Geoenvironmental Engineering* 136, 1555–1564.

Roberts, J., Cabral, O.M.R., Fisch, G., Molion, L.C.B., Moore, C.J. *et al.* (1993) Transpiration from an Amazonian rainforest calculated from stomatal conductance measurements. *Agricultural and Forest Meteorology* 65, 175–196.

Saatchi, S.S., Harris, N.L., Brown, S., Lefsky, M., Mitchard, E.T. *et al.* (2011) Benchmark map of forest carbon stocks in tropical regions across three continents. *Proceedings of the National Academy of Sciences USA* 108, 9899–9904.

Sayama, T., Araki, R., Yamamoto, K. and Apip, A. (2021) Characteristics of soil and hillslope responses in humid tropical forests in Sumatra, Indonesia. *Hydrological Research Letters* 15, 23–30.

Shimizu, A., Kabeya, N., Nobuhiro, T., Kubota, T., Tsuboyama, Y. *et al.* (2007) Runoff characteristics and observations on evapotranspiration in forest watersheds, Central Cambodia. In: Sawada, H., Araki, M., Chappell, N.A., LaFrankie, J.V. and Shimizu, A. (eds) *Forest Environments in the Mekong River Basin*. Springer, Tokyo, pp. 135–146.

Shuttleworth, W.J., Gash, J.H.C., Lloyd, C.R., Moore, C.J., Roberts, J.M. *et al.* (1984) Eddy correlation measurements of energy partition for Amazonian forest. *Quarterly Journal of the Royal Meteorological Society* 110, 1143–1162.

Sidle, R.C. and Ziegler, A.D. (2012) The dilemma of mountain roads. *Nature Geoscience* 5, 437–438.

Sidle, R.C., Ziegler, A.D., Negishi, J.N., Nik, A.R., Siew, R. *et al.* (2006) Erosion processes in steep terrain? truths, myths, and uncertainties related to forest management in southeast asia. *Forest Ecology and Management* 224, 199–225.

Spracklen, D.V., Arnold, S.R. and Taylor, C.M. (2012) Observations of increased tropical rainfall preceded by air passage over forests. *Nature* 489, 282–285.

Takahashi, A., Kumagai, T., Kanamori, H., Fujinami, H., Hiyama, T. *et al.* (2017) Impact of tropical deforestation and forest degradation on precipitation over Borneo Island. *Journal of Hydrometeorology* 18, 2907–2922.

Takamura, N., Hata, Y., Matsumoto, K., Kume, T., Ueyama, M. *et al.* (2023) El Niño-Southern oscillation forcing on carbon and water cycling in a Bornean tropical rainforest. *Proceedings of the National Academy of Sciences USA* 120, e2301596120.

Tanaka, N., Kume, T., Yoshifuji, N., Tanaka, K., Takizawa, H. *et al.* (2008) A review of evapotranspiration estimates from tropical forests in Thailand and adjacent regions. *Agricultural and Forest Meteorology* 148, 807–819.

US Department of Energy (2012) *Research Priorities for Tropical Ecosystems Under Climate Change Workshop Report*. DOE/SC-0153. US Department of Energy, Office of Science, Washington, DC.
van der Ent, R.J., Savenije, H.H.G., Schaefli, B. and Steele-Dunne, S.C. (2010) Origin and fate of atmospheric moisture over continents. *Water Resources Research* 46, W09525.
van der Ent, R.J., Wang-Erlandsson, L., Keys, P.W. and Savenije, H.H.G. (2014) Contrasting roles of interception and transpiration in the hydrological cycle? Part 2: Moisture recycling. *Earth System Dynamics* 5, 471–489.
van der Molen, M.K., Dolman, A.J., Waterloo, M.J. and Bruijnzeel, L.A. (2006) Climate is affected more by maritime than by continental land use change: A multiple scale analysis. *Global and Planetary Change* 54, 128–149.
Vourlitis, G.L., Souza Nogueira, J., Almeida Lobo, F., Sendall, K.M., de Paulo, S.R. *et al.* (2008) Energy balance and canopy conductance of a tropical semi-deciduous forest of the southern Amazon Basin. *Water Resources Research* 44, W03412.
Walsh, R.P.D. (1996) Climate. In: Richards, P.W. (ed.) *The Tropical Rain Forest, An Ecological Study*, 2nd edn. Cambridge University Press, Cambridge, UK, pp. 159–205.
Wohl, E., Barros, A., Brunsell, N., Chappell, N.A., Coe, M. *et al.* (2012) The hydrology of the humid tropics. *Nature Climate Change* 2, 655–662.
Zhang, I., Dawes, W.R. and Walker, G.R. (2001) Response of mean annual evapotranspiration to vegetation changes at catchment scale. *Water Resources Research* 37, 701–708.

10 Hydrology of Forests after Wildfire

P. R. Robichaud*

Rocky Mountain Research Station, USDA Forest Service, Moscow, Idaho, USA

Abstract

Severe wildfires can greatly impact the hydrological response of forested catchments and may result in negative water-quality effects. Peak flows can increase by orders of magnitude leading to soil erosion, major increases in sediment yields and possible debris flows. To understand and model the potential watershed response, several factors need to be understood and evaluated. Fire-induced water repellency reduces infiltration, forest-floor consumption can lead to exposed and disturbed soil that is easily erodible and burned canopy reduces rainfall interception. These conditions are spatially varied due to the mosaic of burn severity conditions, which leads to a varied vegetation recovery. Recent literature reviews provide in-depth discussion on these topics. The watershed response has some degree of predictability via physically based hydrological models. Mitigation treatments may be used to reduce the impacts of high-severity fires, and some models predict the likely impact. This chapter discusses fire's impact on watershed hydrology.

10.1 Introduction

The hydrological response from burned forest watersheds can be some of the most dramatic responses that occur from forested catchments (Bren, 2022; Ebel *et al.*, 2023). High-severity fires may lead to extremely high peak flows, which often strip away easily erodible soil; conversely, low-severity fires may have minimal effect on the watershed response. Most forest watersheds with good hydrological conditions and adequate rainfall sustain stream base flow conditions throughout the year and produce little erosion (DeBano *et al.*, 1998). Wildfire impacts these stable conditions by consuming accumulated forest-floor material, forest vegetation and understorey vegetation (Table 10.1). This vegetation and forest-floor material, when unburned, protects the soil from raindrop impact and overland flow, and promotes infiltration.

10.2 Fire Effects on Soil

10.2.1 Soil infiltration

Water infiltrating the soil is highly dependent on the surface conditions. Runoff from wildfire-burned hillslopes generally increases by one or two orders of magnitude in the first few years after the fire (Moody *et al.*, 2013). The extent of disturbance to the surface organic material is largely the consumption of organic debris (commonly referred to as 'duff' or forest floor) and the fine organic matter that holds soil particles together. The amount of duff consumed during the combustion process is a function of the severity of the fire, including the temperature reached and the duration of heating (Robichaud *et al.*, 2025). The post-fire hydrological response is directly related to the effect of the fire on the soil and duff layers (Robichaud, 1996; Parsons

*Corresponding author: peter.robichaud@usda.gov

 Forest Hydrology: Processes, Management and Assessment (eds Leon J. Bren *et al.*)
DOI: 10.1079/9781800625310.0010

Table 10.1. Hydrological processes affected by wildfire. Specific factors influencing hydrological changes include: soil type and structure; soil cover, vegetation type and regeneration rate, precipitation intensity and frequency, understorey and canopy vegetation cover, and micro- and macro-topography features. Adapted from Neary *et al.* (2005).

Hydrological process	Consequence of high burn severity	
Infiltration	↑ Overland flow ↑ Storm flow ↑ Water repellency	
Soil water storage	↑ Evaporation ↑ Water repellency	↓ Infiltration ↓ Litter absorption
Forest-floor/duff storage	↑ Evaporation ↑ Runoff ↑ Splash erosion ↑ Snow sublimation	↓ Interception
Interception/ evapotranspiration	↑ Water yield ↑ Snowpack	↓ Storage ↓ Evaporation ↓ Transpiration
Surface runoff/ overland flow	↑ Sediment yield ↑ Erosion ↑ Debris flow	
Streamflow	↑ Surface runoff ↑ Snowmelt rate ↑ Erosion	
Peak flow	↑ Volume ↑ Flash flood frequency ↑ Flood levels	
Baseflow	↑ Evapotranspiration	↑ Percolation
Water quality	↑ Suspended sediment ↑ Ash nutrients: Ca^{+}, NO_3–, NH_4+, PO_4^{3-}	

et al., 2010). The post-fire condition of the mineral surface horizons is important because they determine the amount of mineral soil exposed to raindrop splash and overland flow, and the development of water-repellent soil conditions (DeBano, 1981).

10.2.2 Soil water repellency

Soil water repellency has been well documented in burned and unburned soils in forests (Robichaud, 2000; Huffman *et al.*, 2001; Doerr *et al.*, 2006; Jordán *et al.*, 2013; Butzen *et al.*, 2015). Wildfires have often been associated with the formation of water-repellent soil conditions thought to decrease infiltration and increase runoff and soil erosion (Fig. 10.1) (DeBano *et al.*, 1998; Robichaud, 2000). Hydrophobic organic compounds in the litter and topsoil are volatized during combustion and released upwards to the atmosphere and downwards into the soil profile along a temperature gradient. Translocated hydrophobic compounds condense on cooler soil particles below the surface forming water-repellent conditions (DeBano *et al.*, 1976).

Natural water-repellent soil conditions can sometimes occur in unburned forests due to coating of soil particles with hydrophobic compounds leached from organic matter accumulations, by-products of microbial activity, or fungal growth under thick layers of litter and duff material (Savage *et al.*, 1972; DeBano, 2000; Doerr *et al.*, 2000; Butzen *et al.*, 2015). Under unburned conditions, litter and vegetation cover promote water storage and mitigate water-repellency impacts on infiltration and erosion. Fire removes protective organic layers

Fig. 10.1. Water-repellent soil conditions after a wildfire and a 30-min rainfall event showing saturated surface soil and dry soil a few centimeters below the surface.

(litter and duff), exposing the soil to raindrop impact and removing barriers to overland flow (Moffet *et al.*, 2007; Pierson *et al.*, 2008).

Seasonal variability in the presence and strength of soil water repellency under burned and unburned conditions has been observed (Dekker and Ritsema, 2000; Doerr and Thomas, 2000; Huffman *et al.*, 2001). Dekker *et al.* (2001) demonstrated that soil water repellency is a function of soil water content, that critical soil water thresholds demarcate wettable and water-repellent soil conditions, and that the relationship between moisture content and soil water repellency is affected by drying regime. The time since burning was not a significant predictor of soil water repellency in pine forests of the Colorado Front Range (Huffman *et al.*, 2001), as the water-repellent soils became wettable when soil moisture levels exceeded 12–25%. These studies indicate that seasonal variability in site characteristics that influence soil water repellency can confound assessment of long-term soil water-repellency persistence (Doerr *et al.*, 2000).

10.2.3 Soil water storage

Wildfires in a given region often occur after a several-years drought cycle (Westerling *et al.*, 2006). This cycle can affect the soil water storage even before the wildfire starts, with reduced soil water in the soil profile. The same drought cycle that caused the region's wildfire season may persist for several years after the fire. Without the protective layer of duff and forest litter, it is often difficult to recover the soil water 'deficiency' because winter snow may melt but without the 'sponge'-holding effect of the duff, little water remains to replenish the soil profile. It may take several years before antecedent (pre-fire) soil water conditions are reached.

Vegetation recovery after wildfire depends on many factors (e.g. soil burn severity, distance to seed source, fire tolerance of native species), but especially on the precipitation and snowpack (and subsequent melting) in the first post-fire year. There is typically a surge of vegetation growth immediately after a wildfire, with growth increasing at a non-linear rate over the next decade depending on the landscape dynamics. Vegetation depletes the soil water profile via root uptake and evapotranspiration (Niemeyer *et al.*, 2019): however, vegetation also provides important soil stabilization. The stabilizing influence of vegetation is likely more beneficial than any detriment caused by soil water depletion. Both fine and larger roots can also provide infiltration pathways within burned soil profiles that were rendered water repellent from the fire.

Examples of wildfire impacting the water cycle in forested environments include a wildfire in central Washington, which caused an increase in soil water storage as transpiration was reduced from the dead burned trees (Klock and Helvey, 1976). In contrast, in Arizona, a high-severity burned-over ponderosa (*Pinus ponderosa*) forest showed decreased soil water storage due to an increase in overland flow and drying of the bare soil surface (Campbell *et al.*, 1977). These effects often continue until revegetation has occurred.

10.2.4 Forest floor/duff

The hydrological response of the forest soil is influenced by the effects of the wildfire on the organic material (O horizon) found above the mineral soil in the forest floor (Fosberg, 1977; Brown *et al.*, 1985). This organic material commonly has three distinct layers: (i) litter, the top layer, is the undecomposed, unconsolidated material consisting of debris such as twigs, grasses, leaves and needles; (ii) the fermentation layer is below the litter and consists of partially decomposed organic material often bound with fungus; and (iii) humus, the third and deepest organic layer, is extensively decomposed material found just above the A horizon of the mineral soil. In the field, it can be difficult to discern the physical separation between the fermentation and humus layers because humus is usually mixed in varying proportions with partially decomposed organic materials. Forest scientists and fire managers commonly use the term 'duff' to refer collectively to the fermentation and humus layers, while the term forest floor is used to refer to all the surface organic horizons (duff and litter) overlying the mineral soil (DeBano *et al.*, 1998). Although there is commonly a clear division between the mineral soil and overlying duff, site disturbances may mix varying amounts of mineral soil into the duff.

The ground-level effects of wildfires can range from removal of litter to total consumption of the forest floor and alteration of the mineral soil structure below (Wells *et al.*, 1979; Brown *et al.*, 1985; DeBano *et al.*, 1998; Ryan, 2002). Mineral soil that becomes exposed when the forest-floor duff is completely consumed is highly susceptible to erosion (Wells *et al.*, 1979; Soto *et al.*, 1994). Additionally, infiltration and the water-storage capacity of the mineral soil are significantly reduced because the 'sponge' effect of the organic forest-floor material is gone and the mineral soil cannot absorb short-duration, high-intensity rainfall (Baker, 1990).

Any remaining unburned duff below the ash layer can create water-repellent patches when dry, but these can be water-absorbent patches when moist. This patchiness increases the spatial variability of the soil properties and adds complexity to our understanding of post-wildfire runoff and erosion responses. Even when water repellency is extreme, prolonged rainfall can cause the soil to be transformed to a 'normal' wettable state (Doerr *et al.*, 2000; Stoof *et al.*, 2011), but soil can regain its repellent state once dry conditions return (Shakesby and Doerr, 2006).

10.2.5 Soil and spatial variability

Soil properties are naturally highly variable. Soil erosion experiments generally measure standard deviations in erodibility values similar to the means, and coefficients of variation greater than 30% are common (Elliot, 1990). Soils near the tops of ridges tend to be coarser grained and shallower, whereas soils at the bottoms of hillslopes may be finer grained. The disturbance from fire rather than inherent soil properties dominates the erodibility of the soils, and the distribution of the disturbance is seldom uniform (Robichaud *et al.*, 2007).

The combined effects of a mosaic of fire severity and soil variability result in spatial variability of soil erodibility that has some degree of predictability. For instance, the effect of water repellency decreases with an increase in spatial scale (Larson *et al.*, 2009), because water will often find infiltration pathways via natural hillslope or landscape features. There will be areas following wildfire where the fire burned at a higher soil burn severity as defined by Parsons *et al.* (2010), leading to a complete loss of surface cover and most likely the development or enhancement of a water-repellent soil condition. There will be other areas where the fire burned at low soil burn severity, resulting in an area of minimal erosion risk; a large percentage of any fire will generally exhibit a combination of these characteristics (Robichaud *et al.*, 2010b; Wagenbrenner *et al.*, 2010). Spatial variability analyses have shown that following some wildfires, there are definite trends in degree of fire severity, whereas the variability is evenly distributed on a hillslope or watershed following other fires (Robichaud and Miller, 1999).

10.3 Fire Effects on Vegetation: Interception and Evapotranspiration

Wildfire can have a significant effect on the vegetation ranging from complete combustion of the canopy for hundreds of square kilometers to a little charring of needles or leaves. Forests experience a reduction in evaporative losses through precipitation, interception and evapotranspiration, thereby increasing the amount of rain and snow reaching the ground and increasing soil moisture, runoff and streamflow (Neary *et al.*, 2005; Niemeyer *et al.*, 2019). The combustion of the forest canopy has shown a significant effect on interception by decreasing stand rainfall-intercepting capacity. For example, duff and vegetation canopy combustion has been found to decrease water storage or 'hydrological buffering' capacity, especially on north-aspect slopes in dry conifer forests (Ebel, 2013).

Removal of forest canopy by wildfire can also increase accumulation of the snowpack; the difference may be a function of reduced interception and ablation from the tree canopy (Burles and Boon, 2011; Gleason *et al.*, 2013), as canopy removal will reduce interception losses during the spring when precipitation falls as rain. For example, it has been found that a burned forest canopy produced a 4–11% increase in snow water equivalence accumulation as compared with that produced by the mature forest stand. This is similar to other disturbances that remove the canopy such as clearcut logging (Winkler *et al.*, 2010). Canopy reduction in the burned forest stands also resulted in an ablation rate greater than that in the mature forest stand, and burned areas experienced earlier snowmelt than unburned areas in the Cascades Mountains, Washington (Winkler *et al.*, 2010; Burles and Boon, 2011; Gleason *et al.*, 2013).

Decreased canopy cover also generally drives the reduction of evapotranspiration. For example, in a semi-arid pine and mixed-eucalypt forest, Dore *et al.* (2012) found that evapotranspiration decreased immediately after the fire where low or no vegetation cover resulted in a high ratio of evaporation to transpiration. Transpiration decreased the most in dry conifer forest followed by wet conifer forest, deciduous forest and grassland. Fire severity also influences evapotranspiration, as the eucalypt forest experienced 41% lower evapotranspiration after a high-severity burn compared with unburned forest, whereas moderate severity was only 3% lower in the first and second years following the fire. The lower evapotranspiration was offset by regenerating seedlings and forest-floor evapotranspiration and interception loss (Nolan *et al.*, 2014). Recovery from wildfire for evapotranspiration and interception will occur 3–4 years after burning (Soto and Diaz-Fierros, 1997; Nolan *et al.*, 2014), as these processes are correlated with increases in leaf area, canopy density or basal area (Soto and Diaz-Fierros, 1997).

10.4 Fire Effects on Watershed Response

10.4.1 Precipitation

Precipitation patterns in the years following the disturbance are crucial in determining the hydrological response. If precipitation is minimal, there will be little erosion, but there will also be little natural or seeded vegetation regrowth and little soil recovery from water-repellent conditions, meaning that the site can remain susceptible to erosion for several additional years.

If the precipitation comes as short-duration, high-intensity storms, then erosion can be severe. If the weather is very wet, and the soils are water repellent, there is a high likelihood of severe soil erosion, but there may also be rapid vegetation recovery. Runoff and erosion from rainfall or rain-on-snow events will be much greater than runoff from melting snow. Once a site has recovered (e.g. vegetative cover has returned and soils have stabilized), rainfall rates in excess of 50 mm/h, or total rainfall amounts greater than 100 mm within a day, are necessary before any significant upland erosion will occur. This is considered a rare occurrence in forested areas.

10.4.2 Surface runoff/overland flow

When a high-severity fire results in poor hydrological conditions (most precipitation does not infiltrate into the soil and the streamflow response to precipitation is rapid), runoff and peak flows can increase by several orders of magnitude and can cause some of the most extreme hydrological impacts (Neary *et al.*, 2005; Moody and Martin, 2009; Robichaud *et al.*, 2010a; Ebel *et al.*, 2023). These increased watershed responses are typically caused by infiltration-excess and sometimes by saturation-excess overland flow, or a combination of both (Sheridan *et al.*, 2007; Moody *et al.*, 2013). Increased runoff is often attributed to a combination of: development of soil water repellency, the increase in percentage of bare soil, the decrease in canopy interception and the lack of surface water storage. Convective rainfall events are the primary cause of increased runoff.

10.4.3 Stream flow

In general, rainfall-runoff methods assume temporally and spatially uniform rainfall (which is usually not applicable to burned areas in mountainous terrain) and runoff contributions to channel flow from the entire drainage basin area. Depending on the post-wildfire response, hillslope-runoff-generating processes may switch between infiltration-excess and saturation-excess overland flow (Ebel *et al.*, 2012; Moody *et al.*, 2013). Runoff generation by infiltration excess has been found to be more sensitive to the uncertainty associated with precipitation than by saturation excess.

Annual streamflow discharge from a 560 ha burned watershed in the Cascade Range of central Washington increased fivefold relative to a pre-fire streamflow. Differences between the pre- and post-fire streamflow discharge varied from nearly 110 mm in a dry year to about 477 mm in a wet year, as summarized by Neary *et al.* (2005) and Niemeyer *et al.* (2019). Campbell *et al.* (1977) observed a 3.5-fold increase in average annual stormflow discharge from a small 8 ha severely burned watershed following the occurrence of a wildfire in a south-western ponderosa pine forest. Average annual stormflow discharge from a smaller 4 ha moderately burned watershed increased 2.3-fold to almost 15 mm in relation to an unburned (control) watershed. The average runoff efficiency on the severely burned watershed was 350% greater when the precipitation input was rain and 51% less in snowmelt periods. The observed differences during rainfall events were largely due to the lower tree density, a greater reduction in litter cover and a more extensive formation of water-repellent soil, resulting in lower evapotranspiration losses and more stormflow on the severely burned watershed than on the moderately burned watershed. In the spring snowmelt period, the lower tree density of the severely burned watershed allowed more of the snowpack to be lost to evaporation. As a result, less stormflow occurred than on the more shaded, moderately burned watershed.

In the first year after a 150 ha watershed was burned over by a wildfire in southern France, streamflow discharge increased 30% to nearly 60 mm (Lavabre *et al.*, 1993). The pre-fire vegetation on the watershed was primarily a mixture of maquis, cork oak (*Quercus suber*) and chestnut oak (*Quercus prinus*) trees and was attributed to the reduction in evapotranspiration due to the loss of vegetation by the fire.

10.4.4 Peak flow

The effects of wildfire on storm peak flows are highly variable and complex. Some of the most profound impacts besides the wildfire itself can be the post-fire peak flow response (Fig. 10.2) (Neary *et al.*, 2005; Niemeyer *et al.*, 2019). One to three orders of magnitude increases in peak flows are related to the occurrence of short-duration but intense rainfall events, steep watersheds and high-severity burn areas. These peak flows are instrumental in channel formation, sediment transport and sediment redistribution within stream corridors. The timing of these peak flow events is often shorten producing more 'flash floods.' These peak flow events also increase in frequency after the fire.

One consideration of the peak flow is the size of the area (watershed) being affected by the rainfall event, and the burn severity within the watershed (Shakesby and Doerr, 2006). Cannon *et al.* (2001a, b) suggested that areas of about 1 km^2 or less will produce the greatest specific discharge, as this sized area often will have a combination of high soil burn severity, steep slopes and intense rainfall from a single storm cell. Peak flows are also an important consideration in management and design of structures (e.g. bridges, dams, levees, buildings, cultural sites, personnel property).

10.4.5 Base flow

The removal of forest canopy cover decreases interception and transpiration, and this generally increases annual water yields including base flow (MacDonald and Stednick, 2003). The increases in annual water yield following forest harvest are usually assumed to be proportional to the amount of forest cover removed, but at least 15–20% of the trees must be removed to produce a statistically detectable effect; this would be analogous to tree loss from a moderate or severe burn. In areas where the annual precipitation is less than 450–500 mm, removal of the forest canopy is unlikely to significantly increase annual water yields. In drier areas, the decrease in interception and transpiration is generally offset by the increase in soil evaporation, and there is no net change in runoff as long as there is no change in the underlying runoff processes (MacDonald and Stednick, 2003).

Base flows formed from percolated water are often increased after wildfires as evapotranspiration and interception decrease. Local soils and geology play an important role in determining whether the excess water goes to springs, baseflow or groundwater recharge, which are driven by the seasonal patterns of the amount and timing of precipitation (Neary *et al.*, 2005).

10.4.6 Water quality

Fire-impacted watersheds will generally increase their peak flows, which in turn will affect the water quality. Suspended fines (ash and sediment) and bedload material are the most visible effects, often increasing by several orders of magnitude (Fig. 10.3). Sánchez-García *et al.* (2023) described the potential impacts of ash on water quality. Biogeochemical processes can alter the mobility and availability of nutrients and potentially toxic compounds (Santín *et al.*, 2015; Gustine *et al.*, 2022), which can disrupt drinking water supplies (Hohner *et al.*, 2019; Neris *et al.*, 2021).

The review by Smith *et al.* (2011) reported that first-year post-fire suspended sediment exports varied from 0.017 to 50 t/ha/year across a large range of catchment sizes (0.021–1655 km^2). This represented an estimated increase of 1–1459 times unburned exports. Maximum reported concentrations of total suspended solids in streams for the first year after the fire ranged from 11 to 500,000 mg/L.

Fig. 10.2. Channel scour after a high-intensity rainfall event following the 2022 Pipeline Fire in Arizona.

Similarly, there was a large range in first-year post-fire stream exports of total nitrogen (N) (1.1–27 kg/ha/year) and total phosphorus (P) (0.03–3.2 kg/ha/year), representing a multiple change of 0.3–431 times the unburned values, while nitrate (NO_3^-) exports of 0.04–13.0 kg/

Fig. 10.3. Ash and sediment deposits after a high-intensity rainfall event on the 2011 Wallow Fire reburned in the 2021 Horton Fire in Arizona.

ha/year (3–250 times unburned) have been reported. Nitrate nitrogen (NO_3-N), ammonium nitrogen (NH_4-N), phosphate (PO_4^{3-}), potassium (K) and alkalinity increased in stream water following ash input, but concentrations of each returned to pre-fire conditions within 4 months (Earl and Blinn, 2003). Mineral nutrients such as calcium (Ca), magnesium (Mg) and potassium (K) are typically converted to oxides (often a major component of the light-colored ash resulting from complete combustion after fire) that are relatively soluble (Ice *et al.*, 2004). The amount of Ca typically found in ash-contaminated runoff can be used as a marker to define water contaminated with ash runoff. Elevated sodium (Na^+), chloride (Cl^-) and sulfate (SO_4^{2-}) solute yields have been observed soon after a fire in coniferous forests (Smith *et al.*, 2011). The NH_4-N, P and total cyanide (CN^-) concentrations were significantly correlated with Ca concentrations, indicating association of chemicals with ash-related inputs (Crouch *et al.*, 2006).

10.5 Fire Effects on Sediment Yield

10.5.1 Soil erosion

Variability in the post-wildfire erosion response is caused by differences in the runoff and transport processes, which are affected by topography, temporal and spatial variability of fire-affected soils (e.g. water storage, infiltration, fine root breakdown), precipitation and runoff (Moody *et al.*, 2013). Complexity in the erosion response is due to non-uniformity in the spatial distribution of sediment sources and from sediment that is being transported on hillslopes that can change the surface roughness, causing deviations in runoff patterns and sediment transport (Kirkby, 2011). Soil erodibility in post-fire environments can be particularly variable in response to changes caused by heating during wildfires and to changes in soil moisture conditions after wildfires. Thus, the sediment availability, which is a function of the sediment supply, and its associated erodibility and mobility determine the vulnerability of the hillslope to erosion processes (Moody *et al.*, 2013).

Characteristics of climate, topography, soils, vegetation, degree and extent of soil burn severity, and channel proximity create high variability in post-fire responses and recovery rates (DeBano *et al.*, 1998; Robichaud, 2000). More specifically, post-fire runoff, peak flow rates and erosion rates are highly dependent on rainfall intensity and amount, as well as the magnitude and spatial distribution of fire-induced soil disturbances. Fire effects on soil include decreases in soil organic matter and surface litter, reductions in soil aggregates leading to diminished soil structure, loss of interceptive and transpiring vegetation, changes in hydraulic roughness, and alteration or formation of water-repellent soil conditions (DeBano *et al.*, 1998; Certini, 2005; Shakesby and Doerr, 2006). High-severity fires tend to be larger and have more homogeneous patches of soil disturbance than low- or moderate-severity fires. Increased spatial extent, or patch size, of the disturbed soil may result in greater overland flow, increased potential for rilling and larger amounts of sediment transport (Moody *et al.*, 2008).

In severely burned areas, high-intensity, short-duration rain events have increased peak flows 2–2000-fold (DeBano *et al.*, 1998; Neary *et al.*, 2005). Published sediment yields after high-severity wildfires range from 0.01 to over 110 t/ha/year in the first year after burning (Fig. 10.4) (Robichaud *et al.*, 2000; Benavides-Solorio and MacDonald, 2005; Moody and Martin, 2009). In most cases, the decline in soil water repellency and vegetative regrowth means that these large increases in runoff and erosion diminish quite rapidly. Most long-term studies show a return to baseline conditions by about the fourth year after burning (Benavides-Solorio and MacDonald, 2005).

10.5.2 Debris flows

Debris flows are a sediment-laden flow with unconsolidated sediment concentration of 50–77% by volume. These are capable of supporting gravel and boulders while flowing and can occur after wildfires. This is often referred

Fig. 10.4. Sediment and debris after a high-intensity rainfall event after the 2022 Hermit Peak Calf Canyon Fire in New Mexico.

to as 'sediment bulking' (Cannon *et al.*, 2008). Post-wildfire triggering mechanisms for debris flows include progressive entrainment of soil eroded from hillslopes and channels by overland flow (Cannon *et al.*, 2001b; Santi *et al.*, 2008; Rengers *et al.*, 2020). The introduction of ash from fires provides sufficient fine-grained material to support the sediment (Gabet and Sternberg, 2008). The saturation of soil above the fire-induced water-repellent 'layer' initiates 'thin debris flows,' and shallow soil slumps induced by low infiltration into soils increases pore pressure resulting in liquefaction and mobilization.

Cannon *et al.* (2008) suggested that short-duration, high-intensity events in the intermountain west and longer-duration frontal-precipitation events in southern California initiate the majority of the debris flows. In south-eastern Australia, Nyman *et al.* (2011) observed that short-duration, high-intensity rainfall events combined with converging hillslopes with ample available sediment were the cause of observed debris flows.

10.6 Stabilization and Rehabilitation

The US Department of Agriculture (USDA) Forest Service and US Department of Interior land management agencies develop watershed rehabilitation plans with Burned Area Emergency Response teams after severe wildfires to reduce the effects of soil erosion and flooding (Robichaud and Ashmun, 2013). Many factors are taken into consideration such as soil burn severity, climate and values at risk during their assessment. Many post-fire assessment procedures and tools that have been developed for the USA are being used by managers in other countries. Concerted efforts have been made to adapt portions of the US post-fire assessment and treatment recommendation protocol for use within countries throughout the world, including Argentina, Australia, Canada, France, Greece, Israel, Italy, Peru, Portugal, Spain, the UK, Peru and New Zealand.

The general procedure is to map the soil burn severity, determine whether there are values at risk, model the potential increase in watershed response, select treatments and implement these treatments before the first damaging storms. Since 2002, the USDA Forest Service, Geospatial Technology and Applications Center, US Geological Survey, and Earth Resources Observation and Science Center have differenced pre- and post-fire broadband satellite images (10–30m pixels, 5–14 day return period) of the burned area to derive a preliminary classification of landscape change. An image transformation algorithm known as the normalized burn ratio (NBR) which uses a near-infrared (NIR) and a short-wave infrared (SWIR) band is applied to the pre- and post-fire imagery before differencing. The NBR algorithm is:

$$\text{NBR} = (\text{NIR} - \text{SWIR}) / (\text{NIR} + \text{SWIR}) \quad (10.1)$$

The result of the differenced NBR is a continuous raster GIS layer that is classified into four burn severity classes: unburned, low, moderate and high. This map is referred to as the Burned Area Reflectance Classification (BARC) map and is usually the starting point for the soil burn severity map (Parsons *et al.*, 2010). The BARC is validated using a field guide directing the user to make five observations (ground cover, ash color and depth, soil structure, roots and soil water repellency) at various data-collection locations for each site within a fire (Parsons *et al.*, 2010).

Estimation of potential post-fire erosion is often accomplished using the physically based Forest Service Water Erosion Prediction Project (FSWEPP) interfaces (Elliot, 2004; Elliot and Robichaud, 2011), which are adaptations of WEPP for forest and rangeland environments. WEPP and the full suite of FSWEPP interfaces can be accessed online at: https://forest.moscowfsl.wsu.edu/fswepp (accessed 7 May 2025). There are several FSWEPP interfaces that calculate potential post-fire erosion rates. The 'Erosion Risk Management Tool' (ERMiT) of Robichaud *et al.* (2006) is the most common. ERMiT predicts the probability associated with a given hillslope sediment yield (untreated, treated with seeding, dry agricultural straw mulching or erosion barriers) from a single storm in each of 5 years following the wildfire. WEPPcloud is a pre- and post-disturbance (e.g. wildfire) management tool to model forested watersheds throughout the USA, Australia and European Union (Dobre *et al.*, 2022; Lew *et al.*, 2022).

WEPPcloud automatically builds input files from nationally available databases. Model outputs are generated in both tabular and GIS formats that are easy to interpret by users. These tools are used widely for post-wildfire assessments (East *et al.*, 2024).

Alternatively, the Kinematic Runoff and Erosion Model (KINEROS2), a spatially distributed, event-based watershed rainfall-runoff and erosion model, and the companion ArcGIS-based Automated Geospatial Watershed Assessment (AGWA) tool are also used in the post-fire environment (Goodrich *et al.*, 2012). AGWA automates the time-consuming tasks of watershed delineation into distributed model elements and initial parameterization of these elements using commonly available, national GIS data layers. After wildfires, model parameters are changed in the model as a function of burn severity and pre-fire land-cover type. AGWA has a differencing function with which the stored results from pre- and post-fire simulations can be subtracted over all of the spatially distributed model elements. These differences, in absolute or percentage change terms, can then be mapped back into the GIS display to provide a quick visual indication of watershed 'hot spots' where large changes between the two simulations have taken place.

Research syntheses have included measured post-fire erosion rates (Moody and Martin, 2009), fire effects on soils (Cerdá and Robichaud, 2009), effects of fire on soil and water (Neary *et al.*, 2005) and hydrological models (Ebel *et al.*, 2023). Post-fire treatment syntheses and treatment catalogs provide current information on various treatments in formats that are easily used by post-fire assessment teams. Published treatment syntheses (Napper, 2006; Foltz *et al.*, 2009; Robichaud *et al.*, 2010b; Peppin *et al.*, 2011) are available at the BAERTOOLS website (https://forest.moscowfsl.wsu.edu/BAERTOOLS, accessed 7 May 2025).

References

Baker, M.B.J. (1990) Hydrologic and water-quality effects of fire. In: Kramames, J.S. (tech. coord.) *Proceedings: Effects of Fire Management of Southwestern Natural Resources*, General Technical Report RM-GTR-191. US Department of Agriculture, Forest Service, Rocky Mountain Forest and Range Experiment Station Fort Collins, Colorado.

Benavides-Solorio, J. and MacDonald, L.H. (2005) Measurement and prediction of post-fire erosion at the hillslope scale, Colorado front range. *International Journal of Wildland Fire* 14, 457–474.

Bren, L. (2022) *Forest Hydrology and Catchment Management, an Australian Perspective*, 2nd edn. Springer, Dordrecht, Netherlands.

Brown, J.K., Mardsen, M.A., Ryan, K.C. and Reinhardt, E.D. (1985) *Predicting Duff and Woody Fuel Consumed by Prescribed Fire in Northern Rocky Mountains*. Research Paper INT-337. US Department of Agriculture, Forest Service, Intermountain Forest and Range Research Station, Ogden, Utah.

Burles, K. and Boon, S. (2011) Snowmelt energy balance in a burned forest plot, Crowsnest Pass, Alberta, Canada. *Hydrologic Processes* 25, 3012–3029.

Butzen, V., Seeger, M., Marruedo, A., Jonge, L., Wengel, R. *et al.* (2015) Water repellency under coniferous and deciduous forest – experimental assessment and impact on overland flow. *Catena* 133, 255–265.

Campbell, R.E., Baker, M.B.J., Ffolliott, P.F., Larson, F.R. and Avery, C.C. (1977) *Wildfire Effects on a Ponderosa Pine Ecosystem: An Arizona Case Study*. Research Paper RP-191. US Department of Agriculture, Forest Service, Rocky Mountain Forest and Range Experiment Station, Fort Collins, Colorado.

Cannon, S.H., Bigio, E.R. and Mine, E. (2001a) A process for fire-related debris flow initiation, Cerro Grande fire, New Mexico. *Hydrological Processes* 15, 3011–3023.

Cannon, S.H., Kirkham, R.M. and Parise, M. (2001b) Wildfire-related debris-flow initiation processes, Storm King Mountain, Colorado. *Geomorphology* 39, 171–188.

Cannon, S.H., Gartner, J.E., Wilson, R.C. and Laber, J.L. (2008) Storm rainfall conditions for floods and debris flows from recently burned areas in Southwestern Colorado and Southern California. *Geomorphology* 96, 250–269.

Cerdá, A. and Robichaud, P.R. (2009) *Fire Effects on Soils and Restoration Strategies*. Science Publishers, Enfield, New Hampshire.

Certini, G. (2005) Effects of fire on properties of forest soils: A review. *Oecologia* 143, 1–10.

Crouch, R.L., Timmenga, H.J., Barber, T.R. and Fuchsman, P.C. (2006) Post-fire surface water quality: Comparison of fire retardant versus wildfire-related effects. *Chemosphere* 62, 874–889.

DeBano, L.F. (1981) *Water Repellent Soils: A State-of-the Art*. General Technical Report PSW-46. US Department of Agriculture, Forest Service, Pacific Southwest Forest and Range Experiment Station, Berkeley, California.

DeBano, L.F. (2000) The role of fire and soil heating on water repellency in wildland environments: A review. *Journal of Hydrology* 231–232, 195–206.

DeBano, L.F., Savage, S.M. and Hamilton, D.A. (1976) The transfer of heat and hydrophobic substances during burning. *Soil Science Society of America Journal* 40, 779–782.

DeBano, L.F., Neary, D.G. and Ffolliott, P.F. (1998) *Fire's Effects on Ecosystems*. Wiley, New York.

Dekker, L.W. and Ritsema, C.J. (2000) Wetting patterns and moisture variability in water repellent Dutch soils. *Journal of Hydrology* 231–232, 148–164.

Dekker, L.W., Doerr, S.H., Oostindie, K., Ziogas, A.K. and Ritsema, C.J. (2001) Water repellency and critical soil water content in dune sand. *Soil Science Society of America Journal* 65, 1667–1674.

Dobre, M., Srivastava, A., Lew, R., Deval, C., Brooks, E.S. *et al.* (2022) WEPPcloud: An online watershed-scale hydrologic modeling tool. Part II. Model performance assessment and applications to forest management and wildfires. *Journal of Hydrology* 610, 127776. DOI: 10.1016/j.jhydrol.2022.127776.

Doerr, S.H. and Thomas, A.D. (2000) The role of soil moisture in control patterns and moisture variability in water repellent Dutch soils. *Journal of Hydrology* 231–232, 148–164.

Doerr, S.H., Shakesby, R.A. and Walsh, R.P.D. (2000) Soil water repellency: Its causes, characteristics and hydro-geomorphological significance. *Earth-Science Reviews* 51, 33–65.

Doerr, S.H., Shakesby, R.A., Dekker, L.W. and Ritsema, C.J. (2006) Occurrence, prediction and hydrological effects of water repellency amongst major soil and land-use types in a humid temperate climate. *European Journal of Soil Science* 57, 741–754.

Dore, S., Montes-Helu, M., Hart, S.C., Hungate, B.A., Koch, G.W. *et al.* (2012) Recovery of ponderosa pine ecosystem carbon and water fluxes from thinning and stand-replacing fire. *Global Change Biology* 18, 3171–3185.

Earl, S.R. and Blinn, D.W. (2003) Effects of wildfire ash on water chemistry and biota in South-Western U.S.A streams. *Freshwater Biology* 48, 1015–1030.

East, A.E., Logan, J.B., Dow, H.W., Smith, D.P., Iampietro, P. *et al.* (2024) Post-fire sediment yield from a Central California watershed: Field measurements and validation of the WEPP model. *Earth and Space Science* 11, e2024EA003575.

Ebel, B.A., Moody, J.A. and Martin, D.A. (2012) Hydrologic conditions controlling runoff generation immediately after wildfire. *Water Resources Research* 48, W03529.

Ebel, B.A. (2013) Wildfire and aspect effects on hydrologic states after the 2010 Four Mile Canyon Fire. *Vadose Zone Journal* 12, vzj2012.

Ebel, B.A., Shephard, Z.M., Walvoord, M.A., Murphy, S.F., Partridge, T.F. *et al.* (2023) Modeling post-wildfire hydrologic response: Review and future directions for applications of physically based distributed simulation. *Earth's Future* 11, e2022EF003038.

Elliot, W.J. (1990) *A Compendium of Soil Erodibility Data from WEPP Cropland Soil Field Erodibility Experiments 1987 & 88*. Agricultural Research Service, US Department of Agriculture. National Soil Erosion Research Laboratory, Purdue University, West Lafayette, Indiana.

Elliot, W.J. (2004) WEPP internet interfaces for forest erosion prediction. *Journal of American Water Resource Association* 40, 299–309.

Elliot, W.J. and Robichaud, P.R. (2011) Risk-based erosion assessment: Application to forest watershed management and planning. In: Morgan, R.P.C. and Nearing, M.A. (eds) *Handbook of Erosion Modelling*. Wiley-Blackwell, Chichester, UK, pp. 313–323.

Foltz, R.B., Robichaud, P.R. and Rhee, H. (2009) *A Synthesis of Postfire Road Treatments for BAER Teams: Methods, Treatment Effectiveness, and Decision-Making Tools for Rehabilitation*. General Technical Report RMRS-GTR-228. US Department of Agriculture, Forest Service, Rocky Mountain Research Station, Fort Collins, Colorado.

Fosberg, M.A. (1977) *A Synthesis of Postfire Road Treatments for BAER Teams: Methods, Treatment Effectiveness, and Decision-Making Tools for Rehabilitation*. Research Paper RM-195. US

Department of Agriculture, Forest Service, Rocky Mountain Forest and Range Experiment Station, Fort Collins,Colorado.

Gabet, E.J. and Sternberg, P. (2008) The effects of vegetative ash on infiltration capacity, sediment transport, and the generation of progressively bulked flows. *Geomorphology* 101, 666–673.

Gleason, K.E., Nolin, A.W. and Roth, T.R. (2013) Charred forests increase snowmelt: Effects of burned woody debris and incoming solar radiation on snow ablation. *Geophysical Research Letters* 40, 4654–4661.

Goodrich, D.C., Burns, I.S., Unkrich, C.L., Semmens, D.J., Guertin, D.P. *et al*. (2012) KINEROS2/AGWA: Model use, calibration, and validation. *Transactions of the ASABE* 55, 1561–1574.

Gustine, R.N., Hanam, E.J., Robichaud, P.R. and Elliot, W.J. (2022) From burned slopes to streams: How wildfire affects nitrogen cycling and retention in forests and fire-prone watersheds. *Biogeochemistry* 157, 51–68.

Hohner, A.K., Rhoades, C.C., Wilkerson, P. and Rosario-Ortiz, F.L. (2019) Wildfires alter forest watersheds and threaten drinking water quality. *Accounts of Chemical Research Journal* 52, 1234–1244.

Huffman, E.L., MacDonald, L.H. and Stednick, J.D. (2001) Strength and persistence of fire induced soil hydrophobicity under ponderosa and lodgepole pine, Colorado Front Range. *Hydrologic Processes* 15, 2877–2892.

Ice, G.G., Neary, D.G. and Adams, P.W. (2004) Effects of wildfire on soils and watershed processes. *Journal of Forestry* 102, 16–20.

Jordán, A., Zavala, L.M., Mataix-Solera, J. and Doerr, S.H. (2013) Soil water repellency: Origin, assessment and geomorphic consequences. *Catena* 108, 742576.

Kirkby, M.J. (2011) A personal synthesis and commentaries. In: Kirkby, M.J. (ed.) *Benchmark Papers in Hydrology, 6. Hydro-Geomorphology, Erosion and Sedimentation*. International Association of Hydrological Science, pp. 1–36.

Klock, G.O. and Helvey, J.D. (1976) Soil-water trends following a wildfire on the entiat experimental forest. In: *Proceedings of the Tall Timbers Fire Ecology Conference, October 16–17, 1974, Portland, Oregon*, Tall Timbers Research Station, Tallahassee, Florida, pp. 193–200 (Vol. 15).

Larson, I.J., MacDonald, L.H., Brown, E., Rough, D., Welsch, M. *et al*. (2009) Causes of post-fire runoff and erosion; water repellency, cover or soil sealing? *Soil Science Society of American Journal* 73, 1393–1407.

Lavabre, J., Torres, D.S. and Cernesson, F. (1993) Changes in the hydrological response of a small Mediterranean basin a year after a wildfire. *Journal of Hydrology* 142, 273–299.

Lew, R., Dobre, M., Srivastava, A., Brooks, E.S., Elliot, W.J. *et al*. (2022) WEPPcloud: An online watershed-scale hydrologic modeling tool. *Journal of Hydrology* 608, 127603.

MacDonald, L.H. and Stednick, J.D. (2003) *Forests and Water: A State-of-the-art Review for Colorado*. Colorado Water Resources Research Institute Completion Report. Colorado State University, Fort Collins, Colorado.

Moffet, C.A., Pierson, F.B., Robichaud, P.R. and Spaeth, K.E. (2007) Modeling soil erosion on steep sagebrush rangeland before and after prescribed fire. *Catena* 71, 218–228.

Moody, J.A. and Martin, D.A. (2009) Synthesis of sediment yields after wildland fire in different rainfall regimes in the Western United States. *International Journal of Wildland Fire* 18, 96–115.

Moody, J.A., Martin, D.A., Haire, S.L. and Kinner, D.A. (2008) Linking runoff response to burn severity after wildfire. *Hydrological Processes* 22, 2063–2074.

Moody, J.W., Shakesby, R.A., Robichaud, P.R., Cannon, S.H. and Martin, D.A. (2013) Current research issues related to post-wildfire runoff and erosion processes. *Earth-Science Review* 122, 10–37.

Napper, C. (2006) *Burned Area Emergency Response Treatments Catalog*. USDA Forest Service, National Technology and Development Program, Watershed, Soil, Air Management 0625 1801-STTDC, San Dimas, California.

Neary, D.G., Ryan, K.C. and DeBano, L.F. (2005) *Wildland Fire in Ecosystems: Effects of Fire on Soil and Water*. General Technical Report RMRS-GTR-42-V4. General Technical Report RMRS-GTR-42-V4. US Department of Agriculture, Forest Service, Rocky Mountain Research Station, Ogden, Utah.

Neris, J., Santin, C., Lew, R., Robichaud, P.R., Elliot, W.J. *et al*. (2021) Designing tools to predict and mitigate impacts on water quality following the Australian 2019/2020 wildfires: Insights from Sydney's largest water supply catchment. *Integrated Environmental Assessment and Management* 17, 1151–1161.

Niemeyer, R.J., Bladon, K.D. and Woodsmith, R.D. (2019) Long-term hydrologic recovery after wildfire and post-fire forest management in the interior Pacific Northwest. *Hydrological Processes* 34, 1182–1197.

Nolan, R.H., Lane, P.N., Benyon, R.G., Bradstock, R.A. and Mitchell, P.J. (2014) Changes in evapotranspiration following wildfire in resprouting eucalypt forests. *Ecohydrology* 7, 1363–1377.

Nyman, P., Sheridan, G., Smith, H.G. and Lane, P.N.J. (2011) Evidence of debris flow occurrence after wildfire in upland catchments of South-East Australia. *Geomorphology* 125, 383–401.

Parsons, A., Robichaud, P.R., Lewis, S.A., Napper, C. and Clark, J.T. (2010) *Field Guide for Mapping Post-Fire Soil Burn Severity*. General Technical Report RMRS-GTR-243. US Department of Agriculture, Forest Service, Rocky Mountain Research Station, Fort Collins, Colorado.

Peppin, D.L., Fule, P.Z., Sieg, C.H., Beyers, J.L., Hunter, M.E. *et al.* (2011) Recent trends in post-wildfire seeding in western US forests: Costs and seed mixes. *International Journal of Wildland Fire* 20, 702–708.

Pierson, F.B., Robichaud, P.R., Moffet, C.A., Spaeth, K.E., Williams, C.J. *et al.* (2008) Soil water repellency and infiltration in coarse-textured soils of burned and unburned sagebrush ecosystems. *Catena* 74, 98–108.

Rengers, F.K., McGuire, L.A., Oakley, N.S., Kean, J.W., Staley, D.M. *et al.* (2020) Landslides after wildfire: Initiation, magnitude, and mobility. *Landslides* 17, 2631–2641.

Robichaud, P.R. (1996) Spatially-varied erosion potential from harvested hillslopes after prescribed fire in the Interior Northwest. PhD thesis, University of Idaho, Moscow, Idaho.

Robichaud, P.R. (2000) Fire effects on infiltration rates after prescribed fire in Northern Rocky Mountain forests, USA. *Journal of Hydrology* 231–232, 220–229.

Robichaud, P.R. and Miller, S.M. (1999) Spatial interpolation and simulation of post-burn duff thickness after prescribed fire. *International Journal of Wildland Fire* 9, 137–143.

Robichaud, P.R. and Ashmun, L.E. (2013) Tools to aid post-wildfire assessment and erosion-mitigation treatment decisions. *International Journal of Wildland Fire* 22, 95–105.

Robichaud, P.R., Elliot, W.J., Pierson, F.B., Hall, D.E. and Moffet, C.A. (2006) Erosion Risk Management Tool (ERMiT), version 2006.01.18. US Department of Agriculture, Forest Service, Rocky Mountain Research Station, Moscow, ID, USA. Available at: https://forest.moscowfsl.wsu.edu/fswepp/ (accessed 30 April 2025).

Robichaud, P.R., Elliot, W.J., Pierson, F.B., Hall, D.E. and Moffett, C.A. (2007) Predicting postfire erosion and mitigation effectiveness with a web-based probabilistic erosion model. *Catena* 71, 229–241.

Robichaud, P.R., Beyers, J.L. and Neary, D.G. (2000) *Evaluating the Effectiveness of Postfire Rehabilitation Treatments*. General Technical Report RMRS-GTR-63. US Department of Agriculture, Forest Service, Rocky Mountain Research Station, Fort Collins, Colorado.

Robichaud, P.R., Ashmun, L.E. and Sims, B. (2010b) *Post-fire Treatment Effectiveness for Hillslope Stabilization*. General Technical Report RMRS-GTR-240. US Department of Agriculture, Forest Service, Rocky Mountain Research Station, Fort Collins, Colorado.

Robichaud, P.R., Wagenbrenner, J.W. and Brown, R.E. (2010a) Rill erosion in natural and disturbed forests: 1. Measurements. *Water Resources Research* 46, W10506.

Robichaud, P.R., Massman, W.J., Bova, A., Girona-García, A., Alfaro-Leranoz, A. *et al.* (2025) Comparing modelled soil temperature and moisture dynamics during prescribed fires, slash-pile burns, and wildfires. *International Journal of Wildland Fire* 34, WF22082.

Ryan, K.C. (2002) Dynamic interactions between forest structure and fire behavior in boreal ecosystems. *Silva Fennica* 36, 13–39.

Sánchez-García, C., Santín, C., Neris, J., Sigmund, G., Otero, X.L. *et al.* (2023) Chemical characteristics of wildfire ash across the globe and their environmental and socio-economic implications. *Environmental International* 178, 108065.

Santi, P.M., deWolf, V.G., Higgins, J.E., Cannon, S.H. and Gartner, J.E. (2008) Sources of debris flow material in burned areas. *Geomorphology* 96, 310–321.

Santín, C., Doerr, S.H., Otero, X.L. and Chafer, C.J. (2015) Quantity, composition and water contamination potential of ash produced under different wildfire severities. *Environmental Research* 142, 297–308.

Savage, S.M., Osborn, J., Letey, J. and Heaton, C. (1972) Substances contributing to fire-induced water repellency in soils. *Soil Science of America Proceedings* 36, 674–678.

Shakesby, R.A. and Doerr, S.H. (2006) Wildfire as a hydrological and geomorphological agent. *Earth-Science Reviews* 74, 269–307.

Sheridan, G.J., Lane, P.N.J. and Noske, P.J. (2007) Quantification of hillslope runoff and erosion processes before and after wildfire in a wet eucalyptus forest. *Journal of Hydrology* 343, 12–28.

Smith, H.G., Sheridan, G.J., Lane, P.N.J., Nyman, P. and Haydon, S. (2011) Wildfire effects on water quality in forest catchments: A review with implications for water supply. *Journal of Hydrology* 396, 170–192.

Soto, B. and Diaz-Fierros, F. (1997) Soil water balance as affected by throughfall in gorse (*Ulex europaeus* L.) shrubland after burning. *Journal of Hydrology* 195, 218–231.

Soto, B., Basanta, R., Benito, E., Perez, R. and Diaz-Fierros, F. (1994) Runoff and erosion from burnt soils in Northwest Spain. In: Sala, M. and Rubio, J.L. (eds) *Soil Erosion and Degradation as a Consequence of Forest Fires*. Geoforma Ediciones, Logroño, Spain, pp. 91–98.

Stoof, C.R., Moore, D., Ritsema, C.J. and Dekker, L.W. (2011) Natural and fire-induced soil water repellency in a Portuguese shrubland. *Soil Science Society of America Journal* 75, 2283–2295.

Wagenbrenner, J.W., Robichaud, P.R. and Elliot, W.J. (2010) Rill erosion in natural and disturbed forests: 2. Modeling approaches. *Water Resources Research* 46, W10507.

Wells, C.G., Campbell, R.E., DeBano, L.F., Lewis, C.E., Fredrickson, R.L. *et al.* (1979) *Effects of Fire on Soil: A State-of-the-Knowledge Review*. General Technical Report WO-7. US Department of Agriculture, Forest Service, Washington, D.C.

Westerling, A.L., Hidalgo, H.G., Cayan, D.R. and Swetnam, T.W. (2006) Warming and earlier spring increases western U.S. forest wildfire activity. *Science* 313, 940–943.

Winkler, R., Boon, S., Zimonick, B. and Baleshta, K. (2010) Assessing the effects of post-pine beetle forest litter on snow albedo. *Hydrological Processes* 24, 803–812.

11 Hydrology of Taiga Forests at Northern High Latitudes

T. Burenina*, A. Onuchin, I. Danilova, M. Korets and A. Musokhranova
V.N. Sukachev Institute of Forest Siberian Branch of the Russian Academy of Sciences, Krasnoyarsk, Russia

Abstract

The study described in this chapter assessed the influence of the dominating forest vegetation community in the hydrological cycle in river basins draining the territory of Central Siberia and changes in hydrological processes after logging in connection with the change of vegetation cover. The results showed that the formation of the river runoff of Central Siberia is affected by provincial climatic conditions and local river basin features, rather than by zonality. A comparative analysis of moisture consumption for evapotranspiration (ET) with the structure of forest cover in Podkamennaya Tunguska basin showed that an increase in the proportion of dark coniferous species and Scots pine in the forest stand has contributed to an increase in the evaporation of precipitation retained by tree crowns. The dynamics of ET in clear-cut areas at Yenisei Ridge showed that, apart from climatic characteristics, ET is determined by a steadily increasing trend of young and pole stands in these sites. Research in the Angara basin has shown that as a result of logging over the past two decades, a decrease in annual runoff from 1.5 to 4 mm/year has been observed in small and medium-sized tributaries of the Angara. As a result of constant changes in the structure of the vegetation cover, future scenarios of the hydrological regime of rivers after clear-cutting will be determined by both further climate changes and the trajectories of succession of the vegetation cover.

11.1 Introduction

The impact of forests on hydrological processes has long been hotly contested despite the efforts made by many scientists to find answers. Although large arrays of data on forest hydrology have been built over the past few decades, water balance variability caused by local vegetation, climate and type of locality remains the least studied issue. Large differences in results obtained over a range of geographical conditions have led to a number of theories, often conflicting, on the hydrological role of forests. The differences of opinions usually concern the influence of the forest on water yield (volume of flow) (i.e. a quantifiable water protection service related to forest cover). Other forest impacts on river flow, such as maintenance of high water quality and groundwater recharge, have long been agreed upon among forest hydrologists. A special issue in the general forest–water interaction discussion is the impact of forest cutting on the hydrological cycle of a given area. Clear-cutting affects all water balance components. However, water balance structure restores gradually with regenerating forest. Increasing wood resources development in Central Siberia has made it an urgent task to estimate the

*Corresponding author: burenina@ksc.krasn.ru

DOI: 10.1079/9781800625310.0011

hydrological effects of forest cover disturbances and analyze post-cutting and post-fire changes in water balance within geographical zones and provinces.

11.2 Literature Review

Over a fairly long period of time, from late 19th to mid-20th century, water resources were approached by forest policy makers of a number of North American and North Eurasian countries based on a simplified interpretation of forests as increasing water availability, optimally regulating the dynamics of river runoff, and preserving water resources under any hydrological and ecological conditions. Studies conducted in the second half of the 20th and in the early 21st century showed varying concepts about the hydrological role of forests (Hamilton, 2008; Ellison *et al.*, 2012).

Some authors (e.g. Schwarze, 2007) believe that, in contrast to agricultural and other treeless areas, forests increase total moisture evaporation and reduce river flow. However, according to the opposite viewpoint adhered to by, among others, Voronkov (1988) and Hamilton (2008), water yield increases with increasing forest area. Moreover, the evaporation:precipitation ratio between forested and open sites varies considerably from year to year in the same area, depending on season-specific precipitation patterns and heat balance (Onuchin and Burenina, 2008).

According to Onuchin (2015), the differences of opinions on the hydrological role of forests in snow moisture balance in forested and open sites, found under different climatic conditions, has not been duly considered. Where winters are mild, a significant portion of solid atmospheric precipitation is intercepted by the forest canopy. According to research results from Pomeroy and Schmidt (1993), more than 30% of the annual amount of solid precipitation in boreal forests can return to the atmosphere by sublimation. Forest cover removal or thinning under these conditions can result in growing snow accumulation on the ground. In Siberia, particularly in the north, snow water equivalent, in contrast, decreases after clear-cutting (Onuchin, 2015). In regions of sharply continental climate, strong winds cause active snow drifting across large areas after logging; as a result, snow moisture evaporation increases and runoff decreases (Varhola *et al.*, 2010; Onuchin *et al.*, 2021). Thus, it can be argued that there are general patterns of change in the hydrological functions of forests caused by climatic parameters, which gives grounds to talk about the formation of the geographical determinism concept in assessing the hydrological role of forests (Onuchin, 2015). The proposed concept largely removes apparent contradictions, reconciling previously formulated concepts of the general moistening, drying, and uncertain or unstable hydrological role of forests (Voronkov, 1988).

Of all types of forest management in river basins, forest clear-cutting followed by forest regeneration has the highest influence on water balance and runoff. Multiple publications have been devoted to studying the influence of logging and subsequent reforestation on the hydrological regime (e.g. Krestovsky, 1986; Hamilton and Pearce, 1987; Nisbet and McKay, 2002; Grip *et al.*, 2005). It used to be common practice to compare results from paired catchment studies, in which one of the catchments would have the forest clear-cut. Hydrological regime studies covering 100 paired river basins across varying geographical regions showed that the catchments with clear-cutting has considerably increased runoff (Bosch and Hewlett, 1982).

Krestovsky (1986), in his publication on boreal forests of the European part of Russia, showed changes in total evaporation and runoff as forests regenerate. He associated the increase in runoff in the first years after logging with increased snow accumulation in clear-cuts, and the subsequent decrease in runoff 40–50 years after logging with increased evapotranspiration (ET) of closed young coniferous stands.

The hydrological consequences of forest disturbances are determined to a great extent by the characteristics of the stands subjected to cutting, and by local climate and hydrological regimes (Onuchin, 2015; Zhang and Wei, 2021). In recent years, the cumulative effects of forest disturbance on the hydrological cycle of a catchment and the disturbance threshold have aroused special interest (Li *et al.*, 2017; Wei *et al.*, 2022).

Now that remote sensing and global climate and hydrological databases of varying spatiotemporal scales are available, it has become feasible for researchers to analyze the water cycle of any region of the Earth. Estimates of evaporation from land surface based on SEBS (Surface Energy Balance Systems), SEBAL (Surface Energy Balance Algorithm for Land) and METRIC (Mapping Evapotranspiration with Internalized Calibration) data (Gibson *et al.*, 2013; Leng *et al.*, 2017) have been subjected to validation tests in subsatellite experiments of various scales and types in different countries.

The objectives of the study discussed in this chapter were twofold: (i) to assess the influence of forest vegetation on the hydrological regime of the territory in various landscape conditions on the base of the taiga zone of Central Siberia; and (ii) to analyze the change (trends) of water balance elements (ET and runoff) in the river basins of the region disturbed by logging, taking into account their post-logging reforestation.

11.3 Study Sites and Methods

The study sites for this research were the catchments of the Yenisei tributaries (Table 11.1), forming their runoff within four landscape zones: forest-tundra, northern taiga, middle taiga and southern taiga. We used data on the discharge of 39 flow gauging stations for the period 1950–2020 published in hydrological yearbooks (https://gmvo.skniivh.ru/index.php?id=505; accessed 7 May 2025). On the basis of these data, we calculated average annual runoff (mm), runoff volume per unit area (l/s/km²) and the runoff coefficient ($K_{st} = Y/X$, where K_{st} is the runoff coefficient, Y is runoff (mm) and X is precipitation (mm)).

To assess the impact of forest vegetation on the components of the water balance, catchments of medium-sized rivers from Table 11.1 were selected: four in the Angara basin (nos. 30, 31, 36 and 37) and one catchment (no. 28) of the Sukhoy Pit River, a tributary of the Bol'shoy Pit River.

The boundaries of the river basins under study were determined using the watershed calculation tools of ESRI ArcGIS 10, based on USGS EROS Archive – Digital Elevation – Global Multi-resolution Terrain Elevation Data 2010 (GMTED2010, 7.5 arc-second spatial resolution and 66m root mean square error (RMSE) globally compared with the same National Geospatial-Intelligence Agency control points). To refine the catchment shapes for flat areas, we additionally used the ASTER Global Digital Elevation Model V003 (ASTGTM.003, 1 arc-second spatial resolution and vertical accuracy of 20 m). Polygons of the study catchments were delineated from the point location of the hydrological stations.

We used mean monthly air temperature (TMP) and mean monthly precipitation (PRE) data for the 2000–2020 period from the CRU TS 4.01 Spatial Raster Database (30 arc-second spatial resolution, https://crudata.uea.ac.uk/cru/data/hrg/cru_ts_4.01/; accessed 7 May 2025). For each year, we calculated the average annual air temperature and the total of annual precipitation.

ET values were obtained based on the MODIS Global Evapotranspiration Project (MOD16A2, v006, 2000–2020, 8 day, 500m pixel resolution) (https://lpdaac.usgs.gov/products/mod16a2v006; accessed 7 May 2025), which considered the climatic parameters, underlying surface type and vegetation diversity of the area of interest. The 8 day MODIS data were summed for each year (or transpiration season).

The raster data sets (TMP, PRE and ET) were transformed to an Albers Conical Equal Area cartographic projection with 500m spatial resolution and bilinear value interpolation. We then used these data sets to calculate the per-pixel means for TMP, PRE and ET inside the study catchments polygons.

Using zonal statistics methods, we analyzed forest inventory information collected from field trips, forestry data, forest management data and vegetation cover maps (Bartalev *et al.*, 2011).

To analyze the land cover change dynamics (deforestation and reforestation) caused by harvesting and forest fires in the river basins, we used Global Land Cover and Land Use Change (GLAD Dataset 2000–2020, 30m spatial resolution) annual data (Potapov *et al.*, 2022).

We used multiple linear regression and simple linear trend analysis based on the least squares method of calculation with standard R^2 and P values (StatSoft STATISTICA 10 and EXCEL 2007 software).

Table 11.1. Characteristics of the studied catchments.

No.	River/flow gauging station	Geographic coordinates of the flow gauging station		Basin area (km²) (according to DEM SRTM)	Annual runoff (mm)	Runoff volume per unit area (l/s/km²)
		Northern latitude	Eastern longitude			
1	Turukhan/Yanov Stan	65.97	84.32	10,454	347	11.0
2	Eloguy/Kellogg	62.49	86.26	15,821	302	9.6
3	Gorbiachin/Gorbiachin	67.52	88.43	3,065	597	18.9
4	Tukalanda/Tukalanda	68.36	88.26	817	749	23.7
5	Kureika/Svetlogorsk	66.94	88.33	40,966	766	24.3
6	Tembenchi/Tembenchi trading post	64.94	98.88	19,048	434	13.7
7	Erachimo/Bol'shoy Porog	65.63	90.01	9,100	497	15.8
8	Taimura/Kerbo	62.71	101.11	14,859	180	5.7
9	Nizhnyaya Tunguska/3rd Gulmok Creek	57.82	107.29	2,808	153	4.9
10	Nizhnyaya Tunguska-Podvoloshino Podvoloshino Podvoloshino	58.24	108.42	8,270	123	3.9
11	Nizhnyaya Tunguska/Erbogachen	61.27	108.00	77,400	121	3.8
12	Nizhnyaya Tunguska/Kislokan	63.58	103.95	155,000	125	4.0
13	Nizhnyaya Tunguska/Tura	64.26	100.18	268,000	188	5.9
14	Nizhnyaya Tunguska/Bol'shoy Porog	65.63	90.00	447,000	244	7.7
15	Chuna/Strelka Chuna	61.74	102.80	13,195	173	5.5
16	Podkamennaya Tunguska – Chemdalsk	59.6	103.33	16,709	199	6.3
17	Podkamennaya Tunguska/Baikit	61.67	96.35	159,000	201	6.4
18	Podkamennaya Tunguska/Kuz'movka	62.31	92.11	218,000	232	7.4

Continued

Table 11.1. Continued

No.	River/flow gauging station	Geographic coordinates of the flow gauging station: Northern latitude	Eastern longitude	Basin area (km²) (according to DEM SRTM)	Annual runoff (mm)	Runoff volume per unit area (l/s/km²)
19	Teya/Suvorovsky	60.58	92.62	5,626	378	12.0
20	Velmo/7 km below the mouth of the Teya	61.10	92.97	22,900	338	10.7
21	Kas/Aleksandrovsky lock	59.43	89.30	6,995	228	7.2
22	Sym/Sym trading post	60.34	88.34	22,815	322	10.2
23	Kas/Maryina Griva	59.24	88.69	424	188	6.0
24	Dubches/ Sandakches	61.28	88.50	8,568	322	10.3
25	Kem/Mikhailovka	57.95	92.37	2,901	140	4.4
26	Belaya/Belskoye	57.82	92.17	1,344	108	3.4
27	Bol'shoy Pit/Bryanka	59.10	93.47	14,773	280	8.9
28	Sukhoy Pit/Sukhoy Pit	58.80	92.83	1262	452	14.9
29	Bol'shoy Pit/Sukhoy Pit base	58.80	92.83	19,500	333	10.6
30	Karabula/Karabula	58.03	97.38	4,059	89	2.2
31	Irkineeva/Bedoba	58.80	97.19	9,573	163	3.4
32	Taseeva/ Mashukovka	57.79	94.36	116,161	194	6.2
33	Usolka/Troitskoye	57.41	94.83	9,511	82	2.6
34	Kova/Prokopyevo	58.04	100.62	10,193	141	4.5
35	Tatarka/Tatarka	58.17	93.34	1,898	398	12.3
36	Mura/Irba	58.10	98.98	9,566	83	2.6
37	Chadobets/Yarkino	59.12	99.40	14,257	135	4.3
38	Angara/Boguchany	58.38	97.38	81,000	97	3.1
39	Angara/Tatarka	58.18	93.34	252,000	222	7.0

11.4 Forest Characteristics of Forest Cover in Central Siberia

The study region stretches over 2000 km from north to south and encompasses several geographical zones, from tundra to southern taiga. On the Central Siberian Plateau, geographical zones are less pronounced compared with the lowland because of largely rough, often mountainous terrain. These factors account for the heterogeneity of forest cover in Central Siberia.

Forest cover in the Nizhnyaya Tunguska River basin is fairly homogenous, with much of the area being open larch (*Larix sibirica and L. gmelinii*) forest and tundra landscapes. In the Podkamennaya Tunguska River basin, larch is predominant, but the vegetation structure is more complex due to the presence of mixed and Scots pine (*Pinus sylvestris*) stands. In the central part of the river basin, dark-needled conifer species account for 10% of the forest stand composition, while in areas where the Podkamennaya Tunguska River basin is adjacent to the Yenisei River, these species make up 25% of the composition.

Vegetation of the southern taiga, where the entire Angara River basin is situated, is a complex of different vegetation community types. The areas close to the Angara River are mostly mixed Scots pine/larch stands, 42% of which are dominated by Scots pine with 24% being larch dominated. A number of river basins are occupied by pure Scots pine stands. Other conifers are insignificant: spruce (*Picea obovata*), Siberian fir (*Abies sibirica*) and Siberian pine (*Pinus sibirica*)-dominated stands occupy as little as 8%, 4% and 3%, respectively, of the total in the southern taiga zone. In the south of this zone (the basin of the Taseyeva River, the Angara River tributary), the proportion of dark-needled stands made up of Siberian fir and Siberian pine increases. The mountain taiga of the Yenisei Ridge is represented by Siberian fir and Siberian pine.

11.5 Changes in Hydrological Regimes in Central Siberia

A geostatistical analysis (ESRI ArcGIS 10.1) of raster data sets for TMP and PRE (CRU TS 4.01 Spatial Raster Database) and data from the flow gauging stations revealed the spatial variability these parameters for the 39 catchments studied. Precipitation distribution appeared to reflect the regional orographic features and air mass circulation. As water yield is influenced by many factors, we carried out hydrological zoning of the region by one of the river flow indicators, by runoff volume per unit area (l/s/km^2) (Fig. 11.1). As is clear from the map, the hydrological zones identified, except for the basins of the Yenisei left-bank tributaries, reflect the combined influence of geographically specific climate and topography.

It is known that variability of climatic conditions among landscape zones has some impact on the hydrological cycle. The multi-linear regression equations shown in Table 11.2 describe, with an acceptable accuracy, the relationship of mean annual water yield to climatic parameters. The significance of the variable coefficients in the equations (Table 11.2) was estimated based on the T-criterion using a confidence of 95% or a level of 0.05.

The models obtained show that annual runoff depends on the hydrothermal conditions of a given year. For all the rivers studied, water runoff increased with increasing precipitation, especially winter precipitation. River runoff exhibited a complex interaction with air temperature. In the northern Tembenchy and Taimura Rivers, runoff increased with increasing July and September temperatures. A temperature rise in May resulted in decreasing runoff for the Turukhan and Yerachimo River basins.

Fig. 11.2 shows the change in the average long-term runoff coefficient (K_{st}) by landscape zones. According to Fig. 11.2, the formation of river runoff of Central Siberia is affected by provincial climatic conditions and local river basin features, rather than by zonality. As is also clear from Fig. 11.2, runoff coefficients in forest-tundra do not change significantly (0.8–0.9), whereas in the northern taiga zone they range from 0.3 to 0.9, depending on the location, with maximum values being characteristic of rivers flowing from the Putorana Plateau.

For rivers of the middle taiga zone, except for those rivers that form runoff on the Yenisei Mountain Ridge (nos. 19 and 20), the runoff coefficient was 0.4–0.5. In the estuary part of Podkamennaya Tunguska (no. 18), after the confluence of rivers 19 and 20, the runoff coefficient increased to 0.7.

The southern taiga zone is characterized by contrasting features in runoff formation. This is due to the fact that this zone is represented by two different forest formations, with dark coniferous forests (*A. sibirica* and *P. sibirica*) in the mountainous regions of the Yenisei Ridge and light coniferous forests (*P. sylvestris*) on the plateau. Due to the relief features and, accordingly, with the potentially different moisture levels of the territory, the rivers in the southern taiga zone have significant differences in runoff parameters. For example, the runoff coefficients ranged from 0.2 to 0.4 for the lowland rivers of the Angara Region and from 0.6 to 0.7 for the rivers of Yenisei Ridge (nos. 28 and 29).

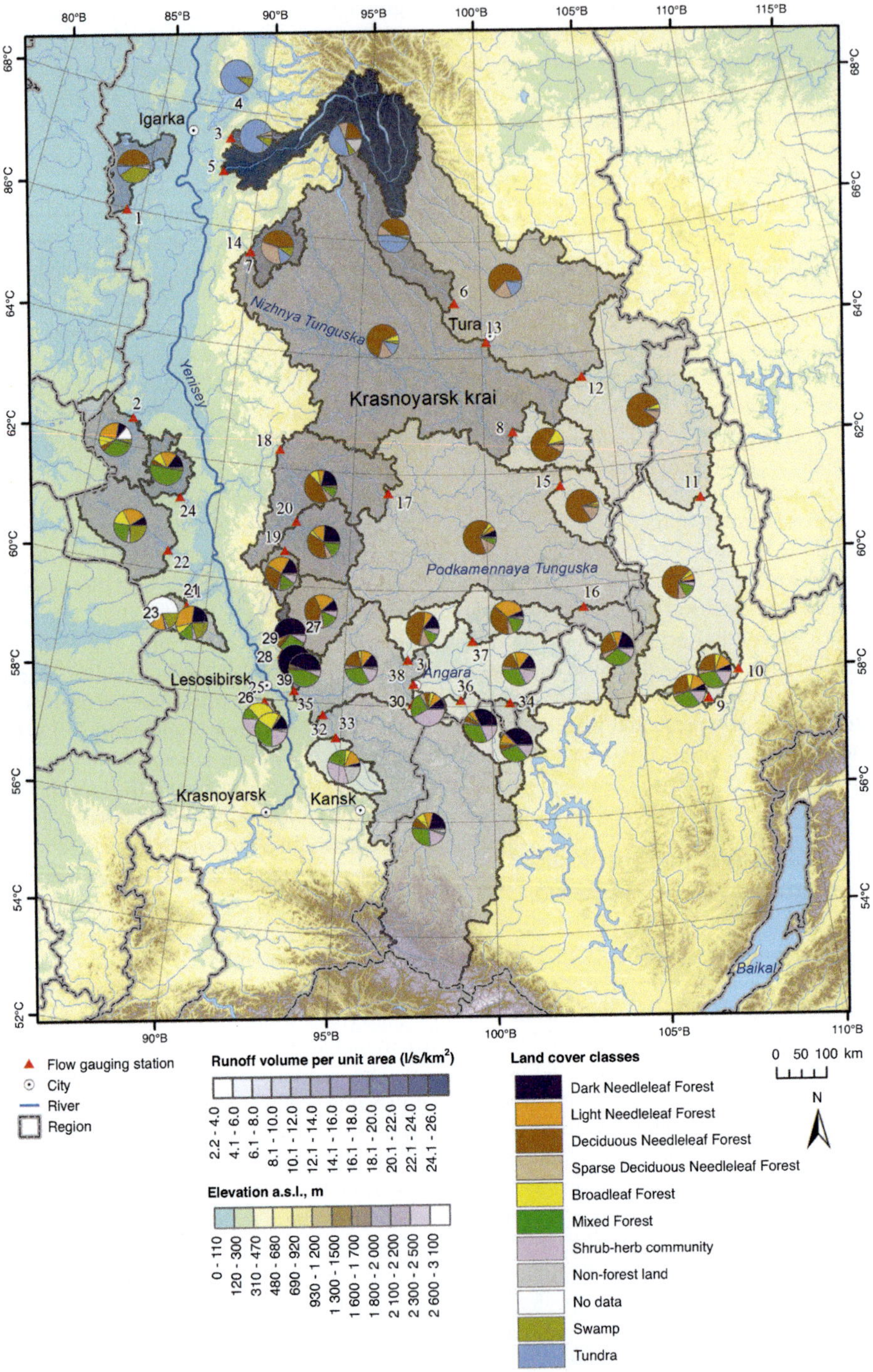

Fig. 11.1. Distribution of land cover classes and runoff volume per unit area across the study catchments (the numbers on the map show the values of runoff volume per unit area). The pie charts show the proportions of landcover class types on the study catchments.

Table 11.2. Regression models of annual runoff (Burenina *et al*., 2015).

River/flow gauging station	Equation	R^2	*G*	*F*
Turukhan/Yanov Stan	$Y = 115.3 + 0.39\,_{Xj} + 0.53\,_{Xt} - 2.69T_{(5+6)}$	0.45	44.0	10.9
Yerachimo/Bol'shoy Porog	$Y = 154.5 + 0.37\,_{Yp} + 0.58\,_{Xj} - 10.1T_5$	0.46	57.5	11.4
Tembenchy/Tembenchy	$Y = -447.7 + 1.5\,_{Xt} + 32.0T_7 + 1.7X_9$	0.57	55.5	10.6
Taimura/Kerbo	$Y = -250.5 + 0.50\,_{Xt} + 10.6T_7 + 0.57X_8$	0.73	22.7	13.7
Podkamennaya Tunguska/Kuz'movka	$Y = 51.3 + 0.15\,_{Xj} + 0.41\,_{Xt} - 3.2T_5$	0.70	19.4	20.3
Karabula/Karabula	$Y = 7.0 + 0.2\,_{Xj} \times X_t/(T_5 \times T_7) + 5.2T_9$	0.47	16.7	15.7
Irkineeva/Bedoba	$Y = 47.3 + 0.31(X_j \times X_t)/(T_5 \times T_7) + 7.0\,_{T9}$	0.40	22.8	11.6
Chadobets/Yarkino	$Y = 69.8 + 0.375(X_j \times X_t)/(T_5 \times T_7)$	0.49	11.8	27.2

Y, annual river runoff (mm); Xj, annual amount of liquid precipitation at a representative weather station (mm); X_t, annual amount of solid atmospheric precipitation (mm); X_8 and X_9, mean monthly precipitation in August and September, respectively; Y_p, runoff of the previous year; T_5, T_6, T_7 and T_9, mean monthly air temperature in May, June, July and September respectively (°C); R^2, coefficient of multiple determination; *G*, standard error of the equation; *F*, Fisher criterion.

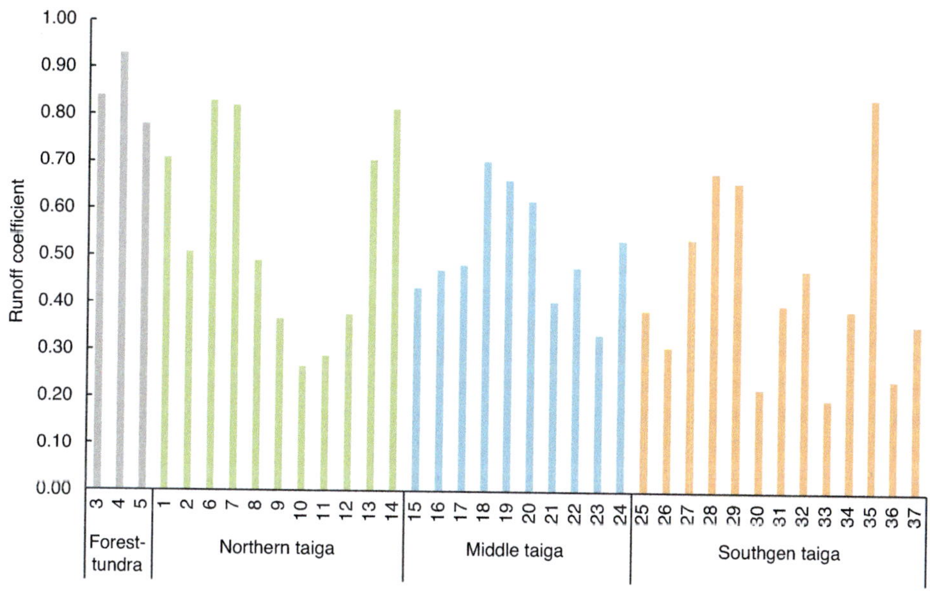

Fig. 11.2. Average long-term runoff coefficient by landscape zone.

Geographical variability in runoff is largely related to changes in ET at both zonal and provincial levels. In the forest-tundra, where the forest cover is represented by sparse forests, transpiration is minimal. According to our previously obtained data (Burenina *et al*., 2002), average long-term ET fluctuates within the range of 184–218 mm, increasing to 236 mm in the western plain part of the studied region (nos 3–5).

In the northern taiga, the forest cover is represented not only by sparse larch forests but also by dark coniferous taiga, which predominates on the left bank of the Yenisei and in the lower reaches of the Nizhnyaya Tunguska River. Annual average ET in dark coniferous stands in catchment areas nos. 1–2 is 323 mm, and in the Nizhnyaya Tunguska River basin, it is up to 285 mm (Burenina *et al*., 2002). The higher moisture consumption by ET in dark coniferous forests is due to the higher evaporation rate of intercepted precipitation.

In the middle and southern taiga, the role of forests in hydrological cycling increases significantly as a result of the increase in the radiation balance and the greater biological productivity of forest stands (Burenina *et al.*, 2002). In the middle taiga, ET fluctuates from 301 to 369 mm, and from 338 to 411 mm in the southern taiga. These differences in ET between the middle taiga zone and the southern taiga are due to both climatic factors and the species composition of the stands (Burenina *et al.*, 2002). The influence of forest productivity and thermal resources on ET has been covered in the works of other authors (Voronkov, 1988; Buttle *et al.*, 2005; Karpechko and Bondarik, 2010).

We studied in more detail the influence of various factors on ET using the example of the Podkamennaya Tunguska River basin. The considerable length of the Podkamennaya Tunguska River (1700 km) and the orientation of its basin from the south-east to the north-west determine the diversity of landscape conditions, which is expressed in the characteristics of the total moisture content and the representation of plant communities. In the west, average long-term precipitation is 666 mm (weather station Kuz'movka), while it is 355 mm (weather station Chemdalsk) in the upper reaches of the river in the south-east. For the studied catchments (nos. 16–18), a fairly close relationship (correlation coefficient 0.56) was found between annual ET and the amount of annual precipitation, and a less pronounced relationship with temperature indicators (Danilova and Burenina, 2021). For other geographic areas, some authors note closer relationship evaporation from summer temperatures and radiation index of aridity (Klaassen, 2001; Zhang *et al.*, 2017).

To estimate influence of dominating forest vegetation community in hydrological cycle based on their contribution to ET, we carried out estimates for landscape differentiation of the basin area. Using vegetation cover maps (Bartalev *et al.*, 2011), we identified seven classes of land cover complexes: 1) birch (*Betula pendula*), 2) dark coniferous (*Abies sibirica*), 3) pine (*Pinus sylvéstris*) + larch pine stands (*Laix sibirica* +*Pinus sylvéstris*), 4) mixed (dark coniferous (*Abies sibirica*) + deciduous), 5) mixed (pine (*Pinus sylvéstris*) + dark coniferous (*Abies sibirica*), 6) pure larch forests (*Laix sibirica* and *L. gmelini*), and 7) burnt areas. Annual ET was calculated for all vegetation classes.

For all vegetation classes ET varied considerably both year to year and geographically. The greatest variability of ET was calculated for birch stands (360–440 mm), both by year (375–440 mm) and territorially (360–405 mm). For larch stands, although it exhibited annual variations, ET tended to increase west-eastward from 365 to 405 mm. A similar tendency was found for Scots pine and mixed Scots pine-larch stands (385–405 mm) (Burenina *et al.*, 2022). According to our data (Danilova and Burenina, 2021; Burenina *et al.*, 2021), in catchment area no. 16 (the upper reaches of the Podkamennaya Tunguska River), significant areas are occupied by dark coniferous species (28%) and larch forests (29%). Pine forests account for a fairly high share in the structure of the forested area (18%). These forest stands take up to 80% of the total annual precipitation as total evaporation. In the central part of the Podkamennaya Tunguska basin (catchment area no. 17), the share of larch forest stands to ET uptake increases to 70%, with a significant share of more than 70% of the total precipitation being up taken as ET by sparse forests and woodlands.

The catchment area no. 18 is characterized by the highest annual precipitation values and more productive forest stands in which the structure of moisture consumption for ET changes. The 43% of larch stands, as well as the dark coniferous stands, which occupy up to 27% of the catchment area, make a significant contribution to ET. The annual ET is about 60% of the total precipitation.

A comparative analysis of moisture consumption for ET with the structure of forest cover in the Podkamennaya Tunguska basin showed that an increase in the proportion of dark coniferous species and Scots pine in the forest stand contributes to an increase in the evaporation of precipitation retained by tree crowns (Danilova and Burenina, 2021; Burenina *et al.*, 2022). In winter, the amount of snow evaporation from tree crowns of dark-needle and Scots pine forest stands is comparable to the evaporation of snow cover in open areas, while in summer, ET in these stands is much higher than in larch forests due to the greater amount of intercepted precipitation by tree crowns.

11.6 Evapotranspiration Affected by Logging in the Mountain Taiga of the Yenisei Ridge

In order to assess the change in ET at sites after forest felling and to analyze its change with gradual clear-cut recovery, we selected the Sukhoy Pit River basin (Fig. 11.1, no. 28). Sukhoy Pit is a tributary of the Bol'shoy Pit River, which flows into the Yenisei River. Sukhoy Pit River is 156 km long, with a basin area of 1262 km^2 and its headwaters on the west-facing macro-slope of the Yenisei Ridge.

Analysis of space imagery of the GLAD (Global Land Cover and Land Use Change) data set showed that 30,835 ha of forest (24.4% of the entire Sukhoy Pit basin) was clear-cut between 2000 and 2020 (Fig. 11.3). Annual data analysis showed that the area of clear-cut varied, with cutting areas of less than 500 ha in some years and up to as much as 3400 ha of forest cut annually during the 2012–2014 and 2018–2020 periods.

The obtained ET values for both the clear-cut areas and areas with undisturbed forest stands were characterized by significant

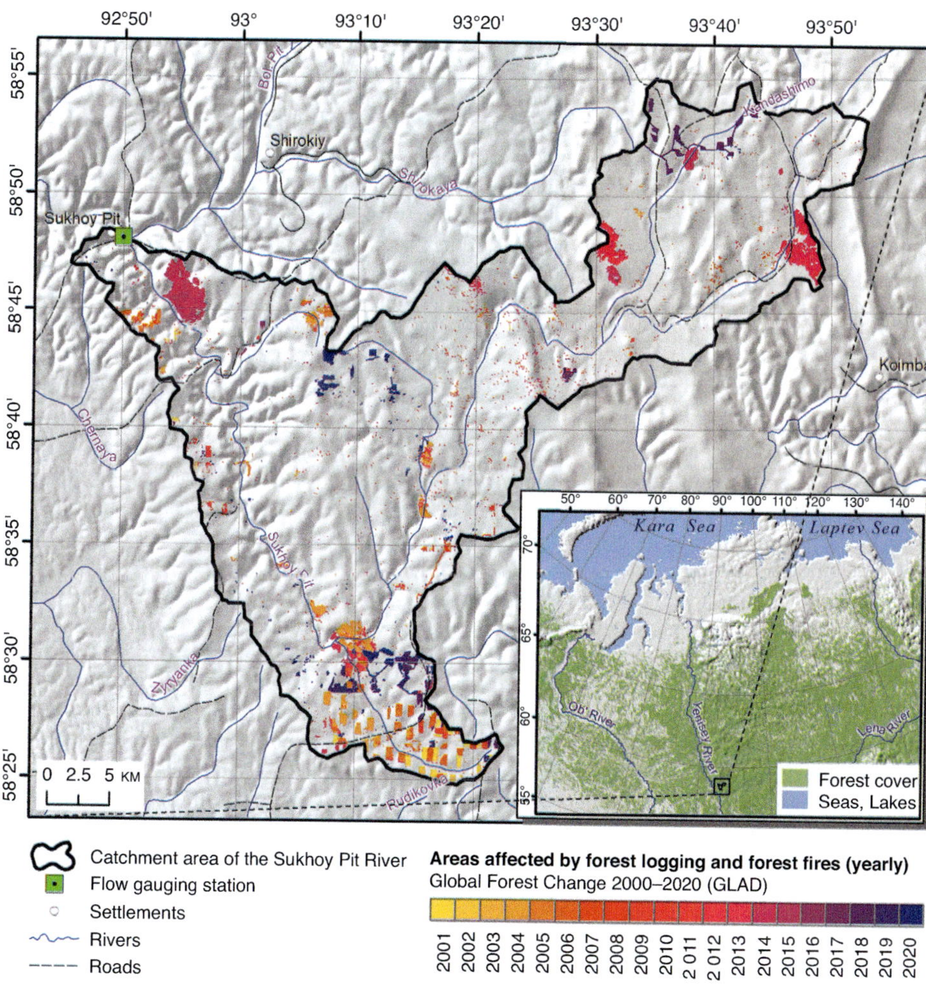

Fig. 11.3. Sukhoy Pit River catchment showing the areas affected by forest logging and forest fires during the period of 2001–2020.

Table 11.3. Annual evapotranspiration (ET) correlation with mean annual temperature, annual precipitation and cumulative clear-cut areas for the Sukhoy Pit River basin for 2001–2020.

Indicator	Correlation coefficient (R)	Significance level (P)
Mean annual air temperature	0.62	0.003
Total annual precipitation and ET of current year (unbiased)	0.36	0.05
Total annual precipitation (1-year prior ET)	0.82	0.00004
Total annual precipitation (2-year prior ET)	0.44	0.04
Cumulative clear-cut area for 2001–2020	0.71	0.0002

variability over time. For example, during the period 2001–2020 in the Sukhoy Pit River basin, overall annual ET varied from 370 to 440 mm. In some years, ET per unit area in cut sites exceeded background values for the areas of no forest change. These ET trends were largely induced by weather conditions, particularly air temperature and precipitation changes (Table 11.3). As mentioned above, the relationship between ET and climate parameters has been noted by most authors who have studied the issues of evaporation by forest vegetation (e.g. Klaassen, 2001; Zhang *et al.*, 2017).

We analyzed ET interaction with mean annual temperature and annual precipitation. The statistical results are presented in Table 11.3. The analysis revealed a lag of 1–2 years between maximum precipitation and maximum ET. For example, 2013–2015 was a period of maximum precipitation and a maximum ET period came later and lasted not less than 3 years (Burenina *et al.*, 2024). As is clear from Table 11.3, the highest correlation coefficient (0.82) was characteristic of current-year ET and the preceding-year precipitation.

A year-to-year comparison of growing season (May–September) ET variability with that of climatic parameters revealed that ET reached its maximum when the combination of air temperature and precipitation was optimal (i.e. when heat was sufficient for precipitated water to evaporate) (Burenina *et al.*, 2024).

According to the plot in Fig. 11.4, the variability of ET in the growing season at clear-cut areas over the years was similar to the variability of ET in areas with undisturbed forest cover. Such linear trends showing an increase in ET since 2001 in both clear-cut areas and undisturbed stands are hypothesized as being reflective of global warming trends.

Our analysis of ET changes in the Sukhoy Pit basin with changing freshly cut area revealed that for maximum freshly cut areas, ET for the entire catchment during growing season was 25–30 mm (8–1%) less than the multi-year mean. Estimating the impact of fresh clear-cutting on ET for the entire basin regardless of other factors is problematic, because clear-cuts of previous years with different vegetation succession stages are quite effective moisture evaporators. To estimate the contribution of postharvest vegetation succession to ET in the basin, we classified cut areas by time since cutting (0, 1, 2–5, 6–10 and ≥11 years), with each class corresponding to a certain vegetation succession stage. Successional changes in clear-cut areas were: (i) 0–1 year after felling: fresh clearings (without vegetation); (ii) 1–5 years: grass–shrub stage; (iii) 6–10 years: young stands; and (iv) 11–20 years: pole stands. We used this classification based on our own observations of successional changes in clear-cut areas in the Sukhoy Pit River basin (Burenina *et al.*, 2014b), and taking into account the characteristics of post-logging reforestation in dark coniferous forests of the region. It is known that in clear-cut areas in the process of afforestation, the biometric indicators of vegetation, in particular the leaf index, change (Sampson *et al.*, 2011).

Due to non-uniform year-to-year cutting, the change of the clear-cut area was wave-like. From 2014, young and pole stands tended to persistently increase in clear-cut sites. Minimal ET values were confined to the largest fresh clear-cuts, whereas maximal ET occurred in

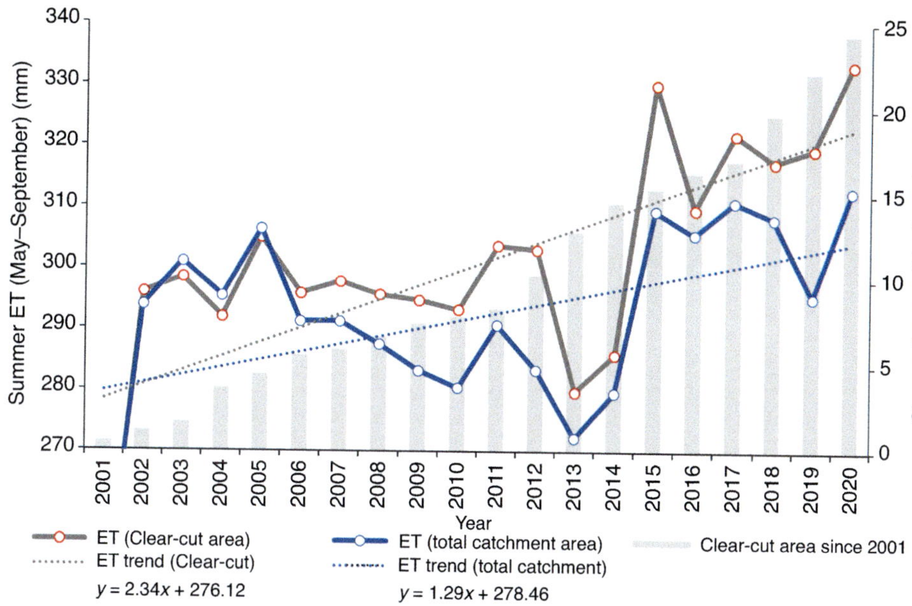

Fig. 11.4. Trends of summer evapotranspiration (ET) for clear-cut areas and the entire catchment area of the Sukhoy Pit River.

2014–2020 with a gradual restoration of clear-cut sites as newly established forest communities enhanced ET (Burenina *et al.*, 2024).

Our analysis of ET values for 2001–2020 years obtained for the Sukhoy Pit River basin (Fig. 11.4) using MODIS data (https://lpdaac.usgs.gov/products/mod16a2v006; accessed 7 May 2025) showed that year-to-year ET variations at the basin scale were primarily due to changing temperature and precipitation combined with the changing size of fresh and recovered clear-cuts. The dynamics of clear-cut versus intact ET showed that, apart from climatic characteristics, ET for this basin was determined by a steadily increasing trend of young and pole stands in old clear-cut sites.

11.7 Water Yield Trends in River Basins Under Large-Scale Clear-Cutting

Among the forest areas of Central Siberia, the Angara River Basin is the area of greatest forest cover disturbance due to catastrophic fires, pest outbreaks, large-scale cutting and industrial activity. All of these factors have led to considerable changes in the regional forest area, structure and species composition.

It is known that from 1957 to 1966, large-scale logging was carried out in the study area, which was replaced by narrow areas of cutting from 1967 to 1988. Since 1989, forest policy has changed in Russia, as a result of which logging has become less controlled by the forest management. In connection with the above, we identified three different periods for studying river flow at four tributaries of the Angara: the Karabula (no. 30), Irkineeva (no. 31), Mura (no. 36) and Chadobets (no. 37) rivers.

Runoff trends were found to vary among the periods of 1957–1966, 1967–1988 and 1989–2017. Runoff tended to decrease for the Irkineeva and Chadobets Rivers in 1957–1966, although the smaller rivers (Mura and Karabula) showed a slight increase in this parameter (Fig. 11.5a). From 1967 to 1988, runoff did not change in the basins of the Mura, Karabula and Irkineeva Rivers, whereas it tended to increase for the Chadobets River (Fig. 11.5b). In 1989–2017, an increase in runoff was observed on the Irkineeva River

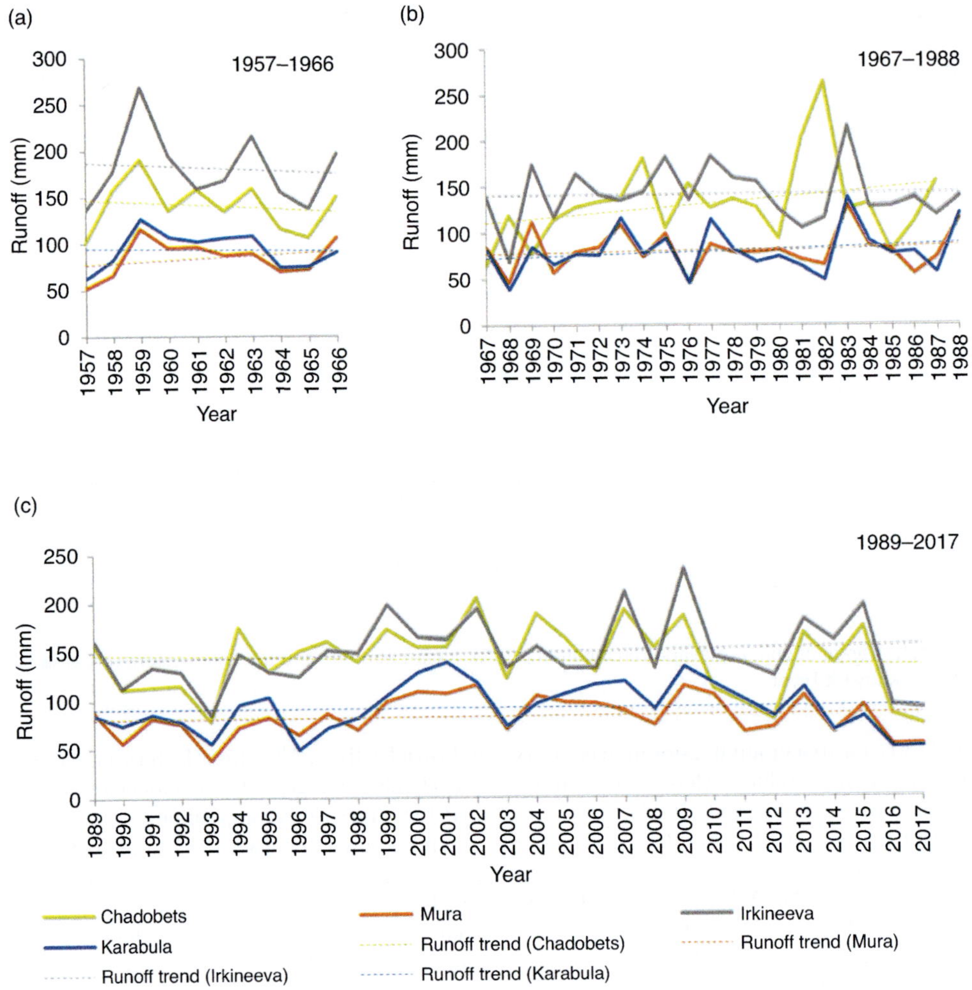

Fig. 11.5. Trends of annual runoff of the study catchments for the periods 1957–1966 (a), 1967–1988 (b) and 1989–2017 (c).

basin, unlike the Chadobets River basin where runoff decreased (Fig. 11.5c).

The authors believe that these differences in the trends of runoff are the result of reforestation, i.e. they were caused by the ratio of the area of treeless clearings, the area of young stands and the area of forest stands with a closed canopy specific to the basins. The analysis of the change in the forested area according to forest management materials showed that in 1966, large areas (up to 20% of the catchment area) were classified as clearings, but by 1988, the clearing areas (despite annual clearings of 1–3%) had decreased to 4.7–8.7% of the river basin area. At the same time, the area of young stands had almost doubled (Table 11.4). As indicated above, the process of reforestation in the Angara region occurs through a change in species, and clearings are overgrown with birch and aspen. Based on this, it can be stated that the composition of plantings in the studied catchments has changed significantly: the area of coniferous forests has decreased, and the area of derivatives represented by deciduous species has increased. As shown in Table 11.4, in 1988, the area of fresh clearings had decreased compared

Table 11.4. Dynamics of the spatial structure of the forested area of the studied catchments from 1966 to 1988.

River/flow gauging station	Catchment area (km²)	1966			1988		
		Clearing area from catchment area (%)	Area of young growth from the catchment area (%)	Forest area from catchment (%)	Clearing area catchment area (%)	Area of young growth from the catchment area (%)	Forest area from catchment (%)
Chadobets/ Yarkino	14,257	21.3	8.9	76	4.7	22.1	84
Irkineeva/ Bedoba	9,573	18.6	8.5	80	8.3	19.4	80
Mura/Irba	9,566	15.1	9.2	83	8.7	15.6	89
Karabula/ Karabula	4,059	15.7	7.9	79	7.9	16.3	85

with 1966, while the total forest cover in the catchments had increased by 6–8% (with the exception of the Irkineeva River catchment) due to the transition of clearings to forested areas.

As shown in Table 11.4, the Irkineeva and Chadobets River basins were subjected to large-scale (concentrated) forest cutting before 1966, and the authors believe that runoff decreased due to highly intensive snow water evaporation. According to Onuchin (2015), strong winds lift dry fine snow up into the air to increase the snow evaporating area to tens to hundreds of times its usual size and, as a result, snow moisture evaporation increases correspondingly.

The positive runoff trend for the Chadobets River basin identified between 1967 and 1988 can be attributed to the fact that the large-scale clear-cuts that existed before this period gradually moved to young growth and then to pole-stand forest regeneration stages. A comparative analysis of snow accumulation and vegetation succession in cut areas revealed that the maximum snow accumulation was in deciduous young and pole stands (Burenina *et al.*, 2014b). The 1967–1988 runoff trend identified for the rivers of interest in this study (Fig. 11.5b) agreed with the results from Burenina and Onuchin (1999) for the Angara region.

A number of studies (Krestovsky, 1986; Burenina *et al.*, 2014a) have found the components of post-cutting water balance changing, depending largely on the age stage of the regenerating vegetation. The water balance of any forested river basin is controlled by the ratio between the area of the major woody species and that of secondary stands. In this respect, one should keep in mind that the distribution of stands differing in age and succession stage across a river basin is never constant. Forest age and inventory characteristics change over time. Therefore, we tried to analyze the runoff trend, taking into account post-logging successional changes.

With the advent of the GLAD data set (Potapov *et al.*, 2022) in 2000, regular data on forest areas disturbed by logging, fires, insects, etc. became available. Using this data set, we calculated the total area disturbed in each river basin of interest (Chadobets, no. 37; Irkineeva no. 31; Mura, no. 36; Karabula, no. 30) in 2001–2020, as well as the cumulative areas falling within the following land disturbance categories: (i) fresh clearcuts and burns; (ii) 1–5 years after disturbance (grass communities); (iii) 6–10 years after disturbance (young growth); (iv) 11–15 years (young growth); and (v) 16–20 years (pole stands). We identified categories of disturbed lands taking into account the characteristics of post-logging reforestation in the pine forests of the study region and using the results of our field observations in the basins of these rivers (Danilova and Burenina, 2025).

We then analyzed the linear trends of river runoff based on long-term runoff observations at flow gauging stations for different stages of reforestation after logging. Fig. 11.6 shows the

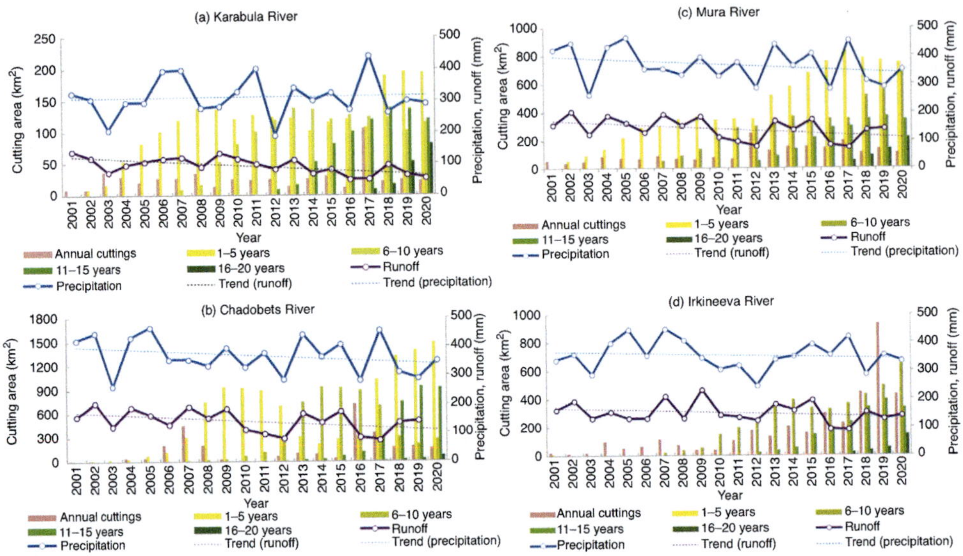

Fig. 11.6. River runoff, precipitation and forest regeneration in clear-cuts and burns in the basins of the Karabula (a), Chadobets (b), Mura (c) and Irkineeva (d) Rivers.

identified changes in annual runoff depending on forest regeneration dynamics for a range of clear-cut stages.

As follows from Fig. 11.6, the influence of both fresh cuttings and various stages of forest restoration on the runoff is manifested differently in the studied river basins depending on their size. It was typical for the Karabula River that, during the grass stages, runoff correlated with precipitation, but since 2009, when the 10-year-old cuttings were overgrown with young growth, the relationship between runoff and precipitation has been disrupted (Fig. 11.6a). In 2017, despite the maximum annual precipitation for the studied period, runoff on the Karabula River was minimal. This can be explained by both the significant area of disturbed forest cover (10%) and the large area of fresh cuttings (>2%).

On the Mura River (Fig. 11.6c), the correlation between precipitation and runoff has been disrupted since 2012, which is obviously related to the onset of intensive logging in the period 2012–2020 (up to 2% of the basin area annually). Since 2012, there has been a sharp increase in the area of grassy communities (Fig. 11.6c), and the total area of disturbed forest cover by 2020 has increased from 10% to 22%.

In the Chadobets River basin with the largest area of all the rivers studied in the Angara region, due to less intensive logging, a fairly close correlation between precipitation and runoff was observed. In 2007 and 2016, logging from 3–5% of the basin area affected the water yield of this river (Fig. 11.6b). In the Irkineeva River basin, pine stands in areas accessible to transport were felled before 1966. Between 2000 and 2016, less than 1% of the catchment area was logging annually, which had little effect on the river runoff (Fig. 11.6d). After 2017, logging was carried out in distant areas (300 km from the Irkineeva River mouth), which in the future will affect the hydrological regime of this territory.

Our study showed that runoff dropped regardless of precipitation only in years of very large-scale clear-cutting (e.g. in 2017 on the Karabula and Irkineeva Rivers) (Fig. 11.6a, d). With annual cutting of less than 2% of the total basin area, runoff fluctuations reflected the variability of annual precipitation. When the annual clear-cut area exceeded 2%, runoff tended to decrease. This is typical for small rivers as Karabula and Mura. For the Chadobets River, the threshold value for forest cover disturbance was 5% of the catchment area (Fig. 11.7) (Burenina and Danilova, 2023).

The results of our study on river flow trends depending on the reforestation tendency are consistent with the literature data (Hornbeck *et al.*, 1993; Oda *et al.*, 2018). Long-term runoff observations in 11 catchments (Hornbeck *et al.*, 1993) showed that changes in species composition after forest cutting on several study catchments eventually resulted in decreased water yields compared with those from uncut, control catchments. Different views can be found in the literature on the consensus of the hydrological regime recovery in river basins subjected to forest clear-cutting. Some experts (Pobedinskiy, 1979; Lebedev, 1982) mention a time range of 10–50 years needed for a hydrological regime to fully recover. For some regions of the USA and Japan, the periods of recovery of hydrological functions of forests after logging differ from the data for the taiga zone of Siberia (Hornbeck *et al.*, 1993; Oda *et al.*, 2018). According to the data obtained by Krestovsky (1986), the recovery of the hydrological regime typical for overmature taiga forest of European Russia can take as long as 100 years. These wide ranges can be attributable to variations in post-cutting forest regeneration and landscape restoration among locations and climates.

11.8 Conclusions

As our results above have shown, changes in climatic conditions in landscape zones have a certain effect on the water regime of watercourses. In general, the region under study (except for the northern rivers on the border of forest-tundra and northern taiga) is characterized by a decreasing trend in annual runoff, which can be explained by both a decrease in precipitation and an increase in forest cover disturbance under the influence of both natural and anthropogenic factors. The increase in the runoff of northern rivers is obviously associated with the influence of global warming, and since 2001–2012, the thawing of permafrost soils has become a more significant factor in the formation of runoff (Burenina *et al.*, 2015).

The obtained results showed that ET of the identified plant communities, which directly affects the total runoff, is characterized by significant variability both in space and time. It is determined by the variability of climatic indicators and the structure of vegetation cover. The observed differences in the mean annual values of total ET in the Podkamennaya Tunguska River basin from east to west, especially in the summer months, are determined by the diversity of the species composition of forest stands and their taxation indicators.

A comparison of ET trends in background areas of the Sukhoy Pit River basin and areas affected by logging showed that changes in ET in the catchment area are influenced not only by climatic characteristics but also by a stable trend of increasing areas of young stands and pole stands in the place of logging in the late 20th and early 21st centuries. Runoff coefficients and changes of clear-cut area conducted for the river basins of the study regions for each year revealed the impact of the clear-cut area on water yield to be markedly dependent on the current-year proportion of the clear-cut area in a basin, and on precipitation of the current and previous years.

Our analysis of the runoff relationship with vegetation succession stages in clear-cuts of the Angara region enabled us to conclude that structural changes in vegetation shift the water balance of a basin toward decreasing or increasing ET, thereby affecting the water yield. For the basins of interest, runoff appeared to become relatively close to the pre-cutting levels with increasing area of 16–20 year-old clearcuts (i.e. with gradual replacement of conifer young growth by pole stands). The manifestation of these patterns is complicated by significant annual variability in precipitation. In the dark-needle forests of Yenisei Ridge, water balance tended to gradually recover back to pre-disturbance levels where dark-needled conifers at clear-cuts had reached the age of over 40 years (the pole-stand forest succession stage).

Our approach to the forest disturbance impact on the river basin hydrological regimes was based on a careful account of the cumulative clear-cutting and fire-disturbed area and the ratio of the clear-cut area to that of the entire basin of interest. Moreover, we paid a close

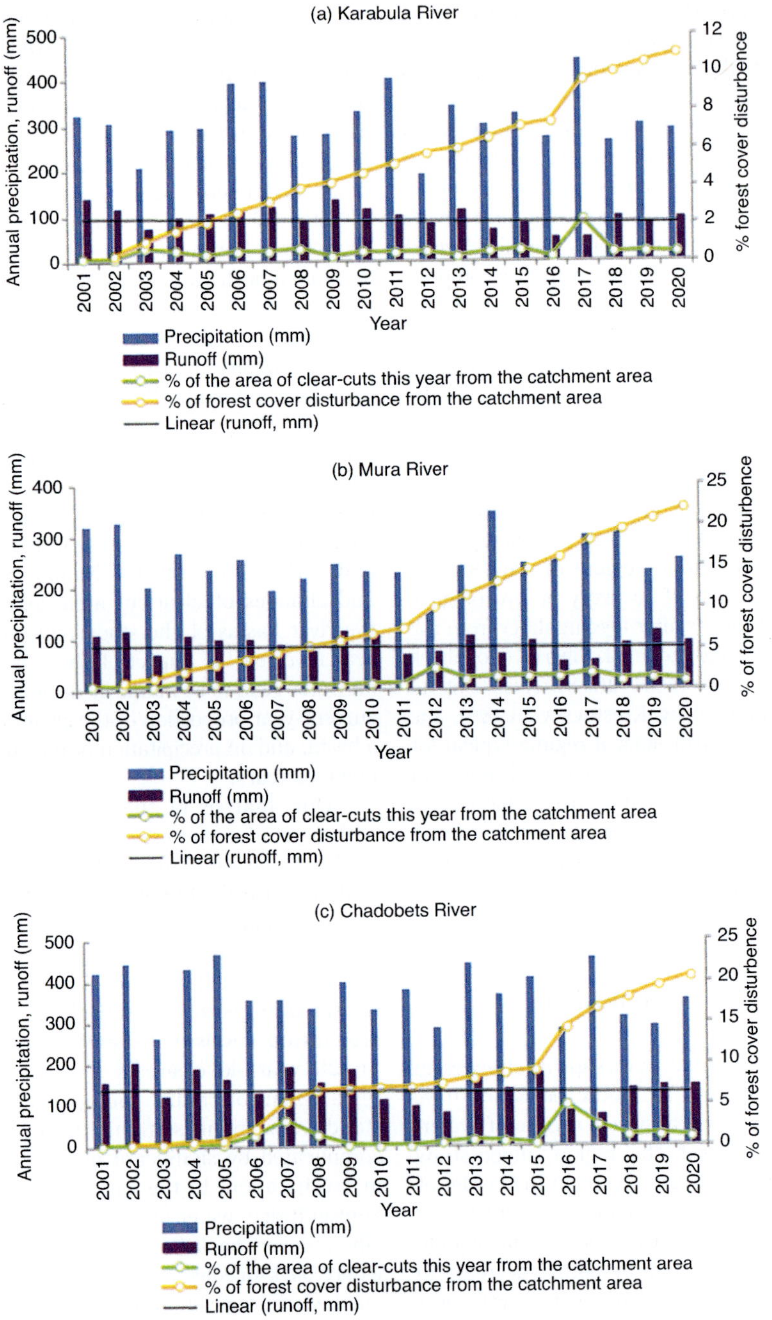

Fig. 11.7. Deviations of the current year's runoff from the long-term average depending on the combined influence of precipitation and forest cover disturbance for the Karabula (a), Mura (b) and Chadobets (c) Rivers.

attention to climatic, landscape and geographical factors, which change the hydrological role of the forest qualitatively. The results of this study confirm the conclusion of the hydrological importance of forest changes geographically, at both zonal and local levels, arrived at by a number of past studies (Idzon and Pimenova, 1975; Onuchin *et al.*, 2021). The landscape-determined concept of the hydrological role of forests suggested by Onuchin and Burenina (2008) considers the mechanisms of moisture cycling in forested basins and thereby explains the influence of forest on water yield variability by geophysical background differences.

Acknowledgements

The chapter was prepared based on the information received in the framework of the scientific project: 'The role of natural and anthropogenic factors in the system of sustainable forest management in Siberia' (2024–2026) FWES-2024-0007.

References

Bartalev, S.A., Egorov, V.A., Ershov, D.V., Isaev, A.S., Lupyan, E.A. *et al.* (2011) Satellite mapping of vegetation cover in Russia based on data spectroradiometer MODIS. *Modern Problem Remote Probing the Earth from Space* 8, 285–302 (in Russian).

Bosch, J.M. and Hewlett, J. (1982) A review of catchment experiments to determine the effect of vegetation changes on water yield and evapotranspiration. *Journal of Hydrology* 55, 3–23.

Burenina, T.A. and Onuchin, A.A. (1999) Dynamics of runoff in the watersheds of the Northern Angara region. *Geography and Natural Resources* 2, 42–46 (in Russian).

Burenina, T.A. and Danilova, I.V. (2023) The influence of forest disturbance on river flow in the Angara River basin: Natural resource potential and environmental rehabilitation of degraded landscapes. In: *Materials of the International Scientific and Practical Conference, Grozny, March 17–18, 2023*, Publishing House of the Kadyrov Chechen State University, Grozny, Russia, pp. 59–66 (in Russian).

Burenina, T., Onuchin, A., Guggenberger, G., Musokhranova, A. and Prisov, D. (2015) Dynamics of hydrological regime in permafrost zone of Central Siberia. In: Liu, J. (ed.) *Environment, Chemistry and Biology IV. International Proceedings of Chemical Biological & Environmental Engineering*, IACSIT Press, Singapore, pp. 125–132 (Vol. 90).

Burenina, T.A., Onuchin, A.A. and Stakanov, B.D. (2002) Geographic variability of forest ecosystem evapotranspiration. In: Pleshikov, F. (ed.) *Forest Ecosystems of the Yenisei Meridian*. Publishing House of SB RAS, Novosibirsk, Russia, pp. 54–57 (in Russian).

Burenina, T.A., Onuchin, A.A. and Fedotova, E.V. (2014a) Hydrological effect of forest logging in boreal zone of Siberia. In: Roberts, N. (ed.) *Forest Ecosystems: Biodiversity, Management and Conservation*. Nova Science Publishers, New York, pp. 117–147.

Burenina, T.A., Musokhranova, A.V. and Suleimanova, Z.R. (2021) Dynamics of the structure of the water balance during reforestation successions in logging sites in the dark coniferous forests of the yenisei ridge. In: *management of forest ecosystems under climate changeProceedings of the International Scientific and Practical Conference Devoted to the 105th Anniversary of the Birth of Doctor of Biological Sciences, Professor, Honored Scientist Peter Alekseevich Gan*, Bishkek, Kyrgyzstan, September 7–9, , pp. 88–92.

Burenina, T.A., Shishikin, A.S. and Onuchin, A.A. (2014b) Snow cover on logged sites in fir and Siberian pine forests of the Yenisei Ridge. *Contemporary Problems of Ecology* 7, 779–788.

Burenina, T.A., Danilova, I.V. and Mikheeva, N.A. (2022) Spatial-temporal dynamics of evapotranspiration in the Podkamennaya Tunguska River Basin. *Contemporary Problems of Ecology* 15, 449–458.

Burenina, T.A., Korets, M.A. and Suleymanova, Z.R. (2024) Regional features of transformation of water balance structure during restorative successions in clearcut areas of dark coniferous forests of the Yenisei Ridge. *Izvestiya of the Russian Academy of Sciences: Geographical Series* 88, 27–40 (in Russian).

Buttle, J.M., Creed, I.F. and Moore, R.D. (2005) Advances in Canadian forest hydrology, 1999–2003. *Hydrological Processes* 19, 169–200.
Danilova, I.V. and Burenina, T.A. (2021) The use of indicative properties of landscapes to assess the hydrological role of forests. In: *International Scientific and Practical Conference "Management of Forest Ecosystems in Conditions of Climate Change". Dedicated to the 105th anniversary of the birth of Doctor of Biological Sciences, Professor, Honored Scientist Peter Alekseevich Gan*, Bishkek, Kyrgyzstan, September 7–9, 2021, pp. 153–158.
Danilova, I.V. and Burenina, T.A. (2025) Dynamics of runoff in catchments with forest cover disturbed by logging in the Angara River basin. *Lesovedenie* 2, 237–251 (in Russian).
Ellison, D., Futter, M.N. and Bishop, K. (2012) On the forest cover-water yield debate: From demand- to supply-side thinking. *Global Change Biology* 18, 806–820.
Gibson, L.A., Jarmain, C., Su, Z. and Eckardt, F.E. (2013) Estimating evapotranspiration using remote sensing and the surface energy balance system – A South African perspective. *Water SA* 39, 477–484.
Grip, H., Fritsch, J.M. and Bruijnzeel, L.A. (2005) Soil and water impacts during forest conversion and stabilization to new land use. In: Bonell, M. and Bruijnzeel, L.A. (eds) *Forests, Water and People in the Humid Tropics*. UNESCO International Hydrology Series, Cambridge University Press, Cambridge, UK, pp. 561–589.
Hamilton, L.S. (2008) *Forests and Water*. FAO Forestry Paper No.155, FAO, Rome.
Hamilton, L.S. and Pearce, A.J. (1987) What are the soil and water benefits of planting trees in developing country watersheds? In: Southgate, D.D. and Disinger, J.F. (eds) *Sustainable Resource Development in the Third World*. Westview Press, Boulder, Colorado, pp. 39–58.
Hornbeck, J.W., Adams, M.B., Corbett, E.S., Verry, E.S. and Lynch, J.A. (1993) Long-term impacts of forest treatments on water yield: A summary for Northeastern USA. *Scientific Research Publishing* 150, 323–344.
Idzon, P.F. and Pimenova, G.S. (1975) *The Influence of Forests on River Flow*. Nauka, Moscow (in Russian).
Karpechko, Y.V. and Bondarik, N.L. (2010) *Hydrological Role of Forestry and Timber Industry Work in the Taiga Zone of the European North of Russia*. Karelian Scientific Center of the Russian Academic Sciences, Institute of Water Problems of the North, Petrozavodsk, Russia (in Russian).
Klaassen, W. (2001) Evaporation from rain-wetted forest in relation to canopy wetness, canopy cover, and net radiation. *Water Resources Research* 37, 3227–3236.
Krestovsky, O.I. (1986) *Logging and Forestation Influence Water-Yield of Rivers*. Hydrometeoizdat, Leningrad, Russia (in Russian).
Lebedev, A.V. (1982) *Hydrological Role of Siberian Mountain Forests*. Nauka, Siberian Branch, Novosibirsk, Russia. (in Russian).
Leng, P., Li, Z.L., Duan, S.B. and Tang, R.L. (2017) A method for deriving all-sky evapotranspiration from the synergistic use of remotely sensed images and meteorological data. *Journal of Geophysical Research* 122, 13263–13277.
Li, Q., Wei, X., Zhang, M., Liu, W., Fan, H. *et al.* (2017) Forest cover change and water yield in large, forested watersheds: A global synthetic assessment. *Ecohydrology* 10, e1838.
Nisbet, T.R. and McKay, H. (2002) Sustainable forestry and the protection of water in great britain. In: *Proceedings of an International Expert Meeting on Forests and Water, Shiga, Japan*, International Forestry Cooperation Office, Forestry Agency, Tokyo, pp. 101–112.
Oda, T., Green, M.B., Urakawa, R., Scanlon, T.M., Sebestyen, S.D. *et al.* (2018) Stream runoff and nitrate recovery times after forest disturbance in the USA and Japan. *Water Resources* 54, 6042–6054.
Onuchin, A. (2015) The reasons for conceptual contradictions in evaluating hydrological role of boreal forests. *Siberian Journal of Forest Science* 2, 41–54.
Onuchin, A. and Burenina, T. (2008) The hydrological role of forests in Siberia. In: Prescott, A. and Barkely, T. (eds) *Trends in Water Resources Research*. Nova Science Publishers, New York, pp. 67–92 (in Russia).
Onuchin, A., Burenina, T., Shvidenko, A., Prysov, D. and Musokhranova, A. (2021) Zonal aspects of the influence of forest cover change on runoff in northern river basins of Central Siberia. *Forest Ecosystems* 8, 45.
Pobedinskiy, A.V. (1979) *Water- and Soil Protective Role of Forests*. Lesnaya Promyshlennost, Moscow (in Russian).

Pomeroy, J.W. and Schmidt, R.A. (1993) The use of fractal geometry in modelling intercepted snow accumulation and sublimation. In: *Proceedings of the 50th Eastern Snow Conference*, Quebec City, Quebec, June 8–10, 1993, pp. 1–10.

Potapov, P., Hansen, M.C., Pickens, A., Hernandez-Serna, A., Tyukavina, A. *et al.* (2022) The global 2000–2020 land cover and land use change dataset derived from the landsat archive: First results. *Frontiers in Remote Sensing* 3, 856903.

Sampson, D.A., Amatya, D.M., Lawson, C.D. and Skaggs, R.W. (2011) Leaf area index (LAI) of loblolly pine and emergent vegetation following a harvest. *Transactions of the ASABE* 54, 2057–2066.

Schwarze, R. (2007) Runoff formation and water balance of forested catchments. In: Puhlmann, H. and Schwarze, R. (eds) *Forests Hydrology: Results of Research in Germany and Russia*. IHP/HWRP-Sekretariat, Koblenz, Germany, pp. 3–22.

Varhola, A., Coops, N.C., Bater, C.W., Teti, P., Boon, S. *et al.* (2010) The influence of ground- and LIDAR-derived forest structure metrics on snow accumulation and ablation in disturbed forests. *Canadian Journal of Forest Research* 40, 812–821.

Voronkov, N.A. (1988) *Significance of Forests for Water Protection*. Gydrometeoizdat, Leningrad, Russia (in Russian).

Wei, X., Giles-Hansen, K., Spencer, S.A., Ge, X., Onuchin, A. *et al.* (2022) Forest harvesting and hydrology in boreal forests under an increased and cumulative disturbance context. *Forest Ecology and Management* 522, 120468.

Zhang, M. and Wei, X. (2021) Deforestation, forestation, and water supply. *Science* 371, 990–991.

Zhang, M., Liu, N., Harper, R., Li, Q., Liu, K. *et al.* (2017) A global review on hydrological responses to forest change across multiple spatial scales: Importance of scale, climate, forest type and hydrological regime. *Journal of Hydrology* 546, 44–59.

12 European Perspectives on Forest Hydrology

C. de Jong*
University of Strasbourg, Strasbourg, France

Abstract
This chapter explores current and potential future issues in European forest hydrology. It addresses the paradigm shift in European forest hydrology hypotheses concerning floods and droughts. Furthermore, it attempts to discuss the impacts of land-use change and European policies and national regulations on the forest water cycle. Lastly, it highlights water-sensitive forest management for improving water quantity and quality under the sword of Damocles of climate and anthropogenic change.

12.1 Introduction

This chapter summarizes different European forest hydrology hypotheses concerning floods and droughts, the impacts of land-use change, the effects of important European policies and national regulations, and water-sensitive forest management geared towards improving water quantity and quality, as well as future challenges linked to climate and anthropogenic change.

One-third of Europe is covered by forests (210 million ha), which provide more than 4 km^3 of water annually to European citizens and host 870,000 km of rivers and almost 33% of European lakes (Zal *et al.*, 2015). About 32% of Europe's forested area is designated for the 'preservation of water resources' (Forest Europe, 2020). However, the condition of European forests is deteriorating due to the increasing frequency of large-scale forest disturbances, including extreme droughts, heat waves, extensive bark beetle outbreaks, and extensive forest fires, some of which hit many catchments simultaneously.

Information on European forest hydrology is highly dispersed and sectoral, mainly because Europe consists of 50 different countries, 28 of which form part of the European Union (EU) with 24 official languages. EU policies are mandatory and are reinforced at the European, national, regional and local levels. European Russia as non-EU covers roughly one-third of Europe's surface but abides by its own laws and regulations. The same is true for Norway, Iceland, Liechtenstein and (not least) Switzerland, an ideal island of comparison within Europe. Considering Europe's wide range of climatic and hydrological regimes, long-term studies are clustered in a surprisingly narrow temperate/alpine belt. Literature on forest hydrology has become increasingly segmented, not least due to the myriad of EU project reports. Furthermore, no single source summarizes the evolution of different EU regulations on hydrological events in forest hydrology. This chapter summarizes European

*Corresponding author: camen.dejong@live-cnrs.unistra.fr

DOI: 10.1079/9781800625310.0012

perspectives on forest hydrology based on EU and non-EU case studies.

European literature on quantitative forest hydrology is limited compared with studies from the USA, Australia and Japan (Schleppi, 2011). This is due to the small-scale ownership structures in Europe (P. Schleppi, Birmensdorf, Switzerland, personal communication, 2015) and the lack of large tracts of government land (L. Bren, Melbourne, Australia, personal communication, 2015) that make basin-wide experiments feasible. Andréassian (2004) corroborated this hypothesis with his observation that the unprecedented development of experimental basins in the 20th century occurred mainly in the USA. Nevertheless, the paired-catchment approach was initiated in Europe in the Swiss Sperbelgraben and Rappengraben areas and then spread to the USA (see Chapter 1, this volume). Later, the concept was taken up in the Plynlimon catchment in Wales, UK (Blackie and Robinson, 2007). Whereas earlier work focused on floods, there has been a recent shift towards climate change, droughts and low flows, as well as water quality and protection of drinking water. However, there is a painful lack of experimental and modelling studies on forest evapotranspiration compared with the USA. Recent European Geosciences Union sessions on 'Advances in Forest Hydrology' further include biogeochemical cycling, interactions with groundwater and impacts of environmental disturbance but not predominantly from Europe. The regional emphasis is also moving from catchment to flood-plain forest hydrology (Schnabel *et al.*, 2022). Globally, studies are increasingly based on remote sensing and modelling but are 'weakened by a lack of ground-truthed data' (Jones *et al.*, 2020). Europe has produced some notable forest hydrologists, but their number is restricted compared with forest biologists, in particular those working in alpine, arid and arctic regions.

The Swiss Sperbelgraben and Rappengraben areas belong to the oldest European experimental forest hydrology sites. They were instrumented as paired watersheds (with nearly 100% and 66% forest cover, respectively) in 1902. Large floods in the 1860s and 1870s were thought to be the result of large-scale deforestation and infrequent high-rainfall events (Keller, 1988). Early Swiss research in the Erlenbach region showed that forests are effective in strongly reducing peak discharge (by up to 50%) for short and intensive rainfall events but that the difference between forested and non-forested catchments diminishes for increasingly strong rainfall events (Fig. 12.1) (Hegg, 2006; Schleppi, 2011). In 1948, discharge and precipitation monitoring was initiated in the Lange Bramke catchment (Harz Mountains) in Germany after heavy deforestation (Herrmann and Schumann, 2009). Although the catchment is now completely afforested, the results showed that forest does not protect against catastrophic floods and can only partially decrease smaller floods.

As early as 1907, the first forest lysimeter station worldwide began investigating water budgets of young trees on the Drachenkopf Mountain, Eberswalde, Germany (Müller and Bolte, 2009). From this, the unique evolution of evapotranspiration and infiltration was available for mixed stands over 80 years. The Wernersbach experimental catchment in the Tharandt Forest (Germany) followed in 1967, accompanied by eddy-covariance flux towers since 1996 (Pluntke *et al.*, 2023). Evapotranspiration is increasing significantly due not only to climate change but also to changes in forest management (e.g. decreased harvesting, change from coniferous to mixed forest) and healthier forest stands.

In his publication on European experiences in long-term hydrology research, Keller (1988) indicated that not all scientific problems have been solved even after nearly 90 years of investigations – in particular, the question of extreme floods in forested areas. In Europe, as for the USA, there was a shift from quantitative to qualitative forest hydrology in the 1980s (Schleppi, 2011). European literature on forest-related water quality is more abundant for obvious reasons. The main problems include suspended sediment concentration from erosion, floods, pollutants and acid deposition.

In recent decades, forest hydrology in Europe has become much more interdisciplinary (with Faculties of Forest, Geological and Hydrological Sciences, Forest and Environmental Sciences, Forest Ecology and Hydrology). Finland hosts the European Forest Institute (EFI) with widespread regional offices. However, experimental forest hydrology research in Europe still stands in the shadow of the USA. This reflects greater

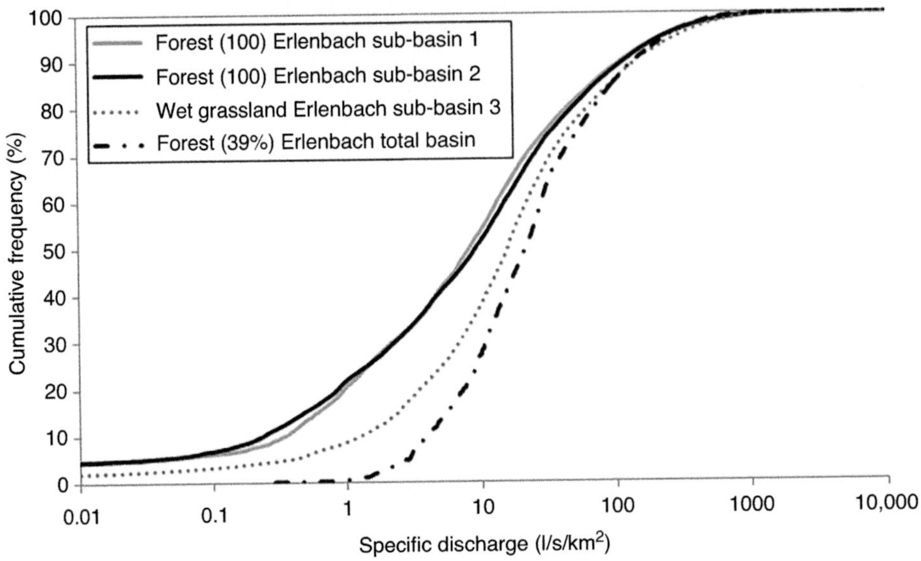

Fig. 12.1. Flow duration curves for forested and non-forested catchments in the Erlenbach region and surrounding three small experimental catchments. Data from Patrick Schleppi.

financial resources devoted to environmental research in the USA, bringing forward 'some of the most noteworthy contributions to catchment area research' worldwide, according to McCulloch and Robinson (1993). Recent decades have witnessed more emphasis on theoretical and modelling work in Europe with a more segmented and specialized, rather than basin-wide, approach towards forest hydrology. In the last decade, there has been less emphasis on water and more on 'climate-smart' projects and the forest–biodiversity–climate–water nexus.

Increasing water yield as a result of deforestation and decreasing water yield by afforestation is still the main hydrological dilemma in European forest management today (Flörke *et al.*, 2011), both in flood and drought risk. The larger the area of deforestation or reforestation, the stronger the effect on the annual water balance, with up to 700 mm difference in total. Compared with deforestation, afforestation experiments are limited, probably due to the long timescales in observation involved (Schleppi, 2011). Long-term simulations over 45–years in the now forested Coalburn catchment of northern England showed that, following afforestation, the development of mature forests has produced a decrease of about 250–300 mm in the annual streamflow compared with the original upland grassland vegetation (Birkinshaw *et al.*, 2014).

Despite the importance of forests, no common forest policy exists in Europe, but a new EU Forest Strategy for 2030 was created in 2021. As 60% of land is private property in EU countries, forest and water management remains complicated. Between 1990 and 2021, forest cover in the EU increased by 10% (145 million ha) through a combination of afforestation and land abandonment. Land is being abandoned in rural and mountain regions (in particular, in the Alps and Pyrenees), despite population increase because of urban growth and has major hydrological impacts (García-Ruiz and Lana-Renault, 2011; Khorchani *et al.*, 2020). However, simultaneously, forests face growing pressure from habitat degradation, overexploitation, fragmentation, expanding urban areas, pollution, invasive species, loss of biodiversity, soil erosion and climate-change consequences such as windthrow, insect outbreaks and fires (EEA, 2019, p. 496). Forest fires largely affect southern Europe, and latterly central and even northern Europe. The future frequency and timing of forest wildfires is expected to alter with drought patterns and significantly change temporal and spatial patterns of forest hydrology. Bark beetle disturbance has doubled its share of the total damage in the last 20 years (Patacca *et al.*, 2023). Despite efforts to halt biodiversity loss, 80% of forest habitat assessments show an unfavourable conservation status, with the boreal zone worst off. The increase in water scarcity has strengthened the focus on of drinking-water provision from forests. More than 20% of European forests (from 12% in the north to 42% in the south-west) are dedicated to protecting water and soils, mainly in mountainous areas. This trend is steadily increasing (Forest Europe, 2020).

12.2 Is Forest Flood Mitigation a Myth?

It is essential to focus on extreme floods in Europe, given their current and projected increase under climate change. While forests and afforestation play an important role in reducing the peaks of small to medium floods, research since the 1990s has shown that this also applies to extreme floods, whatever the catchment size (Andréassian, 2004). The water-retention capacity of forest soils is rapidly exhausted during extreme events and therefore pre-event soil moisture overrides the importance of vegetation type, especially in alpine catchments (Hegg, 2006; Schleppi, 2011). While discharge and peak flow generally increase in deforested catchments, after forest fires or insect outbreaks, Swiss and Swedish experiments have shown that the role of forests in reducing peak flow is proportionally less than for low flows. Comparison of three small Swiss catchments showed no direct relationship between the proportion of forest cover and peak flow (Burch *et al.*, 1996, in Schleppi, 2011). Although forested slopes often have high infiltration rates and peaks are insensitive to short-term rainfall intensities, particularly heavy precipitation events could cause Hortonian flow where infiltration is poor (e.g. shallow soils, fine-textured soils, saturated soils, rock outcrops, roads and ski runs).

In forest hydrology, one of the last remaining challenges is defining a threshold above which forest cover is no longer effective in reducing a flood in terms of precipitation intensity and discharge return intervals. In the European literature, with few exceptions, the quantitative definition of such threshold remains vague. Bathurst (2014) stated that forests do not prevent floods and that they do not appear to affect the magnitude of larger floods. He observes that above a certain magnitude (or frequency) of rainfall event, there is little difference in the peak discharges of forested versus non-forested catchments for those with a surface area larger than 1500 km^2 (Fig. 12.2). Both field data and model studies support the general trend towards either absolute or relative convergence for large events, which seem to lie around the 10-year rainfall return period (Fig. 12.2).

Forest restructuring and afforestation have been recognized as providing mitigation of small floods but no protection against catastrophic flood events (Dittrich *et al.*, 2007; Calder *et al.*, 2008; Hall *et al.*, 2013; Buechel *et al.*, 2022). An intense programme of reforestation believed to reduce floods was implemented in France in the mid-19th century (Mather *et al.*, 1999), but its effectiveness was called into question (Humbert and Najjar, 1992). The Global Forest Resources Assessment also holds natural processes – rather than land management in the upper watershed – responsible for flooding at the macro-scale (Hamilton, 2008). Thus, the importance of forest cover in regulating hydrological flows has been overestimated and the impacts of forest cover removal are evident only at the micro-level and in association with short-duration and low-intensity rainfall events (Flörke *et al.*, 2011). As rainfall duration or intensity increases and the distance down the watershed and river basin becomes greater, other factors such as slope, soil characteristics, altitude and snow dynamics become more influential (Hamilton, 2008; Alewell and Bebi, 2011). Forests may provide protection against peak runoff events mainly during snowmelt.

Substantially increasing forest cover by afforestation in large catchments such as the Inn (nearly 10,000 km^2) in Austria brought no significant reduction in the 100-year flood peak (Blöschl, 2022). Similarly, a simulation for the meso-scale Kamp catchment (622 km^2), Austria, showed that the efficacy of afforestation depends on event magnitude, with negligible peak flow reduction for the largest events. Forest cover in flood management plans in the Scottish

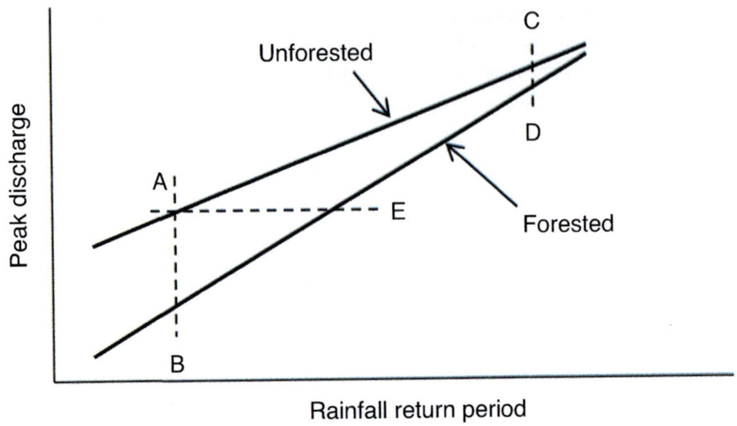

Fig. 12.2. Convergence of peak discharge for forested and non-forested catchments with increasing flood return period. A represents the discharge–rainfall relationship in the unforested catchment and B in the forested catchment. For small but frequent rainfall events, the unforested catchment has a higher peak discharge (line AB). For the forested catchment to achieve the same peak discharge, it would have to be subjected to a rainfall event with a larger return period (line AE). For large but infrequent rainfall events, the two responses converge (line CD) so that the same peak discharge is produced in each catchment by the same rainfall event (Bathurst, 2014).

Borders, UK is considered cautiously as it may even be inversely correlated with mean flow transit times, water storage and groundwater fraction (Peskett *et al.*, 2023). The effectiveness of mountain forest protection laws as a means of flood protection are even being reconsidered in the Upper Rhone, Switzerland (Rüegg *et al.*, 2022).

The lethal, extreme 2021 floods in the Ahr Valley in Germany were predominantly generated in small, forested, headwater catchments unable to buffer exceptional rainfall. In the narrow Schuld meander, the Ahr River rose by nearly 9m following 150mm rainfall in less than 18 hours over only a 500 km^2 catchment. The rapid, extreme flood peak is described as resulting from unhindered surface runoff on clayey soils saturated from pre-event precipitation and two converging forested sub-basins accumulating the highest precipitation rates in the catchment (Dietze *et al.*, 2022; Kahle *et al.*, 2022). Dietze *et al.* (2022) reported that 'ephemeral headwater drainages turned into streams, conveying several decimetres of overland flow' and caused 'overland flow velocity to increase by 1 to 2 orders of magnitude'. Despite the importance of flow connectivity, the literature includes little on the importance of slope gradient with respect to flood generation in forested catchments. In deeply incised low mountain catchments such as the Upper Ahr, lower valley slopes near the river can exceed 35°. Extremely fast and erosive surface flow facilitated by steep slopes and extreme precipitation (85 mm of rain and hail in only 30 min) generated the 1990 catastrophic 150 year flood event in the small, entirely forested alpine Schmiedlaine catchment in Southern Germany (9.4 km^2) (de Jong, 1992). Gradients reach 20° in the upper catchment and exceed 30° along the lower valley slopes. After a dry summer, infiltration was restricted and surface runoff concentrated in hollows on the steep forest floor creating zero-order creeks that actively eroded cobble-sized material and caused headwater incision (Ergenzinger, 1994). Slope steepness and the relief ratio was identified as a trigger for dangerous flash floods in small, partially forested catchments in the French Pyrenees (Garambois *et al.*, 2014). Steep valley slopes channel runoff along drainage flow pathways, favouring rapid concentration times and causing water depth in the drainage network to peak within a few minutes or hours. More experimental analyses on the nature and connectivity of flow pathways during extreme events are necessary. Until the turn of the millennium, these were carried out in small catchments in the Black Forest, Germany (Casper *et al.*, 2003), but were replaced by hydrological modelling.

In the frame of the Water Retention by Land-use (WaReLa project, 2003–2006), water-retention measures were investigated as to their effectiveness in reducing or temporally delaying floods within small forested watersheds in Germany (Schüler, 2006). A positive effect can be achieved if the water-storage capacity of forest sites is not exceeded. Discharge-accelerating linear structures such as roads and ditches should be identified. Runoff increases significantly when road density increases from 20 to 50m/ha. Recommendations included aligning roads parallel to the slope as road orientation is hydrologically significant, reducing the length of logging trails as much as possible to counter soil impermeability, delaying discharge as long as possible in large retention areas such as stream and river valleys, and keeping water courses, riverbeds and banks, and vegetation as natural as possible. More crucially, if all small catchments in a larger watershed are managed with a view to water retention, damaging floods may be reduced. Furthermore, the economic benefits and eco-efficiency of flood-prevention measures should be evaluated. Schüler (2006) recommended that flood prevention should not be restricted to forestry but enlarged to include the water, agriculture and viticulture sectors. Across Europe, the planning and evolution of flood protection in forested catchments has been proceeding at different paces, depending on the political history and status of the countries concerned within the EU. Not all forest types, such as flood-plain and riparian forest, have reached the same level of recognition, despite being essential for flood mitigation. They urgently require protection and restoration. However, according to Serra-Llobet *et al.* (2022), 'Although floodplain forests are protected by the EU Habitat Directive (EEC (European Economic Community), 1992) and are the most species-rich forest type in Central Europe, they have become increasingly endangered and considerably reduced in area.'

In the future, we will have to deal with much stronger effects of anthropogenic changes, such as accelerated or altered flow paths through urbanization and infrastructural developments, in addition to stronger hydrological variability due to climate change in forested catchments.

12.3 Is Forest Drought Mitigation a Reality?

In the past decades, much less consideration was given to the role of European forests in buffering droughts compared with the exposure of forests to drought. Forested catchments in Europe, especially in the mountains, are facing a myriad of pressures (e.g. more frequent storms, parasites) related to climate change, which is decreasing their hydrological resilience to droughts. The water budgets of Central European forests are increasingly endangered by droughts. In north-east Germany, forests are exposed to higher drought risk due to reduced precipitation, groundwater lowering and increased evapotranspiration in response to heatwaves and more erratic precipitation (Ziche *et al.*, 2021). Alpine forests are not exempt from drought either. Relatively small climatic shifts could result in large negative drought-related impacts on forest ecosystem services (Beniston and Stoffel, 2014). In studies on drought stress of European beech, higher-altitude trees were found to be less resistant to water deficit than lower-altitude stands already growing in a drier climate (Pšidová *et al.*, 2015). Drought periods severely affect forest productivity and are an important trigger for forest decline and mortality. A long-term irrigation experiment was started in 2003 on the effects of drought on alpine forest in the Swiss Federal Institute for Forest, Snow and Landscape Research (WSL) Irrigation Experiment (Pfynwald) in the inner-alpine dry Rhone valley near Wallis, Switzerland (d'Odorico *et al.*, 2021). This was expanded in 2024 to VPDrought, the first experiment worldwide to investigate the influence of air dryness and soil drought in young and mature natural forest. Under drought stress, the growth period of non-irrigated trees was 2–5 weeks shorter.

It is generally considered that forests have a negative impact on the water balance during droughts. Several studies have shown that forested areas may produce a lower runoff coefficient and that this may cause water-supply deficiencies in times of increased drought stress under future climate change (Alewell and Bebi, 2011). In alpine catchments, surface runoff is generally lower in forests due to high infiltration rates of humus layers and higher tree evapotranspiration compared with grasslands. Long-term studies from the UK and Ireland have shown that deforestation of conifers substantially increased dry weather baseflow (Xiao *et al.*, 2022). Models show that afforestation in 12 different catchments in the UK systematically decreased median and low flows (Buechel *et al.*, 2022). A study of more than 200 headwater catchments throughout Spain between 1961 and 2013 even attributed the higher frequency, duration and severity of hydrological droughts to an increase in vegetation cover rather than climate change (Pena-Angulo *et al.*, 2021). Similarly, in Czech upland forests, a long-term hydrological simulation since the 1950s showed that water balance shifted from positive to negative at the beginning of this century due to increased transpiration associated with a longer growing season (Kupec *et al.*, 2021). A similarly long hydrological drought analysis in the Bialowieza primeval forest, Poland, also revealed a significant increase in the number of days and aggregated volumes of low flow (Hejduk *et al.*, 2021).

In contrast, for the Czech Ore mountains, a long-term analysis of hydrological droughts demonstrated a significant decrease in average and minimum drought since 1967 attributed solely to climate change under stable land use (Vlach *et al.*, 2021). In forested headwater catchments in Bavaria, Germany, groundwater dominates water fluxes and sustains discharge, even during multi-year droughts (Kaule and Gilfedder, 2021). Under low flow, more than 70% of total discharge is derived from groundwater.

As climate change is accelerating and affecting entire forested catchments, understanding droughts at a seasonal and daily scale is becoming more crucial. Intact forest acts as a sponge by absorbing and storing water and releasing it slowly during the subsequent dry season. RiverDiv is a German–French Interreg project (2023–2025) that is investigating the impacts of climate change, in particular droughts, on the water quality of the forested Lauter River across the Palatinate and North Vosges Forests. Its forested headwater catchment and highly pervious sandstone geology play an essential buffering role in retarding the onset of low flows by several weeks compared with its more exposed downstream reaches (Liu and de Jong, 2024). In addition, hundreds of historic ponds partially damming the Lauter confluences in its forested headwaters (used in the past centuries for creating artificial flood waves for wood rafting) still play an important role in retaining water, limiting both flood peaks and severe low flow.

However, a state-of-the-art review on nature-based solutions in Europe (Yimer *et al.*, 2024) has shown that afforestation is much less commonly implemented to mitigate drought and water scarcity than wetland restoration and urban rainwater harvesting. Surprisingly, enabling policy frameworks in the EU for reducing droughts and floods through nature-based solutions do not yet specifically include forests (Amirzada *et al.*, 2023). European flood-plain forests are more vulnerable to drought due to the combined effects of declining groundwater levels and river regulation. Towards the end of the 2018–2019 drought, tree growth in the Leipzig protected flood-plain forest substantially decreased, indicating that consecutive hotter droughts pose a threat even with relatively high groundwater levels (Schnabel *et al.*, 2022). Similarly, flood-plain forests in the southern Czech Republic are directly threatened both by groundwater modifications and by more frequent droughts (Basu *et al.*, 2024). Nevertheless, riparian forest restoration, combined with river restoration and forest management, are beneficial for increasing resilience against climate-change impacts, improving infiltration, water storage and reducing drought risk (e.g. Evrotas River, Greece, and the Sava and Drina River Corridors Integrated Development Programme, Serbia). The Flemish 'Blue Deal' also aims at restoring forest ecosystems and agroforestry as a lever against water scarcity.

Drought impacts on forests have received less attention than in agriculture as they are less well understood, but their temporal and spatial effects should be integrated in forest-related management. Meteorological droughts have a statistically significant influence on forest wildfires (Ristić *et al.*, 2011; Stagge *et al.*, 2015). Estimates of wildfire severity based on monthly area burned are documented in the European Forest Fire Information System (EFFIS). In the Mediterranean, wildfires dominate as a single, large, peak fire-danger period in late summer whereas the temperate regions of Central Europe produce two distinct fire peaks occurring in the spring and again in late summer. Some alpine forests in the Valais, Switzerland, are even irrigated during summer droughts to prevent wildfires.

It is essential to consider the impacts of forests on evapotranspiration when investigating the water balance. This is particularly so in more complex mountain environments. In the Alps, differences between evapotranspiration of forests and grasslands are greatest in early spring when spruce forests are already transpiring while grasslands are still under closed snow cover and in winter when precipitation interception is effective in forests and/or forests are transpiring during warmer winter days (Alewell and Bebi, 2011).

The alpine forest water balance dilemma still needs to be solved. In the Swiss Alps, Alewell and Bebi (2011) concluded that a complete regrowth of forests in the Reuss catchment, Urseren Valley, within forests' climatic boundaries would only insignificantly influence hydrograph dynamics because of the relatively small area of regrowth compared with the forest-free area above the treeline. Long-term measurements (runoff data since 1904) indicate a 30% reduction in runoff compared with long-term means over the last 60 years (BAFU, 2007). However, it is unclear to what extent these runoff changes are caused by shrub encroachment (30% increase in shrub area between 1959 and 2004), soil degradation, climate change, or changes in the water courses and their management.

12.4 The Bureaucracy of Forest-Relevant EU Policies

12.4.1 Evolution of European forest policies

A wide range of EU policies are concerned with forests. These include the Habitat Directive, the EU Biodiversity Action Plan, the Rural Development Regulation, the Common Agricultural Policy (CAP), the EU Forest Action Plan, Natura 2000 and the Biomass Action Plan. However, forests are not mentioned in the Water Framework Directive (WFD) or the EU Water Scarcity and Drought Policy. It is a challenge to manage forests sustainably across EU member states, local authorities and forest owners. In response to a lack of coherence between forest-related EU policies and national forest policies, a non-binding EU Forest Strategy was created in 1998. In 2013, it was updated due to growing demands on and threats to forests, as well as significant societal and political changes that affected them. Its aim was to protect forests and biodiversity from the transnational effects of storms and fires and increasingly sparse water resources. As the number of targets increases, so does the complexity of competing forest claims and trade-offs at the regional level (Pülzl, 2013).

Another issue has been monitoring and reporting of the status of forests in the EU – knowledge of the pressures they face and ecosystem services they provide is still patchy. Forest data and information are collected at national levels, but this information is not easily available and is seldom comparable from country to country. The 2013 strategy called for coordination of such information and extension of forest-management plans to privately owned forests. One provision was to provide compensation payments for drinking-water and groundwater protection. This was transformed in 2022 into the EU Forest Strategy for 2030. It highlights the importance of European forests as key providers of ecosystem services, in particular for water.

Nearly 90% of the EU's total finance for forestry measures comes from the CAP and the European agricultural fund for rural development (EAFRD). This includes voluntary payment for ecosystem services such as water management, especially for privately owned forests. However, funding for forested areas is much lower than for agriculture and represents less than 1% of the total EU budget, despite forest covering the same area as agriculture. Thus, the European Court of Auditors pointed out in 2021 that the EU has a positive but limited impact on protecting the biodiversity of EU forests and adapting them to climate change.

Next to strategies, the EU also issues green papers and resolutions. In 2010, a *Green Paper on Forest Protection and Information in the EU: Preparing Forests for Climate Change* recognized that forests 'regulate freshwater supplies' and 'play a major role in the storage, purification and release of water to surface water bodies and sub-surface aquifers'. For example, in Belgium, the Ardennes Forest is the principal source of water supply for Brussels and Flanders. In Germany, two-thirds of the 'Wasserschutzgebiete' (drinking-water perimeters) for abstraction of high-quality drinking water are under forest cover, and in Spain, forests in headwater catchments have been given special conservation status because of their capacity to improve drinking-water quality. In 2011, the European Parliament reacted to the Green Paper asking for the compilation and monitoring of indicators relating to the protective functions of forests such as soil retention and water capacity. To achieve the EU 2020 strategy's objectives with regard to national forest action plans, it requested each member state to develop a forest strategy including riverbank reforestation, rainwater capture and selection of traditional tree varieties best adapted to drought.

Forest-management plans based on the principles of sustainable forest management are key instruments in delivering multiple goods and services according to the European Commission. Forest-management plans are at the core of both the EU 2020 Biodiversity Strategy and EU Rural Development funding. Member states should maintain and enhance forest cover to ensure soil protection and water quality and quantity regulation by integrating sustainable forestry practices in the Programme of Measures of River Basin Management Plans under the EU WFD (Water Framework Directive) and in the Rural Development Programmes.

In 2007, a Resolution on Forests and Water was adopted by the Ministerial Conference for the Protection of Forests in Europe. It consists of four parts: (i) sustainable management of forests with relation to water; (ii) coordinating policies on forests and water; (iii) forests, water and climate change; and (iv) economic valuation of water-related forest services. However, progress is insufficient concerning economic valuation (Forest Europe, 2020). The latest United Nations University report on building climate resilience with respect to the 2021 Western Europe flood disaster shows that there are many obstacles to giving higher valuation to ecosystem services concerning forest-based management options in flood-prevention concepts (Hagenlocher *et al.*, 2023). At the same time, the EFI emphasized that 'avoiding deforestation, less frequent or less intensive harvesting, retention of old and dead trees as well as the restoration of peatland hydrology can benefit both forest biodiversity and climate change mitigation' (Verkerk *et al.*, 2022, p. 75). Currently, forest harvesting is an urgent issue, with a nearly 50% increase in harvested forest area over Europe after 2015 compared with the period 2011–2015 (Ceccherini *et al.*, 2020). The highest losses have occurred over the Iberian Peninsula and in the Nordic and Baltic countries. This will not only potentially affect water regulation but also hamper EU forest-based climate mitigation. As such, 'climate-smart forestry' remains a missing component in national strategies (Verkerk *et al.*, 2020).

Achieving policy objectives is challenging due to external pollutants contaminating soil and groundwater, and ageing forests with insufficient protective functions. Furthermore, only a few European countries have 'created framework conditions for the sustainable provision of the qualitatively and quantitatively adequate water supply from protective forest areas and few monitor protective forest areas'. Only one country, Poland, has a specific target for increasing the volume of retained water in forests in lowland areas (by 2.1 million m^3) and in mountainous areas (by 0.4 million m^3) (Forest Europe, 2020). Mountain forests account for one-third of the total forest area in the EU. However, the concept of forest mountain 'water towers' for protecting and providing drinking water for Europe's cities has not yet been addressed in European policies.

12.4.2 Effects of policies on forest hydrology

As yet, there is surprisingly little information available on the effects of policies on forest hydrology in Europe (Meesenburg *et al.*, 2005). Müller (2012) describe the positive hydro-ecological impacts of new guidelines for the creation of stable mixed pine–beech stands and a nature-oriented approach to forest structures by German forestry in the north-eastern lowlands. Stemflow on beech trees in the winter half-year has the advantage of generating additional deep soil seepage and an increased amount of stemflow in the summer half-year linked to more trees with larger diameters increasing topsoil moisture. With less than 600 mm annual precipitation and sandy soils with a low water-storage capacity, water availability is limited and at risk from more frequent future droughts. For some decades, groundwater levels have been falling. Counteracting this trend through targeted forest conversion is a challenge for forestry in north-east Germany (Goral and Müller, 2010). Furthermore, ameliorating the regional water balance could also have positive spin-off effects such as the protection of 'forest mires' (wetlands) and improving drinking-water and groundwater quality in the vicinity of urban agglomerations.

European policies have different hydrological objectives that are not always compatible and can generate conflicting forest hydrological impacts. For example, whereas the EU WFD aims to bring waterbodies into a good ecological status and seeks mainly to combat water pollution, the European Flood Directive focuses specifically on flood prevention through forest management at the catchment scale. As noted by Pülzl *et al.* (2014, p. 57):

> The likely provision of water-related ecosystem services by forests is not clearly recognized in the WFD, and the complex interplay between water protection management and forestry is neglected. While timber-production-oriented forestry is considered a risk in reaching a good ecological water status, especially of local water bodies, the potential benefits of forests and forest management in achieving a good ecological status for waters and their catchments are not recognized.

In 2009, the EFI even suggested establishing an EU Forest Framework Directive to strengthen coordination of forest-related aspects by EU regulation. MOUNTFOR (Preserving and Enhancing the Multifunctionality of Mountain Forests), initiated in 2013 in cooperation with the EFI, was one of the few alpine projects that assesses the potential impacts of forest management and land-use changes on mountain hydrology and the availability/quality of water resources.

A case study from the north-eastern Ore Mountains in Saxony, Germany, carried out in the FLOODsite project evaluated the impacts of the European Flood Directive on forest land management and, in turn, on water dynamics (Wahren and Feger, 2011). Since the disastrous floods of the River Elbe in 2002, a new water law exists for Saxony with regulations concerning flood-originating areas. Natural water retention is to be conserved and improved, and soils are to remain unsealed and afforested, if possible, with compensation measures in case of reduction or loss. Due to its mining history, the area only has 20% forest cover. These catchments traditionally provide drinking water from reservoirs and are known for their lower specific runoff under forest. The competitive goals of future land-use planning such as flood protection, profitable food/wood production, water supply and water protection create conflicts in decision making facing an uncertain future (Wahren and Feger, 2011). Afforestation and 'near-natural' silviculture can increase water retention for smaller flood events, but the quantitative role of non-structural flood-risk management measures with respect to event size remains controversial. While the European Flood Directive does not have clear subsidy policies, flood protection remains at best an additional benefit but not a target (Wahren and Feger, 2011).

Manser (2013) postulated that climate change will proceed at a rate faster than the natural adaptation capacity of forests and that this poses a serious challenge for forest policy. Among future challenges, he identified: (i) understanding the reaction of forest stands to climate change; (ii) its consequence for forest goods and services; (iii) the benefits and risks involved in adaptation strategies and ways to increase the adaptive capacity of forests; (iv) overcoming long research timescales when climate-change strategies require short-term decision making; and (v) how models and tools can be optimized for practitioners.

12.5 Emerging Issues

12.5.1 Drinking-water and groundwater protection

The EU focuses mainly on water-quality issues related to the provision of drinking water from forests (SOER, 2015). Groundwater pollution is identified as a new challenge due to land-use change. In the West Harz mountains, Germany, voluntary drinking-water protection associations have been created at the initiative of forestry (Versuchsanstalt, 2012). Local stakeholders such as foresters and farmers familiar with local conditions and best solutions are empowered to cooperate closely with water managers. The main issue is decreasing suspended sediment concentration in the local drinking-water reservoirs.

In Austria, economic incentives for water management such as transfer payments for water protection by forests around population agglomerations (e.g. Vienna) are provided. Proposed action plans for south-eastern Europe include avoiding clear-cutting within drinking-water protected areas, ensuring a continuous forest cover and stable forest ecosystem, and limiting silviculture-related road construction (CC-Ware, 2014). Often drinking-water protection legislation is not homogeneous, not implemented properly or confusing. In future, an improved applicable legal framework integrating specific subjects of relevance for drinking-water supply, ecosystem services, land use and climate change will be required. Recent EU projects (e.g. Cities4Forests) support forested catchments for urban water supplies and risk reduction (e.g. around Oslo). Schüler *et al.* (2011) identified the prediction of forest management and environmental impacts on groundwater quality as one of our strongest present-day challenges: 'Forest management in a changing environment subject to global warming or air pollution may diminish the protective functions of forests with regard to groundwater quality.' Moreover, interactions between forests and water and impacts of forestry on groundwater quality and runoff remain a scientific grey zone. Forests can concentrate pollutants and decrease water quality by retaining atmospheric pollution (Schleppi, 2011). Forest monoculture such as spruce has been found to have negative impacts on water quality in Central Europe.

12.5.2 Water scarcity

As the Mediterranean region is particularly vulnerable to water scarcity, it may require adaptive forest management in future to adapt forests to water availability by means of artificial regulation of the forest structure and density (González-Sanchis *et al.*, 2015). The European Commission recognizes that forests are particularly important in Mediterranean countries because of their ability to balance the water cycle and therefore considers that reforestation to produce rainwater catchments may be required. To date, there is no clear linkage between forest management and droughts. Pine forests should be managed in order to decrease water interception and plant transpiration (green water) and increase water runoff and/or percolation (blue water).

12.5.3 Flooding triggered by ski runs

Preferential runoff paths that accelerate and increase flood peaks in forested catchments have so far been attributed only to forest roads, timber harvesting and log trails. Ski runs, in addition to new roads constructed to access water reservoirs for snowmaking and constantly expanding ski resorts with their highly reduced infiltration capacity, are rarely mentioned and not sufficiently integrated into modelling approaches (Fig. 12.3).

For the USA, Wemple *et al.* (2007) distinguished four factors associated with ski area development that may affect watershed processes and that are distinct from those associated with traditional forest-management practices: (i) forest clearings created for ski trails are oriented along gravitational flow paths, enhancing the potential for efficient down-slope routing of water, solutes and particulates; (ii) forest clearings for ski trails are intended to persist over time and represent a relatively permanent alteration of the forest landscape; (iii) certain activities associated with ski area development, particularly artificial snowmaking, are not present in traditional forest-management operations; and (iv) other practices, including creation of impervious surfaces and development of drainage infrastructure, are more extensive than those associated with traditional forest-management practices. Little scientific literature specifically addresses the effects of resort development in forests.

More recent studies show that the presence of ski runs significantly changes the hydrology of forested headwater catchments due to

Fig. 12.3. (a) A healthy forest with one ski run in 2014. (b) Pre-flood forest damaged by pine beetle, windthrow and snow load, and the new Marchner ski run in 2020. (c) Post-flood forest severely damaged by disturbance in 2022 with the highly eroded Innerbergbach visible on the photo in Fig. 12.4. From GeoBrowser Maps, South Tyrol, Italy, https://maps.civis.bz.it/ (accessed 8 May 2025).

deforestation and enhanced surface runoff on ski runs and access roads (Fidelus-Orzechowska *et al.*, 2018; Furdada *et al.*, 2020). Torrential flooding as well as excessive erosion, sediment transport and deposition have been monitored as a result of ski resorts in the Spanish Pyrenees, Polish Carpathians, Stara Planina Mountains (Serbia) and Austrian Alps. In the Pyrenees, a strong increase in surface flow and sediment transport has led to gullying, incision, and new channel head development and increasing flood discharge in the main streams. Only 3 years after ski run construction in the Carpathians, up to 1.5m valley incision was recorded and a 26% increase in discharge was modelled for the 10-year event at control points (Fidelus-Orzechowska *et al.*, 2018). In Serbia, ski resorts have also triggered more frequent extreme events in the Zubska River headwater (Ristić *et al.*, 2012).

Catastrophic flooding and debris flows in 2005 and 2015 generated in forested catchments with ski runs in the Paznaun valley in Austria are thought to be linked to loss in soil permeability and vegetation cover and preferential flood routing. Pötzelsberger and Hasenauer (2015) found that ski runs increased runoff considerably in the 10 km^2 Schmittental catchment covered principally by Norway spruce. Land use evolved from sparse to dense forest from 1890 to 1965 but has been increasingly dissected by ski runs since the 1970s. At present, the catchment is covered by 71% forest (520 ha) and 28% pasture (200 ha), of which about 14% (100 ha or 77 km in length) is used as ski slopes. The high average runoff coefficient of 0.74 is attributed to ski slopes with reduced infiltration, in particular when ski runs are groomed and topsoil has been removed for ski run remodelling (Hagen, 2003). Extensive experimental studies also show that ski runs within forests consistently attain the highest runoff coefficients (above 0.5) compared with all other land use (Markart *et al.*, 2017). Furthermore, about 500,000 m^3 of water is introduced into the Schmittental catchment annually from a downstream lake to produce artificial snow on ski runs. This surplus water probably increases snowmelt runoff from ski runs in spring by 20–25% (Pötzelsberger and Hasenauer, 2015). Pumping surplus water from water reservoirs or rivers on to ski runs is common in all ski resorts.

An intense summer hailstorm in 2022 (84 mm of rain in 1 hour, the strongest ever measured locally) caused ski runs on the eastern side of the Kronplatz, Italy, to generate severe overland flow, triggering heavy gullying, stream incision, sediment transport and landslides on the steep, forested slopes, and producing catastrophic flooding in the main valley where a flood-plain forest had been removed (Fig. 12.4). The mountain forest was unable to buffer excess overland flow because the entire summit area above the treeline had been recently mulched and because all the drainage from ski runs on the summit and in the forest were topographically concentrated towards a single, steep stream in the forest. Flow over the ski runs turned into

Fig. 12.4. (a) Extreme hail and rainfall concentrated on the forest-cleared Marchner ski run on the Kronplatz, South Tyrol, Italy, causing (b) severe stream incision and landslides in the Innerbergbach and catastrophic flooding of the Furkelbach River in the main valley on 5 August 2022. Photo courtesy of E. Brunner.

streams (Brunner, personal communication; Fig. 12.4). Flooding was amplified by a debris flow heavily charged with uncleared tree trunks logged due to forest dieback from pine-beetle disease, drought, windfall (2018) and snow load (2019) (Fig. 12.3). Forest cover dominates (72%) but ski runs (22%) largely surpass pasture (2%). This undoubtedly has strong hydrological repercussions, not only on soil water retention and flow but also on evapotranspiration. Other evolving land-use changes that counteract the flood buffering role of forests are the loss of natural water-retention areas such as depressions and wetlands within forests to large impermeable snowmaking water reservoirs and the removal of flood-plain forests in the main valleys to give way to car parks for lifts.

In the European Alps, the extent of ski runs in forested areas has increased substantially and is still increasing. Winter resorts in forested regions should be identified as a strong emerging regional hydrological challenge, in particular because the hydrological resilience of alpine forests is further weakened by other wide-ranging impacts such as pine-beetle attacks, droughts, windfall, snow loads and forest fires.

12.5.4 Climate change

One of the most serious challenges of forest hydrology is coping with climate change, in particular the effects of increased temperature and CO_2 concentration and decreased snow duration (Schleppi, 2011), but also windfall, pests (Andreu *et al.*, 2015) and droughts. As Europe is warming almost twice as fast as the global average (C3S, 2024), European forest hydrology is probably more heavily impacted by climate change than any other region worldwide. European forest transpiration increased by 5% over the 20th century (Frank *et al.*, 2015) but has declined in some regions since 2010 due to drought damage, bark beetle infestation, and events damaging tree crowns such as wind storms and snow load (Pluntke *et al.*, 2023).

A major challenge is combining plot and catchment scale with long-term experiments and unravelling causal factors (Schleppi, 2011). This requires new research questions and adapted methods. Open questions include optimal tree species composition and endemic forest development (Zimmermann *et al.*, 2006). Changes are so fast that existing data is often insufficient. As entire ecosystems cannot be investigated experimentally, researchers need to refer to practitioners and local knowledge. Uncertainty has to be confronted. No single solution is available, but diversifying tree species is essential.

Climate-change impacts on streamflow have been investigated for over 50 years in forested headwater catchments in the Polish Tatras. Despite decreased snow cover and increased temperatures, low flows increase in both summer and winter (Rajwa-Kuligiewicz and Bojarczuk, 2024). The winter low-flow increase is likely due to groundwater storage of snowmelt that also efficiently delays early summer flow. Other rain and snow-dominated forested catchments in southern and Central European experience a decrease in annual streamflow (Zhang *et al.*, 2022). Droughts and water shortage have increased strongly since the turn of the millennium and have triggered an extreme increase in defoliation. Forest fragmentation resulting from such forest dieback will reduce the ability to dampen runoff peaks and impact water quantity and water quality.

12.5.5 Disturbances

European forests are increasingly impacted by storms (windthrow), wildfires and insect outbreaks related to drought and climate change, in particular in mountain catchments. Long-term studies in the Polish Tatra mountains investigated the streamflow response to windstorms (Rajwa-Kuligiewicz and Bojarczuk, 2024). They showed that low flows increased and hydrological droughts decreased after windthrows, mainly due to a decrease in evapotranspiration and reduced catchment storage. A valuable long-term study compared the impact of bark beetle outbreaks causing tree dieback of nearly the entire stand in an unmanaged mountain catchment of the Bohemian Forest, Poland, with a nearby control forest in the Czech Republic (Kopáček *et al.*, 2020). Runoff marginally increased by 6% in the Polish relative to the Czech catchment. The effects of tree dieback on hydrology and microclimate after bark beetle infestation were less pronounced and more variable in unmanaged, naturally disturbed areas than in clear-cut areas.

12.6 Adaptation and Water-Sensitive Forest Management

European projects such as ClimWatAdapt have assessed water-sensitive forest management as an adaptation measure (Flörke *et al.*, 2011). The main climate threats identified were: (i) not enough water (water scarcity and droughts); (ii) too much water (flooding, sea-level rises and coastal erosion); and (iii) deteriorating water quality and biodiversity. Forest-management measures can increase water yield, regulate water flow and reduce drought stress for a forest under low-flow conditions. Some measures put into practice to support the water-regulation role of forests include: (i) reduced density of stand stocking; (ii) shorter length of the cutting cycles; (iii) planting hardwood species; and (iv) regeneration from seedlings rather than sprouts. Afforestation near watercourses benefits flow regulation, water quality and reduces drought severity. Identification and correct application of forest management to reduce water use is a crucial aspect regarding water scarcity. In practice, it is difficult to adapt silviculture management rules to improve tree water balance. Stakeholders pointed out potential limitations, undesired side-effects, and costs. The popular idea that more forests implies more and better water is increasingly challenged.

The EFI Silvistrat project (Response Strategies to Climatic Change in Management of European Forests) has developed adaptive management strategies for sustainable forest management under climate change. It analysed adaptive forest-management strategies to reduce the impacts of drought and other adverse effects of climate change. It suggested substitution of species sensitive to drought and late spring frosts with more drought-tolerant and frost-resistant tree species. Within Central Europe, forests in Slovakia and Hungary were identified as the most drought prone (Hlásny *et al.*, 2014). Adaptive measures included artificial regeneration to enrich local gene pools and increase stand tolerance to droughts. Risk management and disturbance monitoring systems are reinforced. However, adaptation strategies for water retention in drought-endangered forest landscapes lack funding. In the water-scarce Mediterranean Catalan River basins, forest thinning was investigated as a strategy for coping with climate change (Garcia *et al.*, 2024). Coniferous forests respond to thinning with the highest water yield and deciduous forest with the least.

Many recent experimental studies on water-sensitive forest hydrology come from Germany with the aim of creating stable mixed stands and nature-oriented approaches. Forest conversion plays an important role in the regional water budget, water supply and water distribution (J. Müller, personal communication, 2012). Current forest management, which involves replacing Norway spruce with Douglas fir or moving towards mixed spruce–fir–beech stands, has positive impacts on water resources in terms of water-use efficiency, infiltration and drought tolerance, thereby increasing the resilience of European forests (Prietzel *et al.*, 2023).

In Sweden, forests play an increasingly important role in regulating the hydrological cycle, but integration of water management in the day-to-day management of forests is fairly new (Samuelson *et al.*, 2015). Sweden has developed societal strategies to restore and/or maintain forests for the benefit of water-regulation strategies. Bilateral and multi-lateral activities have been initiated to build resilient landscapes and a water-management toolbox developed for forest planners. Forest policies and management strategies are starting to integrate climate-change mitigation and adaptation.

Recently, a catalogue of measures for improving forest water balance was established for low mountain catchments in the Palatinate Forest Biosphere Reserve in south-western Germany (Fig. 12.5) (Müller, 2023). It demonstrates a shift from pure forest protection to a stronger integration of soil protection and creation of water-retention areas. The key challenges identified are: (i) breaking the erosive energy of rain drops through adaptive forest management; (ii) enhancing infiltration through improved soil protection; (iii) distributing overland flow across the surface via ingenious forest development; and (iv) restoring natural habitats by creating retention areas. The overall future message identified under climate

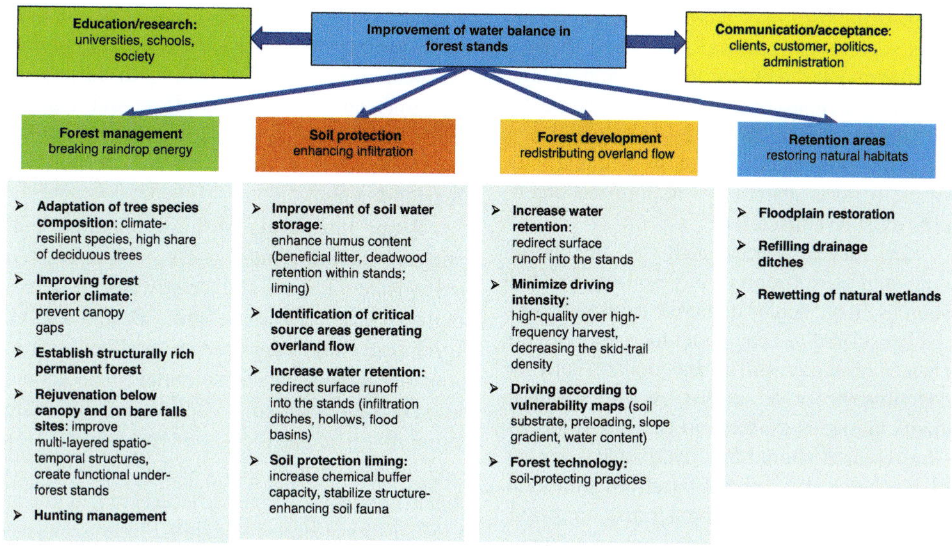

Fig. 12.5. Catalogue of measures for improving water balance in the Palatinate Forest, south-western Germany. Modified from Müller (2023).

change is that water must be retained in the forest, that soil protection means water conservation, and that the forest's interior climate is of central importance and must be preserved in order to use the cooling effect to reduce evaporation (Müller, 2023). Under climate change, especially droughts, forest ecosystem services will become increasingly important for ensuring surface and groundwater recharge. Furthermore, a water-sensitive forest-management adaptation option has been produced by the European Climate Change Adaptation Platform (Climate-ADAPT, 2024).

12.7 Challenges and Future Research Needs

It is difficult to unentangle future research needs on forest hydrology for Europe from those of other continents worldwide in the literature.

In future, a stronger focus is required on the quantitative impacts of climate change on forest hydrology and water resources. As such, a comparative upstream–downstream-coupled discharge–groundwater monitoring approach is advisable in catchments with forested headwaters. The potential sponge role of forests for buffering droughts should be considered, (Peña-Arancibia *et al.*, 2019; Liu and de Jong, 2024). More experimental studies on evaporation and evapotranspiration (including lysimeters and eddy-flux towers) are urgently needed, as this component is increasingly impacted by global warming (Strohmenger *et al.*, 2022) and seasonal differences in snow, soil water and groundwater availability. Forest ecohydrology also requires a more active dialogue between hydrologists and ecophysiologists (Cocozza and Penna, 2022).

The role of forested mountain water towers for providing and protecting drinking water for European cities needs to addressed in research and policies. In general, stronger European laws should be put into place against deforestation, in particular of pristine forests and for infrastructures related to winter tourism. Forest regrowth due to abandonment of pasture in the Alps has reduced discharge from springs, torrents and rivers, and increased evapotranspiration (Dumas, 2011; van den Bergh *et al.*, 2014). Changes are complex, as future forest regrowth might be impeded by increased disturbances from insects, droughts and wildfires. Land abandonment needs to be avoided as it mostly concerns remote headwater catchments acting as essential water towers.

Superimposition of worst-case scenarios in forests such as extreme rainfall, forest dieback and impermeable surfaces from changing land use leading to rapid changes in the water balance and extreme floods should be anticipated. For flood protection, the role of slope gradient with respect to flood generation in forested catchments must be considered.

New developments such as the bioeconomy and substitution of bioresources for fossil resources, in particular in Nordic countries, need to be considered as they could increase pressure on water resources and cause conflicts with the WFD (Rakovic *et al.*, 2020). As the strongest climate change is expected in the boreal forests, future research should link hydrological, energy and biochemical processes (Lindroth and Crill, 2011). A closer look at seasonal transition periods from winter to spring and autumn to winter is necessary. The growing season will be affected by changes in permafrost thaw and snow cover. Shifting patterns of temperature and precipitation will lead to changes in fire frequency and intensity and in drought frequency versus waterlogging.

Since 2009, there has been an expansion of European Critical Zone Observatories focusing on ecohydrological processes in forested catchments (Hissler *et al.*, 2021). Several European projects have dealt with ecosystem experimentation and adaptation and vulnerability to climate change, in particular droughts and the impacts of hydropeaking from dams on riparian forests.

Better integration of forest management, land-use planning and policy measures is a prerequisite for successful forest water resources management (Alewell and Bebi, 2011). Agroforestry and other intersectoral approaches should be supported. Forest management should become decentralized to meet the growing demands on forest ecosystem services such as food, biofuel, timber and disaster protection. A shift in competence from the federal to regional or possibility local level is advisable. Citizen science is a much-needed alternative to enrich knowledge and obtain spatially representative data that are otherwise unavailable.

Another major challenge will be to solve the contradictory approaches of forest hydrology management on afforestation versus forest thinning with respect to droughts and floods.

Acknowledgements

I am grateful to James Bathurst, Newcastle, UK; Patrick Schleppi, Birmensdorf, Germany; Jürgen Müller, Eberswalde, Germany; Friedhard Knolle, Goslar, Germany; and Eva Müller, Trippstadt, Germany, for their help.

References

Alewell, C. and Bebi, P. (2011) Forest development in the European Alps and potential consequences on hydrological regime. In: Bredemeier, M., Cohen, S., Godbold, D.L., Lode, E., Pichler, V. *et al.* (eds) *Forest Management and The Water Cycle: An Ecosystem-Based Approach*. Springer, Dordrecht, Netherlands, pp. 111–126.

Amirzada, Z., Pavlova, I., de Chaisemartin, M., Denoon, R., Kalas, M. *et al.* (2023) Reducing hydrometeorological risks through nature-based solutions: A comprehensive review of enabling policy frameworks in the european union. *Nature-Based Solutions* 4, 100097.

Andréassian, V. (2004) Couvert forestier et comportement hydrologique des bassins versants. *La Houille Blanche* 90, 31–35.

Andreu, J., Solera, A., Paredes-Arquiola, J., Haro-Monteagudo, D. and Lanen, H. (eds) (2015) *Drought: Research and Science–Policy Interfacing*. Taylor & Francis, London.

BAFU (Bundesamt für Unwelt) (2007) *Hydrologisches Jahrbuch der Schweiz 2005*, Umwelt-Wissen, Vol. 713. Bundesamt für Umwelt, Bern.

Basu, S., Stojanović, M., Jevšenak, J., Buras, A., Kulhavý, J. *et al.* (2024) Pedunculate oak is more resistant to drought and extreme events than narrow-leaved ash in Central European floodplain forests. *Forest Ecology and Management* 561, 121907.

Bathurst, J.C. (2014) Comment on "A paradigm shift in understanding and quantifying the effects of forest harvesting on floods in snow environments" by kim C. Green and Younes Alila. *Water Resources Research* 50, 2765–2768.

Beniston, M. and Stoffel, M. (2014) Assessing the impacts of climatic change on mountain water resources. *Science of the Total Environment* 493, 1129–1137.

Birkinshaw, S.J., Bathurst, J.C. and Robinson, M. (2014) 45 years of non-stationary hydrology over a forest plantation growth cycle, coalburn catchment, Northern England. *Journal of Hydrology* 519, 559–573.

Blackie, J. and Robinson, M. (2007) Development of catchment research, with particular attention to plynlimon and its forerunner, the East African catchments. *Hydrology and Earth System Sciences* 11, 26–43.

Blöschl, G. (2022) Three hypotheses on changing river flood hazards. *Hydrology and Earth System Sciences* 26, 5015–5033.

Buechel, M., Slater, L. and Dadson, S. (2022) Hydrological impact of widespread afforestation in great britain using a large ensemble of modelled scenarios. *Communications Earth and Environment* 3, 6.

Burch, H., Forster, F. and Schleppi, P. (1996) Zum Einfluss des Waldes auf die Hydrologie der Flysch-Einzugsgebiete des Alptals [Schweiz]. *Schweizerische Zeitschrift Für Forstwesen* 147, 925–938.

Calder, I., Hofer, T., Vermont, S. and Warren, P. (2008) Towards a new understanding of forests and water. *Unasylva* 229, 3–10.

Casper, M.C., Volkmann, H.N., Waldenmeyer, G. and Plate, E.J. (2003) The separation of flow pathways in a sandstone catchment of the northern black forest using DOC and a nested approach. *Physics and Chemistry of the Earth, Parts A/B/C* 28, 269–275.

CC-Ware (2014) *Towards an Action Plan for the SEE-Region. Guidance for Water Resources under Climate Change.* ETC/SEE Project, Federal Ministry of Agriculture, Forestry, Environment and Water Management (BMLFUW), Forestry Department, Vienna.

Ceccherini, G., Duveiller, G., Grassi, G., Lemoine, G., Avitabile, V. *et al.* (2020) Abrupt increase in harvested forest area over Europe after 2015. *Nature* 583, 72–77.

Climate-ADAPT (2024) *Water Sensitive Forest Management. Climate Adapt, European Environment Agency,* Copenhagen. Available at: https://climate-adapt.eea.europa.eu/en/metadata/adaptation-options/water-sensitive-forest-management (accessed 1 May 2025).

Cocozza, C. and Penna, D. (2022) Towards a more active dialogue between hydrologists and ecophysiologists for interdisciplinary studies in forest ecosystems. *Science of the Total Environment* 807, 150877.

de Jong, C. (1992) A catastrophic flood/multiple debris flow in a confined mountain stream: An example from the Schmiedlaine, Southern Germany. In: Walling, D.E. (ed.) *Erosions, Debris Flows and Environment in Mountain Regions, Proceedings of the Chengdu Symposium*, IAHS Press, Chengdu, China, July 5–9, 1992, pp. 237–245.

Dietze, M., Bell, R., Ozturk, U., Cook, K.L., Andermann, C. *et al.* (2022) More than heavy rain turning into fast-flowing water – A landscape perspective on the 2021 Eifel floods. *Natural Hazards and Earth System Sciences* 22, 1845–1856.

Dittrich, I., Kubatzsch, A. and Münch, A. (2007) *Waldwirkung und Hochwasser. Ein Leitfaden für Landnutzer und Entscheidungsträger*. Initiative Weißeritz-Regio Arbeitsgruppe Wald und Hochwasser, Staatsbetrieb Sachsenforst, Forstbezirk Bärenfels, Altenberg, Germany.

d'Odorico, P., Schönbeck, L., Vitali, V., Meusburger, K., Schaub, M. *et al.* (2021) Drone-based physiological index reveals long-term acclimation and drought stress responses in trees. *Plant, Cell and Environment* 44, 3552–3570.

Dumas, D. (2011) The impact of forests on the evolution of water resources in the mid-altitude Alps from the middle of the 19th century (Chartreuse massif, France). *Journal of Alpine Research* 99, 1–14.

EEA (European Environmental Agency) (2019) *The European Environment – State and Outlook 2020. Knowledge for Transition to a Sustainable Europe*. European Environment Agency, Copenhagen.

Ergenzinger, P. (1994) The susceptibility of valley slopes and river beds to erosion and accretion under the impact of climatic change – Alpine examples. *IAHS Publications, Series of Proceedings and Reports, International Association of Hydrological Sciences* 224, 43–54.

Fidelus-Orzechowska, J., Wrońska-Wałach, D., Cebulski, J. and Żelazny, M. (2018) Effect of the construction of ski runs on changes in relief in a mountain catchment (Inner Carpathians, Southern Poland). *Science of the Total Environment* 630, 1298–1308.

Flörke, M., Wimmer, F., Laaser, C., Vidaurre, R., Tröltzsch, J. *et al.* (2011) *Assessment of Adaptation Measures: Factsheets*. Center for Environmental Systems Research, Kassel, Germany. Available at: www.ecologic.eu/sites/default/files/publication/2016/2331-annex11_climwatadapt_factsheets_0.pdf (accessed 1 May 2025).

Forest Europe (2020) *State of Europe's Forests 2020*. Forest Europe, Jönköping, Sweden.

Frank, D.C., Poulter, B., Saurer, M., Esper, J., Huntingford, C. *et al.* (2015) Water-use efficiency and transpiration across European forests during the anthropocene. *Nature Climate Change* 5, 579–583.

Furdada, G., Victoriano, A., Díez-Herrero, A., Génova, M., Guinau, M. *et al.* (2020) Flood consequences of land-use changes at a ski resort: Overcoming a geomorphological threshold (Portainé, Eastern Pyrenees Iberian Peninsula). *Water* 12, 368.

Garambois, P.A., Larnier, K., Roux, H., Labat, D. and Dartus, D. (2014) Analysis of flash flood-triggering rainfall for a process-oriented hydrological model. *Atmospheric Research* 137, 14–24.

Garcia, X., Estrada, L., Saló, J. and Acuña, V. (2024) Blueing green water from forests as strategy to cope with climate change in water scarce regions: The case of the Catalan River basin district. *Journal of Environmental Management* 353, 120249.

García-Ruiz, J.M. and Lana-Renault, N. (2011) Hydrological and erosive consequences of farmland abandonment in Europe, with special reference to the Mediterranean region – A review. *Agriculture, Ecosystems and Environment* 140, 317–338.

González-Sanchis, M., Campo, A.D., Molina, A.J. and Fernandes, T.J.G. (2015) Modeling adaptive forest management of a semi-arid Mediterranean Aleppo pine plantation. *Ecological Modelling* 308, 34–44.

Goral, F. and Müller, J. (2010) Auswirkungen des waldumbaus im waldgebiet der schorfheide auf die entwicklung der grundwasserhöhen und den zustand der waldmoore. *Naturschutz Und Landschaftspflege in Brandenburg* 19, 4.

Hagen, K. (2003) *Wildbacheinzugsgebiet Schmittenbach (Salzburg): Analyse des Niederschlags-und Abflussgeschehens 1977–1998*. Bundesamt und Forschungszentrum für Wald, Vienna, Austria.

Hagenlocher, M., Okamoto, S., Nagabhatla, N., Dietrich, S., Hassel, J. *et al.* (2023) *Building Climate Resilience: Lessons from the 2021 Floods in Western Europe*. United Nations University (UNU)-Institute for Environment and Human Security, UNU-Comparative Regional Integration Studies and UNU-Maastricht Economic and Social Research Institute on Innovation and Technology.

Hall, J., Arheimer, B., Borga, M., Brázdil, R., Claps, P. *et al.* (2013) Understanding flood regime changes in Europe: A state-of-the-art assessment. *Hydrology and Earth System Sciences Discussions* 10, 15525–15624.

Hamilton, F.C. (2008) Forests and water. In: *A Thematic Study Prepared in the Framework of the Global Resources Assessment 2005*. FAO Forestry Paper No.155, FAO, Rome.

Hegg, C. (2006) Waldwirkung auf Hochwasser. *LWF Wissen* 55, 29–33.

Hejduk, L., Kaznowska, E., Wasilewicz, M. and Hejduk, A. (2021) Hydrological droughts in the Białowieża Primeval Forest, Poland, in the years 1951–2020. *Forests* 12, 1744.

Herrmann, A. and Schumann, S. (2009) Investigations of the runoff formation process as a mechanism for monitoring the basin turnover in the Lange Bramke catchment, Upper Harz Mountains. *Hydrologie Und Wasserbewirtschaftung* 53, 64–79.

Hissler, C., Martínez-Carreras, N., Barnich, F., Gourdol, L., Iffly, J.F. *et al.* (2021) The weierbach experimental catchment in Luxembourg: A decade of critical zone monitoring in a temperate forest-from hydrological investigations to ecohydrological perspectives. *Hydrological Processes* 35, e14140.

Hlásny, T., Mátyás, C., Seidl, R., Kulla, L., Merganičová, K. *et al.* (2014) Climate change increases the drought risk in Central European forests: What are the options for adaptation? *Lesnicky Časopis Forestry Journal* 60, 5–18.

Humbert, J. and Najjar, G. (1992) *Influence de la forêt sur le cycle de l'eau en domaine tempéré: Une analyse de la littérature francophone*. CEREG, Université Louis Pasteur, Strasbourg, France.

Jones, J.A., Wei, X., Archer, E., Bishop, K., Blanco, J.A. *et al.* (2020) Forest–water interactions under global change. In: Levia, D.F., Carlyle-Moses, D.E., Iida, S., Michalzik, B., Nanko, K. *et al.* (eds) *Forest–Water Interactions. Ecological Studies*, Vol. 240. Springer, Cham, Switzerland, pp. 589–624.

Kahle, M., Kempf, M., Martin, B. and Glaser, R. (2022) Classifying the 2021 'Ahrtal' flood event using hermeneutic interpretation, natural language processing, and instrumental data analyses. *Environmental Research Communications* 4, 051002.

Kaule, R. and Gilfedder, B.S. (2021) Groundwater dominates water fluxes in a headwater catchment during drought. *Frontiers in Water* 3, 89.
Keller, H.M. (1988) European experiences in long-term forest hydrology research. In: Swank, W.T. and Crossley, D.A. (eds) *Forest Hydrology and Ecology at Coweeta*. Springer, New York, pp. 407–414.
Khorchani, M., Nadal-Romero, E., Tague, C., Lasanta, T., Zabalza, J. *et al.* (2020) Effects of active and passive land use management after cropland abandonment on water and vegetation dynamics in the Central Spanish Pyrenees. *Science of the Total Environment* 717, 137160.
Kopáček, J., Bače, R., Hejzlar, J., Kaňa, J., Kučera, T. *et al.* (2020) Changes in microclimate and hydrology in an unmanaged mountain forest catchment after insect-induced tree dieback. *Science of the Total Environment* 720, 137518.
Kupec, P., Deutscher, J. and Futter, M. (2021) Longer growing seasons cause hydrological regime shifts in Central European forests. *Forests* 12, 1656.
Lindroth, A. and Crill, P. (2011) Hydrology and biogeochemistry of boreal forests. In: Levia, D.F., Carlyle-Moses, D. and Tanaka, T. (eds) *Forest Hydrology and Biogeochemistry: Synthesis of Past Research and Future Directions*. Springer, Dordrecht, Netherlands, pp. 321–339.
Liu, X. and de Jong, C. (2024) Low-flow similarities between the transboundary Lauter river and Rhine river at Maxau from 1956 to 2022 (France/Germany). *Water* 16, 1584.
Manser, R. (2013) Forest policy in a changing climate – What are we expecting from scientists? In: Wohlgemuth, T. and Priewasser, K. (eds) *ClimTree 2013*. Swiss Federal Research Institute WSL, Birmensdorf, Switzerland, p. 18 (in abstract).
Markart, G., Sotier, B., Stepanek, L., Lechner, V. and Kohl, B. (2017) Waldwirkung auf die Abflussbildung bei Unterschiedlichen Betrachtungsmaßstäben. *Wildbach Lawinenverbau* 180, 100–115.
Mather, A.S., Fairbairn, J. and Needle, C.L. (1999) The course and drivers of the forest transition: The case of France. *Journal of Rural Studies* 15, 65–90.
McCulloch, J.S. and Robinson, M. (1993) History of forest hydrology. *Journal of Hydrology* 150, 189–216.
Meesenburg, H., Jansen, M., Döring, C., Beese, F., Rüping, U. *et al.* (2005) Konzept zur Beurteilung der Auswirkungen Forstlicher Maßnahmen auf den Gewässerzustand nach den Anforderungen der EG-wasserrahmenrichtlinie. In: *Asservorsorge in Bewaldeten Einzugsgebieten*. Eigenverlag der FVA, Freiburg, Germany, pp. 171–180.
Müller, E.V. (2023) *Analysis of Forest-specific Ecosystem Services with Regard to Water Balance Components: Runoff and Groundwater Recharge in the Forest* No. 90/23. Mitteilungen Aus Der Forschungsanstalt Für Waldökologie Und Forstwirtschaft, Trippstadt, Germany, p. 90.
Müller, J. and Bolte, A. (2009) The use of lysimeters in forest hydrology research in North-East Germany. *Landbauforschung Volkenrode* 59, 1–10.
Müller, J. (2012) Auswirkungen von waldstrukturellen veränderungen im zuge des wald-umbaus auf die hydroökologischen bedingungen in den beständen. In: Grünewald, U., Bens, O., Fischer, H., Hüttl, R.F., Kaiser, K. *et al.* (eds) *Wasserbezogene Anpassungsmaßnahmen an den Landschafts- und Klimawandel*. *Schweizerbart*. Stuttgart, Germany, pp. 280–291.
Patacca, M., Lindner, M., Lucas-Borja, M.E., Cordonnier, T., Fidej, G. *et al.* (2023) Significant increase in natural disturbance impacts on European forests since 1950. *Global Change Biology* 29, 1359–1376.
Pena-Angulo, D., Vicente-Serrano, S.M., Domínguez-Castro, F., Noguera, I., Tomas-Burguera, M. *et al.* (2021) Unravelling the role of vegetation on the different trends between climatic and hydrologic drought in headwater catchment of Spain. *Anthropocene* 36, 100309.
Peskett, L.M., Heal, K.V., MacDonald, A.M., Black, A.R. and McDonnell, J.J. (2023) Land cover influence on catchment scale subsurface water storage investigated by multiple methods: Implications for UK natural flood management. *Journal of Hydrology: Regional Studies* 47, 101398.
Peña-Arancibia, J.L., Bruijnzeel, L.A., Mulligan, M. and Dijk, A.I. (2019) Forests as 'sponges' and 'pumps': Assessing the impact of deforestation on dry-season flows across the tropics. *Journal of Hydrology* 574, 946–963.
Pluntke, T., Bernhofer, C., Grünwald, T., Renner, M. and Prasse, H. (2023) Long-term climatological and ecohydrological analysis of a paired catchment–flux tower observatory near dresden (Germany). Is there evidence of climate change in local evapotranspiration? *Journal of Hydrology* 617, 128873.
Pötzelsberger, E. and Hasenauer, H. (2015) Forest–water dynamics within a mountainous catchment in Austria. *Natural Hazards* 77, 625–644.
Prietzel, J., Weis, W., Bachmann, S. and Kern, M. (2023) Effects of douglas fir cultivation in german forests on soil seepage water quantity and quality. *Forest Ecology and Management* 546, 121379.

Pšidová, E., Ditmarová, L., Jamnická, G., Kurjak, D., Majerová, J. *et al.* (2015) Photosynthetic response of beech seedlings of different origin to water deficit. *Photosynthetica* 53, 187–194.

Pülzl, H. (2013) *European Forest Governance: Issues at Stake and the Way Forward*. European Forest Institute, Barcelona, Spain.

Pülzl, H., Hogl, K., Kleinschmit, D., Wydra, D., Arts, B. *et al.* (eds) (2014) *European Forest Governance: Issues at Stake and the Way Forward. What Science Can Tell Us*. European Forest Institute, Joensuu, Finland.

Rajwa-Kuligiewicz, A. and Bojarczuk, A. (2024) Streamflow response to catastrophic windthrow and forest recovery in subalpine spruce forest. *Journal of Hydrology* 634, 131078.

Rakovic, J., Futter, M.N., Kyllmar, K., Rankinen, K., Stutter, M.I. *et al.* (2020) Nordic bioeconomy pathways: Future narratives for assessment of water-related ecosystem services in agricultural and forest management. *Ambio* 49, 1710–1721.

Ristić, R., Radić, B., Vasiljević, N. and Nikić, Z. (2011) Land use change for flood protection: A prospective study for the restoration of the river Jelašnica watershed. *Glasnik Sumarskog Fakulteta* 103, 115–130.

Ristić, R., Kašanin-Grubin, M., Radić, B., Nikić, Z. and Vasiljević, N. (2012) Land degradation at the Stara Planina ski resort. *Environmental Management* 49, 580–592.

Rüegg, J., Moos, C., Gentile, A., Luisier, G., Elsig, A. *et al.* (2022) An approach to evaluate mountain forest protection and management as a means for flood mitigation. *Frontiers in Forests and Global Change* 5, 785740.

Samuelson, L., Bengtsson, K., Celander, T., Johansson, O., Jägrud, L. *et al.* (2015) *Water, Forests, People – Building Resilient Landscapes*. Stockholm International Water Institute, Stockholm.

Schleppi, P. (2011) Forested water catchments in a changing environment. In: Bredemeier, M., Cohen, S., Godbold, D.L., Lode, E., Pichler, V. *et al.* (eds) *Forest Management and the Water Cycle: An Ecosystem-Based Approach*. Springer, Dordrecht, Netherlands, pp. 89–110.

Schnabel, F., Purrucker, S., Schmitt, L., Engelmann, R.A., Kahl, A. *et al.* (2022) Cumulative growth and stress responses to the 2018–2019 drought in a European floodplain forest. *Global Change Biology* 28, 1870–1883.

Schüler, G. (2006) Identification of flood-generating forest areas and forestry measures for water retention. *Forest, Snow, and Landscape Research* 80, 99–114.

Schüler, G., Pfister, L., Vohland, M., Seeling, S. and Hill, J. (2011) Large scale approach to forest and water interactions. In: Bredemeier, M., Cohen, S., Godbold, D.L., Lode, E., Pichler, V. *et al.* (eds) *Forest Management and the Water Cycle: An Ecosystem-Based Approach*. Springer, Dordrecht, Netherlands, pp. 435–452.

Serra-Llobet, A., Jähnig, S.C., Geist, J., Kondolf, G.M., Damm, C. *et al.* (2022) Restoring rivers and floodplains for habitat and flood risk reduction: Experiences in multi-benefit floodplain management from California and Germany. *Frontiers in Environmental Science* 9, 778568.

SOER (2015) *The European Environment – State and Outlook 2015. A Comprehensive Assessment of the European's Environment's State, Trends and Prospects, in A Global Context*. European Environment Agency, Copenhagen. Available at: www.eea.europa.eu/soer/2015 (accessed 1 May 2025).

Stagge, J.H., Dias, S., Rego, F.C. and Tallaksen, L.M. (2015) Integration of climate time series and MODIS data as an analysis tool for forest drought detection. In: Andreu, J., Solera, A., Paredes-Arquiola, J., Haro-Monteagudo, D. and Lanen, H. (eds) *Drought: Research and Science–Policy Interfacing*. Taylor & Francis, London, pp. 51–53.

Strohmenger, L., Ackerer, P., Belfort, B. and Pierret, M.C. (2022) Local and seasonal climate change and its influence on the hydrological cycle in a mountainous forested catchment. *Journal of Hydrology* 610, 127914.

van den Bergh, T., Hiltbrunner, E. and Körner, C. (2014) Land abandonment increases evapotranspiration in mountain grassland. In: Amatya, D., Sun, G., Gowda, P., Jong, C., Nettles, J. *et al.* (eds) *Evapotranspiration: Challenges in Measurement and Modeling from Leaf to the Landscape Scale and Beyond*. American Society of Agricultural and Biological Engineers, St Joseph, Michigan.

Verkerk, H., Delacote, P., Hurmekoski, E., Kunttu, J., Matthews, R. *et al.* (2022) *Forest-based climate change mitigation and adaptation in Europe*. From Science to Policy 14. European Forest Institute, Barcelona, Spain.

Verkerk, P.J., Costanza, R., Hetemäki, L., Kubiszewski, I., Leskinen, P. *et al.* (2020) Climate-smart forestry: The missing link. *Forest Policy and Economics* 115, 102164.

Versuchsanstalt, N.F. (2012) *SILVAQUA-Auswirkungen forstlicher Bewirtschaftungsmaßnahmen auf den Zustand von Gewässern in bewaldeten Einzugsgebieten am Beispiel der Oker im Nordharz*, Vol. 9. University Press Göttingen, Göttingen, Germany.

Vlach, V., Matouskova, M. and Ledvinka, O. (2021) Impacts of regional climate change on hydrological drought characteristics in headwaters of the Ore Mountains. *River Research and Applications* 37, 919–930.

Wahren, A. and Feger, K.H. (2011) Model-based assessment of forest land management on water dynamics at various hydrological scales – A case study. In: Bredemeier, M., Cohen, S., Godbold, D.L., Lode, E., Pichler, V. *et al.* (eds) *Forest Management and the Water Cycle: An Ecosystem-Based Approach*. Springer, Dordrecht, Netherlands, pp. 453–469.

Wemple, B., Shanley, J., Denner, J., Ross, D. and Mills, K. (2007) Hydrology and water quality in two mountain basins of the northeastern US: Assessing baseline conditions and effects of ski area development. *Hydrological Processes* 21, 1639–1650.

Xiao, L., Robinson, M. and O'Connor, M. (2022) Woodland's role in natural flood management: Evidence from catchment studies in Britain and Ireland. *Science of the Total Environment* 813, 151877.

Yimer, E.A., Trift, L., Lobkowicz, I., Villani, L., Nossent, J. *et al.* (2024) The underexposed nature-based solutions: A critical state-of-art review on drought mitigation. *Journal of Environmental Management* 352, 119903.

Zal, N., Bastrup-Birk, A., Bariamis, G., Scholz, M., Tekidou, A. *et al.* (2015) *Water-Retention Potential of Europe's Forests: A European Overview to Support Natural Water-Retention Measures*. European Environment Agency Technical Report 13. European Environment Agency, Copenhagen.

Zhang, M., Liu, S., Jones, J., Sun, G., Wei, X. *et al.* (2022) Managing the forest-water nexus for climate change adaptation. *Forest Ecology and Management* 525, 120545.

Ziche, D., Riek, W., Russ, A., Hentschel, R. and Martin, J. (2021) Water budgets of managed forests in northeast germany under climate change – results from a model study on forest monitoring sites. *Applied Sciences* 11, 2403.

Zimmermann, N.E., Bolliger, J., Gehrig-Fasel, J., Guisan, A., Kienast, F. *et al.* (2006) Wo wachsen die Bäume in 100 Jahren. *Forum Für Wissen* 63–71.

13 Forest Hydrology in China

K. Duan[1]*, M. Liu[1], X. Liu[2], Z. Zhang[3], Y. Wang[4] and G. Sun[5]

[1]School of Civil Engineering, Sun Yat-Sen University, Guangzhou, China; [2]College of Forestry and Landscape Architecture, South China Agricultural University, Guangzhou, China; [3]School of Soil and Water Conservation, Beijing Forestry University, Beijing, China; [4]Ecology and Nature Conservation Institute, Chinese Academy of Forestry, Beijing, China; [5]Eastern Forest Environmental Threat Assessment Center, USDA Forest Service, Research Triangle Park, North Carolina, USA

Abstract

China is a vast nation that has relatively low forest coverage and suffers chronic soil erosion and water shortages due to intensive human activities and natural conditions. Modern forest hydrology research in this populous country started late but has made important advancement during the past 40 years. This chapter provides a summary of the historical development of forest hydrology research in China, highlighting the major research methods, field installation and achievements. We present key research progress in terms of forest canopy interception, evapotranspiration, and the relationships of forests with runoff and soil erosion under the broad background of large-scale ecological restoration campaigns. In addition, we present emerging technologies and techniques such as simulation models and remote sensing, and discussed future opportunities and challenges in forest hydrological research in China.

13.1 Introduction

Traditional Chinese wisdoms value the harmonious relationships between forests and water: known as *Shui Yuan Han Yang* – 'forests retain water' (Sun *et al.*, 2023). However, modern forest hydrology research in China started only around the 1980s in response to the severe lack of knowledge in watershed management for soil erosion control (Ma, 1993; Zhang *et al.*, 2004; Guo *et al.*, 2005). Before the 1980s, limited rigorous studies existed on the role of forests in regulating watershed hydrology in the upper Yellow River basin (Liu and Zhong, 1978) and the upper Yangtze River (Ma, 1993). The last two decades have seen a dramatic increase in Chinese forest hydrology literature, reflecting the tremendous advances in our understanding of the forest–water relationships that are critical to the success of unprecedented ecological restoration in China (Sun *et al.*, 2006, 2023; Wei *et al.*, 2008). Now China has the world's largest plantation forests with total forest coverage reaching more than 24% of the land mass.

This chapter provides a brief summary of the historical development of forest hydrology research in China over the past several decades. Efforts in long-term field study sites and key findings in terms of canopy interception, forest floor interception, forest evapotranspiration

*Corresponding author: duank6@mail.sysu.edu.cn

 Forest Hydrology: Processes, Management and Assessment (eds Leon J. Bren *et al.*)
DOI: 10.1079/9781800625310.0013

(ET), the forest relationship with runoff and soil erosion, emerging technologies and techniques, and future opportunities and challenges are discussed in this chapter.

13.2 Field Studies

13.2.1 National ecosystem research network

Forest ecosystem experimental stations established during the 1950s and 1960s represent much of the early pursuit of forest ecology and hydrology research in China (Zhang *et al.*, 2004), particularly on the nutrient and energy flows at ecosystem scales. For instance, the Changbaishan forest ecosystem research station in north-eastern China joined the international Man and the Biosphere (MAB) program in 1979. In 1982, the Chinese Ministry of Forestry started a forest ecosystem study project and provided continuous financial support to 11 long-term stations. In 1988, the Chinese Academy of Sciences established the Chinese Ecosystem Research Network (CERN) (Fu *et al.*, 2010; Xia *et al.*, 2018). It is a founding member of the International Long-term Ecosystem Research Network (ILTER) and the Global Terrestrial Observation System (GTOS). Since 1998, CERN has conducted long-term observations of typical ecosystems in China in accordance with unified monitoring indicators and technical specifications, obtaining systematic and continuous observation data (e.g. water, soil, atmosphere and biology) and providing shared services. The forestry department in China also gradually established the Chinese Forest Ecosystem Research Network (CFERN) in the 1950s (Niu *et al.*, 2013). Currently, CFERN has developed into a network consisting of 73 forest ecological stations spanning 30° in latitude and covering the typical ecological regions of China.

These long-term monitoring stations aim to monitor the changing hydrological patterns and quantify the forest ecosystem services in different regions of China. According to the national standard 'indicator system for long-term observation of forest ecosystem' issued in 2017 (Wang *et al.*, 2017), important indicators include hydrological processes (e.g. precipitation, throughfall, stemflow, soil moisture, surface runoff, groundwater) and hydrochemical characteristics (e.g. pH, dissolved organic carbon, total nitrogen, total phosphorus). In addition, there are a considerable number of temporary research sites across China for addressing national questions in connection with forestry programs such as the Natural Forests Protection Program, the key shelterbelt system construction projects of 'Three North' and the middle and lower reaches of the Yangtze River, and the Grain for Green Program (GFG) (Zhang *et al.*, 2004). For example, approximately 180 experimental sites dominated by different forest ecosystems have monitored canopy interception for at least 1 year to understand the hydrological functions of forest canopies in precipitation partitioning (Deng *et al.*, 2024).

13.2.2 Hydrometric monitoring at a plot scale

Hydrometric observations at plot scale have been widely used to study hydrological processes, such as canopy interception rate, litter moisture content, soil infiltration and storage capacity, ET and runoff generation. Rain gauges are typically situated at a distance of approximately one to two times the tree's height from the forest edge, while the open space within the forest should encompass a minimum area of 20 × 20 m. Alternatively, these instruments can be elevated above the observation tower for forest microclimate, maintaining a minimum height of 2 m above the forest canopy. The monitoring of throughfall is conducted by five to six cross-shaped or groove-shaped throughfall collectors in the plot (Fig. 13.1a,b). The throughfall collectors are typically positioned beneath the tree canopy and are situated 60 cm above the forest floor. A water outlet connecting to a data collector is incorporated at the base of the throughfall collector.

Stemflow measurements typically involve several standard steps. First, all trees within the observation area are categorized based on their diameter at breast height. Subsequently, two or three standard trees are chosen from each diameter class. In the context of studying forest stemflow, at least eight standard trees are

Fig. 13.1. Observation instruments in the Dinghushan Forest Ecosystem Research Station of China: (a) cross-shaped throughfall collector; (b) groove-shaped throughfall collector; (c) stemflow collector; (d) surface runoff plot. Photos courtesy of Xiaodong Liu (2023).

selected per stand. The methodology for observation entails the use of a polyethylene rubber tube with a diameter of 2–3 cm (Fig. 13.1c). The tube is wrapped spirally around the trunk of the sample tree to create an intercepting channel. After two to three revolutions of the trunk, the tube is secured with nails and the joint is sealed. The end of the polyethylene rubber pipe is connected to a collection device such as a water container or an automatic rain gauge that records stemflow continuously.

Forest litter may absorb significant precipitation and affect water available for infiltration into soils. A change in litter moisture and litter interception is an important process of the water cycle (Liu *et al.*, 2022). The assessment of litter interception is commonly conducted using a funnel-type water measuring device, which consists of stainless-steel funnels with a 50 cm diameter, tapering to a diameter of 3–4 cm at the outlet. They are securely affixed with an iron frame positioned 60 cm above the ground. The outlet of each funnel is directed into a water-collection device on the ground, allowing either manual observation or recording through an automatic rain gauge. Litter moisture content is determined by the oven drying method.

Soil moisture measurements are typically conducted next to a runoff plot (Fig. 13.1d). To ensure comprehensive coverage, a minimum of three observation points is established within each sample plot, evenly distributed along the plot's diagonal axis. The measurement depth is determined based on the maximum soil layer depth, facilitating accurate determination of volumetric soil moisture (%) at varying depths when neutron probes are used. Recent advancements in soil moisture monitoring technology and instrumentation have led to the widespread adoption of time-domain reflectometer methods across various forest ecosystem research stations in China. These methods enable real-time and continuous tracking of soil moisture dynamics.

Fig. 13.2. (a) Sap flow probes used to measure tree transpiration and (b) the eddy flux tower used to measure total water and carbon exchange between forests and the atmosphere in the Huanjiang Observation and Research Station for Karst Ecosystems, southern China. Photo courtesy of Meixian Liu (2023).

Forest ET represents the total flux of water vapor transported to the atmosphere by trees, understory vegetation and the soil. Sap-flow methods are commonly used to monitor transpiration through thermal dissipation probes (Fig. 13.2a) (Du *et al.*, 2021). These probes are inserted into the sapwood to measure temperature differentials caused by sap flow, and the rate of temperature change is used to estimate sap-flow rates. Sensors are typically installed at different depths within the sapwood to capture variations in sap-flow rates across the xylem. Typical depths for sensor placement are 1.5, 2.5 and 3.5 cm from the cambium. Observations are carried out on a variety of tree species, sizes and ages to study differences in sap-flow patterns. To measure the combined water loss from both evaporation and plant transpiration, eddy covariance towers are increasingly used to measure ET rates (Fig. 13.2b) (Liu *et al.*, 2013). Standard tower heights typically range from 1.5 to 30m above the canopy, depending on the ecosystem and the specific research objectives. Instrumentation on the eddy covariance tower includes fast-response sensors for measuring air temperature, humidity, wind speed and CO_2 concentration. Additionally, a three-dimensional sonic anemometer measures wind speed and direction at high frequencies. Eddy covariance systems continuously collect high-frequency data (e.g. 10–20 Hz) on vertical wind speed fluctuations and fluctuations in water vapor density. The covariance between vertical wind fluctuations and water vapor density fluctuations is used to calculate latent heat flux, which represents ET rates.

In lieu of gauged small watersheds, runoff plots are widely used to measure surface runoff and subsurface runoff in China (Fig. 13.1d). Surface runoff plots are usually 5m wide and aligned parallel to contour lines, with a horizontal projection length of 20m and an area of 100 m^2. Along the slope, upper water-retaining walls are erected in parallel from top to bottom, while left and right water-retaining walls are established to form a quadrilateral framework. Fences, 50 cm deep and 25 cm above ground level, are installed, and a water collection channel is positioned at the lower part of the runoff plot, connecting to a collection pool. This pool gathers all effluents from the plot, enabling sampling of runoff amount and water quality.

13.3 Key Findings of Forest Hydrology Research in China

13.3.1 Canopy interception

A large number of studies on canopy interception have been conducted in China (Q. Zhang *et al.*, 2022). Researchers have employed various data sets, including direct measurements, modeling and remote sensing, to quantify interception loss and understand its role in the overall water balance in forested ecosystems. Wen and Liu (1995) found that the canopy interception of different forest ecosystems in different geographical regions of China fell between 11.4% and 34.3%, with variation coefficients of 6.68–55.05%. Averaged annual interception amounts to 134.0–626.7 mm from north to south. Wei *et al.* (2005) reviewed a series of forest hydrological studies conducted from the mid-1980s to the early 2000s. These studies covered the major types of forest in China, including Korean pine (*Pinus koraiensis*), Chinese fir (*Cunninghamia lanceolata*), oak (*Quercus mongolica*), larch (*Larix gmelinii*), Faber's fir (*Abies fabri*), Chinese pine (*Pinus tabuliformis*), Armand pine (*Pinus armandi*), birch (*Betula platyphylla*) and some tropical forests. These studies revealed that forest canopy interception rates varied from 15% to 30% of total rainfall across the country, and stemflow generally accounted for less than 5%, except for the oak forests in north-east China (15%). Q. Zhang *et al.* (2022) compiled a total of 86 publications from 47 different research sites located in 21 provinces or municipalities in China and found that the dominant type of canopy interception in the southern regions was by trees, while in the north it was shrubs. The canopy interception rate varied significantly across regions (~13.6–34.6%) and species (~12.4–47.6%). The regions with the maximum canopy interception rate were Qinghai (34.6%) and Inner Mongolia (29.5%), while the minimum values were found in Shanxi (13.6%) and Ningxia (14.9%). Across different species, the highest and lowest canopy interception rates were recorded in *Myricaria squamosa* (47.6%) and *Cryptomeria fortunei* (42.7%), and *Populus davidiana* (12.4%) and *Robinia pseudoacacia* (13.6%), respectively.

13.3.2 Forest floor interception

The capacity and storage of water intercepted by litter and understory vegetation influences soil water replenishment and plant water availability. The forest floor is usually more porous than bare soil, and the intercepted water evaporates more easily. Forest floor interception capacity can be two- to four times its own weight (Liu, 1996).

Previous studies have examined various forest types across different climatic zones in China, revealing that forest floor interception can vary considerably depending on forest composition, litter characteristics, understory density and climatic conditions. Key findings from these studies highlighted the role of the forest floor in enhancing soil moisture retention, mitigating surface runoff and regulating water balance in forest ecosystems (Li *et al.*, 2013; Sun *et al.*, 2018; Xia *et al.*, 2019; X. Wu *et al.*, 2021). Moreover, researchers have explored the ecological functions of forest floor interception, such as its impact on nutrient cycling, soil erosion control and ecosystem productivity (Tang *et al.*, 2010; Wang *et al.*, 2021). For example, X. Wu *et al.* (2021) used 1045 observational sites of forest ecosystems across China and a random forest model to estimate the spatial distribution of forest water retention. They concluded that litter maximum water-holding capacity contributed 5.37% to the total forest water retention amount in China, while canopy interception and soil water storage accounted for 21.24% and 73.39%, respectively. Sun *et al.* (2018) found that litter interception varied from 0.9 ± 0.3 to 7.7 ± 0.8 mm between 2008 and 2012 across eight representative sites in China. Stand density, Shannon–Wiener index, litter mass and size ratio showed the largest impacts on litter interception, and litter interception was more correlated to undecomposed litter mass than to total litter mass.

13.3.3 Forest ET

Evapotranspiration is the key process in forest hydrology to understand forest influence on water yield and carbon cycles. Common methods adopted in China to measure forest ET include hydrological, micrometeorological and vegetation physiology methods (Tie *et al.*, 2018). Hydrological methods include the water balance method and the lysimeter method, with the former estimating ET by the difference between precipitation and water yield at catchment scales, and the latter usually directly measuring ET beneath forest canopies. Micrometeorological methods include the aerodynamic method, eddy covariance, Bowen's ratio energy balance method and large-aperture scintillometry (S.M. Liu *et al.*, 2011, 2013). These methods are based on the assumption that the coefficients for turbulent diffusion of momentum, energy and water vapor are equivalent, necessitating specific atmospheric and underlying surface conditions (Tang and Wang, 2022). The vegetation physiology methods are mainly the sap-flow method (Du *et al.*, 2021) and stoma meter method, which monitor the transpiration of individual plants or leaves (Wang and Ma, 2002).

The study of forest ET in China has progressed significantly in recent years along with the increasing applications of remote-sensing and eddy covariance flux data sets (Li *et al.*, 2014). Data collected from flux observations across China have shown that annual ET generally declines with increasing latitude. Such latitudinal dependency of ET is controlled by the latitudinal decline in annual temperature and precipitation and the resulting decline in growing season length (Xiao *et al.*, 2013). Recent research emphasizes the impact of vegetation characteristics and changes in climate and land use on ET rates. Synthesized data sets indicate that a multi-decadal and accelerating rise in global ET has been apparent since the 1980s. These changes are largely related to vegetation greening (revealed by increasing leaf area index), while increased precipitation and enhanced atmospheric evaporative demand have secondary roles (Yang *et al.*, 2023). China leads the world in greening through ambitious land-use management programs to conserve and expand forests, aiming to achieve the goals of mitigating land degradation, air pollution and climate change (Chen *et al.*, 2019). A recent assessment using products from the Global Land Evaporation Amsterdam Model (GLEAM), Global Land Data Assimilation System (GLDAS) and Noah Land Surface Model (NOAH) suggest that annual ET increased in 84.82% (52.75% with $P<0.01$) of China's land surface with a countrywide rate of 1.27 ± 0.12 mm/year ($P<0.01$) during 1982–2017 (Yao *et al.*, 2023). Although the greening trends generally had positive effects on ET at the large river basin and national scales, large diversity and fluctuation were also found across the heterogeneous climate and landscapes in China (Li *et al.*, 2017; Ma *et al.*, 2024).

13.3.4 Forest–water yield relationship

The relationship between forests and runoff/streamflow has long been an important research interest in China (Sun *et al.*, 2008; Wei *et al.*, 2008). Forest removal or deforestation generally increases surface runoff and total water yield due to decreased ET and soil degradation (Wang *et al.*, 2008, 2011). However, the effects of increasing vegetation greenness, particularly afforestation, are more complex, uncertain or even controversial than deforestation across basins and spatial scales due to the various climate and basin characteristics, such as topography, soil properties, pre-forestation land cover, tree species and stand structure (Zhou *et al.*, 2010; Zhang and Wei, 2021; B. Zhang *et al.*, 2022a). For example, Sun *et al.* (2006) conducted a national modeling experiment and concluded that average water yield reduction due to forestation may vary from about 50 mm/year (50%) in the semi-arid Loess Plateau region in northern China to about 300 mm/year (30%) in the tropical southern region. Although empirical data are rare and regular streamflow change is influenced by many human influences (e.g. hydraulic engineering, urbanization), Zhou *et al.* (2010) argued that forest recovery in the Guangdong Province of southern China may play a positive role in redistributing water from the wet season to the dry season and thus increase water yield in the dry season. Xue *et al.* (2022) contrasted two dry and two wet basins in China, and found that afforestation explained parts of runoff reduction for dry basins but did not result in significant changes in the total runoff for the two wet basins. Annual runoff is usually more sensitive to forest cover change in water-limited watersheds than in energy-limited watersheds. Given the water shortage in northern China and the negative effects of large-scale vegetation greening on water yield, there are increasing concerns that revegetation programs in semi-arid regions such as the Loess Plateau have caused substantial streamflow reduction that has consequently approached regional sustainable water resource limits (Feng *et al.*, 2016). These studies suggest that afforestation efforts should perhaps focus on southern China with a larger water supply and that a region-specific afforestation policy is crucial to deal with forestation–water quantity tradeoff (Y. Liu *et al.*, 2016; Xue *et al.*, 2022).

Using global or regional land–climate-coupled models at a relatively coarse spatial resolution, recent studies have also explored the potential feedback from vegetation changes on regional precipitation and consequent impacts on the hydrological cycle (Y. Li *et al.*,

2018; Teo *et al.*, 2022). Model simulations incorporating the precipitation response to vegetation increases suggest that increased precipitation could offset some of the negative vegetational impacts on soil moisture and water yield. For example, Y. Li *et al.* (2018) coupled the Laboratoire de Météorologie Dynamique atmospheric general circulation model with zoomed capability (LMDZ, version 5, v2076) and the ORganizing Carbon and Hydrology in Dynamic EcosystEms model (ORCHIDEE, v3035) to investigate vegetation biophysical feedback to regional hydrology in China, and found that the increased precipitation from vegetation greening and the increased forest area supplied enough water to cancel out enhanced ET in North and South-east China in the past three decades. Teo *et al.* (2022) coupled the International Centre for Theoretical Physics (ICTP) Regional Climate Model version 4.7 to the Common Land Model version 4.5 (RegCM4.7-CLM4.5) and concluded that reforestation would increase both ET and precipitation and resulted in a statistically significant increase in water yield for the Loess Plateau–North China Plain, Yangtze Plain and South-east China under future representative concentration pathway scenarios.

13.3.5 Forests and soil erosion

Forest plantations can mitigate soil erosion by reducing surface runoff and river peak flow (Gong *et al.*, 2022). Guo *et al.* (2015) analyzed soil loss rates of five water erosion regions in China based on 2823 plot-year data from 73 different sites, and found small regional differences in the soil loss rate of forest, shrub and grasslands, which were 0–1.89, 0.28–8.06 and 3.98–1.57 t/ha/year, respectively. Meanwhile, the soil loss rate was 30.87–107.44 t/ha/year on fallow land and 7.65–49.38 t/ha/year on farmland under conventional tillage, showing large regional differences across the country.

Dams and reservoirs have played a dominant role in reducing sediments in major Chinese rivers in the second half of the 20th century (Chu *et al.*, 2009; T. Li *et al.*, 2018). Since 1999, China's vegetation cover has increased and the water erosion area began to decrease significantly. The country's water erosion area decreased from $1.83 \times 10^6 km^2$ in 1996 to $1.29 \times 10^6 km^2$ in 2011 according to the first national water census (T. Li *et al.*, 2018). In the Loess Plateau in particular, the average vegetation coverage of the 12 main Yellow River subcatchments increased from 25% in 1978 to 29% in 1998, and then to 46% in 2010. Between 1980–1999 and 2000–2010, a decrease of 26% in water yield and 21% in sediment concentration could be attributed to vegetation restoration, and about 57% (0.23 Gt/year) of the reduction in average sediment transport between these two time periods resulted from the increased vegetation coverage (Wang *et al.*, 2015). As a result of the restoration of soil-stabilizing forests and grasslands on barren or farmed land, sediment discharge into the Yellow River declined to levels close to those estimated for the agriculture period of the Tang dynasty in 740 AD (Chen *et al.*, 2015).

13.4 Emerging Technologies and Techniques

13.4.1 Forest hydrology modelling tools

Various models have been used in China ranging from simple water balance equations (Wang *et al.*, 2011) to more complex landscape distributed models (Zhang *et al.*, 2008; Zhou *et al.*, 2011) with different assumptions, spatiotemporal scales, structural complexities, parameter transferability and data requirements. The Budyko framework is widely used to explain the large-scale, long-term hydrological response to climate and watershed characteristics across China (Zhou *et al.*, 2021). Models including SWAT (Zhou *et al.*, 2011), MIKE-SHE (Zhang *et al.*, 2008; Wang *et al.*, 2012), TOPOG (Yu *et al.*, 2010), GBHM (Gao *et al.*, 2015), RHESSys (Peng *et al.*, 2016), HSPF (Zhou *et al.*, 2017) and WaSSI (J. Zhang *et al.*, 2021; X. Wang *et al.*, 2022) have also been applied in different basins to study the forest water mechanism. As well as process-based ecohydrological models, watershed-scale hydrological models with improvements in vegetation and terrain representations are increasingly used to address hydrological responses to climate and land-cover changes in the hydrology communities (Gao *et al.*, 2015; Zhou *et al.*, 2017).

13.4.2 Remote sensing of precipitation

In contemporary times, the detection of regional precipitation relies heavily on weather radar and satellite-based remote sensing. Weather radar uses the principle of backscattering to detect and quantify precipitation in the atmosphere. When the emitted electromagnetic wave pulse from a weather radar encounters precipitation particles, a fraction of its energy is scattered in various directions. Through empirical and physical models correlating echo power, radar parameters and target characteristics, the radar system can be used to analyze the intensity, location and type of precipitation from the back-scattered signals. Besides conventional radar, Doppler weather radars not only gauge precipitation intensity but also track the radial velocity of target objects along its path. Compared with traditional rain gauges, radar-based precipitation measurement offers substantial advantages in terms of scale (up to kilometer-level coverage), high resolution and real-time performance. This translates into the provision of more comprehensive, precise and timely precipitation data. China has taken significant strides in this realm, establishing a national Doppler weather radar network comprising 236 radar stations as of 2022 (http://en.weather.com.cn/radar/; accessed 9 May 2025). These radar stations, operating on S-band, C-band and X-band frequencies, are strategically spaced approximately 150–300 km apart based on terrain, precipitation characteristics and radar capabilities. S-band radars are predominantly deployed in areas susceptible to intense convective weather, while C-band radars cover arid and semi-arid regions. X-band radars fill the coverage gaps left by the aforementioned radar types.

Satellite-based remote sensing of precipitation involves using satellites equipped with sensors (e.g. visible light or reflected infrared sensors, thermal infrared sensors and microwave sensors) to detect and measure various parameters related to precipitation (Y.B. Liu *et al.*, 2011). China's achievements in this domain include the launch of 20 Fengyun (FY) meteorological satellites since the 1970s, such as geostationary satellite FY-4B and polar orbit satellite FY-3E. Notably, the successful launch of FY-3G satellite in April 2023 marked a significant milestone. Equipped with a dual-frequency active precipitation-measurement radar, combined with passive microwave and optical remote-sensing capabilities, this satellite offers enhanced monitoring of precipitation-formation processes. It can detect relevant atmospheric factors such as clouds, water vapor, aerosols and wind fields, accurately sensing precipitation-intensity changes as subtle as 0.2 mm/h (N. Li *et al.*, 2018; Xue *et al.*, 2021; H. Wang *et al.*, 2022).

13.4.3 The stable isotope method in ET partitioning

Stable isotope techniques have been increasingly applied to study ET partitioning in forest ecosystems in China (Liu and Li, 2008; Wu *et al.*, 2014). This involves using isotopic signatures to differentiate between soil/canopy evaporation and plant transpiration.

The isotopic compositions of water are typically characterized by the relative deviation of their isotope ratio (*R*, the ratio of deuterium:hydrogen (D:H) or ^{18}O:^{16}O) from that of the Vienna Standard Mean Ocean Water (V-SMOW) reference standard. These deviations are commonly expressed as:

$$\delta = \left[R_{\text{Sample}}/R_{\text{V-SMOW}} - 1\right] \times 1000\% \tag{13.1}$$

where $R_{\text{V-SMOW}} = 0.0020052$ for ^{18}O:^{16}O and 0.00015576 for D:H (Zhang *et al.*, 2010). Differences in isotopic compositions ($\delta^{18}O$ and δD) of water fluxes from foliage and soil water are used to partition the ecosystem ET (Liu *et al.*, 2017; Han *et al.*, 2022; Ren *et al.*, 2022). For example, Peng *et al.* (2020) used the Craig–Gordon model, isotope steady-state assumption and the Keeling plot model to partition ET within the Qinghai spruce forest ecosystem; Y.X. Wu *et al.* (2021) employed the Keeling curves alongside the Craig–Gordon model, incorporating both isotopic steady-state and non-steady-state models to simulate ET, evaporation and transpiration water vapor $\delta^{18}O$ in the *Platycladus orientalis* ecosystem, thereby comparing the ET classification outcomes derived from these two models. The stable isotope method provides a promising alternative for accurate partitioning of ET components, requiring only small water samples. However, isotope analysis equipment and data-processing tools can be costly, and field campaigns involving isotopic sampling and modeling require significant labor and resources.

13.4.4 Contactless technology in discharge monitoring

Traditional discharge monitoring relies on stream gauges that convert water level to discharge with an established stage-discharge curve, or information on stable cross-sections and flow velocity obtained from various flow velocimeters, such as a rotor current meter, acoustic Doppler current profiler, radar velocimeter and other ultrasonic devices (Kasuga *et al.*, 2003). However, these approaches require substantial investment in equipment implementation, personnel training with specialized expertise and ongoing maintenance (Yorke and Oberg, 2002; Fujita *et al.*, 2007; Czuba *et al.*, 2017). To overcome the limitations of traditional methods, a few image-based approaches have been introduced for water stage, flow velocity and discharge measurement in rivers (Leduc *et al.*, 2018; Noto *et al.*, 2022). Image-based approaches rely solely on capturing digital images of streams using inexpensive commercial cameras, presenting a promising alternative for continuous, non-invasive and low-cost streamflow monitoring. These methods usually assume that the surface characteristics of a river mirror the motion pattern of its water body and utilize image-processing techniques to analyze river-surface images for flow measurement. The velocity–area method is then used to indirectly deduce discharge after identifying the stage and average velocity. Image-based approaches or algorithms such as large-scale particle image velocimetry, particle tracking velocimetry, space–time image velocimetry and optical tracking velocimetry have increasingly been implemented (Zhang *et al.*, 2024). For example, the AiFlow system developed by the Intelligent Water Research Institute of Wuhan University and Wuhan Dashuiyun Technology Co. have been applied for stage and streamflow monitoring in over ten hydrological stations across China, while Fang *et al.* (2024) developed a procedure to extract information on flow patterns and volumes from high-frequency RGB (red, green, blue) images with deep convolutional neural networks and achieved satisfactory accuracy in discharge measurement at a steep mountain stream in southern China.

13.4.5 Unmanned aerial vehicle technology

Recent developments in unmanned aerial vehicle (UAV) technologies have drawn the attention of many researchers in the ecohydrology community. UAV-based remote sensing stands out due to its agility, operational flexibility, ability to carry multiple payloads and high-resolution surveying accuracy (J.X. Zhang *et al.*, 2021). Typically, UAVs are outfitted with visible light sensors, LiDAR (light detection and ranging) and hyperspectral sensors, each serving distinct purposes. UAVs equipped with visible light sensors are prevalently employed in terrain analysis and related domains such as soil erosion, surface rock outcrop examination, and land-use and vegetation categorization owing to their prowess in capturing high-resolution topographical data. Meanwhile, UAVs integrated with LiDAR are instrumental in geological and forestry surveys, canopy structure analysis, understory vegetation monitoring and vegetation biomass quantification due to their high structural measurement accuracy and the penetrability of laser scanning beams. Furthermore, UAVs with hyperspectral sensors are extensively utilized in vegetation growth monitoring and crop pest detection. In forest hydrology studies, UAV applications encompass the monitoring of ET and soil moisture (Luo *et al.*, 2019), soil erosion (Li *et al.*, 2024), forest canopy structure (e.g. leaf area index) (Lu *et al.*, 2020), topography (Du *et al.*, 2023), biomass (Mao *et al.*, 2023) and runoff (Yang *et al.*, 2021). For example, Q. Liu *et al.* (2016) used UAV-borne LiDAR and hyperspectral sensors to delineate the types and height profiles of dominant tree species in subtropical forests of Hubei Province, while Yang *et al.* (2021) developed a new UAV-based method for river flow estimation employing the high-precision Digital Orthophoto Map, the Digital Surface Model and the classic Manning formula river flow algorithm.

13.5 Unique Opportunities and Challenges Brought by Massive Reforestation

Zhang and Song (2006) characterized the change of forest cover in China into three stages: (i) the transition stage in 1949–1981; (ii) the slow-increase stage in 1982–1993; and (iii) the rapid-increase stage after 1994. In the 20th century, intensive deforestation and land reclamation on steep slopes and sandy lands caused severe soil erosion and environmental damage in China. The government has since heavily invested in various ecological restoration projects, such as the Three North Shelterbelt Program, the GFG, and rocky desertification control projects across the country (Shao *et al.*, 2022). In 1999, the Returning Farmland to Forest Program was launched in the provinces of Sichuan, Shaanxi and Gansu in north-eastern and south-western China. This initiative marked the beginning of the world-renowned GFG program. From 1999 to 2008, the GFG program covered a total of 26.9 million ha of land, including 9.3 million ha of reforestation on farmland and 17.6 million ha of afforestation on bare lands. Alongside the GFG program, the Ministry of Water Resources launched the Soil and Water Conservation Ecological Restoration Program. The first demonstration plots, initiated in 2002, covered a total area of 58.9×10^3 km^2 across 27 provinces. The second demonstration plots, initiated in 2006, covered 95 counties in 28 provinces. The main objective of this program is to increase vegetation cover to a rate higher than 75%, and to convert all sloping farmland with gradients higher than 25°, as well as regions prohibited from cultivation, to forest or grassland.

Over the past 20 years, China has continued to introduce policies and regulations to further promote ecological restoration and ecological civilization construction. Since 2016, the government has invested approximately 10 billion Chinese Yuan (approximately US$1.4 billion) annually in the implementation of 51 'Integrated Protection and Restoration Projects for Mountains, Rivers, Forests, Fields, Lakes, Grasslands, and Sand' across the country. It is projected that from 2021 to 2030, the total investment will exceed 250 billion Chinese Yuan and revitalize 10 million ha of land. China's planted forests in 2020 had a total area of 0.77 million km^2, accounting for 31.30% of the world's total planted forests (Cheng *et al.*, 2023). Over three-quarters (77.45 %) of China's planted forests were located in the eastern, center-south and south-western regions (Tong *et al.*, 2023).

China's massive forestation effort serves as a giant 'land manipulation experiment,' providing a rare opportunity to rigorously study how forest cover change can affect hydrological processes and water balance at a much bigger spatial level than traditional paired-watershed experiments (Sun *et al.*, 2006; Wei *et al.*, 2008). In the past four decades, water and carbon fluxes in China have exhibited an overall increasing trend, with significant growth in ET and gross primary productivity, and divergent changes in runoff (Fig. 13.3). Besides the generally warming and wetting climate, researchers have shown that vegetation greening played a dominating role in enhancing ET (Yang *et al.*, 2023; Yao *et al.*, 2023), altering the magnitudes and seasonality of water yield (Zhou *et al.*, 2010; Teo *et al.*, 2022; Xue *et al.*, 2022), and reducing soil erosion and sediment discharges (Chu *et al.*, 2009; T. Li *et al.*, 2018).

Recent vegetation expansion in China has also raised concerns on the inherent tradeoffs between water yield, ecosystem carbon sequestration, soil erosion controls and sediment delivery (Chen *et al.*, 2015; Feng *et al.*, 2016; Yu *et al.*, 2024). Although these tradeoffs are well recognized, our understanding of their extent and mechanisms is still limited. It is particularly challenging for China to balance the conflicting demands for water between the ecosystem and humans in fragile ecosystems such as the Loess Plateau. The changing climate and extensive afforestation have reduced water availability for humans due to enhanced vegetation ET. Meanwhile, the changing climate and

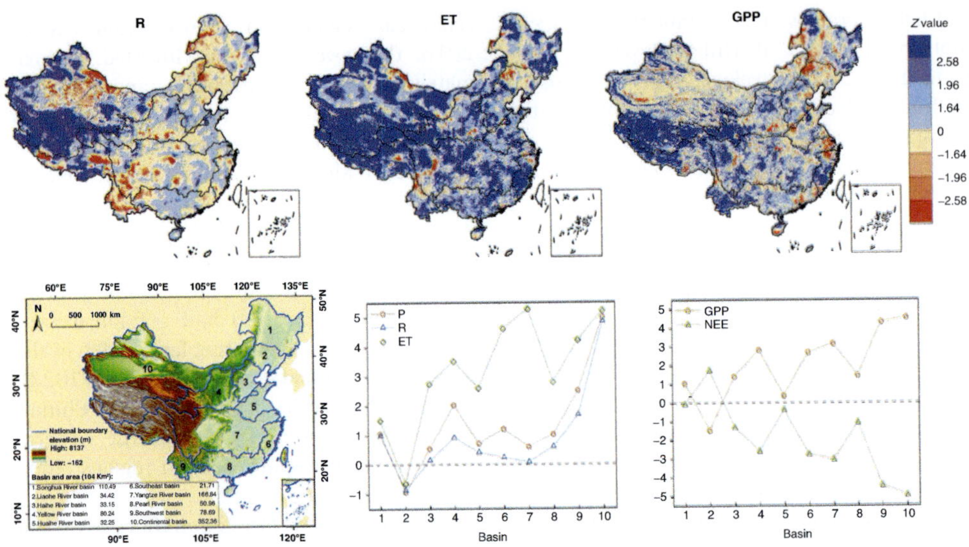

Fig. 13.3. Trends in water and carbon fluxes in the ten basins of China in 1982–2018 derived from the WaSSI model. Trends in precipitation (P), evapotranspiration (ET), runoff (R), gross primary productivity (GPP) and net ecosystem exchange (NEE) were evaluated by the *Z* value of the Mann–Kendall method. Positive and negative *Z* values indicate increasing and decreasing trends, respectively.

afforestation have used up much of the Loess Plateau's limited water and reduced soil moisture, which has made the fragile loess soil more vulnerable to wind and water erosion and has caused vegetation degradation in some areas (Fu *et al.*, 2017). The reduction in total sediment flux delivered to the sea, driven by soil erosion control and reservoir operation, has also resulted in challenges for the sustainable management of river deltas (Chu *et al.*, 2009).

13.6 Conclusions

Although forest hydrology research in China had a late start, great progress has been made in understanding forest–water relationships in the past decades thanks to the large investments in ecological restoration and relevant studies. The 21st century has seen a dramatic rise of Chinese forest hydrology literature, reflecting our advances in understanding the role of forests in water yield regulation, flood/drought mitigation, soil erosion control and ecosystem conservation.

The developments in forest hydrology research have largely relied on the advancements in data acquisition and processing technology. Continuous satellite monitoring of underlying characteristics and the water cycle since the early 1980s has facilitated the development of long-term high-resolution data sets of forest resources and key ecohydrological processes such as ET. The integration of multiple data sources based on on-site observations, physical algorithms, machine learning and remote sensing has provided new opportunities to enhance our understanding of forest hydrology. Nevertheless, evidence derived from multi-source data highlights the uncertainty in the effects of increasing vegetation greenness across basin characteristics, spatial scales and plant species.

The lessons learned from China can provide much-needed guidance to other developing countries or regions with similar challenges in the management of water and land resources. However, challenges remain for scientists and decision makers to fully understand the complexities of forest–water relationships in the face of climatic changes and intensive anthropogenic disturbances. The long-lasting debates over the role of forests or forestation in reducing floods and water yield across the country's different hydro-climatic and geophysical regions remain due to the lack of long-term experimental research and the ambiguity in scale effects. Stable support for basic forest ecosystem research with long-term commitment is crucial to fill the gaps between China's forest hydrological research and that of other developed countries. Rigorous data quality control and further interdisciplinary research are needed to better understand the individual and combined effects of forest management, climate change and other anthropogenic factors (e.g. reservoir regulation, water withdrawal, irrigation) on water resources and ecosystem functions.

References

Cheng, K., Su, Y., Guan, H., Tao, S., Ren, Y. *et al.* (2023) Mapping China's planted forests using high resolution imagery and massive amounts of crowdsourced samples. *ISPRS Journal of Photogrammetry and Remote Sensing* 196, 356–371.

Chen, Y., Wang, K., Lin, Y., Shi, W., Song, Y. *et al.* (2015) Balancing green and grain trade. *Nature Geoscience* 8, 739–741.

Chen, C., Park, T., Wang, X., Piao, S., Xu, B. *et al.* (2019) China and India lead in greening of the world through land-use management. *Nature Sustainability* 2, 122–129.

Chu, Z.X., Zhai, S.K., Lu, X.X., Liu, J.P., Xu, J.X. *et al.* (2009) A quantitative assessment of human impacts on decrease in sediment flux from major Chinese rivers entering the Western Pacific Ocean. *Geophysical Research Letters* 36, L19603.

Czuba, J.A., Foufoula-Georgiou, E., Gran, K.B., Belmont, P. and Wilcock, P.R. (2017) Interplay between spatially explicit sediment sourcing, hierarchical river-network structure, and in-channel bed material sediment transport and storage dynamics. *Journal of Geophysical Research: Earth Surface* 122, 1090–1120.

Deng, Y., Zhao, X., Cui, Z., Feng, Y., Zhang, W. *et al.* (2024) Canopy rainfall interception characteristics of forest ecosystems in China. *Acta Ecologica Sinica* 44, 2981–2992.

Du, H., Zeng, F., Song, T., Liu, K., Wang, K. *et al.* (2021) Water depletion of climax forests over humid karst terrain: Patterns, controlling factors and implications. *Agricultural Water Management* 244, 106541.

Du, P., Xu, Y., Guo, Y. and Li, H. (2023) Assessing Loess landslide volume using high-precision UAV-derived DEM: A case study of the 15 March 2019 landslide in zaoling township, Xiangning County In North China. *Natural Hazards Research* 3, 640–645.

Fang, C., Yuan, G., Zheng, Z., Zhong, Q. and Duan, K. (2024) Monitoring discharge of mountain streams by retrieving image features with deep learning. *Hydrology and Earth System Sciences* 28, 4085–4098.

Feng, X., Fu, B., Piao, S., Wang, S., Ciais, P. *et al.* (2016) Revegetation in China's Loess Plateau is approaching sustainable water resource limits. *Nature Climate Change* 6, 1019–1022.

Fu, B., Li, S., Yu, X., Yang, P., Yu, G. *et al.* (2010) Chinese ecosystem research network: Progress and perspectives. *Ecological Complexity* 7, 225–233.

Fu, B., Wang, S., Liu, Y., Liu, J., Liang, W.C. *et al.* (2017) Hydrogeomorphic ecosystem responses to natural and anthropogenic changes in the Loess Plateau of China. *Annual Review of Earth and Planetary Sciences* 45, 223–243.

Fujita, I., Watanabe, H. and Tsubaki, R. (2007) Development of a non-intrusive and efficient flow monitoring technique: The space-time image velocimetry (STIV). *International Journal of River Basin Management* 5, 105–114.

Gao, B., Qin, Y., Wang, Y., Yang, D. and Zheng, Y. (2015) Modeling ecohydrological processes and spatial patterns in the upper Heihe Basin in China. *Forests* 7, 10.

Gong, C., Tan, Q., Liu, G. and Xu, M. (2022) Impacts of mixed forests on controlling soil erosion in China. *Catena* 213, 106147.

Guo, M., Wang, Y. and Yu, P. (2005) A review of forest hydrology studies. *World Forestry Research* 18, 6–11.

Guo, Q., Hao, Y. and Liu, B. (2015) Rates of soil erosion in China: A study based on runoff plot data. *Catena* 124, 68–76.

Han, J., Tian, L., Cai, Z., Ren, W., Liu, W. *et al.* (2022) Season-specific evapotranspiration partitioning using dual water isotopes in a *Pinus yunnanensis* ecosystem, Southwest China. *Journal of Hydrology* 608, 127672.

Kasuga, K., Hachiya, H. and Kinosita, T. (2003) Quantitative estimation of the ultrasound transmission characteristics for river flow measurement during a flood. *Japanese Journal of Applied Physics* 42, 3212–3215.

Leduc, P., Ashmore, P. and Sjogren, D. (2018) Technical note: Stage and water width measurement of a mountain stream using a simple time-lapse camera. *Hydrology and Earth System Sciences* 22, 1–11.

Li, G., Zhang, F., Jing, Y., Liu, Y. and Sun, G. (2017) Response of evapotranspiration to changes in land use and land cover and climate in China during 2001–2013. *Science of the Total Environment* 596–597, 256–265.

Li, N., Zhang, S. and He, J. (2018) Simulation and precipitation detection in the tropical cyclones based on the microwave humidity and temperature sounder onboard the Fengyun-3C Satellite. *Progress In Electromagnetics Research C* 88, 269–283.

Li, P., Li, D., Hu, J., Fassnacht, F.E., Latifi, H. *et al.* (2024) Improving the application of UAV-lidar for erosion monitoring through accounting for uncertainty in DEM of difference. *Catena* 234, 107534.

Li, X., Niu, J. and Xie, B.J.P. (2013) Study on hydrological functions of litter layers in North China. *PLOS ONE* 8, e70328.

Li, X., Liang, S., Yuan, W., Yu, G., Cheng, X. *et al.* (2014) Estimation of evapotranspiration over the terrestrial ecosystems in China. *Ecohydrology* 7, 139–149.

Li, Y., Piao, S., Li, L.Z., Chen, A., Wang, X. *et al.* (2018) Divergent hydrological response to large-scale afforestation and vegetation greening in China. *Science Advances* 4, eaar4182.

Li, T., Wang, S., Liu, Y., Fu, B. and Zhao, W. (2018) Driving forces and their contribution to the recent decrease in sediment flux to ocean of major rivers in China. *Science of the Total Environment* 634, 534–541.

Liu, C. and Zhong, J. (1978) The influence of forest cover upon annual runoff in the Loess Plateau of China. *Acta Geographica Sinica* 33, 112–126.

Liu, H. and Li, J. (2008) Application of stable isotopic carbon in studies of plant water use efficiency. *Journal of Northwest Forestry University* 23, 54.

Liu, L., Jia, G., Yu, X. and Zhang, Y. (2017) Daily variations of δ^{18}o and its quantitative distinction in evapotranspiration components of *Platycladus orientalis* plantation during the rapid growth season in the mountainous area of Beijing. *Journal of Beijing Forestry University* 39, 61–70.

Liu, Q., Tan, B. and Hu, K. (2016) The remote sensing experiment on airborne lidar and hyperspectral integrated system for subtropical forest estimation. *Chinese High Technology Letters* 26, 264–274.

Liu, S. (1996) *Study on Hydrology of Forest Ecosystems in China*. Chinese Forestry Publishing House, Beijing.

Liu, S.M., Xu, Z.W., Wang, W.Z., Jia, Z.Z., Zhu, M.J. *et al.* (2011) A comparison of eddy-covariance and large aperture scintillometer measurements with respect to the energy balance closure problem. *Hydrology and Earth System Science* 15, 1291–1306.

Liu, S.M., Xu, Z.W., Zhu, Z.L., Jia, Z.Z. and Zhu, M.J. (2013) Measurements of evapotranspiration from eddy-covariance systems and large aperture scintillometers in the Hai River Basin, China. *Journal of Hydrology* 487, 24–38.

Liu, X., Feng, Y., Liu, P., Zhang, Q., Njoroge, B. *et al.* (2022) Soil moisture dominated the temporal dynamics of litter moisture content in subtropical forests: A 7-year observation in South China. *Journal of Hydrology: Regional Studies* 41, 101102.

Liu, Y., Xiao, J., Ju, W., Xu, K., Zhou, Y. *et al.* (2016) Recent trends in vegetation greenness in China significantly altered annual evapotranspiration and water yield. *Environmental Research Letters* 11, 094010.

Liu, Y.B., Fu, Q.N., Song, P., Zhao, X.S. and Dou, C.C. (2011) Satellite retrieval of precipitation: An overview. *Advances in Earth Sciences* 26, 1162–1172.

Lu, J., Wang, H., Qin, S., Cao, L., Pu, R. *et al.* (2020) Estimation of aboveground biomass of *Robinia pseudoacacia* forest in the Yellow River delta based on UAV and backpack LiDAR point clouds. *International Journal of Applied Earth Observation and Geoinformation* 86, 102014.

Luo, W., Xu, X., Liu, W., Liu, M., Li, Z. *et al.* (2019) UAV based soil moisture remote sensing in a karst mountainous catchment. *Catena* 174, 478–489.

Ma, X. (1993) *Forest Hydrology*. China Forestry Publishing House, Beijing.

Ma, L., Yu, G., Chen, Z., Yang, M., Hao, T. *et al.* (2024) Cascade effects of climate and vegetation influencing the spatial variation of evapotranspiration in China. *Agricultural and Forest Meteorology* 344, 109826.

Mao, P., Jiang, B., Shi, Z., He, Y., Shen, T. *et al.* (2023) Effects of UAV flight height on biomass estimation of desert shrub communities. *Ecological Indicators* 154, 110698.

Niu, X., Wang, B. and Wei, W.J. (2013) Chinese forest ecosystem research network: A platform for observing and studying sustainable forestry. *Journal of Food, Agriculture and Environment* 11, 1008–1016.

Noto, S., Tauro, F., Petroselli, A., Apollonio, C., Botter, G. *et al.* (2022) Low-cost stage-camera system for continuous water-level monitoring in ephemeral streams. *Hydrological Sciences Journal* 67, 1439–1448.

Peng, H., Tague, C. and Jia, Y. (2016) Evaluating the eco-hydrologic impacts of reforestation in the Loess Plateau, China, using an eco-hydrologic model. *Ecohydrology* 9, 498–513.

Peng, W.L., Zhao, L.J., Xie, C., Dong, X.Y., Liu, Q.Y. *et al.* (2020) Evapotranspiration partitioning of the *Picea crassifolia* forest ecosystem in the upper reaches of the Heihe river. *Journal of Glaciology and Geocryology* 42, 629–640. (in Chinese with English abstract).

Ren, X., Jia, J., Chen, Y., Hu, Y., Wang, Y. *et al.* (2022) Comparison of two isotope-based methods used in determining forest evapotranspiration partitioning. *Ecological Indicators* 139, 108937.

Shao, Q., Liu, S., Ning, J., Liu, G., Yang, F. *et al.* (2022) Assessment of ecological benefits of key national ecological projects in China in 2000–2019 using remote sensing. *Acta Geographica Sinica* 77, 2133–2153.

Sun, G., Zhou, G., Zhang, Z., Wei, X., McNulty, S.G. *et al.* (2006) Potential water yield reduction due to forestation across China. *Journal of Hydrology* 328, 548–558.

Sun, G., Liu, S., Zhang, Z. and Wei, X. (2008) Forest hydrology in China: Introduction to the featured collection. *Journal of the American Water Resources Association* 44, 1073–1075.

Sun, G., Zhang, L. and Wang, Y. (2023) On accurately defining and quantifying the water retention services of forests. *Acta Ecologica Sinica* 43, 9–25.

Sun, J., Yu, X., Wang, H., Jia, G., Zhao, Y. *et al.* (2018) Effects of forest structure on hydrological processes in China. *Journal of Hydrology* 561, 187–199.

Tang, J.Q. and Wang, C.J. (2022) Research progress of surface water and heat flux observation with large aperture scintillometer. *Advances in Meteorological Science and Technology* 12, 37–42.

Tang, J.-W., Cao, M., Zhang, J.-H. and Li, M.-H. (2010) Litterfall production, decomposition and nutrient use efficiency varies with tropical forest types in Xishuangbanna, SW China: A 10-year study. *Plant and Soil* 335, 271–288.

Teo, H.C., Raghavan, S.V., He, X., Zeng, Z. and Cheng, Y. (2022) Large-scale reforestation can increase water yield and reduce drought risk for water-insecure regions in the Asia-Pacific. *Global Change Biology* 28, 6385–6403.

Tie, Q., Hu, H., Tian, F. and Holbrook, N.M. (2018) Comparing different methods for determining forest evapotranspiration and its components at multiple temporal scales. *Science of the Total Environment* 633, 12–29.

Tong, X., Brandt, M., Yue, Y., Zhang, X., Fensholt, R. *et al.* (2023) Reforestation policies around 2000 in Southern China led to forest densification and expansion in the 2010s. *Communications Earth and Environment* 4, 260.

Wang, B., Niu, X., Jiang, Y., Guo, Q., Liu, S. *et al.* (2017) *Indicators System for Long-term Observation of Forest Ecosystem (GB/T35377–2017)*. Standard Press, Beijing.

Wang, B., Liu, D., Yang, J., Zhu, Z., Darboux, F. *et al.* (2021) Effects of forest floor characteristics on soil labile carbon as varied by topography and vegetation type in the Chinese Loess Plateau. *Catena* 196, 104825.

Wang, H. and Ma, L. (2002) Measurement of whole tree's water consumption with thermal dissipation sap flow probe (TDP). *Acta Phytoecologica Sinica* 26, 661–667.

Wang, H., Zhang, P., Yin, D., Li, Z., Shang, H. *et al.* (2022) Shortwave infrared multi-angle polarization imager (MAPI) onboard Fengyun-3 precipitation satellite for enhanced cloud characterization. *Remote Sensing* 14, 4855.

Wang, S., Zhang, Z., Sun, G., Strauss, P., Guo, J. *et al.* (2012) Multi-site calibration, validation, and sensitivity analysis of the MIKE SHE model for a large watershed in Northern China. *Hydrology Earth System Sciences* 16, 4621–4632.

Wang, S., Fu, B., Piao, S., Lü, Y., Ciais, P. *et al.* (2015) Reduced sediment transport in the Yellow River due to anthropogenic changes. *Nature Geoscience* 9, 38–41.

Wang, X., Duan, K. and Wei, L. (2022) Simulation of water and carbon coupling of the Pearl River basin based on the WaSSI model. *Journal of Applied Ecology* 33, 1377–1386.

Wang, Y., Yu, P., Xiong, W., Shen, Z., Guo, M. *et al.* (2008) Water-yield reduction after afforestation and related processes in the semiarid Liupan Mountains, Northwest China. *Journal of the American Water Resources Association* 44, 1086–1097.

Wang, Y., Yu, P., Feger, K.H., Wei, X., Sun, G. *et al.* (2011) Annual runoff and evapotranspiration of forestlands and non-forestlands in selected basins of the Loess Plateau of China. *Ecohydrology* 4, 277–287.

Wei, X., Liu, S., Zhou, G. and Wang, C. (2005) Hydrological processes in major types of Chinese forest. *Hydrological Processes* 19, 63–75.

Wei, X., Sun, G., Liu, S., Jiang, H., Zhou, G. *et al.* (2008) The forest-streamflow relationship in China: A 40-year retrospect. *Journal of the American Water Resources Association* 44, 1076–1085.

Wen, Y. and Liu, S. (1995) Quantifying forest canopy interception of major chinese forest ecosystems. *Scientia Silvae Sinicae* 31, 289–298.

Wu, J.E., Liu, W.J. and Zhu, C.J. (2014) Application of stable isotope techniques in the study of plant water sources and use efficiency. *Journal of Southwest Forestry University* 34, 103–110.

Wu, X., Shi, W. and Tao, F. (2021) Estimations of forest water retention across China from an observation site-scale to a national-scale. *Ecological Indicators* 132, 108274.

Wu, Y.X., Zhang, Y.E., Jia, G.D., Wang, Y.S. and Yu, X.X. (2021) Quantitative separation of evapotranspiration components of *Platycladus orientalis* ecosystem based on multiple isotope models. *Chinese Journal of Applied Ecology* 32, 1971–1979. (in Chinese with English abstract).

Xia, L., Song, X., Fu, N., Cui, S., Li, L. *et al.* (2019) Effects of forest litter cover on hydrological response of hillslopes in the Loess Plateau of China. *Catena* 181, 104076.

Xia, S., Liu, Y., Yu, X. and Fu, B. (2018) Challenges in coupling LTER with environmental assessments: An insight from potential and reality of the Chinese Ecological Research Network in servicing environment assessments. *Science of the Total Environment* 633, 1302–1313.

Xiao, J., Sun, G., Chen, J., Chen, H., Chen, S. *et al.* (2013) Carbon fluxes, evapotranspiration, and water use efficiency of terrestrial ecosystems in China. *Agricultural and Forest Meteorology* 182–183, 76–90.

Xue, B., A, Y., Wang, G., Helman, D., Sun, G. *et al.* (2022) Divergent hydrological responses to forest expansion in dry and wet basins of China: Implications for future afforestation planning. *Water Resources Research* 58, e2021WR031856.

Xue, M., Hang, R., Liu, Q., Yuan, X.-T. and Lu, X. (2021) CNN-based near-real-time precipitation estimation from Fengyun-2 Satellite over Xinjiang, China. *Atmospheric Research* 250, 105337.

Yang, S.T., Wang, P.F., Wang, J., Lou, H.Z. and Gong, T.L. (2021) River flow estimation method based on UAV aerial photogrammetry (in Chinese with English abstract). *National Remote Sensing Bulletin* 25, 1284–1293.

Yang, Y., Roderick, M.L., Guo, H., Miralles, D.G., Zhang, L. *et al.* (2023) Evapotranspiration on a greening Earth. *Nature Reviews Earth and Environment* 4, 626–641.

Yao, T., Lu, H., Yu, Q., Feng, S., Xue, Y. *et al.* (2023) Uncertainties of three high-resolution actual evapotranspiration products across China: Comparisons and applications. *Atmospheric Research* 286, 106682.

Yorke, T.H. and Oberg, K.A. (2002) Measuring river velocity and discharge with acoustic Doppler profilers. *Flow Measurement and Instrumentation* 13, 191–195.

Yu, P., Wang, Y., Wu, X., Dong, X., Wei, X. *et al.* (2010) Water yield reduction due to forestation in arid mountainous regions, northwest China. *International Journal of Sediment Research* 25, 423–430.

Yu, Z., Liu, S., Li, H., Liang, J., Liu, W. *et al.* (2024) Maximizing carbon sequestration potential in Chinese forests through optimal management. *Nature Communications* 15, 3154.

Zhang, B., Tian, L., Yang, Y. and He, X. (2022a) Revegetation does not decrease water yield in the Loess Plateau of China. *Geophysical Research Letters* 49, e2022GL098025.

Zhang, J., Zhang, Y., Sun, G., Song, C., Dannenberg, M.P. *et al.* (2021) Vegetation greening weakened the capacity of water supply to China's South-to-North Water Diversion Project. *Hydrology and Earth System Sciences* 25, 5623–5640.

Zhang, J.X., Liu, F. and Wang, J. (2021) Review of the light-weighted and small UAV system for aerial photography and remote sensing. *National Remote Sensing Bulletin* 25, 708–724. (in Chinese with English abstract).

Zhang, M. and Wei, X. (2021) Deforestation, forestation, and water supply. *Science* 371, 990–991.

Zhang, Q., Lv, X., Yu, X., Ni, Y., Ma, L. *et al.* (2022) Species and spatial differences in vegetation rainfall interception capacity: A synthesis and meta-analysis in China. *Catena* 213, 106223.

Zhang, S., Wen, X., Wang, J., Yu, G. and Sun, X. (2010) The use of stable isotopes to partition evapotranspiration fluxes into evaporation and transpiration. *Acta Ecologica Sinica* 30, 201–209.

Zhang, Y. and Song, C. (2006) Impacts of afforestation, deforestation, and reforestation on forest cover in China from 1949 to 2003. *Journal of Forestry* 104, 383–387.

Zhang, Z., Wang, L. and Wang, S. (2004) Forest hydrology research in China. *Science of Soil and Water Conservation* 2, 68–73.

Zhang, Z., Wang, S., Sun, G., McNulty, S.G., Zhang, H. *et al.* (2008) Evaluation of the MIKE SHE model for application in the Loess Plateau, China. *Journal of the American Water Resources Association* 44, 1108–1120.

Zhang, Z., Jiang, T., Zhao, L. and Cheng, Z. (2024) Camera calibration using cross-section waterline orientation for video-based flow measurement. *Journal of Electronics and Information Technology* 46, 1428–1437. (in Chinese with English abstract).

Zhou, G., Wei, X., Luo, Y., Zhang, M., Li, Y. *et al.* (2010) Forest recovery and river discharge at the regional scale of Guangdong Province, China. *Water Resources Research* 46, W09503.

Zhou, G., Wei, X., Wu, Y., Liu, S., Huang, Y. *et al.* (2011) Quantifying the hydrological responses to climate change in an intact forested small watershed in Southern China. *Global Change Biology* 17, 3736–3746.

Zhou, G., Xia, J., Zhou, P., Shi, T. and Li, L. (2021) Not vegetation itself but mis-revegetation reduces water resources. *Science China Earth Sciences* 64, 404–411.

Zhou, Z., Ouyang, Y., Li, Y., Qiu, Z. and Moran, M. (2017) Estimating impact of rainfall change on hydrological processes in Jianfengling rainforest watershed, China using BASINS-HSPF-CAT modeling system. *Ecological Engineering* 105, 87–94.

14 Forests for Municipal Water Supply Watersheds

L.J. Bren[1]* and A.A. Webb[2]

[1]*The University of Melbourne, Creswick, Victoria, Australia;* [2]*WaterNSW, Tamworth, New South Wales, Australia*

Abstract

It is probably impossible to overstate the importance of forested municipal watersheds in sustaining towns and cities large and small around the world. They are a first line of defence in providing clean, disease-free water. In a mature system, this then is stored and sterilized by time in a reservoir, 'treated' to improve properties and sent to consumers through a network of pipes. Relative to their importance, there is surprisingly little literature on their attributes, valuation and management. The vision presented is one of financial valuation based on alternative replacement cost, benchmarking of hydrological and forest characteristics using forest hydrology science, comparisons with 'best performance' ideals and the necessary capital expenditure to achieve such performance. The chapter examines factors such as time, competing water and conservation demands, political pressures relating to forest management and forest recreation, and questions of 'resource ownership'. It is believed that new developments in forest hydrological monitoring using remote sensing to monitor rainfall and watershed storage plus a relaxation of 'keep-out' restrictions on watershed access will lead to a new and exciting age of forested watershed management.

14.1 Introduction

To the authors, it is a miracle of nature that areas of forest that might be described as 'scruffy' or 'nondescript' provide a sustained outflow of clean, high-quality water. This ability has long been noted by water supply managers. This chapter looks at why this is, the pros and cons of forested watersheds for water supplies, examines how forest hydrology science can add to the science of their management, considers how they should be valued (at least at their replacement cost) and discusses how their use might be optimized in a period when growing urbanization is putting increased supply demand on these areas. We have relied on work mainly from the USA and Australia but believe that the findings have a more general applicability.

The provision of water to towns both large and small has always had problematic elements. In 1854, a brilliant English medical practitioner, John Snow, traced an outbreak of cholera in London to contamination of a water supply pipe by faecal matter. This made the world recognize the need for good water supplies to cities to ensure the health of city dwellers; this message was particularly heeded by the cohort of growing towns in the 'new worlds' of the American continents, Africa and Australia. Since then, provision of a clean, plentiful water supply has been associated with good community governance. Some cities reserved large areas for municipal watersheds, while others met this challenge by a mixture

*Corresponding author: lbren@iinet.net.au

 Forest Hydrology: Processes, Management and Assessment (eds Leon J. Bren *et al.*)
DOI: 10.1079/9781800625310.0014

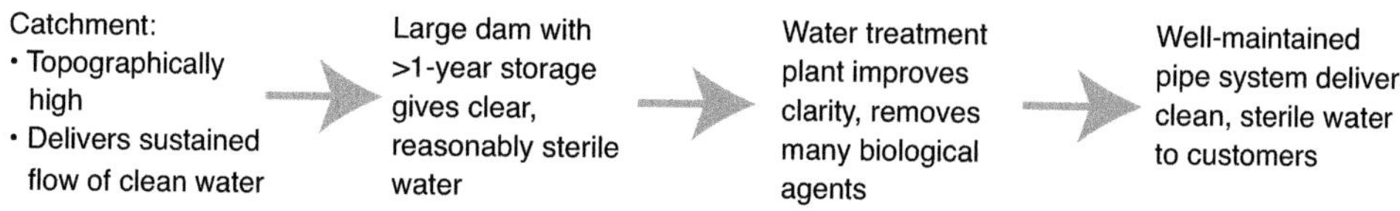

Fig. 14.1. Schematic of a basic but safe water supply with multiple levels of protection for a small town or city.

of structures and increased regulations on residents and users.

A prosaic description of the task of a watershed might specify the volume of water of a given standard to be supplied to a town over a set period. The usual approach to town water supplies uses sequential levels of protection; these are illustrated for a small but sophisticated and well-capitalized water supply in Fig. 14.1. Thus, the town receives its water supply from a topographically high, unpolluted watershed. If the watershed slopes are forested, then much of the rainfall infiltrates through the forest floor to recharge hillslope groundwater. The passage of most of the water through the soil effectively 'filters' the water of particulates and biological organisms and reduces the length of time needed in reservoir storage (e.g. McGinnis and Kerans, 2013). The water passes into a dam where it is stored for a year or more, giving sterile, clear water. A water treatment plant may further remove some turbidity, colour and other pollutants, and modify some physical, biological and chemical characteristics of the water. This then passes clean, sterile water into a well-maintained network of pipes that delivers uncontaminated water to the customers. Larger water supplies may have many dams and a management regime that allows water to bypass problematic reservoirs or supply intakes. Sometimes the water is mixed with water from other reservoirs or different reservoir depths to give a water supply of constant taste and appearance. Lately, some towns have been installing desalination plants that may (possibly) allow cities to exist without municipal watersheds. It is assumed that the watershed exists in a planning environment that effectively 'guards' or 'protects' the purity of the streams and dams (commonly by restricting building developments and roads), although fires can usually breach these safeguards.

14.1.1 An historical view of the excellence of forested watersheds

A link between forests and water has been acknowledged by human societies for centuries (Andréassian, 2004). In the 1st century, Pliny the Elder wrote in his *Natural History*: 'Often, after woods have been cut down, springs on which trees used to feed emerge... Often, disastrous torrents are formed after the felling of mountain woods...' His quote established that humans were aware over 2000 years ago that forests played a role in the hydrological cycle. The Romans put this knowledge to good use by building impressive aqueducts (some of which still stand) to carry water to their cities from distant, high-elevation forested watersheds.

During the Middle Ages, France recognized the link between forests and water with King Phillippe Auguste enacting the royal ordinance of the 'Waters and Forests' in 1219 (Kittredge, 1948). To this day, the French have retained this association between 'Eaux et Forets' (Water and Forests) in relation to forest management (Andréassian, 2004). Indeed, France's main forest management agency is the Agence Nationale des Eaux et Forets (National Agency for Water and Forests).

In the USA, in response to issues of flooding and sedimentation, National Forest reserves were established in the late 19th century with a major aim of protecting and enhancing watershed values (Sedell *et al.*, 2000). The Organic Administration Act of 1897 stated: 'The Act provides that no national forest may be established except to improve and protect the forest, or to secure favorable conditions of water flows, and to furnish a continuous supply of timber.' In the contiguous USA, forest comprises about 29% of the total land area but contributes 46% of the surface water supply (see Liu *et al.*, 2022, for a statistical characterization of the importance of forests for the conterminous USA). Much of the focus of forest hydrology in the first half of the 20th century was on water quantity. It was not until the 1950s that more studies began to be concerned with forests and water quality.

Paired watershed studies played a significant role in advancing our knowledge of the role of forests and their management in driving water quality outcomes. At Hubbard Brook in New Hampshire, Likens *et al.* (1970) reported on an experiment where a watershed of northern hardwood trees was cut but where the timber was left in place where it fell, the slash was distributed across the hillslopes, no roads were made and care was taken to minimize soil erosion. Herbicides were then applied for a period of 3 years to prevent forest regrowth. The results of the experiment demonstrated that a reduction in assimilation of nitrogen and rapid decomposition of organic matter, when compared with an undisturbed forest, led to the export of high levels of nitrates and accompanying cations in stream water that exceeded safe drinking water standards.

Subsequent studies at Hubbard Brook (e.g. Hornbeck *et al.*, 1986) showed that, by comparison, commercial logging operations had a lower effect on water quality due to natural forest regrowth retaining higher levels of nutrients. Further research identified lower impacts on stream nutrient levels following forest harvesting when riparian buffer strips were retained (Hornbeck *et al.*, 1987). New lines of investigation at Hubbard Brook have built on the long legacy of watershed research resulting in a shift from comparing inputs and outputs and quantifying pools and fluxes to a more mechanistic understanding of ecosystem processes within watersheds. For example, hydropedological studies have shed light on linkages between hydrological flow paths and soil development that provide valuable perspectives for managing forests and understanding stream water quality (Campbell *et al.*, 2020). An earlier published study at the Fernow Experimental Forest in West Virginia found much lower impacts on water quality following logging, which highlighted the importance not only of using riparian buffer strips but also of ensuring that roads and skid trails were designed to minimize the delivery of turbid water to streams (Aubertin and Patric, 1974). These early studies laid the foundation for multiple studies into the effects of forestry activities, largely on soil erosion and suspended sediment concentrations in streams (e.g. Beschta, 1978; Binkley and Brown, 1993; Macdonald *et al.*, 2003; Chappell *et al.*, 2004; Gomi *et al.*, 2005; Ide *et al.*, 2009; Basher *et al.*, 2011; Webb *et al.*, 2012; Zimmermann *et al.*, 2012).

A summary of the linkages between forest soils and water quality and quantity was provided by Neary *et al.* (2009), who noted the following:

- Forest soils feature litter layers and high organic content, both of which contribute to an abundant and diverse micro- and macro-fauna.
- Root systems under forests are extensive and deep compared with agricultural lands and grasslands.
- These biological conditions create soils with high macroporosity, low bulk density and high saturated hydraulic conductivities and infiltration rates.
- Surface runoff is rare in forest environments, and most rainfall moves to streams by subsurface flow pathways where ‘filtration’, nutrient uptake, cycling and contaminant sorption processes are rapid.
- Because of the dominance of subsurface flow processes, peak flows are moderated and baseflows are prolonged. This is particularly evident in continued low flows during prolonged dry periods.

Forests also lack the pollutant inputs associated with agriculture and developed lands (Caldwell *et al.*, 2023). It is for these reasons that the best physical quality water in the world is sourced from forested watersheds (Dissmeyer, 2000). When forest is converted to other land uses, commonly the effect on source water quality is detrimental (Barten and Ernst, 2004).

14.2 A Look at Watershed Properties

14.2.1 Some quantitative properties of watersheds

For most cities, the historic criterion of watershed selection was that a substantial stream reach carrying relatively pure water emanating from a forest close to the city was available for diversion to the city. Often such land was naturally forested (Duffy *et al.*, 2020). Commonly, the watershed forests were held by the government so that there was little cash outlay in proclaiming these as municipal watersheds. Usually, little thought was given to determining or optimizing the quantitative properties of these watersheds for many decades. More recently, there has been a formal selection process (e.g. New York, as described by Pires, 2004) as cities have attempted to secure a high-quality 'natural' water supply to guarantee their future. Pires (2004) noted that 'the City of New York acquired the 'political and financial clout' required to tap this most essential of natural resources from the far reaches of its rural hinterland'.

The literature on water supplies is surprisingly mute on the desirable quantitative properties of the watersheds that supply them, and there is little discussion on how such properties of existing watersheds might be 'improved'. We have listed a few key parameters below, but it is not hard to devise others.

Watershed efficiency

This is the ratio of water leaving the watershed to water entering the watershed (generally through precipitation). The concept is widespread in hydrological literature and may have other names such as 'runoff ratio' or 'runoff efficiency'. Figure 1.5 (Chapter 1, this volume) presented the runoff curves of Zhang *et al.* (2001) showing mean annual streamflow (mm) as a function of mean annual precipitation for both forest and pasture on a worldwide basis. These can be used to compare various important characteristics of forested watersheds compared with agricultural watersheds. It is a simple transformation to derive from these the annual percentage efficiency of runoff (ratio of annual water yield to annual rainfall) of forested and agricultural watersheds (Fig. 14.2); the marked reduction in water yield due to the greater

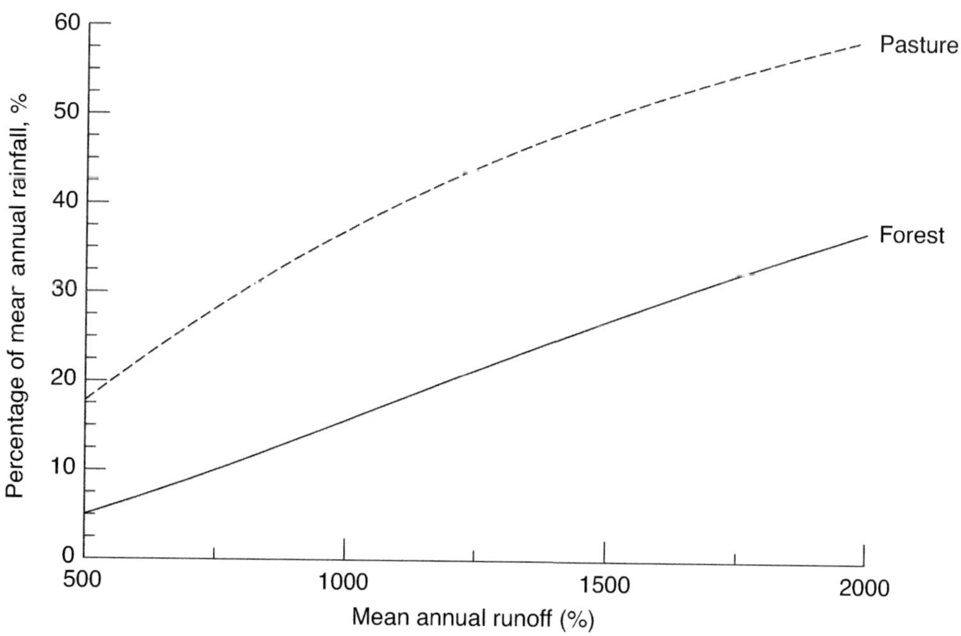

Fig. 14.2. Watershed efficiency (%) of forested and pastured watersheds as a function of mean annual rainfall using the curves of Zhang *et al.* (2001). Efficiency is defined as the percentage ratio of streamflow to rainfall. The illustration does mask much variation encountered by Zhang *et al.* (2001) in the preparation of the curves.

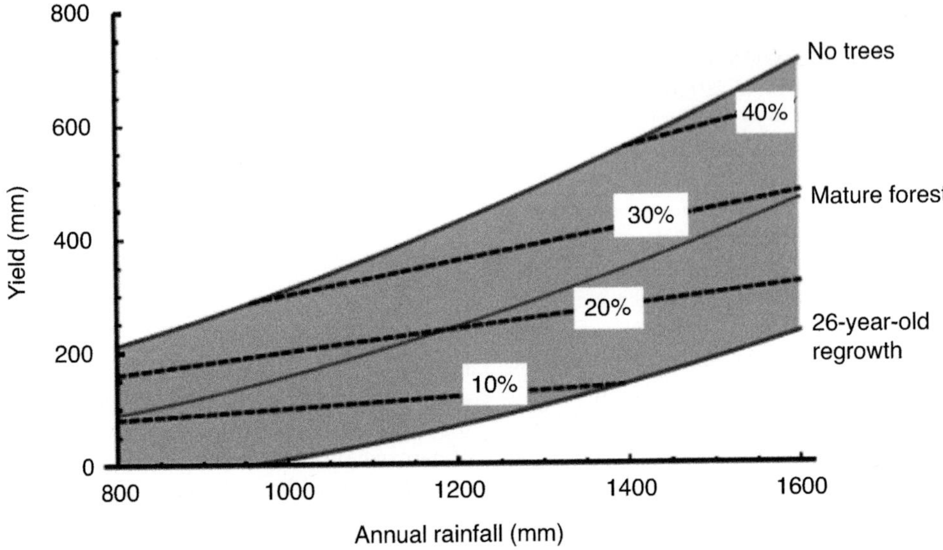

Fig. 14.3. Exploration of water yield efficiency on an Australian eucalypt catchment of mountain ash (*Eucalyptus regnans*) with different forest densities, ranging from 'no trees' to dense 26-year-old regrowth, as a function of mean annual rainfall. The blue line is the 'forest line' of Zhang *et al*. (2001). The percentage values are the watershed efficiency. Relatively sparse forest in the highest rainfall areas gives the highest efficiency.

transpiration of trees compared with grass cover is obvious. It should be noted that there is much variation inherent in such curves, so that they should be viewed as 'generalizations'.

There are substantial gains in both water yield and watershed efficiency with increasing annual rainfall. Annual watershed efficiency provides the most basic (but rarely discussed) 'benchmark' for watershed performance. Barten *et al.* (2012) cited the Quabbin Watershed of Boston, Massachusetts, as having a greater than 50% efficiency; this would be high by world standards and reflects a high annual precipitation, low potential evapotranspiration, excellent capture and transmission of water, deciduous tree species in a high winter rainfall area and good management practices.

Although achieving the highest efficiency possible is desirable, most managers only have a limited range of improvement achievable by skilled vegetation management. As an example, Fig. 14.3 is a 'domain diagram' produced as an adjunct of the work of Bren *et al.* (2010) showing the range of efficiency from a municipal watershed forested with native mountain ash (*Eucalyptus regnans*) regrowth in Victoria, Australia; this combines information from an early 'runoff curve' of Kuczera (1987) and the curves of Zhang *et al.* (2001). There is a restricted range of runoff that can be produced from the catchment by forest manipulation for a given annual rainfall. The influence of dense regrowth forest in reducing streamflow from the watershed is evident. It is likely that, in the absence of forest management aimed at keeping the forest density at the lower end of the range, a low overall water supply efficiency will follow. We found it difficult to find estimates of the water supply efficiency of municipal watersheds around the world for comparative purposes; this alone would make an excellent study. The influence of forest density on water yields in these Australian forests has been particularly highlighted by the exceptional work of Benyon *et al.* (2023).

Runoff elasticity

'Rainfall elasticity' (sometimes called 'runoff elasticity') is defined as the ratio of the

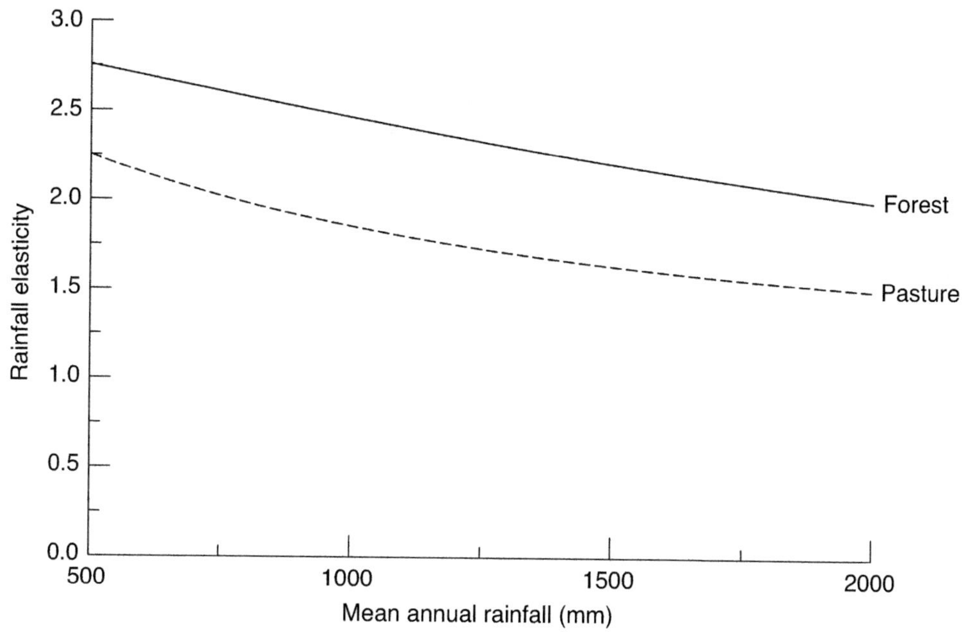

Fig. 14.4. Elasticity of runoff from forest and pasture, derived using the curves of Zhang *et al.* (2001).

proportional change in runoff for a given proportional change in rainfall. It is computed from the curves of Fig. 1.4 (Chapter 1, this volume) by differentiation (Bren, 2023). Fig. 14.4 shows the corresponding plots of rainfall elasticity for pasture and forests. Chiew (2006) looked at rainfall elasticity of streamflow in Australian watersheds and found it ranged from 2 to 3.5 for 70% of the watersheds. Thus, a 5% decrease in rainfall may lead to a 10% reduction in streamflow in an agricultural watershed and a 15% reduction in a forested watershed. The high rainfall elasticity of forested watersheds is occasionally cited as a negative consequence of forested watersheds because, it is argued, downstream communities are subject to a greater flow change for a given change in rainfall than they would be if the watersheds were not forested. del Campo *et al.* (2019) showed that thinning semi-arid forests reduced the elasticity of the watersheds, making them less susceptible to drought. Ervinia *et al.* (2019) thought that rainfall elasticity could be used as a good indicator of watershed reliability and/or resilience. In some circumstances, adequate dam storage could overcome elasticity issues for a municipal supply. The plot does highlight the major drawback of municipal watersheds – if rain is low for 2 or 3 years, the town runs out of water fast.

How many people can be supplied?

The curves of Fig. 1.4 (Chapter 1, this volume) can be used to provide an interesting visualization of how much watershed area per head of population is needed, and how many citizens can be supplied from a given forest area. The citizens of the City of Melbourne (Victoria, Australia) are being encouraged to reduce water consumption to 150 l/day, although current consumption is marginally higher. Using this value, it is a simple matter (Bren, 2023) to transform the runoff curves of Zhang *et al.* (2001) in a way that shows the watershed area per head and the number of people per square kilometre that can be supplied as a function of mean annual rainfall (Fig. 14.5a,b). Fig. 14.5c uses the same approach to estimate the forested watershed area necessary to supply 5, 7, and 9 million people as a function of mean annual rainfall. Currently, Melbourne's population is 5.5 million and is expected to pass 8 million in 2050, according to the Australian Bureau of Statistics (www.abs.gov.au/, accessed 12 May

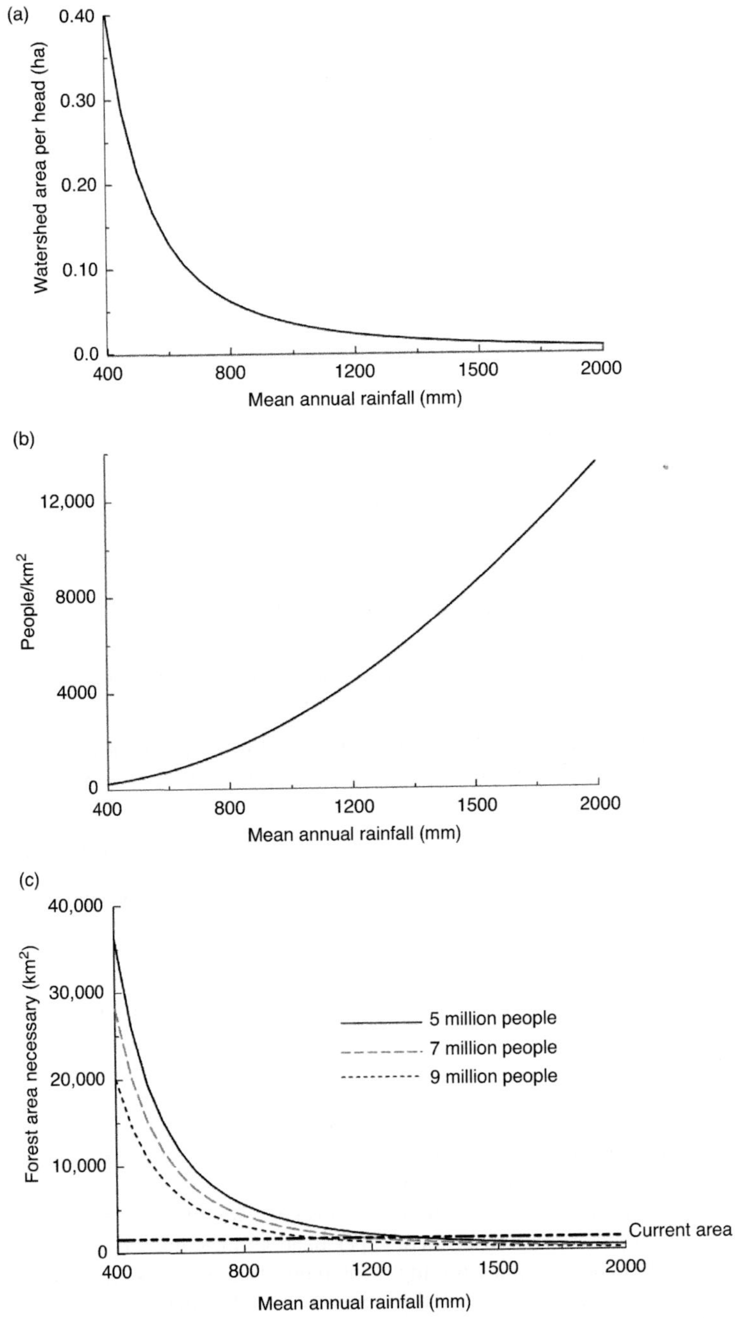

Fig. 14.5. Use of the curves of Zhang *et al*. (2001) to present watershed information (based on Bren, 2023). (a) Area of watershed per person as a function of mean annual rainfall for a forested watershed supplying 150 l/day for each person. (b) Number of people per square kilometre who can be supplied at 150 l/day by a forested watershed as a function of mean annual rainfall. (c) Area of forested watershed necessary to supply a city of 5, 7 and 9 million people as a function of mean annual rainfall. The area of Melbourne's watersheds is indicated.

2025). The plot does highlight the impact of high rainfall elasticity in forests – a relatively small change in mean annual rainfall has a large impact on the number of people who can be supplied by the municipal watershed.

The application of such an approach to any watershed of known rainfall can give an approximate idea of the maximum number of people that could be supplied from that watershed; this in turn allows focus on extending or 'improving' the watershed if the population is growing. Using such plots, given Melbourne's forested water catchment area of 163,000 ha and assuming a mean average rainfall of 1200 mm means that their forested watersheds could supply about 5–7 million people on average. At the time of writing, there has been a run of bounteous years and a build-up of reservoir storage. The Melbourne experience has been that sooner or later a run of drought years will deplete the water storages. Although such an approach is not perfect, it does allow an approximate quantification of the upper limits of supply of forested watersheds. In this case, the city of Melbourne's options are to modify demand, enhance watershed efficiency, find more watersheds, or find additional distant sources of supply to allow coping with sustained runs of below-average rainfall and constant growth in population.

Slope water storage

An often recognized but surprisingly unquantified attribute of upland forests compared with other land uses is the slope water storage (see Chapter 2, this volume, for a discussion of how slopes store water). Upland forests maintain a highly infiltrating slope surface and rainfall usually passes through this to underlying soils/rock and associated groundwater. The 'soil' can be many metres deep and may pass into fractured or oxidized rock zones, which store and transmit water. A muted version of the same process exists in flat-land forests, with the added process of bank storage/return flow due to height variations in stream water levels. The depth of these subsurface strata is hard to measure, but in the Cropper Creek Project Area of south-eastern Australia (e.g. Bren and Hopmans, 2007), seismic work showed these zones to be around 200m depth below the ground surface. The annual slope hydrological cycle can be viewed as porous media storing water and then slowly discharging it. The processes of water retention and release give a sustained, steady outflow of 'filtered water', which is substantially unaffected by short-term rainfall fluctuations. From a water supply perspective, this makes the catchment outflow much more predictable and reduces the need for large reservoirs to buffer storage or to stop potential 'drinking water' passing downstream.

A replicated 'classic' experiment at the Coweeta Hydrologic Laboratory, North Carolina, gave some insight into these properties. Hewlett and Hibbert (1963) measured the outflow of water from a constructed sealed 'box' of reconstituted forest soil that had been saturated with water. Outflow continued for around 145 days. Some 55 years later, the experiment was repeated using the same *in situ* box (Lee *et al.*, 2020). The high-flow results were unchanged but the low flow results were very different due to a leak in the box passing water to 'deep storage' in the watershed. Thus, outflow the second time only occurred for around 14 days, with much of the water passing through the leak to subsoil or bed-rock storage. The direct inference is that these deeper storage zones account for the long-term recession behaviour of watersheds in the absence of rainfall. This matched the senior author's experience on the Australian Croppers Creek paired catchment project (Bren, 1980). On these slopes, water infiltrated very fast to many metres in depth. The water storage and transmission in the saprolite/weathered rock zone dominated the hydrology during non-rainfall periods – the soil acted as a fast transmission conduit to take water to greater depths. A corollary was that the soil moisture levels in the top few metres were not sensitive measures of the watershed water storage.

The buffer storage of forested slopes has been critical to 'see-you-through' storage sustaining communities in times of little rainfall. Nippgen *et al.* (2016) examined the 'watershed memory' of forested watershed at the Coweeta Hydrologic Laboratory. They found that the outflow response of the watershed to rainfall was affected by the legacy of past hydroclimatic conditions. A severe drought would cause a lingering impact of unresponsiveness due to a lack of buffer storage. Similarly, a wet period would lead to enhanced responsiveness for up to a year after the rainfall. Severe drought or heavy vegetation water use can deplete such buffer storages to the extent that overlying vegetation suffers and the responsiveness of the watershed to rainfall is diminished until the buffer is refilled. This influence was discussed by Peterson *et al.* (2021) who noted that, 7 years after the cessation of the long Australian 'millennium drought', the watersheds had not returned to their pre-drought responsiveness to rain. Similar results have been found when plantations replaced low-water-use native vegetation. Scott and Lesch (1997) found in the Mokobulaan paired catchment experiment of South Africa that pine and eucalypt plantations caused the streams to dry up in the ninth year after planting. Streamflow did not return to the pre-planting levels until 5 years after clear-falling of the plantations. Webb and Kathuria (2012) in the Red Hill paired catchment experiment in Australia found that the hydrological impact of thinning was delayed by the onset of drought reducing catchment storage. The hypothesis in each case was that the buffer storage in the watershed slopes was so reduced that the 'normal catchment hydrology' could not function until this deficit had been made up. It is likely that, in such a case, dense forest cover on the slopes would delay recovery, so thinning of the forest would be a reasonable recovery strategy.

McNamara *et al.* (2011) suggested that catchment storage is a useful 'metric' for characterizing watersheds but note that it is seldom done and that tracking the dynamics of storage is even rarer. Although we can find examples of the storage retention/release characteristic being quantified for small watersheds (e.g. Sayama *et al.*, 2011; Nanda and Safeeq, 2024), we are unaware of studies that have quantified this for major watersheds. This storage gives the manager a major cost-saving in avoiding or deferring major reservoir structures. The academic task in such studies is to go from the abstract concept of millimetres of storage to volumetric flow measurements (cubic metres per second) for the entire watershed.

The common approach to monitoring slope storage has been measuring 'intensive' correlated parameters such as soil moisture levels or phreatic pressures, but innovations in high-resolution soil moisture sensors in the last few decades and recent new remote-sensing developments have exciting potential for monitoring, although assimilation of such information presents challenges for the moment. These give the possibility of measuring the absolute amount of watershed storage rather than the relative change from some past base level. Thus, the older generation of 'Terrestrial Water Storage' data recoverable from the Gravity Recovery and Climate Experiment (GRACE) and Global Land Data Assimilation System (GLDAS) satellites and associated technology literally allows weighing of the Earth's surface (by deflection of a satellite, as measured by a following satellite) to determine moisture storage (e.g. Moghim, 2020; Soltani *et al.*, 2021). Newer-generation endeavours are covered in Chapter 16 (this volume). Merging of this information with hydrological models ('data assimilation') will allow a far more comprehensive monitoring in the future than has ever been achieved in the past. The glittering model is that once the overhead of assimilation is provided, the water resource manager would receive periodic updates of the storage data with little additional field-work required.

Given the reduced flow why are forested watersheds favoured?

Figures 14.2–14.5 suggest that where rainfall is relatively low (<800 mm), a forested watershed may not be a good choice for a municipal watershed because the water yield will be lower than agricultural watersheds, although probably of higher quality. In subhumid environments, riparian forest will sometimes be removed ('phreatophyte removal') to enhance streamflow. Forests in these areas will often be heavily thinned to maximize streamflow. Francis *et al.* (2017) showed that more complex forest boundaries enhance water quality for the same percentage of forest cover. Forest management for municipal watersheds in low-rainfall areas should favour open forest species with low water use if possible. English *et al.* (2024) provided a case study for the south-eastern USA, suggesting that a low-density fire-maintained woodland ecosystem provides distinct ecosystem services (which, if necessary, can be valued) for watersheds in this area.

In most parts of the world, there is little choice in the watershed land use because of population pressures and the history of settlement. Where there is some choice (and the rainfall is adequate), water managers usually prefer a forested watershed because of the sustained flow, reduced peak flows, better water quality and relatively easy management. Even when the watershed is predominantly agricultural, watershed authorities commonly attempt limited reforestation along stream banks or on vacant land in the hope of gaining some of these benefits. As well as these, there are other aspects:

1. Forests tend to have fewer humans domiciled within them, avoiding water contamination issues associated with human activity (particularly faecal contamination associated with domestic dwellings).
2. Forest cover precludes domestic animals. A common Australian ranking is that pigs > cows > sheep >native fauna in their ability to transmit waterborne diseases to humans (but see Miller *et al.*, 2017, for a more scientific approach to this large, important subject). Thus, watersheds with substantial populations of common domestic animals pose a direct threat to water consumers if there is a failure in watershed management, dam capacity, dam residence time or water treatment protection. In many forests, populations of feral animals can pose a similar hazard. In Australia, there are feral species of deer, horses, pigs, goats, dogs, cats and foxes commonly found; most countries have a similar list of 'difficult animals'. Hampton and Davis (2020) and Bennett *et al.* (2015) provided a discussion of the impact of feral deer in Melbourne's watersheds. Wild pigs can be particularly problematic (see Bolds *et al.*, 2021, for a discussion on these and recommendations for control). Animal trails can sometimes lead to gully erosion, impairing turbidity. Feral animal control can be a major task, sometimes necessarily involving distastefully bloody (and politically unpopular) methods of killing the animals.
3. Forest cover usually precludes cultivation. Cultivation breaks down the soil structure, and runoff from cultivated land can lead to highly turbid water passing into the stream (e.g. Bren and Ryan, 2024). This turbidity may be associated with clay dispersion, which 'colours' the water of dams. This can usually be removed by 'water treatment' (e.g. flocculation, polyphosphate treatment), but this may be expensive.

14.2.2 Some economics of selling water from forested watersheds

Our forest hydrology runoff curves can give some indication of the financial aspects of operating forested watersheds. These aspects show that, where water is priced, selling of clean water from forested watersheds can indeed be profitable (or at least generate a high revenue). Fig. 14.6a uses the 2022 retail price of water for the City of Melbourne to compute the annual water value of runoff from a forested watershed as a function of annual rainfall. This has the unrealistic assumption that every drop is sold to Melbourne consumers and ignores the cost of storage and distribution, but it does show that producing clean water has the potential to be a lucrative exercise; indeed, analyses such as that of Vardon *et al.* (2019) show that the designation 'cash-cow' can be an apt term applied to municipal watersheds. Bren (2023) carried this analysis further by computing the discounted value of such watersheds into the future (Fig. 14.6b) and showed a discounted capital worth of A$1–2 million/ha (depending on the discount rate chosen) for such high rainfall watersheds supplying major cities.

Given that water collection may be such a profitable task, one might ask whether collecting larger volumes of water from agricultural land and 'treating it' to bring it to the generally higher-quality standard of a forested watershed would be a successful economic strategy. This has two problems. First, the cost of water treatment for expected pollutants (turbidity, bacteria, nitrogen) may be too high to realize any gains from the enhanced volume. Second, the possibilities of unexpected pollutants or contamination from agricultural chemicals (or as a by-product of the additional treatment) is very real. The experience of Sydney, Australia, in 2000 when

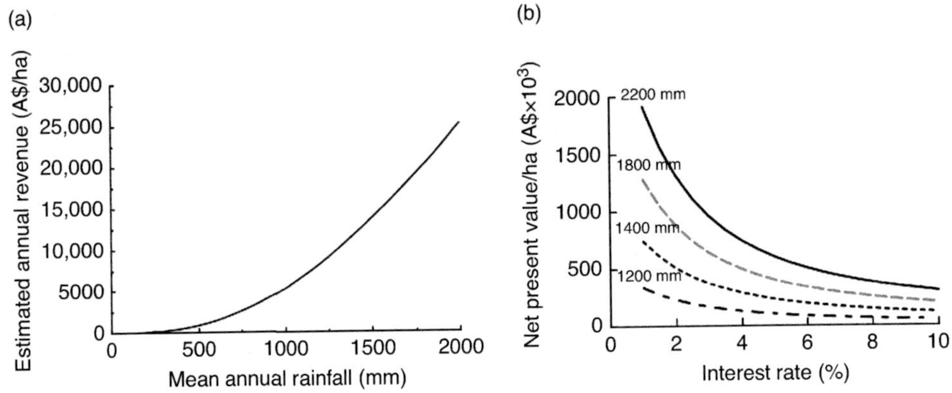

Fig. 14.6. (a) Estimated annual revenue (in Australian dollars) per hectare of watershed if all water was sold as a function of mean annual rainfall, and (b) capitalized value of this for different annual rainfalls as a function of the interest rate adopted.

the city's water supply was possibly contaminated by organisms associated with domestic animal activity in the watersheds showed that the judgement of consumers towards the water authority (and its officers) was not kind (e.g. Carson and White, 1998). The relative economics of treating agricultural runoff to bring it to a higher standard have been well discussed in the literature (e.g. Cunha *et al.*, 2016; Price and Heberling, 2018). In general, small gains in the quantity of water from agricultural land do not compensate for the cost and difficulty associated with impaired water quality. Warziniack *et al.* (2017) found that increasing forest cover by 1% in municipal watersheds decreased turbidity by 3%, and that increased grazing led to increase turbidity. Having forested watersheds gives clear economic savings.

The ability of forested watersheds to generate large gross revenues has received surprisingly little discussion in the literature, with most financial focus going on the cost of storage, distribution or water treatment structures. A corollary of this is that if a watershed requires expenditure to protect or enhance it, one might reasonably expect that such money should be freely available, as the watershed is the source of all this wealth. Observations in Australia suggest that all too often this is often not the case. Such expenditure may be begrudged because of 'competing demands for capital', structures may provide better 'ribbon-cutting photo opportunities', the assumption that the watershed will 'look after itself' and perhaps that the managers are unclear of what forest management course to take. In some cases, the managers may wish to avoid possible controversy associated with 'new' forest management strategies (particularly if forest cutting is involved).

In general, high rainfall land is productive of other goods (particularly wood products). The question of whether community wealth can be maximized by producing usufructs other than water (as well as water) gives rise to vigorous debate in Australia (e.g. Creedy and Wurzbacher, 2001; Taylor *et al.*, 2019; Bren, 2024). A common conflict is whether logging should be allowed in municipal watersheds. The wood may be a needed community resource, and its appropriate cutting may help the quest to have healthy, resilient forests and an acceptable watershed efficiency. The sale of wood products gives a healthy revenue stream, but poorly managed logging can impact on water quality. Harvesting may lead to vigorous regrowth, with water yield implications. In general, the results are sensitive to the assumptions made and the long-term discount rates applied. It has been noted that communities located in or near watersheds commonly express generational anger when their economic use of 'traditional' forest lands are curtailed because the water from their local areas is piped to distant cities that have, otherwise, shown no interest in these communities.

14.3 Management Issues of Forested Watersheds

An unromantic view of forested watersheds is that they are there solely to supply the needs of a downstream town. Hence, anything that enhances the quality and quantity of water yielded is good and anything that diminishes the quantity and quality of water is bad. A basis for valuation and expenditure is what would the community have to do (or spend) if the watersheds did not exist; this gives an immediate basis for valuation of such watersheds in terms of 'replacement costs'. Such a prosaic view may cause conflict with other 'management aims' for such forests. Commonly, there is no feasible 'replacement'.

Fires are a major issue in large areas of the world and the impact of these is covered in Chapter 10 (this volume). Forested municipal watersheds ideally should have a capital investment of fire protection installations (roads, water storages, helipads, fire lines, fuel-reduced areas, operational bases, fire tankers, trained fire crews, fire detection systems, agreements with neighbouring brigades, and fire-fighting procedures to deal with fires and disasters). These are costly and may have adverse impacts on water quality because of construction, the presence of compacted areas, disturbance due to vehicle and tractor movements, and possibilities of contamination from long-lasting pollutant chemicals used in firefighting. Necessary fuel reduction strategies can be controversial on ecological grounds (e.g. Berlinck and Batista, 2020) or smoke nuisance grounds (Schollaert *et al.*, 2024). Worldwide experience is that these are necessary investments for large areas of forested watershed but require careful management. Fig. 14.7 shows the impact of wild-fire burning on sediment in an Australian municipal watershed (see Dragovich *et al.*, 2012, for an example).

Fig. 14.7. Sediment deposited on the bed of a stream in the Cotter River watershed following the 2003 Canberra fires, Australia. Photo courtesy of Ashley A. Webb.

In the absence of water treatment plants, this turbidity pollution can put dams out of service for some years (e.g. Jackson *et al.*, 2024).

From the forest management point of view, the major question is the type of forest on the watershed and the appropriate management. Two diverse Australian examples are:

1. The City of Ballarat (Victoria, Australia). This has had almost 150 years of managing commercial *Pinus radiata* plantations on their watersheds, with smaller areas of native *Eucalyptus* forest in less accessible areas. These emanated from decisions made in the 1870s to get a good water supply for this thriving gold-mining city by reforesting mined-over land. Currently, the watersheds are a mixture of agricultural land and forest land owned by the water authority. The forests have provided a flow of revenue to supplement water revenues, and the associated management provides services such as weed control on forested land and riparian areas. The forests have a good road network providing both internal and external access. The water supply is protected by good watershed management, dams, water treatment plants, interlinking of supplies and a well-maintained pipe network. Bren (2023) noted that this has been 'uncontroversial' and going on for so long that it is rarely noticed or commented on by residents. There are restrictions on public entry into the watersheds, although the needs of recreational fishing are catered for. Runoff from native eucalypt forest in this area can give a 'tannin' colour to water, which effectively requires water treatment to remove.
2. The City of Melbourne (Victoria, Australia). This has around 163,000 ha of native forest, most of which is in various state-owned land tenures and managed jointly for water and conservation. Expansion of the watershed areas in recent decades to meet the growth of Melbourne meant the inclusion of state forests with continued forest harvesting commitments, but these were controversial (see Taylor *et al.*, 2019, and Bren, 2024 for examples of arguments against forest harvesting). The water supply has been augmented by interconnection with the wider Victorian irrigation supply network and construction of a large desalination plant to meet the challenges of occasional drought. There is a strong 'citizen's lobby' totally and actively opposed to any commercial forest management or even forest thinning on these watersheds, which has caused the state government to cease native hardwood logging in its areas (e.g. Fig. 14.8 for an example of anti-logging propaganda). Many of the watersheds are 'closed' to the public for access, and this has historically been supported by city-dwelling consumers.

The forests have given a clean and reliable water supply, but Melbourne has historically had limitations on domestic water consumption due to an inadequate catchment area and/or rainfall. Hence, water consumption by vegetation has been of importance. More recently, there has been concern that fires burning in the incised lower rainfall areas near major rivers and dams will lead to landslips, passing sediment directly into these structures (Nyman *et al.*, 2021). Much of Melbourne's water has been pure enough to avoid water treatment in the past. Indeed, a major rationale of management has been avoiding the capital expenditure on water treatment plants. Hence (other than by bypassing) there is no easy way to deal with such needed dams contaminated by turbidity.

A number of specific forest management issues commonly arise. These include the following:

1. Native forests can carry high populations of native and introduced fauna (e.g. deer and feral hogs in the USA, kangaroos and introduced deer in Australia). These can lead to faecal pollution (as indicated by *Escherichia coli* levels), which requires water treatment (commonly chlorination) to bring the water back to acceptable bacterial levels for human consumption. The 'chlorine taste' of the water can be unattractive to consumers. In some cases (particularly *Giardia* and *Cryptosporidium* spp.), the organisms can be difficult to remove, and this has political ramifications (e.g. see Carson and White, 1998, on the public reaction, political opportunism and retribution when Sydney, Australia, possibly had these in its water supply). Reduction of

Fig. 14.8. An example of anti-logging propaganda about logging on Melbourne's water catchments in 2000. This was a postcard to be mailed to the Premier. There is no evidence of forest harvesting perturbing drinking water quality or risking public health as implied. The postcard is now public domain.

animal populations may involve sophisticated management, chasing away fauna and various forms of wildlife 'culling'. These may not accord with other land-management values, particularly if the watersheds are also conservation reserves. Lee *et al.* (1970) examined changes in the bacterial quality of streamflow as forests acquired human inhabitants. They found that, for low human density, the bacterial contamination of the streams was similar to that of the undisturbed forest from native fauna. As humans moved in, the bacterial loads of the streams increased, and the nature of the contamination increased the possibilities of cross-infections to humans who drank the water.

2. The forests may transpire a considerable amount of water, and it is argued that this reduces the outflow (and hence the value to cities) of the watersheds. Much of Melbourne's watersheds has mountain ash (*E. regnans*), a tall, fast-growing native species. This is an even-aged forest that regenerates fast after logging or burning (which kills the old forest). Until recently, it was claimed that these forests had an age-related characteristic water use (Kuczera, 1987). More recently, Benyon *et al.* (2023) showed that major fires in 1939 burnt relatively open mountain ash forest and replaced these with dense thickets of regrowth. This led to a noticeable reduction in the water yield into the dams. The diminished yield is better viewed as a special case of replacing a sparse, mature forest with dense regrowth. A corollary of this is that such actively growing forests are either inappropriate for the watershed or require active thinning if water yield from the watershed is low relative to demand and other water alternatives are not easily available. Thinning operations effectively have a 'time window', as in dense stands only smaller trees can be removed without unacceptable damage to retained trees.
3. To log or not to log? Forest harvesting produces wood products, which are often viewed as 'essential goods' and also produce a substantial financial return to the watershed owners that helps defray the cost of management. In more affluent societies, native-forest logging is often viewed with disfavour, and sometimes with accusations that it leads to excessive erosion or water quality degradation. This was shown by 'The Battle of Bull Run' in Oregon. This is US Forest Service land providing water for the major city of Portland. In 1952, logging commenced in this 25,000 ha forest. This began a battle (described by Larson, 2009, as 'Machiavellian', 'long-running' and 'bitter') between the forces for and against logging. The major argument against logging was the impact on water quality. The US Forest Service cited a need to improve forest resilience. There was an agreement between all parties that any decision to log or not to log must be science based, but there the agreement ended. Larson (2009) noted that 'scientific data supporting one side's position was summarily rejected by the other as inconclusive or incorrect'. Finally, in 1996, logging, mining and public recreation in the watershed was banned by a US federal law. Since then, the watershed has been 'protected' (i.e.no tree cutting) by the city of Portland (City of Portland, 2024). The conflict illustrates the basic conservatism of affluent communities when it comes to their watersheds.

14.4 'Closed' or 'Open' Watersheds: Is It an Issue?

A 'closed watershed' is one in which the only land use allowed is water harvesting and in which entry to the watershed is limited to management personnel. In their purest form, closed watersheds are found only in the uncommon circumstance of the managing authority owning or controlling large tracts of land and having the power to exclude people. The City of Melbourne has a substantial area of closed watershed, and Portland's 'Bull Run' watershed is arguably closed. Around the world, municipal watersheds are effectively 'closed' by having restrictions on public access (e.g. fences, locked gates, signs noting that 'Trespassers will be prosecuted'). Notwithstanding this *de facto* closure, the municipality may have no 'proclaimed policy' on the matter. Having a closed watershed does have the following advantages:

1. It may avoid pollution by uncontrolled human activity.
2. The concept sometimes gains (strong) support from distant city dwellers as 'protecting 'their' water supply'.
3. The management authority avoids scrutiny of its actions because the public cannot get in to see. This can be particularly useful if poor land management, noxious weeds, erosion, animal culling, forest harvesting, unsightly construction work or unhealthy forests are involved.

Equally, the concept has a number of drawbacks:

1. It alienates local residents who view forest usufructs as being 'stolen' from them, or 'the closure' as a device to keep them out of 'their forests'. Often the closure can be vigorously resisted (by trespassing), and this may require unpopular and costly enforcement.
2. Closed watersheds dislocate transport networks, which then 'go around' the closed watersheds, imposing loads on other areas (e.g. Evans, 2005).
3. The management authority bears the full cost of land management including control of weeds and feral animals.
4. It reduces the prospects for 'local ecotourism', particularly as the watersheds may have major scenic attractions such as waterfalls or forest glades but visitors are not allowed.
5. Alienation of the local population makes it hard for the watershed 'owner' to engender goodwill for their activities.

Allowing access to the public for recreation in watershed areas sometimes appears to cross deeply held management tenets, yet limited managed access (e.g. walking tracks, picnic areas) seems to pose little threat, particularly when the water is stored in large reservoirs and the water supply has 'treatment plants'. Such multiple uses add greatly to the valuation of the watersheds by the public. It also breaks the 'lost waters' nexus of diminished access to waterways associated with the development of municipal water supplies articulated by Nathan (2004).

The view of the senior author is that having closed watersheds is an unsophisticated and unrewarding approach that does not maximize the value of the watershed forests to the community.

14.5 Creating Forested Watersheds

Generally, the land use of a municipal watershed is viewed as 'fixed', and the management must do as best they can. But occasionally a vision (or visionary) is persuasive enough to take others with them, and large areas of hitherto agricultural land can be converted to forest. This requires both money and about a century to come to fruition. Land is acquired and planted up as forest, requiring relocation of settlers and even the removal of entire towns. Pires (2004) gave an account of the city of New York and its drive to obtain adequate watershed protection for their watersheds, which include elements of reforestation. A striking example of creating a forested watershed out of what was then agricultural land is that of Quabbin Dam and the Swift River watershed in the USA. Steinberg and Clark (1999) referred to a 'critical water resource narrative' between a powerful and 'superior' place (i.e. the city) and a 'subordinate place' (i.e. the watershed and surrounds) and how the former would extract water from the latter by 'physically uprooting its population, flooding its land, and destroying the livelihoods of its people'. Others (e.g. Baram, 2019) have been kinder to such concepts as 'providing essential infrastructure'. In a children's book (Yolen, 1992), Sally Jane mourns what was lost in flooding the Swift River Valley but ultimately accepts 'the triumph of modernity over her rural community'.

14.5.1 Quabbin Reservoir and Watershed Development (1930–present)

The Quabbin Reservoir drains the Swift River watershed of about 40,000 ha and supplies water to Boston, Massachusetts, and nearby towns. The land underwent major clearing for sheep farming around 1850, but by 1900 much of the land had been abandoned. When the large Quabbin Dam was being constructed (1930–1939), the state government used a mixture of private purchases and compulsory acquisition to acquire most of the watershed; the basic rationale was to obtain the best water quality possible. Over time, this land has been reforested by a combination of natural regeneration and planting, and now forms a large forest reserve. This project has been described as a 'stunning destination for outdoor enthusiasts, surrounded by forested hills, hiking trails and an abundance of wildlife'.

Overall, the watershed now has 87% forest cover and 6% wetland cover, with the balance being agriculture, residential or other uses. The forest lands are neither wilderness nor unplanned but represent years of effort by the city to acquire land in a semi-agricultural watershed as it became available. In recent years, there has been criticism of the mandatory historic land purchases (at 'depression-era prices'), but a doctoral thesis (Egan, 2018) concluded that the residents at the time were objective about the project and some 'took the remarkably selfless view that their eviction was a noble sacrifice for the greater good of the Commonwealth'.

Forest management on the watershed is 'multi-faceted' and includes wildlife habitat management, access control, thinning, harvesting, prescribed fire, regeneration and stand improvement. Silviculture occurs on about 600–800 ha/year, including regeneration, thinning, enrichment plantings and forest tending. For any given stand, the frequency of return cutting is about 15–30 years. No measurable water quality degradation has been associated with the practices by the intensive monitoring program. Wildlife management has become of concern due to increasing awareness of the potential of wild populations of animals and birds to transmit pathogens – particularly *Salmonella*, *Cryptosporidium* and *Giardia* spp. A 'bird harassment' program aimed at modifying nesting behaviour has been effective but time-consuming. Controlled hunting has been used to reduce the deer population. Active visitor management (under the supervision of a Citizen's Advisory Committee) is used to both maintain good public relations and to avoid water supply risks associated with forest recreation.

In 2012, the managing agency announced a relaunch of its watershed forestry programme after a harvesting moratorium (Barten *et al.*, 2012). This programme is aiming to build age and species diversity into the Quabbin Forest. They noted the unease with which forest harvesting was viewed by members of the general public. Barten *et al.* (2012, p.69) commented that 'This heightened attention and concern happened to coincide with the recent transition from relatively innocuous thinning treatments to much less photogenic regeneration treatments. Most people are unpleasantly surprised by any kind of logging, especially when its purpose is unknown, unclear, or misrepresented. Few people have the opportunity to observe forest change over time.'

Recreation at Quabbin has been a popular experience and the conservation value of the watershed resource has been fully recognized. In recent years, there have been calls for the designation of the Quabbin Forest as some sort of National Park, biosphere reserve or 'wildland'. The proposals do not explain how this designation would equal or exceed the efficacy of current management. Such proposals assert that active land management is, at best, unnecessary – that a decision to 'let nature takes its course' is the most prudent way of management. Barten *et al.* (2012) noted that this idea is appealing to the public and many policy makers, but that this *laissez-faire* approach is at variance with the time-tested principles and practices of water supply management, and the risk-management principles that have guided most other areas of economic endeavour.

The Quabbin experience shows that, given political will, large agricultural watersheds can be converted to forested watersheds, with corresponding gains in water supply quality, but that the experience has its own unique issues and trade-offs, requires a large amount of political will and capital, and can only be achieved over many decades.

14.6 Difficult Questions on Forested Municipal Watersheds

14.6.1 Benchmarking watershed forests

Benchmarking is defined as evaluating something by comparison with external criteria (e.g. Berg, 2007; Hanna *et al.*, 2018). For a forested catchment, benchmarking may include comparisons with other watersheds, idealized performance criteria or some completely different water supply approach. It is suggested that achievable benchmarks (water yield, water quality, flow duration, comparison with other approaches, financial viability) be prescribed and that management should attempt to achieve these benchmarks, given their importance to towns and cities. A widespread adoption of benchmarking would allow the introduction of many aspects of forest hydrology science into their management. It would also provide a focus on areas where changed forest management may give methods of improving performance.

14.6.2 Sharing of forested municipal watersheds

For major cities, municipal forested watersheds can be large areas of more-or-less 'natural' land close to their population base. This carries the risk that they are classed as 'conservation reserves', with an unclear question of who is in charge, what the dominant land aims are, and whether conservation or water supply takes priority. It may well be that the 'park' and the 'watershed management' aims are congruent, but it is likely that, in the long term, one will dominate over the other, with a resultant compromise necessary. Difficulties may include whether thinning of dense forests to enhance water yield is permissible, culling of native fauna species to improve bacterial quality of streamflow, and providing active fire-fighting programmes (including fuel reduction burning) and facilities. It is suggested that development of proper benchmarks will help point the way for management to go in the right direction. If cash flow is included in the benchmarking, the revenue generated by water production usually dominates over other uses, reflecting the importance of the watershed to the economic welfare of communities and that water is sold in a cash transaction.

14.6.3 Can afforesting watersheds make them better?

We present forested watersheds as having better properties than agricultural ones, provided the rainfall is adequate to give the required water yields. Suppose one defines the properties of the ideal watershed. Could, then, a program of afforestation transform a problematic watershed into one with desirable properties? Bren (2023) discussed this question in some detail and concluded that it is possible to move towards this ideal, but it would require perhaps a century to bring to fruition. Afforestation is not a panacea, but it can be a useful, self-sustaining tool to achieve 'better' watersheds. The answer to the question is a guarded 'yes'. In many countries, agricultural land will become forested without management action, but the forest may not meet the criteria for watershed management.

14.6.4 Is there a future for natural watersheds in water supply?

As demand for water resources increases, freshwater is becoming scarcer. To offset this scarcity, cities have been turning to alternative sources of water. One source is salty or brackish water, which must be treated via desalination. There are now more than 15,000 desalination plants worldwide, with concentrations in the USA, China, Australia, Europe, North Africa and the Middle East (Jones *et al.*, 2019). While desalination can produce drinking water of high quality, the various technologies and processes involved are energy intensive and expensive (Curto *et al.*, 2021). El Saliby *et al.* (2009) estimated that desalination plants cost five times that of forested watersheds per unit volume of water. Disposal of the saline waste stream gives many problems. If forested municipal watersheds are still the lowest cost option for providing high-quality drinking water, how do we ensure that managers maintain their economic 'edge' in the face of climate and other risks? Answers reasonably include the following:

- Development of knowledge of their performance by measurement of flows with plots and single-watershed measurements, or even paired-watershed experiments, and measurements of water quality. This allows quantitative comparisons with alternatives.
- Forest management to enhance water production, to maintain forest health, to ensure acceptable animal populations, and

to ensure best practice is carried out on the watershed. This might commonly include thinning of dense forests to acceptable forest densities to enhance water yields but should also include skilled monitoring of forest health and water quality, prescribed burning, wild-fire protection and protection from other hazards.

- Achievement of benchmarking goals for water quantity, quality and price. These give an immediate focus on areas that must be improved and allow managers to achieve the best standards of water quantity and quality, and to combine different approaches to give the best economic outcome for the community.

Postel and Thompson (2005) noted that 'healthy watersheds' are essential for the future of cities and presented a 'spectrum of institutional mechanisms' for safeguarding these services, using examples from Ecuador, Costa Rica and New York City. They noted that the loss of natural watershed functions plays an important part in driving water supply costs up and gave examples of a wide variety of institutional mechanisms for watershed protection.

If the rainfalls are low or the water use of the vegetation high, then it is possible that water supply authorities may use options such as desalination plants, groundwater extraction or pipelines to other available resources. This is exactly the situation in the water supply of the city of Perth in Western Australia. The city has had decades of low rainfall blamed on climate change. Thus, a 20% reduction in annual rainfall has translated to a 50% reduction in dam inflows (reflecting the rainfall elasticity) and the city has developed alternative water resources that do not use their historical forested watersheds. Harper *et al.* (2019) suggested that here 'water supply has been decoupled from forests as a result of management and climate change'. Reduced rainfall on the watersheds coupled with dense forests of jarrah (*Eucalyptus marginata*) have led to very low stream flows into reservoirs. Thinning of these forests has been resisted by community groups on the grounds of conservation (e.g. Calver and Wardell-Johnson, 2004). Forested watersheds are now not seen as significant to supplying water to Perth. They may have value for conservation or ecological services, but these do not attract financial revenue streams in the way that water does. Do these watersheds face a future of obsolescence? We suggest that this is an unusual combination of circumstances, but clearly forested watersheds may become obsolete, particularly when rainfall is low. In such a circumstance, it is appropriate that the watersheds are retired to other land uses.

14.6.5 New forms of measurement and hydrological inventory

The past decade has seen a revolution in remote hydrological measurements due to the Global Precipitation Measurement Mission of NASA and other remote-sensing initiatives. New satellites offer access to watershed monitoring never before available, including mapping of rainfall, evaporation and soil moisture, monitoring of forest health and measurement of slope storage by literally 'weighting the watersheds'. The data may well be free and the inventory/measurement is done in the office, so is very economical. This should allow a new age of quantification and monitoring of the functioning and performance (technical and financial) of municipal watersheds.

14.6.6 Who 'owns' the water?

Ownership is the legal possession and control over property (tangible or intangible). The usual legal position is that the world's water is a 'common good' and not owned by anyone. Notwithstanding this, water emanating from watersheds is often 'claimed' as an entitlement by the owners of water supply dams, which appears to have generally been supported by the legal system in most countries, as the need for a water supply is commonly recognized. Occasionally, this will be questioned; thus, Nathan (2004) examined the diversion of water from watersheds outside of Ballarat (see Section 14.3) and the issues associated with this. Nathan noted that 'the process, whereby Ballarat water interests incrementally extended across the watersheds, had social consequences that polarized and entrenched the water conflict in the region' but noted that viewing it as a resource battle between urban recipients and rural donors oversimplified the matter. One can find many examples, big and small, of conflicts over the question of water ownership, but a municipality 'owning' its watersheds (if possible) helps avoid these.

14.6.7 Recognizing the effects of time on forests

Many of the watersheds cited in this chapter have been providing hydrological services for between one and two centuries. The difficulty is that the forests change over time and the newer forests may not have the desirable properties of the older forests for water supply or other uses. In the 'nature taking its course' and/or nature-based solutions approach, forests may age, and trees and stands may senesce, to be replaced by a different type. In Australia, it is likely that a major forest fire will pass across the area; for at least some species, this will kill the mature forest and create a new, even-aged 'regrowth forest'. Other catastrophic events that may reasonably be expected over long periods of time include windthrow and insect attack. Even if nothing happens, the ageing forest may not meet the modern criterion of 'resilience'.

A perfect solution to what should be done in such cases does not exist, but with knowledge of the forest runoff as a function of stand density and time, besides precipitation and soils, and the value of water, the damage could be valued and the 'repair' treated systematically in the same way that ageing and damage to other major assets (e.g. a building or reservoir) would be approached. This would usually include forest remedial work such as thinning or densification, and removal/replacement of particularly vulnerable trees or stands. One might reasonably aim at obtaining a diversity of forest types and age classes such that large contiguous areas of particularly vulnerable forest are not present. It is suggested that community consultation may be appropriate to reduce the potential for conflict.

14.6.8 How might we 'do better' with our forested watersheds?

Given their economic importance, there sometimes seems to be little concern about the performance of municipal watersheds; perhaps this reflects their reliable performance in the past. Taking the watershed as an asset and then 'doing better' might reasonably follow the steps of any asset improvement plan, as follows:

1. Obtain benchmark indicators of the watershed performance.
2. Identify ideal key performance indicators.
3. Define the gap and make an evaluation of deficiencies in performance, which may be converted to financial terms by looking at the 'replacement cost'.
4. Form an asset improvement plan to close the gap.
5. Implement the plan.

Caveats for a forested watershed may be recognition of the time periods involved, that any path will be a compromise between many factors, and that in some cases there may be elements of uncertainty.

The 'procedures' are best viewed as developing 'best management practices' (Pennington *et al.*, 2003; Hubbart, 2021). These could cover many routine operations including road building, forest harvesting, other silvicultural practices, fire protection and various land-use practices. Particular attention might be paid to the 'upper limits' of performance, for example how many people the watershed can serve with water supply into the future.

'Future-gazing' would involve projecting forest growth over one or more centuries, including harvesting, likely urbanization threats or needs, the probability of and the realistic management of forest health (including fire issues), and the impacts of these on reservoir inflows.

14.7 Conclusions

Forested municipal watersheds are of great utility to human society. Because they have been so effective and reliable at producing regulated flows of pure water, particularly in high rainfall areas, they appear not to have had as much intellectual scrutiny on aspects of performance or valuation as they warrant. New remote-sensing tools are likely to change this because of their ability to measure hydroclimatic and ecological variables related to the hydrological cycle. Current forest hydrology knowledge and the development of new hydrology tools offer opportunities for improved hydrological management. The authors believe that we are on the cusp of an exciting new age of quantification, applications of forest hydrology science, and monitoring and modelling to optimize the use of these natural resources and evaluate the impacts they might have from population growth, urbanization and climate change.

References

Andréassian, V. (2004) Waters and forests: From historical controversy to scientific debate. *Journal of Hydrology* 291, 1–27.

Aubertin, G.M. and Patric, J.H. (1974) Water quality after clearcutting a small watershed in West Virginia. *Journal of Environmental Quality* 3, 243–249.

Baram, M. (2019) Resilience and essential public infrastructure. In: Wiig, S. and Fahlbruch, B. (eds) *Exploring Resilience. SpringerBriefs in Applied Sciences and Technology*. Springer, Cham, Switzerland, pp. 33–40.

Barten, P.K. and Ernst, C.E. (2004) Land conservation and watershed management for source protection. *Journal of the American Water Works Association* 96, 121–135.

Barten, P.K., Ashton, M.S., Boyce, J.K., Brooks, R.T., Buonaccorsi, J.P. *et al.* (2012) *Review of the Massachusetts DWSP Watershed Forestry Program*. DWSP Science and Technical Advisory Committee, Massachusetts Division of Water Supply Protection, Department of Conservation and Recreation, Boston, Massachusetts.

Basher, L.R., Hicks, D.M., Clapp, B. and Hewitt, T. (2011) Sediment yield response to large storm events and forest harvesting, Motueka River, New Zealand. *New Zealand Journal of Marine and Freshwater Research* 45, 333–356.

Bennett, A., Haydon, S., Stevens, M. and Coulson, G. (2015) Culling reduces faecal pellet deposition by introduced sambar (*Rusa unicolor*) in a protected water catchment. *Wildlife Society Bulletin* 39, 268–275.

Benyon, R.G., Inbar, A., Sheridan, G.J., Lyell, C.S. and Lane, P. (2023) Variable self-thinning explains hydrological responses to stand replacement in even-aged forests. *Journal of Hydrology* 618, 129–157.

Berg, S.V. (2007) Conflict resolution: Benchmarking water utility performance. *Public Administration and Development* 27, 1–11.

Berlinck, C.N. and Batista, E.K. (2020) Good fire, bad fire: It depends on who burns. *Flora* 268, 151610.

Beschta, R.L. (1978) Long-term patterns of sediment production following road construction and logging in the Oregon Coast Range. *Water Resources Research* 14, 1011–1016.

Binkley, D. and Brown, T.C. (1993) Forest practices as nonpoint sources of pollution in North America. *Journal of the American Water Resources Association* 29, 729–740.

Bolds, A.A., Lockaby, B.G., Ditchkoff, S.S., Smith, M.D. and VerCauteren, K.C. (2021) Impacts of a large invasive mammal on water quality in riparian ecosystems. *Journal of Environmental Quality* 50, 441–453.

Bren, L.J. (1980) The hydrology of a small, forested catchment. *Australian Forestry Research* 10, 39–51.

Bren, L.J. (2023) *Forest Hydrology and Watershed Management: An Australian Perspective*, 2nd edn. Springer, Dordrecht, Netherlands.

Bren, L.J. (2024) A review of scientific and linked media allegations that "Melbourne's water catchments are under threat" from logging in the Thomson catchment, Victoria, Australia. *Science of the Total Environment* 954, 176343.

Bren, L.J. and Hopmans, P. (2007) Paired catchment observations on the water yield of mature eucalypt and immature radiata pine plantations in Victoria, Australia. *Journal of Hydrology* 336, 416–429.

Bren, L.J. and Ryan, M. (2024) An examination of stream water quality data from monitoring of forest harvesting in the eastern highlands of Victoria. *Land* 13, 1217.

Bren, L.J., Lane, P. and Hepworth, G. (2010) Longer-term water use of native eucalyptus forest after logging and regeneration: The Coranderrk Experiment. *Journal of Hydrology* 384, 52–64.

Caldwell, P.V., Martin, K.L., Vose, J.M., Baker, J.S., Warziniack, T.W. *et al.* (2023) Forested watersheds provide the highest water quality among all land cover types, but the benefit of this ecosystem service depends on landscape context. *Science of the Total Environment* 882, 163550.

Calver, M. and Wardell-Johnson, G. (2004) Sustained unsustainability? An evaluation of evidence for a history of overcutting in the Jarrah forests of Western Australia and its consequences for fauna conservation. In: Lunney, D. (ed.) *Conservation of Australia's Forest Fauna*. Royal Zoological Society of New South Wales, Mosman, Australia, pp. 94–114.

Campbell, J.L., Rustad, L.E., Bailey, S.W., Bernhardt, E.S., Driscoll, C.T. *et al.* (2020) Watershed studies at the Hubbard Brook Experimental Forest: Building on a long legacy of research with new approaches and sources of data. *Hydrological Processes* 35, e14016.

Carson, L. and White, S. (1998) Sydney water contamination crisis: Manufacturing dissent. *Scientific and Public Policy* 254, 265–271.

Chappell, N.A., Douglas, I., Hanapi, J.M. and Tych, W. (2004) Sources of suspended sediment within a tropical watershed recovering from selective logging. *Hydrological Processes* 18, 685–701.

Chiew F. (2006) Estimation of rainfall elasticity of streamflow in Australia. *Hydrologic Sciences Journal* 51, 613–625.

City of Portland (2024) *How Bull Run is Protected*. Portland.gov, Portland, Oregon. Available at: www.portland.gov/water/about-portlands-water-system/how-bull-run-protected (accessed 1 May 2025).

Creedy, J. and Wurzbacher, A.D. (2001) The economic value of a forested watershed with timber, water and carbon sequestration benefits. *Ecological Economics* 38, 71–83.

Cunha, D.G.F., Sabogal-Paz, L.P. and Dodds, W.K. (2016) Land use influence on raw surface water quality and treatment costs for drinking supply in Sao Paulo State (Brazil). *Ecological Engineering* 94, 516–524.

Curto, D., Franzitta, V. and Guercio, A. (2021) A review of the water desalination technologies. *Applied Sciences* 11, 670.

del Campo, A.D., González-Sanchis, M., Molina, A.J., García-Prats, A., Ceacero, C.J. *et al.* (2019) Effectiveness of water-oriented thinning in two semiarid forests: The redistribution of increased net rainfall into soil water, drainage and runoff. *Forest Ecology and Management* 438, 163–175.

Dissmeyer, G.E. (2000) *Drinking Water from Forests and Grasslands*. Southern Research Station, USDA Forest Service, Ashville, North Carolina.

Dragovich, D., Webb, A.A. and Jamshidi, R. (2012) Suspended sediment yield following wildfires in a mixed species eucalypt forest, Southeastern Australia. In: Stone, M., A., C. and M, T. (eds) *Wildfire and Water Quality: Processes, Impacts and Challenges*. IAHS Publication No.354. IAHS, Wallingford, UK, pp. 17–24.

Duffy, C., O'Donoghue, C., Ryan, M., Kilclinec, K., Upton, V. *et al.* (2020) The impact of forestry as a land use on water quality outcomes: An integrated analysis. *Forest Policy and Economics* 116, 102185.

Egan, J. (2018) Watershed decisions: An environmental history of Boston's Quabbin Reservoir 1870–1988. PhD thesis, University of Connecticut, Storrs, Connecticut.

El Saliby, I., Okour, Y., Shon, H.K., Kandasamy, J. and Kim, I.S. (2009) Desalination plants in Australia, review and facts. *Desalination* 247, 1–14.

English, C.J., Younger, S.E., Cannon, J.B., Brantley, S.T., Markewitz, D. *et al.* (2024) Forest management for water yield: Assessing the barriers and impacts of privately-owned open pine woodlands in the Southeastern United States. *Trees, Forests and People* 17, 100600.

Ervinia, A., Huang, J., Huang, Y. and Lin, J. (2019) Coupled effects of climate variability and land use pattern on surface water quality: An elasticity perspective and watershed health indicators. *Science of the Total Environment* 693, 133592.

Evans, P. (2005) The Great Wall of China: Catchment policy and the forests beyond the Yarra Watershed 1850–1950. In: Calver, M., Bigler-Cole, H., Bolton, G., Gaynor, A., Horwitz, P. *et al.* (eds) *A Forest Consciousness: Proceedings of the Sixth National Conference of the Australian Forest History Society*, Millpress Science Publishers, Rotterdam, Netherlands, pp. 613–622.

Francis, C., Ruiz, J., Rodríguez, M.A., Blais, B. and Campeau, S. (2017) Landscape diversity and forest edge density regulate stream water quality in agricultural catchments. *Ecological Indicators* 72, 627–639.

Gomi, T., Moore, R.D. and Hassan, M.A. (2005) Suspended sediment dynamics in small forest streams of the Pacific Northwest 1. *Journal of the American Water Resources Association* 41, 877–898.

Hampton, J.O. and Davis, N.E. (2020) Impacts of introduced deer in Victoria. *Victorian Naturalist* 137, 276–281.

Hanna, S.M., Thompson, M.J., Dahab, M.F., Williams, R.E. and Dvorak, B.I. (2018) Benchmarking the energy intensity of small water resource recovery facilities. *Water Environment Research* 90, 738–747.

Harper, R., Smettem, K.R.J., Ruprecht, J.K., Dell, B. and Liu, N. (2019) Forest–water interactions in the changing environment of south-Western Australia. *Annals of Forest Science* 76, 95.

Hewlett, J.D. and Hibbert, A.R. (1963) Moisture and energy conditions within a sloping soil mass during drainage. *Journal of Geophysical Research* 68, 1081–1087.

Hornbeck, J.W., Martin, C.W., Pierce, R.S., Bormann, F.H., Likens, G.E. *et al.* (1986) Clearcutting northern hardwoods: Effects on hydrologic and nutrient ion budgets. *Forest Science* 32, 667–686.

Hornbeck, J.W., Martin, C.W., Pierce, R.S., Bormann, F.H., Likens, G.E. *et al.* (1987) *The Northern Hardwood Forest Ecosystem: Ten Years of Recovery from Clearcutting*. US Department of Agriculture, Forest Service, Northeastern Forest Experiment Station, Broomall, Pennsylvania.

Hubbart, J.A. (2021) Improving best management practice decisions in mixed land use and/or municipal watersheds: Should approaches be standardized. *Land* 10, 1402.

Ide, J.I., Kume, T., Wakiyama, Y., Higashi, N., Chiwa, M. *et al.* (2009) Estimation of annual suspended sediment yield from a Japanese cypress (*Chamaecyparis obtusa*) plantation considering antecedent rainfalls. *Forest Ecology and Management* 257, 1955–1965.

Jackson, R., Krishna, K.B., Li, M., Sathasivan, S. and Senevirathna, L. (2024) The influence of recent bushfires on water quality and the operation of water purification systems in regional NSW. *Scientific Reports* 14, 16222.

Jones, E., Qadir, M., Vliet, M.T., Smakhtin, V. and Kang, S.M. (2019) The state of desalination and brine production: A global outlook. *Science of the Total Environment* 657, 1343–1356.

Kittredge, J. (1948) *The Effects of Woody Vegetation on Climate, Water and Soil, With Applications to the Conservation of Water and the Control of Floods and Erosion*. McGraw Hill, New York.

Kuczera, G. (1987) Prediction of water yield reductions following a bushfire in ash-mixed species eucalypt forest. *Journal of Hydrology* 94, 215–236.

Larson, D.W. (2009) The battle of bull run. *American Scientist* 97, 182–184.

Lee, R.D., Symons, J.M. and Robeck, G.C. (1970) *Watershed human use levels and water quality.* American Water Works Association, Denver, Colorado.

Lee, R.M., McGuire, K.J., Strahm, B.D., Knoepp, J.D., Jackson, C.R. *et al.* (2020) Revisiting the Hewlett and Hibbert (1963) hillslope drainage experiment and modelling effects of decadal pedogenic processes and leaky soil boundary conditions. *Water Resources Research* 56, e2019WR025090.

Likens, G.E., Bormann, F.H., Johnson, N.M., Fisher, D.W. and Pierce, R.S. (1970) Effects of forest cutting and herbicide treatment on nutrient budgets in the Hubbard Brook watershed-ecosystem. *Ecological Monographs* 40, 23–47.

Liu, N., Dobb, G.R., Caldwell, P.V., Miniat, C.F., Sun, G. *et al.* (2022) Quantifying the Role of the National Forest System and Other Forested Lands in Providing Surface Drinking Water Supply for the Conterminous United States. General Technical Report WO-100. Quilcene, Washington: US Forest Service.

Macdonald, J.S., Beaudry, P.G., MacIsaac, E.A. and Herunter, H.E. (2003) The effects of forest harvesting and best management practices on streamflow and suspended sediment concentrations during snowmelt in headwater streams in sub-boreal forests of british Columbia, Canada. *Canadian Journal of Forest Research* 33, 1397–1407.

McGinnis, S. and Kerans, B.L. (2013) Land use and host community characteristics as predictors of disease risk. *Landscape Ecology* 28, 29–44.

McNamara, J.P., Tetzlaff, D., Bishop, K., Soulsby, C., Seyfried, M. *et al.* (2011) Storage as a metric of catchment comparison. *Hydrological Processes* 25, 3364–3371.

Miller, R.S., Sweeney, S.J., Slootmaker, C., Grear, D.A., Di Salvo, P.A. *et al.* (2017) Cross-species transmission potential between wild pigs, livestock, poultry, wildlife, and humans: Implications for disease risk management in North America. *Scientific Reports* 7, 7821.

Moghim, S. (2020) Assessment of water storage changes using GRACE and GLDAS. *Water Resources Management* 34, 685–697.

Nanda, A. and Safeeq, M. (2024) Nonlinear storage–discharge dynamics of forested headwater catchment: A hysteresis index approach. *Hydrological Processes* 38, e15201.

Nathan, E. (2004) Lost waters of West Moorabool: A history of a community and its catchment. PhD thesis, School of Anthropology, Geography and Environmental Studies, The University of Melbourne, Parkville, Australia.

Neary, D.G., Ice, G.C. and Jackson, C.R. (2009) Linkages between forest soils and water quality and quantity. *Forest Ecology and Management* 258, 2269–2281.
Nippgen, F., McGlynn, B.L., Emanuel, R.E. and Vose, J.M. (2016) Watershed memory at the Coweeta Hydrologic Laboratory: The effect of past precipitation and storage on hydrologic response. *Water Resources Research* 52, 1673–1695.
Nyman, P., Yeates, P., Langhans, C., Noske, P.J., Peleg, N. *et al.* (2021) Probability and consequence of postfire erosion for treatability of water in an unfiltered supply system. *Water Resources Research* 57, 2019WR026185.
Pennington, S.R., Kaplowitz, M.D. and Witter, S.G. (2003) Re-examining best management practices for improving water quality in urban watersheds. *Journal of the American Water Resources Association* 39, 1027–1041.
Peterson, T.J., Saft, M., Peel, M.C. and John, A. (2021) Watersheds may not recover from drought. *Science* 372, 745–749.
Pires, M. (2004) Watershed protection for a world city: The case of New York. *Land Use Policy* 21, 161–175.
Postel, S.L. and Thompson, B.H., Jr (2005) Watershed protection: Capturing the benefits of nature's water supply services. *Natural Resources Forum* 292, 98–108.
Price, J.I. and Heberling, M.T. (2018) The effects of source water quality on drinking water treatment costs: A review and synthesis of empirical literature. *Ecological Economics* 151, 195–209.
Sayama, T., McDonnell, J.J., Dhakal, A. and Sullivan, K. (2011) How much water can a watershed store? *Hydrological Processes* 25, 3899–3908.
Schollaert, C.L., Jung, J., Wilkins, J., Alvarado, E., Baumgartner, J. *et al.* (2024) Quantifying the smoke-related public health trade-offs of forest management. *Nature Sustainability* 7, 130–139.
Scott, D.F. and Lesch, W. (1997) Streamflow responses to afforestation with *Eucalyptus grandis* and *Pinus patula* and to felling in the Mokobulaan experimental catchments, South Africa. *Journal of Hydrology* 199, 360–377.
Sedell, J., Sharpe, M., Apple, D.D., Copenhagen, M. and Furniss, M. (2000) *Water and the Forest Service*. Publication FS-660. US Department of Agriculture, Forest Service.
Soltani, S.S., Ataie-Ashtiani, B. and Simmons, C.T. (2021) Review of assimilating GRACE terrestrial water storage data into hydrological models: Advances, challenges and opportunities. *Earth-Science Reviews* 213, 103487.
Steinberg, P. and Clark, G. (1999) Troubled water? Acquiescence, conflict, and the politics of place in watershed management. *Political Geography* 18, 477–508.
Taylor, C., Blair, D., Keith, H. and Lindenmayer, D. (2019) Modelling water yields in response to logging and representative climate futures. *Science of the Total Environment* 688, 890–902.
Vardon, M., Keith, H. and Lindenmayer, D. (2019) Accounting and valuing the ecosystem services related to water supply in the Central Highlands of Victoria, Australia. *Ecosystem Services* 39, 101004.
Warziniack, T., Sham, C.H., Morgan, R. and Feferholtz, Y. (2017) Effect of forest cover on water treatment costs. *Water Economics and Policy* 3, 1750006.
Webb, A.A. and Kathuria, A. (2012) Response of streamflow to afforestation and thinning at Red Hill, Murray Darling Basin, Australia. *Journal of Hydrology* 412, 133–140.
Webb, A.A., Dragovich, D. and Jamshidi, R. (2012) Temporary increases in suspended sediment yields following selective eucalypt forest harvesting. *Forest Ecology and Management* 283, 96–105.
Yolen, J. (1992) *Letting Swift River Go*. Little Brown Books, New York.
Zhang, L., Dawes, W.R. and Walker, G.R. (2001) Response of mean annual evapotranspiration to vegetation changes at catchment scale. *Water Resources Research* 37, 701–708.
Zimmermann, A., Francke, T. and Elsenbeer, H. (2012) Forests and erosion: Insights from a study of suspended-sediment dynamics in an overland flow-prone rainforest catchment. *Journal of Hydrology* 428, 170–181.

15 Forest Hydrology Modeling: Processes, Methods and Example Models

H.E. Golden[1*], D.M. Amatya[2] and G. Sun[3]

[1]Office of Research and Development, Center for Environmental Measurement and Modeling, US Environmental Protection Agency, Cincinnati, Ohio, USA; [2]Santee Experimental Forest, Southern Research Station, USDA Forest Service, Cordesville, South Carolina, USA; [3]Eastern Forest Environmental Threat Assessment Center, USDA Forest Service, Research Triangle Park, North Carolina, USA

Abstract

This chapter provides a brief overview of why and how forest hydrological models are applied to answer important global research and management questions. It steps through physical functions that most forest hydrological models represent, how to best select the model for a particular goal, and how to think about model parameterization and calibration. Finally, a suite of example forest hydrology models focused on different model domains (i.e. plots, watersheds, groundwater and coupled surface–subsurface) are discussed.

15.1 Introduction

Characterizing and quantifying interactions among components of the forest hydrological cycle is complex and usually requires a combination of field monitoring and modeling approaches (Weiler and McDonnell, 2004; National Research Council, 2008). In the past decade, integrated watershed management approaches have required improved understanding of soil, vegetation, and water quality and quantity responses to disturbances such as wildland fires (Hallema *et al.*, 2018; Beyene *et al.*, 2021; Paul *et al.*, 2022; Povak *et al.*, 2022) and climate change (Zema and Lucas-Borja, 2022; Zhang *et al.*, 2022), as well as practices such as urban forestry (Department of Environmental Resources, 1999), green infrastructure (Ercolani *et al.*, 2018; Coville *et al.*, 2020), forest thinning (Lundquist *et al.*, 2013; Sun *et al.*, 2022) and reforestation (Li *et al.*, 2018; Peña-Arancibia *et al.*, 2019). Modern forest hydrological models are important tools for testing related hypotheses, understanding hydrological processes, filling knowledge gaps and synthesizing forest experimental data (Sun *et al.*, 2011; Komatsu and Kume, 2020). Sun *et al.* (2023) provided a recent, concise, integrative overview of the status of forest hydrological modeling.

A well-calibrated and verified model that incorporates the general principles of forest hydrology can supplement field measurements (e.g. Hydrograph Separation Program, HYSEP; Sloto and Crouse, 1996; Barlow *et al.*, 2015). In turn, these measurements can provide data to improve a model and its performance, particularly when long-term data become available (Tian *et al.*, 2012a; b). Forest hydrology models

*Corresponding author: golden.heather@epa.gov

DOI: 10.1079/9781800625310.0015

can also simulate water quantity and quality in forested watersheds (i.e. catchments) and larger river basin scales with limited recorded measurements, such as stream discharge (Sivapalan, 2003), groundwater levels and evapotranspiration (ET) (Sun *et al.*, 2011). Many forest hydrology models can also quantify forest biogeochemical cycling as well as surface-water quality in watersheds (DeWalle, 2003; Amatya *et al.*, 2013; Nelitz *et al.*, 2013), and all in response to modern pressures such as climate change and urbanization.

However, improved models that project the hydrological responses to forest management practices are needed (Amatya *et al.*, 2011; Vose *et al.*, 2011; Tian *et al.*, 2012b; Komatsu and Kume, 2020). This requires a suite of approaches, incorporating decades of research, on the response of processes regulating hydrological dynamics to forest management (Jones *et al.*, 2009; Buttle, 2011; Tian *et al.*, 2012b). Forest hydrology models are a necessity to project beyond current hydrological conditions and from young stands to forests with full canopied watersheds (e.g. quantifying the effects of forest site preparation, forest growth and silvicultural techniques on the hydrological cycle). This involves simulating changes in components of the forest or the forest watershed's water balances, including runoff, ET, snow accumulation/melt, melting permafrost and the cumulative effects of these changes on stream, river and lake processes (Beckers *et al.*, 2009; Sun *et al.*, 2011; Tian *et al.*, 2012a). Projecting such shifts in the forest hydrological cycle requires numerical modeling methods because of their ability to conduct time-stepped simulations of specific hydrological processes within a given domain and to scale data to broader spatial extents using physically or process-based approaches (National Research Council, 2007).

This chapter provides a brief overview of forest hydrological modeling for answering important global research and management questions. Many hundreds of hydrological models have been applied globally across multiple decades to represent and predict forest hydrological processes (Beckers *et al.*, 2009; Amatya *et al.*, 2013; Nelitz *et al.*, 2013), and these have been comprehensively discussed in a recent review paper (Sun *et al.*, 2023). The focus of this chapter is to highlight a sampling of these process-based models and approaches. We specifically focus on 'forest hydrology models': physically based simulation tools that quantify compartments of the forest hydrological cycle. Physically based models can be considered those that describe the conservation of mass, momentum and/or energy (Beckers *et al.*, 2009). While we provide minimal emphasis on empirical modeling methods (those that use measured data to derive relationships between variables, e.g. linear regression models), these approaches can be embedded within physically based models.

We focus on the watershed scale, which includes multiple drainage areas ranging across various orders of magnitude (e.g. 0.1–1000 km^2), based on Golden *et al.* (2014) and consistent with Wei and Zhang (2011). Temporal scales of each model are associated with the time step the modeler selects to solve the governing equations within the model based on the specific processes of interest, typically hourly for streamflow hydrograph predictions and flow rates in urban settings, daily for most hydrological applications, monthly for forested and/or large-scale ecosystem models, and annually for the transient groundwater flow models.

15.2 Model Functionality and Complexity

Forest hydrology models can range in both functionality and complexity. Each model's functionality and complexity results, in part, from the hydrological processes embedded in the model's structure, the mathematical equations representing these processes, and how the spatial extent of the model domain is discretized (Beckers *et al.*, 2009). These models also include the interactions among processes and across scales (Clark *et al.*, 2017). Most models simulate, at minimum, a basic water balance that includes moisture inputs (e.g. rainfall, snow and/or snowmelt) and water flux outputs via ET including canopy and soil/understory evaporation and runoff as a combination of surface and subsurface flows (Fig. 15.1).

How the water balance and water fluxes are simulated and what are required as model inputs vary widely based on how models describe the hydrological processes at different spatial/temporal scales. Typically, many models consider

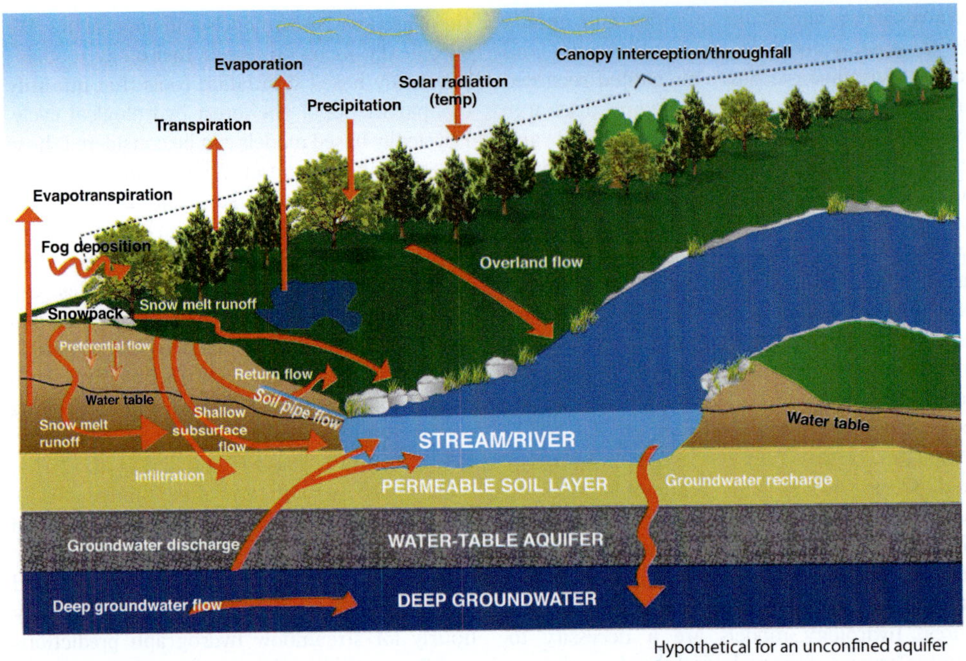

Fig. 15.1. Processes that can be modeled with forest hydrology simulation tools.

the forest ecosystem as mature (e.g. static), while other select models (e.g. DRAINMOD-FOREST; Tian *et al.*, 2012a) explicitly simulate forest physiological and phenological dynamics and how these dynamics affect a forest stand's water balance. Some models simulate the interactions among the soil, vegetation and atmosphere that affect the soil moisture dynamics and water-use efficiency of vegetation in forests (e.g. PnET-N-DNDC simulations of N_2O and NO emissions from forest soils; Li *et al.*, 2000; Stange *et al.*, 2000). Simulated outputs are diverse across models as well, but generally include peak flow, low flow, total streamflow/water yield, ET, and/or changes in soil moisture or groundwater table elevations over time.

15.2.1 Rainfall–runoff functions

The initiation of overland flow or subsurface flow in models that include rainfall–runoff dynamics on hillslope and river channels (i.e. 'watershed models') occurs when a threshold soil moisture level, such as soil field capacity, is reached. Each model calculates a threshold value and runoff-generating processes differently. Forest hydrology models will typically represent one (but sometimes more than one) of these runoff-generating processes, which might include variable source area (VSA) dynamics (e.g. TOPMODEL, Beven and Kirkby, 1979), the USDA Natural Resources Conservation Service curve number method (e.g. Soil and Water Assessment Tool, Neitsch *et al.*, 2011), Green-Ampt infiltration processes (e.g. Hydrologic Simulation ProgramFORTRAN (HSPF)), Hooghoudt's equation for subsurface drainage and Kirkham's equation for ponding conditions for shallow water table and drainage rates (e.g. DRAINMOD-FOREST, Tian *et al.*, 2012a), soil moisture response function (e.g. Visualizing Ecosystems for Land Management Assessment (VELMA), Abdelnour *et al.*, 2011), three-dimensional finite difference or linear reservoir approach for saturated zones, the two-dimensional Richard's equation or two-layer water balance models (e.g. MIKE SHE, Graham and Butts, 2005; Distributed Hydrology Soil Vegetation Model (DHSVM), Wigmosta *et al.*, 1994) or soil moisture balance

(Water Supply Stress Index (WaSSI), Sun *et al.*, 2011) approaches, depending on the temporal scale of simulation. Additional details on these processes are covered in associated chapters in this volume.

For forest hydrology models that explicitly simulate rainfall–runoff processes, runoff may be initiated in the model of interest via infiltration-excess runoff (Horton, 1933), saturation-excess overland flow (Dunne and Black, 1970), including VSA dynamics, or a combination of both overland flow types. Subsurface storm flow (Hursh and Brater, 1941), including preferential flows, may also be implemented in the model's water balance routine, as well as return flows (i.e. flow that travels through the shallow subsurface before reissuing to the land surface). Surface depressional storage capacity, Manning's overland surface runoff coefficient, land slope, and other landscape and surface/litter vegetation characteristics are then used to store or route the excess rainfall after soil saturation to the nearest stream.

15.2.2 Evapotranspiration functions

To effectively model forest hydrological responses to changes in climate, management, fires and other land cover practices, ET must be simulated. Evapotranspiration of a forest ecosystem is the sum of canopy, soil, water litter and understory vegetation surface evaporation, as well as stand transpiration, and is a major component of the forest hydrological cycle (Sun *et al.*, 2016; Komatsu and Kume, 2020; Sun *et al.*, 2023). Plot and watershed models typically estimate subdaily, daily or monthly potential (as the upper limit of ET; Jovanovic and Israel, 2012) via one (or more) of three basic ET formulas: empirical Hamon (Hamon, 1963), energy-based Priestley–Taylor (Priestley and Taylor, 1972) and/or the fully process-based Penman–Monteith (McNaughton and Black, 1973), in order of their complexity. Recently, some traditionally non-forest watershed-scale models have used the ET process and ET-related parameters embedded in these formulations to expand representations of the forest hydrological cycle across commonly used watershed models (e.g. Soil and Water Assessment Tool (SWAT),Haas *et al.*, 2022a, b). Komatsu and Kume (2020) have provided a thorough background of how forest hydrologists have modeled ET through time.

15.2.3 Model parameterization

Depending on the study or management objectives, the complexity of the model and the forest composition, the number and breadth of parameters required to simulate key processes may vary substantially. In addition to several previously mentioned soil and runoff related parameters, parameters related to simulation of forest ET may include rooting depth and distribution, leaf area index, canopy density, canopy structure and interception capacity, stomatal or canopy conductance, and other biophysical characteristics including those that represent the understory type/species (additional details are included Chapter 3, this volume). A recent review of current hydrological models indicated that ET is not well quantified for forested watersheds due to lack of data related to vegetation-specific parameters (Sun *et al.*, 2023). Overall, a key modeling challenge is the availability and quality of spatially explicit data to help put values on model parameters (Clark *et al.*, 2017).

Anthropogenic processes may also be incorporated into the functionality of many forest hydrological models. Some models can project variations in hydrological processes in response to climate change, management scenarios, wildfire, insect and disease outbreaks, and shifts in land cover/land use. For example, studies have applied modeling approaches to project the effects of forest harvesting and management on peak flows in British Columbia (e.g. Whitaker *et al.*, 2003; Thyer *et al.*, 2004), portions of South America (Bathurst *et al.*, 2011; Birkinshaw *et al.*, 2011), China (Sun *et al.*, 2006; Cao *et al.*, 2021), north-western USA (Stednick, 2008; Abdelnour *et al.*, 2011; Yan *et al.*, 2021) and forest management practices in California (Cristea *et al.*, 2014), among others. In a synthesis study, Amatya *et al.* (2013) outlined eight criteria for an ideal forest hydrology model that can describe the impacts of forest fertilization on southern US forested landscapes.

15.2.4 Balancing model functionality and complexity

Model functionality and complexity go hand in hand: typically, the greater number of functions the model simulates, the more complex the model (Golden *et al.*, 2017). Forest hydrology models can vary considerably in complexity from simple empirical models (not discussed here) to process-based models that cover a range of low (Forest Hydrology Model (ForHyM), Arp and Yin, 1992) to medium (VELMA, Abdelnour *et al.*, 2011) to highly complex (e.g. HydroGeoSphere, Brunner and Simmons, 2012; MIKE SHE, Graham and Butts, 2005) hydrological representations: that is, from simple bucket-type models to models that implement multiple soil–water interactions and transport processes. Model complexity can also vary with spatial scale: highly complex and computationally intensive models often function best at finer spatial scales; as the spatial scale expands, the resolution of the modeled system necessarily needs to coarsen to decrease computational demands.

Forest hydrology process-based studies coupled with modeling approaches have most commonly been approached from a small watershed scale (National Research Council, 2008) using paired-watershed approaches (Von Stackelberg *et al.*, 2007) starting as early as 1909–1928 in North America (Bates and Henry, 1928). However, more recent studies have expanded their spatial scales toward the scale of management (e.g. large watersheds, regions, nations, globally) and have used generalizing principles derived from finer-scale studies.

Models can be discretized in different ways based on the model structure, the spatial scale of interest, and the management or research question. Typically, models are spatially discretized in one of three ways: lumped, semi-distributed or fully distributed. Spatially lumped models represent the model domain as one unit, without any heterogeneity in its characteristics (e.g. all soils and land cover are the same throughout the domain). Semi-distributed models carve the watershed into similarly functioning units, such as hydrological response units (similar combinations of land cover, soils and slope) (Neitsch *et al.*, 2011), field management units, subbasins or finite difference grids. Fully distributed models have varying parameters and processes across the model domain (Kampf and Burges, 2007; Arnold *et al.*, 2016). Forest hydrology models can also vary temporally, with some operating on a continuous time step (e.g. daily, monthly and/or annually).

15.3 Model Selection

A forest hydrology model is a simplification of reality. This is an important consideration when selecting the appropriate model for the forest hydrological management and/or research question being addressed. The current state of the science remains limited on insights into choosing the most appropriate spatial resolution to represent the hydrological processes of a specific system. Of utmost importance is the development of a perceptual hydrological model of the study area (i.e. a conceptualization of the forested watershed and its hydrological processes) (McMillan *et al.*, 2023), based on tying together information provided by spatial data based on remote sensing (e.g. light detection and ranging (LiDAR), Landsat, moderate resolution imagery spectroradiometer (MODIS)), a Geographic Information System (GIS), monitoring (e.g. precipitation, stream flow, snowpack depths), past modeling efforts and professional knowledge. This will help to: (i) determine the most important hydrological processes of the study area; (ii) select a model that can simulate these dominant processes; and (iii) determine whether the simplifying hydrological assumptions in the chosen model (e.g. spatial discretization – breaking the model into grids or subbasins or another spatial unit – and temporal resolution) is valid for the system (Golden *et al.*, 2017).

When selecting a model, various factors should be considered. For example, a lumped parameter model might be appropriate (compared with a spatially explicit model) where spatial heterogeneity is low or the spatial scale of the study area is broad (e.g. region, national), or there is a combination of the two. Model selection must also consider the management or research questions, the hydrological processes important to those questions and what future projections need to be simulated, such as climate change or forest management scenarios that vary in complexity.

Practical considerations for choosing a forest hydrology model include the modeling objective, input data needs and parameter availability, computational time and cost–benefits of model complexity. For example, most watershed-based forest hydrology models require an accurate digital elevation model and stream network layers as base data for watershed delineation and drainage network configuration, and spatial data on soils and land cover, in addition to measurements of hydrometeorological variables (e.g. precipitation, temperature), ET (e.g. eddy covariance flux, remote sensing), snowpack depths, water-table variations and streamflow measurements at the watershed outlet.

Depending on the study question, additional information may be needed. For example, water-level data from groundwater wells, piezometers and other surface-water features in the watershed (e.g. wetlands, lakes, dams) may be required to parameterize the model and quantify the full water balance. Furthermore, model set-up, implementation and spin-up (the initial period taken for the model to equilibrate) times, as well as the skill level required to execute the model, all increase with model complexity. Therefore, considering the balance of benefits associated with minimizing model uncertainty versus the increased computational intensity costs associated with added model complexity – including spatial, temporal and process complexity – is imperative (Dehotin and Braud, 2008; Clark *et al.*, 2015, 2017).

15.4 Model Diagnostics and Evaluation

Model evaluation needs to be conducted to determine whether a forest hydrology model is characterizing the system appropriately. In the most general sense, model calibration is a process by which model parameters are adjusted so that the simulated model output matches an observed set of data within a predefined acceptable range. Model parameters are often fitted using measured data or calibrated within reasonable ranges, as determined by the system and/or literature values when data are lacking. However, careful consideration of equifinality (Beven, 1993), which describes the process of arriving at the same simulated model output using a variety of different model parameter sets or structures – without knowing which one might be closest to 'reality' – is important.

Traditionally, watershed models are calibrated by varying parameter values so that simulated streamflow closely matches observed stream gauge data at the watershed outlet. However, models can be improved by also calibrating to spatially distributed observed water table depths, soil moisture, ET and other water balance components. In forest hydrology models that incorporate plant growth components, model calibration to measured leaf area index, total biomass, rooting depth and/or ecosystem productivity, alongside hydrological variables, will also improve the model.

Parameter estimation programs (e.g. PEST, Doherty and Johnston, 2003; OSTRICH, Matott, 2005) that generate best-fit parameter sets to match simulated results to observed data and assess accuracy based on one or more objective functions (optimization; Price *et al.*, 2012) should be selected to generate the best-fit parameter set to match simulated results to observed data (Boyle *et al.*, 2003). A multi-objective framework reduces the problems associated with calibrating to local objective function minima and avoids subjectivity and information loss in model acceptability criteria by simultaneously minimizing observed and simulated differences of multiple functions (Doherty and Johnston, 2003; Flerchinger *et al.*, 2012). Arnold *et al.* (2016) recommended a diagnostic approach that looks at signature patterns of behavior in the model outputs to determine which processes, and thus parameters, need further adjustment during calibration. In a companion study, Malone *et al.* (2015) developed parameterization guidelines and considerations for hydrological models. Additionally, McMillan *et al.* (2017) and Yan *et al.* (2023) provided guidelines for hydrological signatures that could be evaluated in this process.

Validation of the model outputs traditionally suggests the successful testing of simulated outputs against observed data using an input data set different from that of the calibration data set. A split-sample approach is one popular example whereby calibration and validation are conducted during a sequential set of years: one continuous set is used for calibration, the other for validation (Klemes, 1986). More recent work detailed obstacles involved in using split-sample approaches and argued for applying full data sets for calibration only, rather than splitting them with model validation (Arsenault *et al.*, 2018).

15.4.1 Uncertainty and sensitivity analysis

Uncertainties in forest hydrology models must be accounted for in some capacity. Model uncertainties can take the form of parameter uncertainties, input data uncertainties, process uncertainties and predictive uncertainties, among others. For example, McMillan *et al.* (2018) categorized hydrological model uncertainties into measurement, derived data, interpolation, scaling, and data management uncertainties. Whatever the focus is, the first step in uncertainty analysis typically involves quantifying simulation output uncertainty by propagating uncertainties throughout the model and generating a probabilistic distribution of simulated outputs. How uncertainties should be handled in hydrological modeling is a debate that has continued for decades (Matott *et al.*, 2009; Pianosi and Wagener, 2016; Herrera *et al.*, 2022), and new developments, such as those in the realm of Bayesian frameworks for uncertainty analyses, have widened the debate (Tasdighi *et al.*, 2018).

Beven and Young (2013) suggested that uncertainties in hydrological models can be aleatory (irreducible) or epistemic (reducible) in nature. Aleatory uncertainties are random and can be treated probabilistically in the model, while epistemic errors are associated with current lack of knowledge of processes operating within the system. Regardless of the form of uncertainties in the model, it is appropriate to detail the assumptions underlying these uncertainties and quantify them, where appropriate and feasible. Sensitivity analysis is one way of estimating the output uncertainties caused by changes in values model parameters. Sensitivity analysis can determine which parameters assert the most quantifiable control over model outputs, that is, the analysis can quantify which model parameters produce a change in simulated outputs based on incremental changes in a parameter values.

15.5 Examples of Forest Hydrology Models

We grouped our example forest hydrology models into three categories, each based on the model's focal domain: (i) plot and watershed models; (ii) groundwater models; and (iii) coupled surface–subsurface hydrology models. Stand-level models are a specific type of plot-scale model. While many models, with varying discretization approaches, can be applied to questions in forest hydrology (Fig. 15.2), here we expound on a handful of these.

Watershed models – also called catchment models or basin-scale models – refer to physically based models that simulate processes contributing to the forest hydrology water balance (e.g. surface and subsurface runoff, ET), including overland and channel routing to a surface-water system. Watershed models use topographically defined watersheds as boundaries and simulate surface and shallow subsurface processes. Unlike groundwater models, watershed models quantify groundwater as part of the watershed's water balance; the groundwater flow equation is not explicitly solved. Therefore, the deep groundwater system is considered a hydrological 'sink.' The contributions of lentic surface-water features such as lakes, ponds and wetlands to the water balance are also often not explicitly simulated.

Some watershed models exemplified here focus heavily on ecosystem processes, such as forest growth and productivity (e.g. DHSVM), while others are more hydrologically focused (e.g. TOPMODEL) without forest stand dynamics included. All watershed models exemplified here include some means of streamflow routing – from conceptual (e.g. all excess runoff is transported to the subbasin outlet to simple time of concentration) to more complex kinematic wave routing, so that streamflow at the watershed outlet can be simulated.

Groundwater models focus on solving groundwater flow equations specifically. Coupled surface–subsurface models (e.g. GSFLOW)

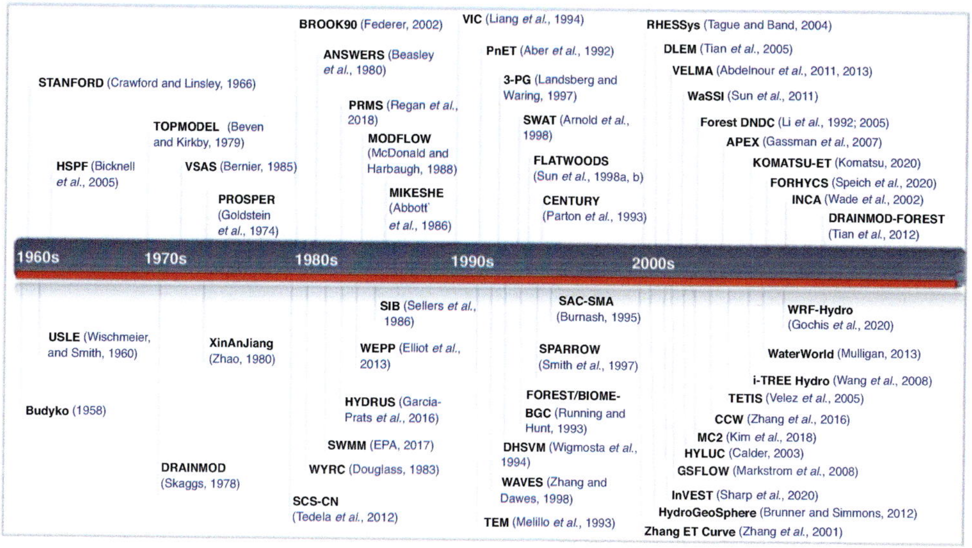

Fig. 15.2. A host of models can be applied to questions related to forest hydrology. We focus on a sample of these models. Modified from Sun *et al.* (2023).

are typically highly complex and involve both process-based surface and subsurface hydrological processes.

15.5.1 Plot and watershed models

PnET-BGC (Gbondo-Tugbawa *et al.*, 2001) and CENTURY (Parton *et al.*, 1993; Parton, 1996) are plot-scale models that simulate forest hydrological processes across a forest stand. PnET is a lumped parameter, monthly or daily time step, and stand-level (more generally called plot-scale) model that quantifies carbon and water dynamics in mature forests. Hydrological processes simulated by the PnET (photosynthetic/ET) model include canopy interception, plant transpiration, macropore flow, lateral flow and deep percolation to the aquifer. CENTURY is a plot-scale terrestrial biogeochemical model that operates at a monthly time step (Parton *et al.*, 1993; Parton, 1996). The model is composed of linked submodels representing forest production, grassland and crop production, soil organic matter and water balance. The simplified water balance submodel simulates monthly evaporation, transpiration, soil water content, snow water content and saturated flow between soil layers.

Several examples of watershed models that can be applied to forest hydrology include ForHyM (Arp and Yin, 1992; Meng *et al.*, 1995), TOPMODEL (Beven and Kirkby, 1979), iTree Hydro (Wang *et al.*, 2008), VELMA (Abdelnour *et al.*, 2011, 2013), Agricultural Policy/Environmental eXtender (APEX, Williams and Izaurralde, 2005; Gassman *et al.*, 2007), Precipitation-Runoff Modeling System (PRMS, Leavesley *et al.*, 1983, 2005), DHSVM (Wigmosta *et al.*, 1994, 2002), BROOK90 (Federer *et al.*, 2003), Variable Infiltration Capacity (VIC, Liang *et al.*, 1994, 1996), Integrated Nitrogen Catchment Model (INCA, Wade *et al.*, 2002), Forest-BGC (Running and Gower, 1991), Regional Hydro-Ecological Simulation System (RHESSys, Band *et al.*, 1993; Tague and Band, 2004) and WaSSI (Sun *et al.*, 2011; Caldwell *et al.*, 2012; Sun *et al.*, 2015). A brief synopsis of each is included below, and additional details are given in Table 15.1.

TOPMODEL is a semi-physically based flexible watershed hydrological mass balance modeling tool that is particularly robust in forested watersheds with shallow soils (Beven and Kirkby, 1979). Flow routing in TOPMODEL is driven by VSA dynamics and includes both saturated flow and infiltration-excess overland flow.

Table 15.1. Example plot-scale and watershed models for forest hydrology applications.

Model[a]	Hydrological approach	Simulation outputs: hydrology	Time steps	Spatial scale	Level of complexity[b]	Appropriate regions of application	Model files publicly accessible	Available online user manual and website[c]
iTree-Hydro	Six main routines for rainfall runoff processes: interception, impervious surface, soils, evaporation and transpiration, routing and pollution; uses time-area delay function or one-parameter diffusion-based exponential function for constructing downstream hydrograph	Streamflow, water balance	Daily	Multi-scale watersheds and plots (i.e. city or parcel)	Medium	Multiple: can be applied to watersheds with different rainfall–runoff mechanisms; recent cold region module development (version 2; Yang *et al.*, 2011)	Yes, by request	Yes; www.itreetools.org/hydro/
PnET (all)	Lumped-parameterized one-dimensional water balance model from canopy to soil	ET, watershed runoff	Daily to monthly	Plot to regional	Low	All forest ecosystem, both upland and lowland	Yes	Yes; www.pnet.sr.unh.edu/
CENTURY	Simplified water balance incorporating evapotranspiration, soil water content and saturated flow	Water routing from landscape with carbon, nitrogen, phosphorus and sulfur loadings	Monthly	Plot	Low	Temperate and tropical forests	Yes	Yes; www.nrel.colostate.edu/projects/century/
TOPMODEL	Semi-distributed rainfall-runoff model; variable source area dynamics but simulates both saturated excess and infiltration excess overland flows; assumes water table follows topography	Various water balance components including streamflow, overland flow (saturated and infiltration excess), subsurface flow, and return flow	Daily	Multi-scale watersheds	Medium	Multiple: typically best in systems with moderate to steep topography and shallow soils	Yes, as an ArcGIS extension, R Statistical Package code or the original code	Yes, in the dynamic TOPMODEL R statistical version: https://cran.r-project.org/web/packages/dynatop/index.html

VELMA	Spatially distributed ecohydrological model; simulates daily infiltration, evapotranspiration and surface/subsurface runoff through four soil layers	Various water balance components including streamflow, overland flow (saturated and infiltration excess), and subsurface flows through four user-defined soil layers	Daily	Mult -scale watersheds	Medium	Multiple: developed in small forested watersheds of Pacific north-west USA but applied/tested elsewhere	Yes	www.epa.gov/water-research/visualizing-ecosystem-land-management-assessments-velma-model
APEX	Curve number method (five options) and Green-Ampt infiltration (four options), subsurface drainage, and five options for PET for ET estimates	Precipitation, snowfall, interception, surface runoff, subsurface flow, snowmelt, irrigation, total water yield, PET, ET	Daily, monthly or annual	Field/small watershed, grid-based	Medium	Upland agricultural and forested fields or watersheds	Yes	Yes; https://epicapex.tamu.edu/ and http://epicapex.tamu.edu/apex/
PRMS	Semi-distributed processed-based rainfall–runoff model for multiple spatial and temporal scales	Various water balance components including evaporation, transpiration, runoff and infiltration, and considers the interaction with the forest/plant canopy, snowpack dynamics and soil hydrological processes	Daily to centuries	Multi-scale watersheds	Medium	Multiple	Yes	Yes; www.usgs.gov/software/precipitation-runoff-modeling-system-prms

Continued

Table 15.1. Continued

Model[a]	Hydrological approach	Simulation outputs: hydrology	Time steps	Spatial scale	Level of complexity[b]	Appropriate regions of application	Model files publicly accessible	Available online user manual and website[c]
DHSVM	Saturation-excess infiltration; Darcy's law for unsaturated and kinematic for saturated subsurface	Surface runoff, snow and snowmelt, soil moisture and ET, streamflow	Subdaily	Watershed	Medium	Applied worldwide across various climate regimes, although originally developed for mountainous terrains	Yes	Yes; www.pnnl.gov/projects/distributed-hydrology-soil-vegetation-model; GITHUB: https://github.com/pnnl/DHSVM-PNNL
BROOK90	One-dimensional water balance with canopy and multi-layer soil profile	Soil water content, overland flow, bypass flow from soil layers, groundwater flow	Daily	Plot	Low	Applied worldwide although designed for forests within north-east USA	Yes	Yes; www.ecoshift.net/brook/brook90.htm
VIC	Grid-cell based water balance with canopy, and three-layer soil profile	Watershed water balance, streamflow with accompanying routing model	Daily to monthly	Regional, global	Medium	Any, as intended to accompany large-scale general circulation models	Yes	https://vic.readthedocs.io/en/master/
INCA	Semi-distributed water balance with canopy, soil and riparian components	Streamflow, watershed water balance	Daily	Watershed	Medium	No limitations; applied extensively in Europe	No, commercial product	www.watres.com/software/INCA/
SWAT	Semi-distributed processed-based rainfall-runoff model for multiple spatial and temporal scales	Streamflow, watershed water balance	Daily	Watershed	Medium	Originally for agricultural watersheds but recent developments to improve forest parameterization	Yes	https://swat.tamu.edu/

WEPP and GeoWEPP	Distributed rainfall runoff and erosion simulation model	Runoff from field and plots, sediment delivery	Multiple	Hillslope and small watersheds (<2.6 km^2)	Medium	Managed forests, roads, wildfire areas	Yes	www.nrcs.usda.gov/resources/tech-tools/water-erosion-prediction-project, https://fargo.nserl.purdue.edu/geowepp/ and https://doc.wepp.cloud/
FOREST-BGC	One-dimensional water balance with forest canopy and soil surface	Outflow, ET	Daily	Watershed to regional	Medium	Any forest watersheds	Yes	https://daac.ornl.gov/OTTER/guides/Runnings_Forest_BGC_Model.html
BIOME BGC	Ecosystem model simulating hydrological across multiple scales based on one-dimensional water balance concepts	ET, soil water, snow depth, outflow	Daily	Multiple scale	Medium	Terrestrial ecosystem consisting of one or multiple biomes	Yes	www.umt.edu/numerical-terradynamic-simulation-group/project/biome-bgc.php
RHESSys	Semi-distributed water balance incorporating TOPMODEL, CENTURY, DHSVM and BIOME BGC	Streamflow, carbon, and nutrient dynamics at watershed outlet and within model components	Daily	Watershed	Medium	Mountainous watersheds	Yes	Yes; http://fiesta.bren.ucsb.edu/~rhessys/
WaSSI	Distributed by HUC-12 watersheds at the continental scale	Water yield, ET, ecosystem productivity, water supply stress index	Monthly	Watershed	Low	Continental	Yes	https://web.wassiweb.fs.usda.gov/

[a]See text for definitions of model names.
[b]Complexity levels refer to the balance of spatial and temporal model discretization, parameterization, and computational burden of the model. 'High complexity' means all complexity sources are demanding.
[c]All accessed 13 May 2025.
ET, evapotranspiration; PET, potential evapotranspiration.

iTree Hydro (previously called UFORE-Hydro; Wang *et al.*, 2008) is a physically based, semi-distributed urban forestry hydrological model that simulates runoff volume and water quality across different urban land covers. Simulations in iTree Hydro are at a daily time step and can operate at multiple watershed or plot (i.e. city, parcel) scales. A user can simulate the effects of various urban impervious and vegetation cover scenarios on the urban forest water balance, including interception, ET, infiltration and runoff.

The VELMA model is a fully distributed (i.e. it captures spatial variability in physical properties of a watershed) ecohydrological model initially developed for forested watersheds, particularly in the Pacific north-west USA (Abdelnour *et al.*, 2011; Abdelnour *et al.*, 2013). VELMA can simulate multiple parts of the forest hydrological cycle (e.g. daily infiltration and redistribution, ET, surface and subsurface runoff) using a four-layer soil column structure.

The semi-distributed APEX model (Williams and Izaurralde, 2005; Gassman *et al.*, 2007) was developed to evaluate land management impacts of hydrology, water and soil quality, and vegetation growth and competition in upland watersheds. The forestry version includes rainfall interception by canopy/litter, silvicultural practices and subsurface flow that includes deep percolation and lateral seepage using storage routing and pipe flow equations (Saleh *et al.*, 2004; Williams and Izaurralde, 2005).

PRMS is a semi-distributed processed-based watershed model that simulates components of the water balance, including evaporation, transpiration, runoff and infiltration, and quantifies interactions with the forest/plant canopy, snowpack dynamics and soil hydrological processes (Leavesley *et al.*, 1983; Leavesley *et al.*, 2005). PRMS has been applied across many landscape types and broad spatial scales. At broader spatial scales, PRMS is often calibrated in forested headwaters.

BROOK90 is a one-dimensional process-based hydrologic model that operates on a daily time step and was originally developed for forested watersheds in north-eastern USA (Federer *et al.*, 2003). The model includes components for interception by a single-layer canopy, snow accumulation and melt, direct evaporation from soil and snow, transpiration from a single-layer canopy and multi-layered soil, and multi-layered soil water movement.

DHSVM is a watershed-scale hydrological model that operates at subdaily time steps (Wigmosta *et al.*, 1994; Wigmosta *et al.*, 2002; Perkins *et al.*, 2019). The model is composed of seven modules representing ET, snowpack accumulation and melting, canopy snow interception and release, unsaturated subsurface flow, saturated subsurface flow, surface overland flow and channel flow. DHSVM is frequently applied to evaluate forest management hydrological effects across a variety of physiographical settings (Storck *et al.*, 1998; Bowling and Lettenmaier, 2001; Cristea *et al.*, 2014; Yan *et al.*, 2021; Sun *et al.*, 2022).

The VIC model is a macroscale hydrological model that operates at daily to monthly time steps; it complements global-scale general circulation models used for climate simulations and weather prediction (Liang *et al.*, 1994; Liang *et al.*, 1996). The model includes forest ET, canopy storage, surface and surface runoff, aerodynamic flux, and snow accumulation and melt.

INCA is a semi-distributed process-based watershed model that operates at a daily time step and is popularly used in Western European forested watershed studies (Whitehead *et al.*, 1998a, b; Wade *et al.*, 2002). The INCA hydrological module simulates soil moisture, storage and evaporation, topographic impacts on flow, and streamflow, and can be applied to assess the effects of forest management on watershed-scale hydrology and biogeochemical cycling.

SWAT, a widely used watershed-scale, semi-distributed model (Arnold *et al.*, 1998) originally developed for upland agricultural landscapes, has been tested, modified and updated for its application to large landscapes containing large portions of forest lands (von Stackelberg *et al.*, 2007; Watson *et al.*, 2009; Parajuli, 2010; Amatya and Jha, 2011), and has recently been improved for forested systems (Haas *et al.*, 2022a, b).

The Watershed Erosion Prediction Project (WEPP; Foster and Lane, 1987), its geospatial interface GeoWEPP (Renschler, 2003), and the most recent watershed-scale version (Brooks *et al.*, 2016; Lew *et al.*, 2022) focus on simulating the water balance and sediment delivery in hillslope and small watersheds ($<36\,km^2$), areas within managed forests, along forests with roads and in wildfire areas, as well as rangelands and agricultural systems.

The FOREST-BGC model (Running and Gower, 1991) and its successors, such as the BIOME-BGC model (White *et al.*, 2000) and other BGC family models, are process-based, watershed-scale ecosystem-focused models that can be spatially aggregated and averaged to a per-unit area basis. FOREST-BGC's water balance is simulated at a daily time step and includes evaporation, transpiration, rainfall interception, throughfall, soil moisture, snow water equivalent depth and soil outflow of water.

RHESSys is a semi-distributed hydrological model that operates at a daily time step and is used to simulate mountainous watersheds (Band *et al.*, 1993; Tague and Band, 2004). The hydrological component of the model simulates atmospheric processes and soil hydrological and transport processes including vertical seepage, soil evaporation, lateral flow, and canopy radiative and moisture processes.

Finally, the WaSSI model is a relatively low-complexity, integrated, lumped parameter (limited spatial variations in physical characteristics within a subwatershed), semi-process-based model that solves water balances at broad spatial scales (Fig. 15.3) (Sun *et al.*, 2011, 2015; Caldwell *et al.*, 2012; Liu *et al.*, 2022). It operates on a monthly time step and simulates the full monthly water balance (ET, streamflow, ecosystem productivity and soil moisture storage) for each land cover class at a user-defined watershed scale.

15.5.2 Groundwater models

Groundwater models explicitly simulate groundwater table dynamics and flows, and are needed

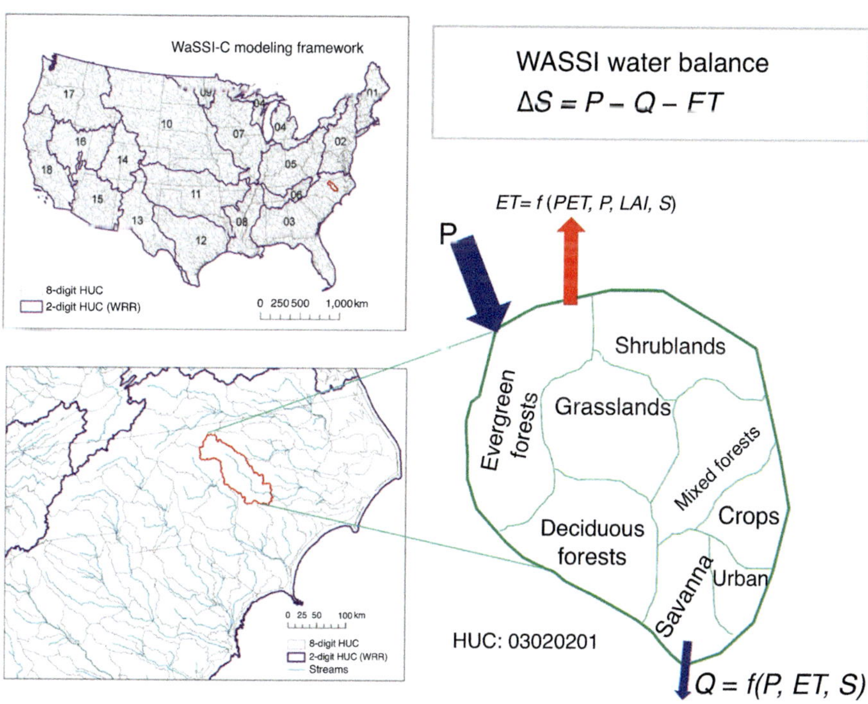

Fig. 15.3. WaSSI: an example of a relatively low-complexity, large-scale ecosystem model that can be applied to forest hydrology management and research questions. *ET*, evapotranspiration; HUC, hydrologic unit code (a watershed identifier); *LAI*, leaf area index; *P*, precipitation; *PET*, potential evapotranspiration; *Q*, watershed runoff; Δ*S*, change in water storage at the watershed scale; WRR, water resource region,

to answer research and management questions related to the movement and transport of subsurface flows and solutes through saturated porous media. Groundwater models are typically bounded by deep subsurface flow networks that reach across multiple watershed boundaries and use the groundwater flow equation (i.e. mass balance and Darcy's law) to estimate deep groundwater transport, which is based on relationships among hydraulic conductivity, specific yield, hydraulic gradient, fluid flow rates and the contributing model domain area. Surface-water flows and features (e.g. lakes, ponds, wetlands, streams, rivers) are not explicitly modeled in groundwater simulations and are considered boundary conditions. Two example groundwater models that could be used for simulating forest hydrological systems with a strong groundwater component are MIKE SHE (Abbott *et al.*, 1986a, b) and DRAINMOD-FOREST (Tian *et al.*, 2012a) (Table 15.2).

The MIKE SHE model (Abbott *et al.*, 1986a, b) is a physically based, fully distributed watershed-scale hydrological modeling system (without streamflow routing unless linked with the MIKE11 module) that was designed to describe the full hydrological cycle in a watershed. The model simulates the hydrological processes of canopy interception, soil evaporation, transpiration, infiltration, overland flow, unsaturated flow in soils, groundwater flow in aquifers and channel flows in rivers.

DRAINMOD-FOREST (Tian *et al.*, 2012a) is a field-scale, fully process-based and integrated model for simulating hydrology, soil carbon and nitrogen cycles, and vegetation growth in lowland forests under various climate conditions and silvicultural practices. Hydrological processes in DRAINMOD-FOREST are simulated on a daily or hourly basis and include ET, rainfall interception, infiltration, subsurface drainage, surface runoff, deep seepage and soil water dynamics in the unsaturated zone.

15.5.3 Coupled surface–subsurface hydrology models

Coupled surface–subsurface hydrology models are highly complex modeling systems that link surface and groundwater models by dividing surface and subsurface flow into different regions or domains, and solve the governing equations in each region using iterative solution methods (e.g. Markstrom *et al.*, 2008) or simultaneously solve the governing equations for surface and subsurface flows (e.g. Panday and Huyakorn, 2004). These models consider feedback among various components of the surface and subsurface water balances (e.g. runoff, groundwater flows and ET) and are thus extremely complex and computationally arduous. Two examples of such models that can be used to address forest hydrological management and research-related questions are HydroGeoSphere (Therrien *et al.*, 2010; Brunner and Simmons, 2012) and GSFLOW (Markstrom *et al.*, 2008) (Table 15.3).

HydroGeoSphere is a fully integrated, physically based numerical model that simulates, at a variety of time steps, coupled-surface (in two dimensions) and subsurface (in three dimensions) hydrological processes so that all primary components of the hydrological cycle are modeled (i.e. overland flow, streamflow, evaporation, transpiration, groundwater recharge, subsurface discharge into surface-water bodies) (Therrien *et al.*, 2010; Brunner and Simmons, 2012; Cornelissen *et al.*, 2014; de Schepper *et al.*, 2015).

GSFLOW is a high-complexity coupled surface–subsurface hydrological model that operates at a daily time step (Fig. 15.4) (Markstrom *et al.*, 2008). The model integrates the surface-water PRMS (Leavesley *et al.*, 1983; Leavesley and Stannard, 1995) and the Modular Groundwater Flow Model (MODFLOW) (Harbaugh *et al.*, 2000; Harbaugh, 2005). PRMS (Table 15.1) simulates land-surface hydrologic processes in ET, runoff, infiltration and interflow, plant canopy interception and storage, and snowpack. MODFLOW is a popular grid-based groundwater flow model that simulates three-dimensional saturated groundwater flow and storage, one-dimensional unsaturated flow, and groundwater interaction with streams.

15.5.4 Emerging modeling methods

Machine learning (ML) is a subfield of artificial intelligence that enables computers to recognize patterns and make predictions, and deep

Table 15.2. Examples of groundwater models for forest hydrology applications.

Model	Hydrological approach	Simulation outputs: hydrology	Time steps	Spatial scale	Level of complexity	Appropriate regions of application	Model files publicly accessible	Available online user manual and website
MIKE SHE	Fully distributed	ET, streamflow, soil moisture content, water table level	Subdaily, daily	Variable	High	All	No; this is a commercial product	Yes; http://www.mikepoweredbydhi.com/products/mike-she (accessed 13 may 2025)
DRAINMOD-FOREST	One-dimensional water balance of canopy, the soil surface and along a soil column	ET, outflow, soil water, seepage, water table level	Subdaily, daily	Field scale	Medium	Lowland	Yes	No

Complexity levels refer to the balance of spatial and temporal model discretization, parameterization and thereby computational burden of the model. High complexity means all complexity sources are demanding.
ET, evapotranspiration.

Table 15.3. Examples of coupled surface–subsurface models for forest hydrology applications.

Model	Hydrological approach	Simulation outputs: hydrology	Time steps	Spatial scales	Level of complexity	Appropriate regions of application	Model files publicly accessible	Available online user manual and website
HydroGeoSphere	Surface domain represented as two-dimensional overland flow; subsurface represented as three-dimensional unsaturated/saturated flow	All primary components of the hydrological cycle are modeled (i.e.overland flow, streamflow, evaporation, transpiration, groundwater recharge, subsurface discharge to surface-water bodies)	Daily to centuries	Multi-scale domains (e.g.plots, watersheds)	High	Multiple	No, this is a commercial product	No, but information can be found at: www.aquanty.com/hydrogeosphere/ (accessed 13 May 2025)
GSFLOW	Coupled surface and groundwater model; PRMS for surface; finite difference equation for subsurface	HRU water balance, streamflow, groundwater dynamics	Daily	Watershed, regional	High	Any	Yes	Yes; http://wwwbrr.cr.usgs.gov/projects/SW_MoWS/GSFLOW.html (accessed 13 May 2025)

Complexity levels refer to the balance of spatial and temporal model discretization, parameterization and thereby computational burden of the model. High complexity means all complexity sources are demanding.
HRU, hydrological response unit; PRMS, Precipitation-Runoff Modeling System.

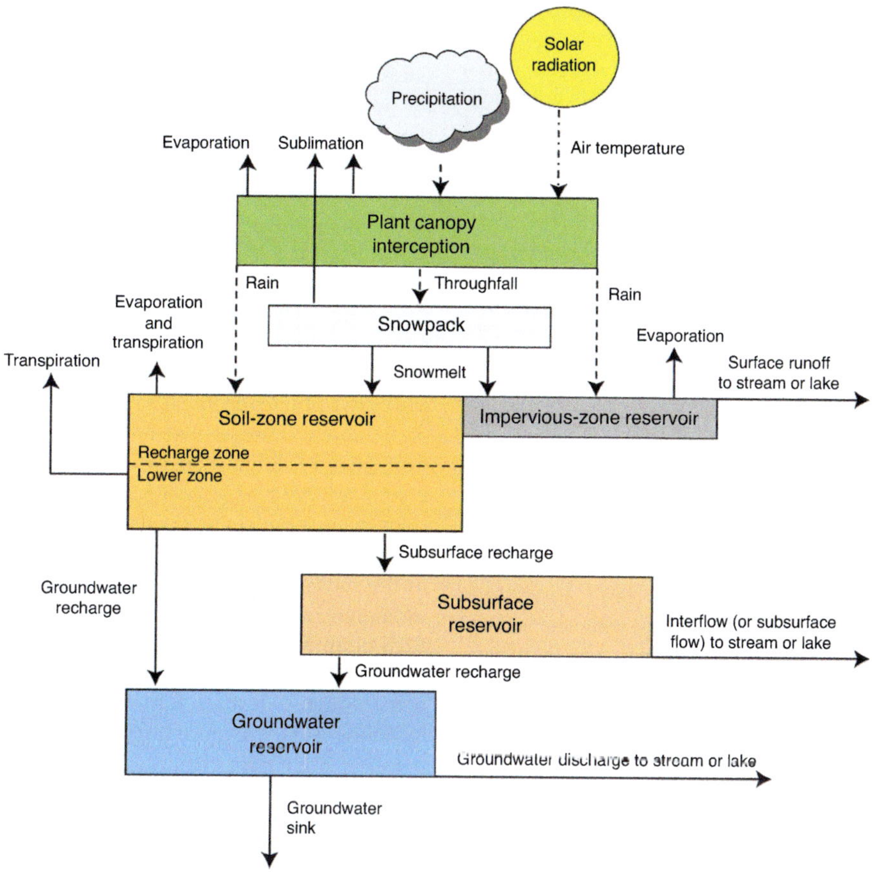

Fig. 15.4. GSFLOW model structure: an example of the complex coupled surface-subsurface modeling system. From Markstrom *et al.* (2008), with permission; S. Markstrom, US Geological Survey, personal communication (2015).

learning (DL) (LeCun *et al.*, 2015) is a type of ML that applies neural networks with multiple layers to understand data patterns. While ML has gained popularity in multiple disciplines in the past decades, applications in hydrology span less than a decade (Lange and Sippel, 2020). The specific use of DL in hydrology has been even more recent (Shen, 2018; Shen *et al.*, 2018; Gauch *et al.*, 2021). However, DL is on a rapid upward trajectory in the hydrological sciences (Shen and Lawson, 2021).

Several ML modeling approaches have recently been applied to test hypotheses and pose questions regarding forest hydrological processes. For example, Schmidt *et al.* (2020) applied a combined neural network approach to estimate forest soil water content across space and time. Random forest applications have also been applied to estimate annual runoff in response to forest losses (Li *et al.*, 2020), to predict inundation dynamics and hydroperiods (Riley and Stillwell, 2023), to understand woody wetland loss due to sea-level rise (He *et al.*, 2023), to track forest soil moisture dynamics (in comparison with other ML methods; de Oliveira *et al.*, 2021) and to transfer output, via a random forest-based tool, from a process-based model focused on forest thinning impacts on snowpack to other nearby forested watersheds (Krogh *et al.*, 2020).

The prediction capacity of ML models often matches or goes beyond that of process-based models (Kratzert *et al.*, 2019). The downside is that they provide lower interpretability regarding forest hydrological processes. To improve the interpretability of these modeling methods, process-guided DL has been introduced (Willard *et al.*, 2022). In this approach, the processes embedded in a physically based model can 'guide' the ML or DL model toward better process understanding for the modeler. Relatedly, explainable deep learning is another method to better understand drivers and patterns in DL models (Zhi *et al.*, 2024), such as that used to discover mechanisms causing flood patterns across the USA (Jiang *et al.*, 2022). Furthermore, emerging fusions of process-based models with DL methods, such as differentiable process-based models (Feng *et al.*, 2022; Shen *et al.*, 2023) and neural ordinary differential equations (Höge *et al.*, 2022), integrate the power of DL into process-based models for hydrological predictions.

Another major development for forest hydrological modeling has been integrating remotely sensed (satellite and airborne; optical or sensor) data into process-based forest hydrology models to better calibrate and 'capture' forest hydrological processes and related parameters, such as ET, leaf area index, canopy leaf temperature and albedo. Remote sensing enables viewing of large land expanses instantaneously and integrating heterogeneous landscapes at a range of spatial resolutions (Sass and Creed, 2011). Use of data such as remotely sensed ET from MODIS (Mu *et al.*, 2011) can be directly integrated into watershed ecohydrology models to improve their accuracy (Rajib *et al.*, 2018).

Overall, the rapid development of supercomputing, accessible 'big' spatial and temporal data stored – and being computed – across multiple cloud-based platforms, and remote-sensing technology is changing how we think about, and model, forest hydrological systems. However, distributed watershed forest hydrological modeling, which allows detailed representations of forest hydrological processes, has not kept pace: the developments have remained stable for over a decade since 2010 (Sun *et al.*, 2023). Therefore, given recent technological advances in data availability and computing platforms, there is still much room for growth in efficient forest hydrological modeling, potentially with reduced prediction uncertainties.

15.6 Conclusions

Forest hydrology models are important tools for developing a clearer understanding of the dominant hydrological processes of a forest stand or watershed (i.e. catchmen) and the process-based hydrological responses to future forest impacts, such as silvicultural practices, implementation of management activities and climate change, on water resources and other ecosystem services. These models can vary widely in complexity; therefore, clarity regarding research or management questions in addition to the conceptual hydrological model of the forest stand or watershed is imperative for model selection. Model evaluation, including uncertainty and sensitivity analyses, is a primary approach to determine whether hydrological processes of interest and/or importance in the modeled system are well characterized, ultimately reporting output variables of interest with quantified prediction uncertainties. With technological and high-speed computing developments in recent years, future forest hydrology modeling work will move further toward incorporating innovative remote sensing, ML, and geophysical and biogeochemical methods for improved parameterization and process understanding. Furthermore, empirical methods (e.g. tracer and isotopic studies for hydrograph separation) and statistical approaches should continue to be integrated into mechanistic modeling structures. Finally, the development of new, simplified, but physically based models might be most appropriate in some forested systems (Sidle *et al.*, 2011), improving our general understanding and guiding forest management.

15.7 Disclaimer

The views expressed in this chapter are those of the authors and do not necessarily represent the views or policies of the US Environmental Protection Agency and the US Department of Agriculture Forest Service.

References

Abbott, M.B., Bathurst, J.C., Cunge, J.A., O'Connell, P.E. and Rasmussen, J. (1986a) An introduction to the European Hydrological System – Systeme Hydrologique European, "SHE", 1: History and philosophy of a physically-based, distributed modelling system. *Journal of Hydrology* 87, 45–59.

Abbott, M.B., Bathurst, J.C., Cunge, J.A., O'Connell, P.E. and Rasmussen, J. (1986b) An introduction to the European Hydrological System – Systeme Hydrologique European, "SHE", 2: Structure of a physically-based, distributed modelling system. *Journal of Hydrology* 87, 61–77.

Abdelnour, A., Stieglitz, M., Pan, F. and McKane, R. (2011) Catchment hydrological responses to forest harvest amount and spatial pattern. *Water Resources Research* 47, e2024WR039854.

Abdelnour, A., McKane, R., Stieglitz, M., Pan, F. and Cheng, Y. (2013) Effects of harvest on carbon and nitrogen dynamics in a pacific-northwest forest catchment. *Water Resources Research* 49, 1292–1313.

Aber, J.D. and Federer, C.A. (1992) A generalized, lumped-parameter model of photosynthesis, evapotranspiration and net primary production in temperate and boreal forest ecosystems. *Oecologia* 92, 463–474.

Amatya, D.M. and Jha, M.K. (2011) Evaluating SWAT model for a low gradient forested watershed in coastal South Carolina. *Transactions of the ASABE* 54, 2151–2163.

Amatya, D.M., Douglas-Mankin, K.R., Williams, T.M., Skaggs, R.W. and Nettles, J.E. (2011) Advances in forest hydrology: Challenges and opportunities. *Transactions of the ASABE* 54, 2049–2056.

Amatya, D.M., Rossi, C.G., Dai, Z., Williams, R., Saleh, A. *et al.* (2013) Review of nitrogen fate models applicable to forest landscapes in the Southern US. *Transactions of the ASABE* 56, 1731–1757.

Arnold, J.G., Srinivasan, R., Muttiah, R.S. and Williams, J.R. (1998) Large-area hydrological modeling and assessment: Model development. *Journal of the American Water Resources Association* 34, 73–89.

Arnold, J.G., Youssef, M.A., Yen, H., White, M.J., Sheshukov, A. *et al.* (2016) Hydrological processes and model representation: Impact of soft data on calibration. *Transactions of the ASABE* 58, 1637–1660.

Arp, P.A. and Yin, X. (1992) Predicting water fluxes through forests from monthly precipitation and mean monthly air temperature records. *Canadian Journal of Forest Research* 22, 864–877.

Arsenault, R., Brissette, F. and Martel, J.-L. (2018) The hazards of split-sample validation in hydrological model calibration. *Journal of Hydrology* 566, 346–362.

Band, L.E., Patterson, P., Nemani, R. and Running, S.W. (1993) Forest ecosystem processes at the watershed scale: Incorporating hillslope hydrology. *Agricultural and Forest Meteorology* 63, 93–126.

Barlow, P.M., Cunningham, W.L., Zhai, T. and Gray, M. (2015) *Groundwater Toolbox, A Graphical and Mapping Interface for Analysis of Hydrologic Data (Version 1.0) – User Guide for Estimation of Base Flow, Runoff, and Groundwater Recharge from Streamflow Data. Techniques and Methods, Book 3, Chapter B10.* US Geological Survey, US Department of Interior, Reston, Virginia.

Bates, C.G. and Henry, A.J. (1928) *Forest and Stream-flow Experiment at Wagon Wheel Gap, Colorado.* US Weather Bureau Monthly Weather Review, Supplement No. 30.

Bathurst, J.C., Birkinshaw, S.J., Cisneros, F., Fallas, J., Iroumé, A. *et al.* (2011) Forest impact on floods due to extreme rainfall and snowmelt in four Latin American Environments 2: Model analysis. *Journal of Hydrology* 400, 292–304.

Beasley, D.B., Huggins, L.F. and Monke, A. (1980) ANSWERS: A model for watershed planning. *Transactions of the ASAE* 23, 938–944.

Beckers, J., Smerdon, B. and Wilson, M. (2009) *Review of Hydrologic Forest Models for Forest Management and Climate Change Applications in British Columbia and Alberta.* FORREX Series Report No. 25. FORREX Forum for Research and Extension in Natural Resources Society, Kamloops, British Columbia, Canada.

Bernier, P.Y. (1985) Variable source areas and storm-flow generation: An update of the concept and a simulation effort. *Journal of Hydrology* 79, 195–213.

Beven, K. (1993) Prophecy, reality and uncertainty in distributed hydrological modeling. *Advances in Water Resources* 16, 41–51.

Beven, K. and Kirkby, M. (1979) A physically-based, variable contributing area model of basin hydrology. *Hydrological Sciences Bulletin* 24, 43–69.

Beven, K. and Young, P. (2013) A guide to good practice in modeling semantics for authors and referees. *Water Resources Research* 49, 5092–5098.

Beyene, M.T., Leibowitz, S.G. and Pennino, M.J. (2021) Parsing weather variability and wildfire effects on the post-fire changes in daily stream flows: A quantile-based statistical approach and its application. *Water Resources Research* 57, e2020WR028029.

Bicknell, B.R., Imhoff, J.C., Kittle, J.L., Jobes, T.H. and Donigian, A.S. (2005) *Hydrological Simulation Program-Fortran: HSPF Version 12.2 User's Manual*. Environmental Research Laboratory Office of Research and Development, US Environmental Protection Agency, Athens, Georgia.

Birkinshaw, S.J., Bathurst, J.C., Iroumé, A. and Palacios, H. (2011) The effect of forest cover on peak flow and sediment discharge – An integrated field and modelling study in central–Southern Chile. *Hydrological Processes* 25, 1284–1297.

Bowling, L.C. and Lettenmaier, D.P. (2001) The effects of forest roads and harvest on catchment hydrology in a mountainous maritime environment. In: Wigmosta, M.S. and Burges, S.J. (eds) *Land Use and Watersheds: Human Influence on Hydrology and Geomorphology in Urban and Forest Areas*. American Geophysical Union, Washington, D.C, pp. 145–164. DOI: 10.1029/WS002p0145.

Boyle, D.P., Gupta, H.V. and Sorooshian, S. (2003) Multicriteria calibration of hydrologic models. In: Duan, Q., H.V., Gupta., Sorooshian, S., Rousseau, A.N. and Turcotte, R. (eds) *Calibration of Watershed Models*. American Geophysical Union Publications, Washington, DC, pp. 185–196.

Brooks, E.S., Dobre, M., Elliot, W.J., Wu, J.Q. and Boll, J. (2016) Watershed-scale evaluation of the Water Erosion Prediction Project (WEPP) model in the Lake Tahoe Basin. *Journal of Hydrology* 533, 389–402.

Brunner, P. and Simmons, C.T. (2012) HydroGeoSphere: A fully integrated, physically based hydrological model. *Ground Water* 50, 170–176.

Burnash, R.J.C. (1995) The NWS river forecast system – catchment modeling. In: Singh, V.P. (ed.) *Computer Models of Watershed Hydrology*. Water Resources Publication, Littleton, Colorado, pp. 311–366.

Buttle, J.M. (2011) The effects of forest harvesting on forest hydrology and biogeochemistry. In: Levia, D.F., Tanaka, T. and Carlyle-Moses, D. (eds) *Forest Hydrology and Biogeochemistry: Synthesis of Past Research and Future Directions*. Springer, Dordrecht, Netherlands, pp. 659–677.

Calder, I.R. (2003) Assessing the water use of short vegetation and forests: Development of the Hydrological Land Use Change (HYLUC) model. *Water Resources Research* 39, 1318.

Caldwell, P.V., Sun, G., McNulty, S.G., Cohen, E.C. and Myers, M.J.A. (2012) Impacts of impervious cover, water withdrawals, and climate change on river flows in the conterminous US. *Hydrology and Earth System Sciences* 16, 2839–2857.

Cao, W., Zhang, Z., Liu, Y., Band, L.E., Wang, S. *et al.* (2021) Seasonal differences in future climate and streamflow variation in a watershed of Northern China. *Journal of Hydrology: Regional Studies* 38, 100959.

Clark, M.P., Fan, Y., Lawrence, D.M., Adam, J.C., Bolster, D. *et al.* (2015) Improving the representation of hydrologic processes in Earth System Models. *Water Resources Research* 51, 5929–5956.

Clark, M.P., Bierkens, M.F.P., Samaniego, L., Woods, R.A., Uijlenhoet, R. *et al.* (2017) The evolution of process-based hydrologic models: Historical challenges and the collective quest for physical realism. *Hydrology and Earth System Sciences* 21, 3427–3440.

Cornelissen, T., Diekkrüger, B., Bogena, H.R. (2014) Significance of scale and lower boundary condition in the 3D simulation of hydrological processes and soil moisture variability in a forested headwater catchment. *Journal of Hydrology* 516, 140–153.

Coville, R., Endreny, T. and Nowak, D.J. (2020) Modeling the impact of urban trees on hydrology. In: Levia, D.F., Carlyle-Moses, D.E., Iida, S., Michalzik, B., Nanko, K. *et al.* (eds) *Forest–Water Interactions*, Vol. 240. Ecological Studies, Springer, Cham, Switzerland, pp. 459–487.

Crawford, N.H. and Linsley, R.K. (1966) *Digital Simulation in Hydrology: Stanford Watershed Model IV*. Technical Report no.39. Department of Civil and Environmental Engineering, Stanford University, Stanford, California.

Cristea, N.C., Lundquist, J.D., Loheide, S.P., Lowry, C.S. and Moore, C.E. (2014) Modelling how vegetation cover affects climate change impacts on streamflow timing and magnitude in the snowmelt-dominated upper Tuolumne Basin, Sierra Nevada. *Hydrological Processes* 28, 3896–3918.

de Oliveira, V.A., Rodrigues, A.F., Vieira Morais, M.A., de Castro Nunes Santos Terra, M., Guo, L. *et al.* (2021) Spatiotemporal modelling of soil moisture in an Atlantic forest through machine learning algorithms. *European Journal of Soil Science* 72, 1969–1987.

de Schepper, G., Therrien, R., Refsgaard, J.C. and Hansen, A.L. (2015) Simulating coupled surface and subsurface water flow in a tile-drained agricultural catchment. *Journal of Hydrology* 521, 374–388.

Dehotin, J. and Braud, I. (2008) Which spatial discretization for distributed hydrological models? Proposition of a methodology and illustration for medium to large-scale catchments. *Hydrology and Earth System Sciences* 12, 769–796.

Department of Environmental Resources (1999) *Low-Impact Development Design Strategies; an Integrated Design Approach*. US Environmental Protection Agency Washington, DC. Available at: https://nepis.epa.gov/Exe/ZyPURL.cgi?Dockey=20004JX4.TXT (accessed 2 May 2025).

DeWalle, D.R. (2003) Forest hydrology revisited. *Hydrological Processes* 17, 1255–1256.

Doherty, J. and Johnston, J.M. (2003) Methodologies for calibration and predictive analysis of a watershed model. *Journal of the American Water Resources Association* 39, 251–265.

Douglass, J.E. (1983) The potential for water yield augmentation from forest management in the Eastern United States 1. *Journal of the American Water Resources Association* 19, 351–358.

Dunne, T. and Black, R.D. (1970) Partial-area contributions to storm runoff in a small New England watershed. *Water Resources Research* 6, 1296–1311.

Elliot, W.J. (2013) Erosion processes and prediction with WEPP technology in forests in the northwestern US. *Transactions of the ASABE* 56, 563–579.

EPA (US Environmental Protection Agency) (2017) *Storm Water Management Model Reference Manual Volume II – Hydraulics*. EPA/600/R-17/111. US Environmental Protection Agency, Washington, DC.

Ercolani, G., Chiaradia, E.A., Gandolfi, C., Castelli, F. and Masseroni, D. (2018) Evaluating performances of green roofs for stormwater runoff mitigation in a high flood risk urban catchment. *Journal of Hydrology* 566, 830–845.

Federer, C.A. (2002) BROOK 90: A simulation model for evaporation, soil water, and streamflow. Available at: www.ecoshift.net/brook/brook90.htm (accessed 13 May 2025).

Federer, C.A., Vorosmarty, C. and Fekete, B. (2003) Sensitivity of annual evaporation to soil and root properties in two models of contrasting complexity. *Journal of Hydrometeorology* 4, 1276–1290.

Feng, D., Liu, J., Lawson, K. and Shen, C. (2022) Differentiable, learnable, regionalized process-based models with multi-physical outputs can approach state-of-the-art hydrologic prediction accuracy. *Water Resources Research* 58, e2022WR032404.

Flerchinger, G.N., Caldwell, T.G., Cho, J. and Hardegree, S.P. (2012) Simultaneous heat and water (SHAW) model: Model use, calibration, and validation. *Transactions of the ASABE* 55, 1395–1411.

Foster, G.R. and Lane, L.J. (1987) *User Requirements: USDA-Water Erosion Prediction Project (WEPP)*. NSERL Report No.1. National Soil Erosion Research Laboratory, USDA ARS, West Lafayette, Indiana.

Garcia-Prats, A., Campo, A.D. and Pulido-Velazquez, M. (2016) A hydroeconomic modeling framework for optimal integrated management of forest and water. *Water Resources Research* 52, 8277–8294.

Gassman, P.W., Reyes, M.R., Green, C.H. and Arnold, J.G. (2007) The soil and water assessment tool: Historical development, applications, and future directions. *Transactions of the ASABE* 50, 1211–1250.

Gauch, M., Kratzert, F., Klotz, D., Nearing, G., Lin, J. *et al.* (2021) Rainfall–runoff prediction at multiple timescales with a single long short-term memory network. *Hydrology and Earth System Sciences* 25, 2045–2062.

Gbondo-Tugbawa, S.S., Driscoll, C.T., Aber, J.D. and Likens, G.E. (2001) Evaluation of an integrated biogeochemical model (PnET-BGC) at a Northern Hardwood Forest Ecosystem. *Water Resources Research* 37, 1057–1070.

Gochis, D.J., Barlage, M., Cabell, R., Casali, M., Dugger, A. *et al.* (2020) The WRF-Hydro® modeling system technical description (version 5.1.1). National Center for Atmospheric Research, Boulder, Colorado. Available at: https://ral.ucar.edu/projects/wrf_hydro/documentation/wrf-hydro-v511-documentation (accessed 13 May 2025).

Golden, H.E., Lane, C.R., Amatya, D.M., Bandilla, K.W., Raanan Kiperwas, H. *et al.* (2014) Hydrologic connectivity between geographically isolated wetlands and surface water systems: A review of select modeling methods. *Environmental Modelling and Software* 53, 190–206.

Golden, H.E., Creed, I.F., Ali, G., Basu, N.B., Neff, B.P. *et al.* (2017) Integrating geographically isolated wetlands into land management decisions. *Frontiers in Ecology and the Environment* 15, 319–327.

Goldstein, R.A., Mankin, J.B. and Luxmoore, R.J. (1974) *Documentation of PROSPER. A Model of Atmosphere–Soil–Plant Water Flow*. No. EDFB-IBP-73-9Oak Ridge National Laboratory, Oak Ridge, Tennessee.

Graham, D.N. and Butts, M. (2005) Flexible, integrated watershed modeling with MIKE SHE. In: Singh, V.P. and Frevert, D.K. (eds) *Watershed Models*. CRC Press, Boca Raton, pp. 242–272.

Haas, H., Kalin, L. and Srivastava, P. (2022b) Improved forest dynamics leads to better hydrological predictions in watershed modeling. *Science of the Total Environment* 821, 153180.

Haas, H., Reaver, N.G.F., Karki, R., Kalin, L., Srivastava, P. *et al.* (2022a) Improving the representation of forests in hydrological models. *Science of the Total Environment* 812, 151425.

Hallema, D.W., Sun, G., Caldwell, P.V., Norman, S.P., Cohen, E.C. *et al.* (2018) Burned forests impact water supplies. *Nature Communications* 9, 1307.

Hamon, W.R. (1963) Computation of direct runoff amounts from storm rainfall. *International Association of Science Hydrology Publishing* 63, 52–62.

Harbaugh, A.W. (2005) *MODFLOW-2005, the US Geological Survey Modular Ground-water Model: The Ground-water Flow Process*. US Department of the Interior, US Geological Survey, Reston, Virginia.

Harbaugh, A.W., Banta, E.R., Hill, M.C. and McDonald, M.G. (2000) *MODFLOW-2000, The US Geological Survey Modular Ground-water Model: User Guide to Modularization Concepts and the Ground-water Flow Process*. US Geological Survey, Reston, Virginia.

He, K., Li, W., Zhang, Y., Sun, G., McNulty, S.G. *et al.* (2023) Identifying driving hydrogeomorphic factors of coastal wetland downgrading using random forest classification models. *Science of the Total Environment* 894: 164995.

Herrera, P.A., Marazuela, M.A. and Hofmann, R. (2022) Parameter estimation and uncertainty analysis in hydrological modeling. *WIREs Water* 9: e1569.

Höge, M., Scheidegger, A., Baity-Jesi, M., Albert, C. and Fenicia, F. (2022) Improving hydrologic models for predictions and process understanding using neural ODEs. *Hydrology and Earth System Sciences* 26, 5085–5102.

Horton, R.E. (1933) The role of infiltration in the hydrologic cycle. *Transactions of the American Geophysical Union* 14, 446–460.

Hursh, C.R. and Brater, E. (1941) Separating storm hydrographs from small drainage areas into surface and subsurface flow. *Transactions of the American Geophysical Union* 22, 863–871.

Jiang, S., Zheng, Y., Wang, C. and Babovic, V. (2022) Uncovering flooding mechanisms across the contiguous united states through interpretive deep learning on representative catchments. *Water Resources Research* 58, e2021WR030185.

Jones, J.A., Achterman, G.L., Augustine, L.A., Creed, I.F., Folliott, P.F. *et al.* (2009) Hydrologic effects of a changing forested landscape-challenges for the hydrological sciences. *Hydrological Processes* 23, 2699–2704.

Jovanovic, N. and Israel, S. (2012) Critical review of methods for the estimation of actual evapotranspiration in hydrological models. In: Irmak, A. (ed.) *Evapotranspiration: Remote Sensing and Modeling*. IntechOpen, London. Available at: www.intechopen.com/books/200 (accessed 2 May 2025).

Kampf, S.K. and Burges, S.J. (2007) A framework for classifying and comparing distributed hillslope and catchment hydrologic models. *Water Resources Research* 43, W05423.

Kim, J.B., Kerns, B.K., Drapek, R.J., Pitts, G.S. and Halofsky, J.E. (2018) Simulating vegetation response to climate change in the blue mountains with MC2 dynamic global vegetation model. *Climate Services* 10, 20–32.

Klemes, V. (1986) Operational testing of hydrological simulation models. *Hydrological Sciences Journal* 31, 13–24.

Komatsu, H. (2020) Modeling evapotranspiration changes with managing Japanese cedar and cypress plantations. *Forest Ecology and Management* 475, 118395.

Komatsu, H. and Kume, T. (2020) Modeling of evapotranspiration changes with forest management practices: A genealogical review. *Journal of Hydrology* 585, 124835.

Kratzert, F., Klotz, D., Shalev, G., Klambauer, G., Hockreiter, S. *et al.* (2019) Towards learning universal, regional, and local hydrological behaviors via machine learning applied to large-sample datasets. *Hydrology and Earth System Sciences* 23, 5089–5110.

Krogh, S.A., Broxton, P.D., Manley, P.N. and Harpold, A.A. (2020) Using process based snow modeling and lidar to predict the effects of forest thinning on the northern sierra nevada snowpack. *Frontiers in Forest and Global Change* 3, 21.

Landsberg, J.J. and Waring, R.H. (1997) A generalised model of forest productivity using simplified concepts of radiation-use efficiency, carbon balance and partitioning. *Forest Ecology and Management* 95, 209–228.

Lange, H. and Sippel, S. (2020) Machine learning applications in hydrology. In: Levia, D.F., Carlyle-Moses, D.E., Iida, S., Michalzik, B., Nanko, K. *et al.* (eds) *Forest--Water Interactions*, Vol. 240. Ecological Studies, Springer, Cham, Switzerland, pp. 223–257.

Leavesley, G.H., Lichty, R.W., Troutman, B.M. and Saindon, L.G. (1983) *Precipitation-Runoff Modeling System; User's Manual*. Water-Resources Investigations Report 83-4238. US Geological Survey, Denver, Colorado. Available at: http://pubs.usgs.gov/wri/1983/4238/report.pdf (accessed 2 May 2025).

Leavesley, G.H. and Stannard, L.G. (1995) The precipitation-runoff modeling system PRMS. In: Singh, V.P. (ed.) *Computer Models of Watershed Hydrology*. Water Resources Publications, Highlands Ranch, Colorado, pp. 281–310.

Leavesley, G.H., Markstrom, S.L., Viger, R.J. and Hay, L.E. (2005) USGS modular modeling system (MMS) – precipitation-runoff modeling system (PRMS). In: Singh, V. and Frevert, D. (eds) *Watershed Models*. CRC Press, Boca Raton, Florida, pp. 159–177.

LeCun, Y., Bengio, Y. and Hinton, G. (2015) Deep learning. *Nature* 521, 436–444.

Li, C., Frolking, S. and Frolking, T.A. (1992) A model of nitrous oxide evolution from soil driven by rainfall events: 1. Model structure and sensitivity. *Journal of Geophysical Research: Atmospheres* 97, 9759–9776.

Li, C., Aber, J., Stange, F., Butterbach-Bahl, K. and Papen, H. (2000) A process-oriented model of N_2O and NO emissions from forest soils: 1. Model development. *Journal of Geophysical Research: Atmospheres* 105, 4369–4384.

Li, C., Trettin, C., Sun, G., McNulty, S. and Butterbach-Bahl, K. (2005) Modeling carbon and nitrogen biogeochemistry in forest ecosystems. In: Zhu, Z., Minami, K. and Xing, G. (eds) *3rd International Nitrogen Conference, 12–16 October 2004, Nanjing, China*, Science Press, Monmouth Junction, New Jersey, pp. 893–898.

Li, M., Zhang, Y., Wallace, J. and Campbell, E. (2020) Estimating annual runoff in response to forest change: A statistical method based on random forest. *Journal of Hydrology* 589, 125168.

Li, Y., Piao, S., Li, L.Z.X., Chen, A., Wang, X. *et al.* (2018) Divergent hydrological response to large-scale afforestation and vegetation greening in China. *Sciences Advances* 4, aar4182.

Lew, R., Dobre, M., Srivastava, A., Brooks, E.S. and Elliot, W.J. (2022) WEPPcloud: An online watershed-scale hydrologic modeling tool. Part I. Model description. *Journal of Hydrology* 608, 127603.

Liang, X., Lettenmaier, D.P., Wood, E.F. and Burges, S.J. (1994) A simple hydrologically based model of land surface water and energy fluxes for general circulation models. *Journal of Geophysical Research* 99, 415–528.

Liang, X., Wood, E.F. and Lettenmaier, D.P. (1996) Surface soil moisture parameterization of the VIC-2L model: Evaluation and modification. *Global and Planetary Change* 13, 195–206.

Liu, N., Caldwell, P.V., Miniat, C.F., Sun, G., Duan, K. *et al.* (2022) *Quantifying the Role of National Forest System and Other Forested Lands in Providing Surface Drinking Water Supply for the Conterminous United States*. General Technical Report WO-100. US Department of Agriculture, Forest Service, Washington DC.

Lundquist, J.D., Dickerson-Lange, S.E., Lutz, J.A. and Cristea, N.C. (2013) Lower forest density enhances snow retention in regions with warmer winters: A global framework developed from plot-scale observations and modeling. *Water Resources Research* 49, 6356–6370.

Malone, R.W., Yagow, G., Baffaut, C., Gitau, M., Qi, Z. *et al.* (2015) Parameterization guidelines and considerations for hydrologic models. *Transactions of the ASABE* 58, 1681–1703.

Markstrom, S.L., Niswonger, R.G., Regan, R.S., Prudic, D.E. and Barlow, P.M. (2008) *GSFLOW, coupled ground-water and surface-water flow model based on the integration of the precipitation-runoff modeling system (PRMS) and the modular ground-water flow model (MODFLOW-2005)*, Chapter 1 of Section D, Ground-Water/Surface-Water, Book 6, Modeling Techniques. US Deptartment of the Interior, US Geological Survey, Reston, Virigina. Available at: http://pubs.usgs.gov/tm/tm6d1/pdf/tm6d1.pdf (accessed 2 May 2025).

Matott, L.S. (2005) *OSTRICH: An Optimization Software Tool, Documentation and User's Guide, Version 1.6*. University at Buffalo, Buffalo, New York. Available at: www.eng.buffalo.edu/~lsmatott/Ostrich/OstrichMain.html (accessed 2 May 2025).

Matott, L.S., Babendreier, J.E. and Purucker, S.T. (2009) Evaluating uncertainty in integrated environmental models: A review of concepts and tools. *Water Resources Research* 45, W06421.

McDonald, M.G. and Harbaugh, A.W. (1988) *A Modular Three-Dimensional Finite-Difference Ground-Water Flow Model*. US Geological Survey, Reston, Virginia.

McMillan, H., Westerberg, I. and Branger, F. (2017) Five guidelines for selecting hydrological signatures. *Hydrological Processes* 31, 4757–4761.

McMillan, H., Westerberg, I.K. and Krueger, T. (2018) Hydrological data uncertainty and its implications. *WIREs Water* 5, e1319.

McMillan, H., Araki, R., Gnann, S., Woods, R. and Wagener, T. (2023) How do hydrologists perceive watersheds? A survey and analysis of perceptual model figures for experimental watersheds. *Hydrological Processes* 37, e14845.

McNaughton, K.G. and Black, T.A. (1973) A study of evapotranspiration from a Douglas fir forest using the energy balance approach. *Water Resources Research* 9, 1579–1590.

Melillo, J.M., McGuire, A.D., Kicklighter, D.W., Moore, B., Vorosmarty, C.J. *et al.* (1993) Global climate change and terrestrial net primary production. *Nature* 363, 234–240.

Meng, F.R., Bourque, C.P.A., Jewett, K., Daugharty, D. and Arp, P.A. (1995) The Nashwaak Experimental Watershed Project – analyzing effects of clearcutting on soil-temperature, soil-moisture, snowpack, snowmelt and stream-flow. *Water, Air, and Soil Pollution* 82, 363–374.

Mu, Q., Zhao, M. and Running, S. (2011) Improvements to a MODIS global terrestrial evapotranspiration algorithm. *Remote Sensing of Environment* 115, 1781–1800.

Mulligan, M. (2013) WaterWorld: A self-parameterising, physically based model for application in data-poor but problem-rich environments globally. *Hydrology Research* 44, 748–769.

National Research Council (2007) *Models in Environmental Decision Making*. National Academies Press, Washington, DC.

National Research Council (2008) *Hydrologic Effects of a Changing Forest Landscape*. National Academies Press, Washington, DC. Available at: www.nap.edu/catalog/12223.html (accessed 2 May 2025).

Neitsch, S.L., Arnold, J.G., Kiniry, J.R. and Williams, J.R. (2011) *Soil and Water Assessment Tool: Theoretical Documentation Version 2009, TR-2011*. Texas Water Resources Institute, College Station, Texas.

Nelitz, M., Boardley, S. and Smith, R. (2013) *Tools for Climate Change Vulnerability Assessments for Watersheds*. PN 1494. ESSA Technologies Ltd, prepared for Canadian Council of Ministers of the Environment.

Panday, S. and Huyakorn, P.S. (2004) A fully coupled physically-based spatially-distributed model for evaluating surface/subsurface flow. *Advances in Water Resources* 27, 361–382.

Parajuli, P.B. (2010) Assessing sensitivity of hydrologic responses to climate change from forested watershed in mississippi. *Hydrologic Processes* 24, 3785–3797.

Parton, W.J., Scurlock, J.M.O., Ojima, D.S., Gilmanov, T.G., Scholes, R.J. *et al.* (1993) Observations and modeling of biomass and soil organic matter dynamics for the grassland biome worldwide. *Global Biogeochemical Cycles* 7, 785–809.

Parton, W.J. (1996) The CENTURY model. In: Powlson, D., Smith, P. and Smith, J. (eds) *Evaluation of Soil Organic Matter Models*. Springer, Berlin/Heidelberg, pp. 283–291.

Paul, M.J., LeDuc, S.D., Lassiter, M.G., Moorhead, L.C., Noyes, P.D. *et al.* (2022) Wildfire induces change in receiving waters: A review with considerations for water quality management. *Water Resources Research* 58, e2021WR030699.

Perkins, W.A., Duan, Z., Sun, N., Wigmosta, M.S., Richmond, M.C. *et al.* (2019) Parallel distributed hydrology soil vegetation model (DHSVM) using global arrays. *Environmental Modelling & Software* 122, 104533.

Peña-Arancibia, J.L., Bruijnzeel, L.A., Mulligan, M. and Dijk, A.I. (2019) Forests as 'pumps' and 'sponges': Assessing the impact of deforestation on dry-season flows across the tropics. *Journal of Hydrology* 574, 946–963.

Pianosi, F. and Wagener, T. (2016) Understanding the time-varying importance of different uncertainty sources in hydrological modelling using global sensitivity analysis. *Hydrological Processes* 30, 3991–4003.

Povak, N.A., Furniss, T.J., Hessburg, P.F., Salter, R.B., Wigmosta, M. *et al.* (2022) Evaluating basin-scale forest adaptation scenarios: Wildfire, streamflow, biomass, and economic recovery synergies and trade-offs. *Frontiers in Forests and Global Change* 5, 805179.

Price, K., Purucker, S.T., Kraemer, S.R. and Babendreier, J.E. (2012) Tradeoffs among watershed model calibration targets for parameter estimation. *Water Resources Research* 48, W10542.

Priestley, C.H.B. and Taylor, R.J. (1972) On assessment of surface heat flux and evaporation using large-scale parameters. *Monthly Weather Review* 100, 81–92.

Rajib, A., Evenson, G.R., Golden, H.E. and Lane, C.R. (2018) Hydrologic model predictability improves with spatially explicit calibration using remotely sensed evapotranspiration and biophysical parameters. *Journal of Hydrology* 567, 668–683.

Regan, R.S., Markstrom, S.L., Hay, L.E., Viger, R.J., Norton, P.A. *et al.* (2018) Description of the national hydrologic model for use with the precipitation-runoff modeling system (PRMS). In: *Techniques and Methods*. Book 6, Chapter B9, US Geological Survey, Reston, Virginia. DOI: 10.3133/tm6B9. (accessed 2 May 2025).

Renschler, C.S. (2003) Designing geo-spatial interfaces to scale process models: The GeoWEPP approach. *Hydrological Processes* 17, 1005–1017.

Riley, J.W. and Stillwell, C.C. (2023) Predicting inundation dynamics and hydroperiods of small, isolated wetlands using a machine learning approach. *Wetlands* 43, 63.

Running, S.W. and Gower, S.T. (1991) Forest-BGC, a general-model of forest ecosystem processes for regional applications. 2. Dynamic carbon allocation and nitrogen budgets. *Tree Physiology* 9, 147–160.

Running, S.W. and Hunt, E.R., Jr (1993) Generalization of a forest ecosystem model for other biomes, BIOME-BGC, and an application for global-scale models. In: Ehleringer, J.R. and Field, C.B. (eds) *Scaling Physiological Processes: Leaf to Globe*. Academic Press, London, pp. 141–158.

Saleh, A., Williams, J.R., Wood, J.C., Hauck, L.M. and Blackburn, W. (2004) Application of Apex for forestry. *Transactions of the ASAE* 47, 751–765.

Sass, G.Z. and Creed, I.F. (2011) Bird's-eye view of forest hydrology: Novel approaches using remote sensing techniques. In: Levia, D.F., Carlyle-Moses, D. and Tanaka, T. (eds) *Forest Hydrology and Biogeochemistry: Synthesis of Past Research and Future Directions*. Springer, Dordrecht, Netherlands, pp. 45–68.

Schmidt, A., Mainwaring, D.B. and Maguire, D.A. (2020) Development of a tailored combination of machine learning approaches to model volumetric soil water content within a mesic forest in the Pacific Northwest. *Journal of Hydrology* 588, 125044.

Sellers, P., Mintz, Y., Sud, Y.E.A. and Dalcher, A. (1986) A simple biosphere model (SiB) for use within general circulation models. *Journal of the Atmospheric Sciences* 43, 505–531.

Sharp, R., Douglass, J., Wolny, S., Arkema, K., Bernhardt, J. *et al.* (2020) *InVEST 3.8.5 User's Guide*. The Natural Capital Project. Stanford University, University of Minnesota, The Nature Conservancy, and World Wildlife Fund.

Shen, C. (2018) A transdisciplinary review of deep learning research and its relevance for water resources scientists. *Water Resources Research* 54, 8558–8593.

Shen, C., Laloy, E., Elshorbagy, A., Albert, A., Bales, J. *et al.* (2018) Opinions: Incubating deep-learning-powered hydrologic science advances as a community. *Hydrology and Earth System Sciences* 22, 5639–5656.

Shen, C. and Lawson, K. (2021) Applications of deep learning in hydrology. In: Camps-Valls, G., Tuia, D., Zhu, X.X. and Reichstein, M. (eds) *Deep Learning for the Earth Sciences: A Comprehensive Approach to Remote Sensing, Climate Science, and Geosciences*. Wiley, Chichester, UK, pp. 283–297.

Shen, C., Appling, A.P., Gentine, P., Bandai, T., Gupta, H. *et al.* (2023) Differentiable modelling to unify machine learning and physical models for geoscience. *Nature Reviews Earth & Environment* 4, 552–567.

Sidle, R.C., Kim, K., Tsuboyama, Y. and Hosoda, I. (2011) Development and application of a simple hydrogeomorphic model for headwater catchments. *Water Resources Research* 47, W00H13.

Sivapalan, M. (2003) Prediction in ungauged basins: A grand challenge for theoretical hydrology. *Hydrological Processes* 17, 3163–3170.

Skaggs, R.W. (1978) *A Water Management Model for Shallow Water Table Soils*. Raleigh, North Carolina: Technical Report no.134. North Carolina State University, Water Resources Research Institute.

Sloto, R.A. and Crouse, M.Y. (1996) *HYSEP: A Computer Program for Streamflow Hydrograph Separation and Analysis. Water-Resources Investigations Report 96-4040*. US Geological Survey, Reston, Virginia. Available at: http://water.usgs.gov/software/HYSEP/ (accessed 2 May 2025).

Smith, R.A., Schwarz, G.E. and Alexander, R.B. (1997) Regional interpretation of water-quality monitoring data. *Water Resources Research* 33, 2781–2798.

Speich, M.J., Zappa, M., Scherstjanoi, M. and Lischke, H. (2020) FORests and hydrology under climate change in Switzerland v1. 0: A spatially distributed model combining hydrology and forest dynamics. *Geoscientific Model Development* 13, 537–564.

Stange, F., Butterbach-Bahl, K., Papen, H., Zechmeister-Boltenstern, S., Li, C. *et al.* (2000) A process-oriented model of N_2O and NO emissions from forest soils: 2. Sensitivity analysis and validation. *Journal of Geophysical Research: Atmospheres* 105, 4385–4398.

Stednick, J. (ed.) (2008) *Hydrological and Biological Responses to Forest Practices*, Ecological Studies, No. 190. Springer, New York.

Storck, P., Bowling, L., Wetherbee, P. and Lettenmaier, D. (1998) Application of a GIS-based distributed hydrology model for prediction of forest harvest effects on peak stream flow in the Pacific Northwest. *Hydrological Processes* 12, 889–904.

Sun, G., Riekerk, H. and Comerford, N.B. (1998a) Modeling the hydrologic impacts of forest harvesting on Florida Flatwoods. *Journal of the American Water Resources Association* 34, 843–854.

Sun, G., Riekerk, H. and Comerford, N.B. (1998b) Modeling the forest hydrology of wetland-upland ecosystems in Florida. *Journal of the American Water Resources Association* 34, 827–841.

Sun, G., Zhou, G., Zhang, Z., Wei, X., McNulty, S.G. *et al.* (2006) Potential water yield reduction due to forestation across China. *Journal of Hydrology* 328, 548–558.

Sun, G., Caldwell, P., Noormets, A., Cohen, E., McNulty, S. *et al.* (2011) Upscaling key ecosystem functions across the conterminous United States by a water-centric ecosystem model. *Journal of Geophysical Research* 116, G00J05.

Sun, G., Caldwell, P.V. and McNulty, S.G. (2015) Modeling the potential role of forest thinning in maintaining water supplies under a changing climate across the conterminous United States. *Hydrological Processes* 29, 5016–5030.

Sun, G., Domec, J.-C. and Amatya, D. (2016) Forest evapotranspiration: Measurements and modeling at multiple scales. In: Amatya, D.M., Williams, T.M., Bren, L. and Jong, C. (eds) *Forest Hydrology: Processes, Management and Assessment*. CAB International, Wallingford, UK, pp. 32–50.

Sun, G., Wei, X., Hao, L., Gonzalez Sanchis, M., Hou, L. *et al.* (2023) Forest hydrology modeling tools for watershed management: A review. *Forest Ecology and Management* 530, 120755.

Sun, N., Yan, H., Wigmosta, M.S., Lundquist, J., Dickerson-Lange, S. *et al.* (2022) Forest canopy density effects on snowpack across the climate gradients of the western United States mountain ranges. *Water Resources Research* 58, e2020WR029194.

Tague, C.L. and Band, L.E. (2004) RHESSys: Regional hydro-ecologic simulation system – An object-oriented approach to spatially distributed modeling of carbon, water, and nutrient cycling. *Earth Interactions* 8, 1–42.

Tasdighi, A., Arabi, M., Harmel, D. and Line, D. (2018) A bayesian total uncertainty analysis framework for assessment of management practices using watershed models. *Environmental Modelling and Software* 108, 240–252.

Tedela, N.H., McCutcheon, S.C., Rasmussen, T.C., Hawkins, R.H., Swank, W.T. *et al.* (2012) Runoff curve numbers for 10 small forested watersheds in the mountains of the eastern United States. *Journal of Hydrologic Engineering* 17, 1188–1198.

Therrien, R., McLaren, R.G., Sudicky, E.A. and Panday, S.M. (2010) *HydroGeoSphere: A Three-dimensional Numerical Model Describing Fully-integrated Subsurface and Surface Flow and Solute Transport*. Groundwater Simulations Group, University of Waterloo, Waterloo, Ontario.

Thyer, M., Beckers, J., Spittlehouse, D., Alila, Y. and Winkler, R. (2004) Diagnosing a distributed hydrologic model for two high-elevation forested catchments based on detailed stand- and basin-scale data. *Water Resources Research* 40, W01103.

Tian, H.Q., Liu, M.L.S., Xu, X.F. and Lu, C.Q. (2005) *DLEM – The Dynamic Land Ecosystem Model, User Manual*. Ecosystem Science and Regional Analysis Laboratory, Auburn University, Auburn, Alabama.

Tian, M.A., Youssef, M.A., Skaggs, R.W., Amatya, D.M. and Chescheir, G.M. (2012a) DRAINMOD-FOREST: Integrated modeling of hydrology, soil carbon, and nitrogen dynamics, and plant growth for drained FOREST. *Journal of Environmental Quality* 41, 764–782.

Tian, M.A., Youssef, M.A., Skaggs, R.W., Amatya, D.M. and Chescheir, G.M. (2012b) Modeling water, carbon and nitrogen dynamics for two drained pine plantations under intensive management practices. *Forest Ecology and Management* 264, 20–36.

Vélez, J., Francés, F. and Vélez, I. (2005) TETIS: A catchment hydrological distributed conceptual model. *Geophysical Research Abstracts* 7, 03503.

Von Stackelberg, N.O., Chescheir, G.M., Skaggs, R.W. and Amatya, D.M. (2007) Simulation of the hydrologic effect of afforestation in the Tacuarembo River Basin, Uruguay. *Transactions of the ASABE* 50, 455–468.

Vose, J.M., Sun, G., Ford, C.R., Bredemeier, M., Otsuki, K. *et al.* (2011) Forest ecohydrological research in the 21st century: What are the critical needs? *Ecohydrology* 4, 146–158.

Wade, A., Durand, P., Beaujouan, V., Wessel, W., Raat, K. *et al.* (2002) A nitrogen model for European catchments: INCA, new model structure and equations. *Hydrology and Earth System Sciences* 6, 559–582.

Wang, J., Endreny, T.A. and Nowak, T.J. (2008) Mechanistic simulation of tree effects in an urban water balance model. *Journal of the American Water Resources Association* 44, 75–85.

Watson, B.M., McKeown, R.A., Putz, G. and MacDonald, J.D. (2009) Modification of SWAT for modelling streamflow from forested watersheds on Canadian boreal plain. *Journal of Environmental Engineering and Science* 7, 145–159.

Wei, A. and Zhang, M. (2011) A review on research methods for assessing the impacts of forest disturbance on hydrology at large-scale watersheds. In: Li, C., Lafortezza, R. and Chen, J. (eds) *Landscape Ecology and Forest Management: Challenges and Solutions in a Changing Globe*. Springer, New York, pp. 119–147.

Weiler, M. and McDonnell, J. (2004) Virtual experiments: A new approach for improving process conceptualization in hillslope hydrology. *Journal of Hydrology* 285, 3–18.

Whitaker, A., Alila, Y., Beckers, J. and Toews, D. (2003) Application of the distributed hydrology soil vegetation model to redfish Creek, British Columbia: Model evaluation using internal catchment data. *Hydrological Processes* 17, 199–224.

White, M.A., Thornton, P.E., Running, S.W. and Nemani, R.R. (2000) Parameterization and sensitivity analysis of the BIOME-BGC terrestrial ecosystem model: Net primary production controls. *Earth Interactions* 4, 1–85.

Whitehead, P., Wilson, E. and Butterfield, D. (1998a) A semi-distributed integrated nitrogen model for multiple source assessment in catchments (INCA): Part I – model structure and process equations. *Science of the Total Environment* 210, 547–558.

Whitehead, P., Wilson, E., Butterfield, D. and Seed, K. (1998b) A semi-distributed integrated flow and nitrogen model for multiple source assessment in catchments (INCA): Part II – Application to large river basins in South Wales and Eastern England. *Science of the Total Environment* 210, 559–583.

Wigmosta, M.S., Vail, L.W. and Lettenmaier, D.P. (1994) A distributed hydrology–vegetation model for complex terrain. *Water Resources Research* 30, 1665–1679.

Wigmosta, M.S., Nijssen, B., Storck, P. and Lettenmaier, D. (2002) The distributed hydrology soil vegetation model. In: Singh, V.P. and Frevert, D.K. (eds) *Mathematical Models of Small Watershed Hydrology and Applications*. Water Resources Publications, Littleton, Colorado, pp. 7–42.

Willard, J., Jia, X., Xu, S., Steinback, M. and Kumar, V. (2022) Integrating scientific knowledge with machine learning for engineering and environmental systems. *ACM Computing Surveys* 55, 1–37.

Williams, J.R. and Izaurralde, R.C. (2005) The APEX model. In: Singh, V.P. and Frevert, D.K. (eds) *Watershed Models*. CRC Press, Boca Raton, Florida, pp. 437–482.

Yan, H., Sun, N., Fullerton, A. and Baerwalde, M. (2021) Greater vulnerability of snowmelt-fed river thermal regimes to a warming climate. *Environmental Research Letters* 16, 054006.

Yan, H., Sun, N., Eldardiry, H., Thurber, T.B., Reed, P.M. *et al.* (2023) Large ensemble diagnostic evaluation of hydrologic parameter uncertainty in the community land model version 5 (CLM5). *Journal of Advances in Modeling Earth Systems* 15, e2022MS003312.

Wischmeier, W.H. and Smith, D.D. (1960) A universal soil-loss equation to guide conservation farm planning. In: *Transactions of the 7th International Congress of Soil Science, 14–24 August 1960, Madison, Wisconsin*. International Society of Soil Science, Madison, Wisconsin, pp. 418–425.

Yang, Y., Endreny, T.A. and Nowak, D.J. (2011) ITree-hydro: Snow hydrology update for the urban forest hydrology model. *Journal of the American Water Resources Association* 47, 1211–1218.

Zema, D.A. and Lucas-Borja, M.E. (2022) The impact of climate change on forest hydrology. In: *The Handbook of Environmental Chemistry*. Springer, Berlin/Heidelberg.

Zhang, L. and Dawes, W. (1998) WAVES – an integrated energy and water balance model. *CSIRO Land Water Technical Report* 31, 98.

Zhang, L., Dawes, W. and Walker, G. (2001) Response of mean annual evapotranspiration to vegetation changes at catchment scale. *Water Resources Research* 37, 701–708.

Zhang, M., Liu, S., Jones, J., Sun, G., Wei, X. *et al*. (2022) Managing the forest–water nexus for climate change adaptation. *Forest Ecology and Management* 525, 120545.

Zhang, Y., Song, C., Sun, G., Band, L.E., McNulty, S. *et al*. (2016) Development of a coupled carbon and water model for estimating global gross primary productivity and evapotranspiration based on eddy flux and remote sensing data. *Agricultural and Forest Meteorology* 223, 116–131.

Zhao, R. (1980) The Xinanjiang model. In: *Proceedings of the Oxford Symposium on Hydrological Forecasting*, IAHS Publication 129, pp. 351–356.

Zhi, W., Appling, A.P., Golden, H.E., Podgorski, J. and Li, L. (2024) Deep learning for water quality. *Nature Water* 2, 228–241.

16 Geospatial Technology Applications in Forest Hydrology

S.S. Panda[1]*, E. Masson[2], M.S. Rana[3], S. Sen[4] and H.W. Kim[5]

[1]Institute of Environmental Spatial Analysis, University of Georgia, Gainesville, Georgia, USA; [2]Université Lille 1 Sciences and Technologies, Geography Department, Cedex, France; [3]Graduate Student, Civil Engineering, Indian Institute of Technology, Guwahati, Assam, India; [4]Civil Engineering Department, Indian Institute of of Technology, Uttarakhand, India; [5]Department of Environmental and Energy Engineering, Anyang University, Anyang, Gyeonggi-do, Korea

Abstract

Geospatial technology in its various forms offers excellent prospects for forest hydrology management. By their nature, forests are inherently large and variable. Remote sensing offers great advantages in collecting and classifying information. New advances allow this to be used in a multitude of ways, from global classifications of forests to detailed mapping of local forests. Specialized approaches allow production of a diverse range of information including biomass, erosion and landslide hazards, flooding hazards, real-time mapping of weather, fire danger predictions and ongoing-fire predictions. Recent advances allow direct input of remotely sensed data into real-time models. Newer technology is allowing the exploration of groundwater (and related effects) on a regional or national scale; this can be particularly important in low-lying wetland forests adjacent to large water bodies (e.g. mangroves). The relationship between ground-measured hydrological data and remotely sensed hydrological data is not always simple. Geospatial technologies have tremendous potential for decision support in forest hydrology management.

16.1 Introduction

Two separate disciplines, hydrology and forestry, together constitute 'forest hydrology.' It is the study of how water moves through forested landscapes, exploring the complex interactions between precipitation, vegetation, soil and groundwater, and how these interactions influence surface and groundwater quantity and quality. Forests help reduce peak flows to mitigate floods by regulating the timing and amount of water entering streams while improving water quality by filtering pollutants and sediments. It is of note that all features that are part of forest hydrology constitute geospatial characteristics that are fundamental to understanding its dynamics and its impact on the environment. The geospatial nature of the features supports efficient environmental management decision system design with geospatial artificial intelligence (GeoAI) for proactive management support.

'Forest' and 'water' are two standardized land-cover mapping classifications of the National Land Cover Database used by

*Corresponding author: jami.nettles@weyerhaeuser.com

 Forest Hydrology: Processes, Management and Assessment (eds Leon J. Bren *et al.*)
DOI: 10.1079/9781800625310.0016

environmental planners in the USA (USGS Land Cover Institute, 2015), CORINE (Co-ORdinated INformation on the Environment) data sets (EEA, 2006) established and used by the European Community and other countries' national land-cover mapping systems for developing an environmental management decision support system (DSS). Due to the spatial nature of forest hydrology, a GeoAI-supported geospatial modeling approach and map algebraic analyses can help in making informed decisions about prudent forest-management practices to protect water resources.

Due to its spatial nature and management needs, forest hydrology ideally should be managed with a sound DSS based on a geospatial technology (GT) application such as remote sensing (RS), geographic information systems (GIS), the global navigation satellite system (GNSS), and information technology (IT). It is humanly impossible to deal with or analyze features of larger areas (e.g. forests or their smaller fragments) for accurate management decision making, because site-specific forest management decision support (SSFMDS) based on scouting only would take years. Management of forests to support silviculture involves large-scale spatial and tabular (attribute) data (gigabytes or even terabytes) and numerous SSFMDS parameters including soil, climatological, hydrological and crop growth attributes. This inherent data volume and intricacy related to SSFMDS, and especially forest hydrology phenomena can be effectively and efficiently monitored using GT for site-specific forest management (von Gadow and Bredenkamp, 1992; Panda *et al.*, 2010). In fact, GT, especially GIS, has become a fundamental part of forestry management in many commercial forestry enterprises (Austin and Meyers, 1996). Currently, applications of advanced RS technologies such as ultra-high (<1 m) spatial resolution ortho- or satellite images, hyperspectral images and radio detection and ranging (radar) data have been extremely useful in the effective management of forest hydrology. Unmanned aerial vehicles (UAVs) and unmanned aircraft systems (UASs) are making hydrological management more efficient through the acquisition of centimeter-scale spatial resolution images with user-specified bandwidths, thus helping in SSFMDS by mapping soil moisture, plant stomatal conductance, canopy temperature and leaf area index (LAI) to measure forest evapotranspiration (ET) and monitor forest fires (Grenzdörffer *et al.*, 2008).

GT, through raster imagery acquisition and mapping, has the ability to depict accurate pixel-based analysis of larger areas using several parameters such as land use/land cover, soil, elevation/topography, hydrology, transportation, population density and adjacency, and climate/weather. RS technology helps in surveying the entire earth with unprecedented regularity; thus, major global forest cover change can be discovered or monitored efficiently to provide insight into forest hydrology management. The Shuttle Radar Topography Mission satellites obtain global elevation data from which earth topographical changes can be monitored proficiently, suggesting changes to forest hydrology. In the current decade, with the introduction of Light Detection and Ranging (LiDAR) and UAV/UAS technology, earth elevation including tree heights in forests is being monitored with centimeter accuracy for forest biomass estimation and ET assessment (Zarco-Tejada *et al.*, 2014; Khosravipour *et al.*, 2015). Currently, weather satellites monitor global atmospheric conditions hourly, including water vapor in the atmosphere on a spatial basis (Panda *et al.*, 2015). RS imagery provides information on drought, vegetation vigor, flood damage, forest fires, deforestation and other natural disasters that are directly or indirectly influenced by forest hydrology (Panda *et al.*, 2015). D'Urso and Minacapilli (2006) used a semi-empirical approach for forest surface soil water content estimation using radar data. Potential RS systems, such as color infrared aerial photography, most other multi-spectral scanners (MSSs) (e.g. Landsat, QuickBird), and hyperspectral systems (e.g. AVIRIS, HyMap, CASI), bathymetric LiDAR, MISR, Hyperion, TOPEX/Poseidon, MERIS, AVHRR (advanced very high-resolution radiometer) and CERES, are being used by scientists to remotely estimate the hydrological flux on the earth's surface, including forest land cover (Panda *et al.*, 2015).

Newer specialized satellites such as SMAP, ECOSTRESS (Liu *et al.*, 2021; Panda *et al.*, 2024a), Global Ecosystem Dynamics Investigation (GEDI) and others are helping management of forest hydrology very efficiently. These NASA satellites provide valuable and already processed global soil moisture (rootzone, surface and average), ET and biomass amount on a spatial basis with emphasis on forest hydrology remote estimation.

GIS provides the tools to accurately map this information globally and locally, including the development of automated geospatial models for precise and proficient forest hydrology management decision support. In recent times, the most widely used global positioning system (GPS) technology (a part of GNSS) accurately tracks the position of environmental disasters such as forest fires, mud slides and other phenomena related to forest hydrology. IT helps improve DSS development and popularize these fascinating but sometimes challenging tools (Panda *et al.*, 2004b).

16.2 Geospatial Technology Application in Forest Hydrological Processes Management

GT has become an invaluable tool in forest hydrology management by enhancing our understanding and developing strategies for management of forested landscapes. Opportunities include precise forest mapping, inventorying forest resources, monitoring forest health through detection of forest disturbances, assessing forest growth and productivity, and watershed modeling to determine water quantity and quality with forest as a positive-impact land cover to develop sustainable management plans. These strategies are aimed at improving the long-term sustainability of forest ecosystems. Satellite imageries provide synoptic views of large areas, allowing monitoring of forest cover, vegetation health and water resources emanating from the forest areas (Khan *et al.*, 2010; Panda *et al.*, 2016a, 2018; Muwamba *et al.*, 2019). LiDAR technology generates ultra-high spatial resolution digital elevation models and detailed forest biomass information through height detection to support forest hydrological modeling and other associated assessments (Amatya *et al.*, 2011, 2013; Dassot *et al.*, 2011; Panda *et al.*, 2016b, 2024b; San Juan and Domingo-Santos, 2018; Schumacher and Christiansen, 2020). The latest advance in UAV imaging offers ultra-high spatial resolution imageries in forest cover for very detailed mapping and monitoring of small-scale hydrological features such as LAI, stomatal conductance and forest speciation (Panda *et al.*, 2016a, b, 2018; Sankey and Tatum, 2022). Forest hydrologists and forest managers can make informed decisions by leveraging the power of RS to promote sustainable forest management.

The most important aspect of forest management is that forest cover provides a cleaner and more dependable supply of water compared with most other land covers (Richmond, 2007). The basic forest hydrological concept is that the first element of the hydrological process, interception of raindrops by the plant canopy or the forest canopy, occurs in abundance – around 25–30% of total precipitation (Zinke, 1967) – and with higher infiltration and lower runoff than any other land-cover types except wetlands due to supportive forest soil texture and structure (Zinke, 1967). People have been observing the link between forests and water for thousands of years (Amatya *et al.*, 2015). Historically, the study of forests, water and climate was referred to as 'forest influences.' This is still a useful term and a meaningful concept (Barten, 2006). Globally, the forest flourishes when precipitation is much greater than potential ET, the growing season is long, the climate is moderate and the frequency of natural disturbance is low (Barten, 2006).

16.2.1 Forest cover mapping and change analysis

The forest cover supports climate stabilization, biodiversity preservation, soil enrichment for agricultural lands, erosion control, clean water supply and its cycle regulation, bioenergy production, and fodder and timber supply for human and animal sustenance. Forest plant litters decompose and recycle nutrients through the shedding of leaves and seeds with the support of forest hydrological cycles, thus enriching the soil (Osman, 2013). These enriched soils from higher-elevation forest cover can be washed to more flat topographic agricultural land and help in higher crop production (Osman, 2013). The tree roots and soil binding in the forest reduce excessive soil erosion (Kittredge, 1948). Forest

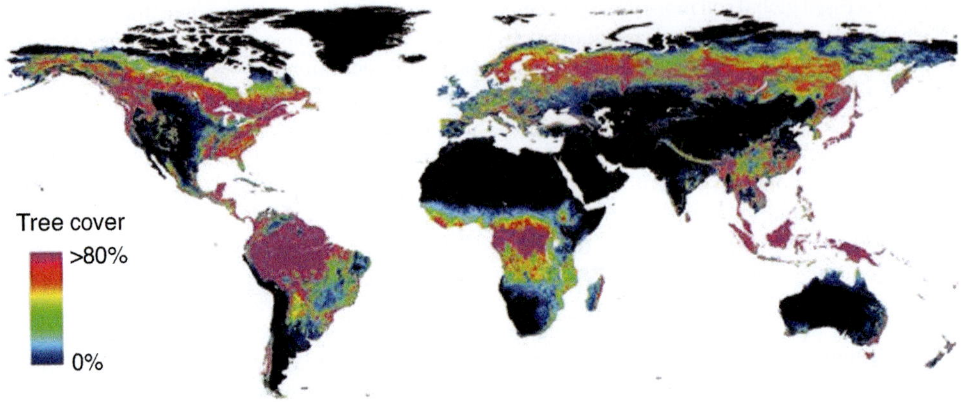

Fig. 16.1. High-resolution global map of 21st century forest cover density. Modified from Hansen *et al.* (2013).

cover regulates the water cycle by absorbing and redistributing rainwater equally to every species living within its range (Perry *et al.*, 2008). Moreover, riparian forest has proved its potential to clean surface water and reduce nitrate accumulation in soils and river flows (Lowrance, 1992; Pinay *et al.*, 1993). Additionally, mapping the forest plant canopy can help to quantify interception and subsequent evaporation. Therefore, efficient mapping and analysis of forest cover with GT supports a better management DSS.

The Food and Agriculture Organization of the United Nations (FAO) monitors global forest cover with 250m resolution Moderate Resolution Imaging Spectroradiometer (MODIS) data. The National Oceanic and Atmospheric Administration (NOAA) uses 1 km resolution AVHRR satellite imagery to constantly monitor global vegetation changes over time. Fig. 16.1 provides a high-resolution global map of 21st-century forest cover density developed by Hansen *et al.* (2013) with the use of 30m spatial resolution Landsat data. Lepers *et al.* (2005) used remotely sensed imagery to construct a temporal change analysis of global forest land cover between 1981 and 2000 (Fig. 16.2). The map and their study (Lepers *et al.*, 2005) provided an insight into worldwide forest loss. Areas in the map (Fig. 16.2) are defined as hotspots when deforestation rates exceed threshold values, as estimated from available deforestation data or from expert opinion. The National Land Cover Database classifies three prominent forest land covers excluding forested wetlands. Medium resolution (30 m) Landsat (five MSSs, seven Enhanced Thematic Mapper Plus (ETM+)) images are used to classify the land cover of the USA on a temporal basis (based on the fly-over cycle of the satellites), and the Anderson land cover classification scheme includes deciduous forest (#41), evergreen forest (#42) and mixed forest (#43) as forest categories. The NASA-GAP project develops US land cover maps at regular intervals (i.e. 1974, 1985, 1992, 2001, 2005 (few states), 2011, and 2021). The National Land Cover Database helps in studying the US land-cover change, especially forest-cover change.

The conversion of natural land cover into human-dominated land-use types such as forest harvesting, deforestation, urbanization and agricultural intensification continues to be a change of global proportions with many environmentally unfriendly consequences for local climate, energy, hydrology and water balance, biogeochemistry and biodiversity (Potter *et al.*, 2007). Deforestation allows soil erosion, causing nutrient-rich soils to be lost into rivers, lakes and oceans (Panda *et al.*, 2004a). According to Samec (2023), the global tropical deforestation rate is around 8% of the current tropical forest inventory per decade. In the Indian subcontinent, Asia and Africa, shifting cultivation (e.g. slash and burn agriculture) is a prime example of mismanagement of forest resources (forest soils and water) (Panda *et al.*, 2004a).

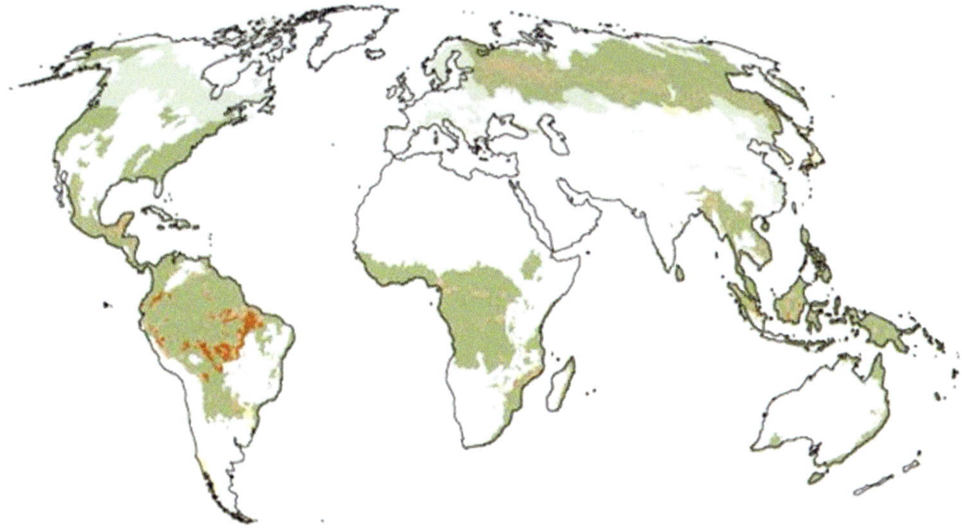

Fig. 16.2. Forest cover change between 1981 and 2010 using remote sensing data and expert opinion. From Lepers *et al.* (2005), published with permission.

The global inventory of tropical land under shifting cultivation (including fallow) was 3 million km^2 by the 1980s (Fig. 16.2) (Samec, 2023). Shifting cultivation in high-elevation forest lands adds to forest degradation by reducing the fertility level of the soil, which is accelerated by soil erosion due to land mass exposure. In the areas under shifting cultivation, nutrient losses occur through leaching, runoff and erosion, making the land uncultivable after two or three cropping seasons (Szott *et al.*, 1999; Panda *et al.*, 2005). Due to the changing dynamics of forest hydrology from deforestation, it is almost impossible for regeneration of the forest in the area under shifting cultivation (Szott *et al.*, 1999; Samec, 2023).

Potter *et al.* (2007) assessed land-cover change detection in California, USA, using the MODIS 250m resolution time series of the Enhanced Vegetation Index. The authors reported that areas affected by forest management and encroachment of residential development into natural vegetation zones should be prime locations for applications of land-cover change detection. Goward *et al.* (2008) reported that research projects within the North American Carbon Program are combining RS and forest inventory data to map the extent and rate of forest disturbance in the conterminous USA. Given that disturbance processes vary in their extent, duration and intensity, the authors suggested a multi-pronged approach with different satellite technologies targeted toward different space and time scales. For example, NOAA's AVHRR and NASA's MODIS are being used to map transient phenomena occurring at the coarsest spatial scales, including insect outbreaks, drought stress and storm damage, and for estimating fire emissions and global mapping of active fires and burned areas.

Since 1994, the CORINE land-cover data has provided land-cover levels in Europe at a 25 ha minimum surface unit, and 5 ha change detection between each data version, according to four forest classes (EEA, 2006). The European program used SPOT (Satellite Pour l'Observation de la Terre), MSS, TM, ETM+ and IPR6 data for the 1994, 2000 and 2006 data sets. The 2012 update was released in September 2015 with significant improvement and reliability according to the RS data used (i.e. SPOT-4 and IRS-P6 data) for a spatial coverage of 39 European countries. An example of forest and other natural cover changes in Europe between 2000 and 2006 as developed by European Environmental Agency is shown on the website (www.eea.europa.eu; accessed 14 May 2025), showing the distribution and the intensity of selected land-cover flows in forest area in Europe using CORINE data. CORINE land-cover classified data is comparable to the Landsat satellite-based land-cover classified map (National Land Cover Database) of the USA that is available through the Multi-Resolution Land Characteristics Consortium (www.mrlc.gov/, accessed 14 May 2025).

Kim *et al.* (2015) studied the land-cover change in North Korea from 2001 to 2014. They found consistent decreases in normalized difference vegetation index (NDVI) values for 14 years, but observed a 4% increase in forest land covers that included evergreen needleleaf forest, evergreen broadleaf forest, deciduous needleleaf forest, deciduous broadleaf forest, mixed forest, closed shrublands, open shrublands and woody savannahs (Fig. 16.3). This is a positive development by forest managers in North Korea. Globally, two aspects of climate change (i.e. temperature and precipitation) affect the photosynthesis of forest ecosystems. High-elevation tropical forests of five continents have been experiencing higher 'browning,' (i.e. forests are losing foliage) and less photosynthetic activities (Krishnaswamy *et al.*, 2014).

Mangrove forests are declining in many parts of the world and the decrease is more marked than in inland tropical forests. Giri *et al.* (2007) used RS data and geospatial mapping to study the hydrodynamics of the mangrove forests of the Sundarbans of Bangladesh from the 1970s to the 2000s. They found the mangrove forest and its inter-tidal zone hydrology processes were constantly changing due to erosion, aggradation, deforestation and mangrove rehabilitation programs.

16.2.2 Forest soil water/moisture estimation and forested wetlands analysis

In general, forest soils naturally consist of high levels of organic matter with high porosity and permeability, allowing high infiltration and low runoff (Pritchett, 1979; Osman, 2013). Soils in forested wetlands, found mostly in coastal and lower elevation flat areas, in general are saturated in nature with abundant availability of soil water. Upland forests also hold high amounts of soil moisture (Jipp *et al.*, 1998). Therefore, forest soils are the nexus for many ecological processes, such as energy exchange, water storage and movement, nutrient cycling, plant growth and carbon cycling at the base of the food web (Johnson *et al.*, 2000). According to the FAO soil

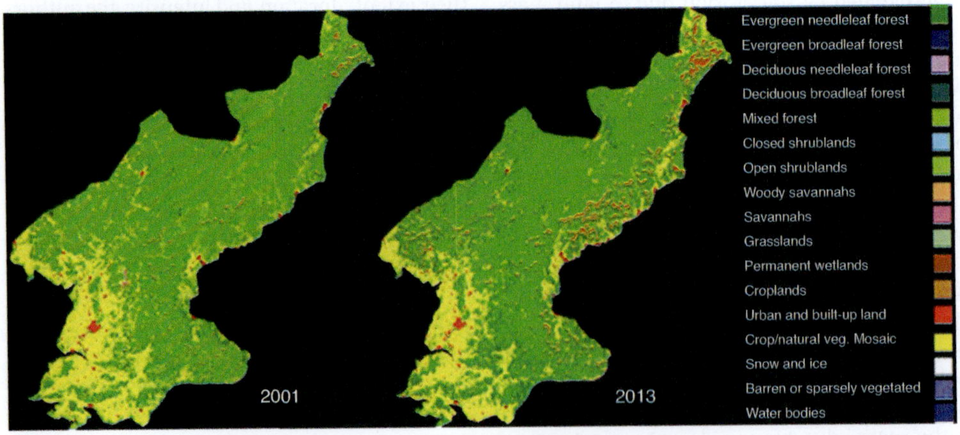

Fig. 16.3. Land-cover (including forest cover) change analysis maps for 2001 and 2013 in North Korea completed using MODIS land-cover product data (Kim *et al.*, 2015).

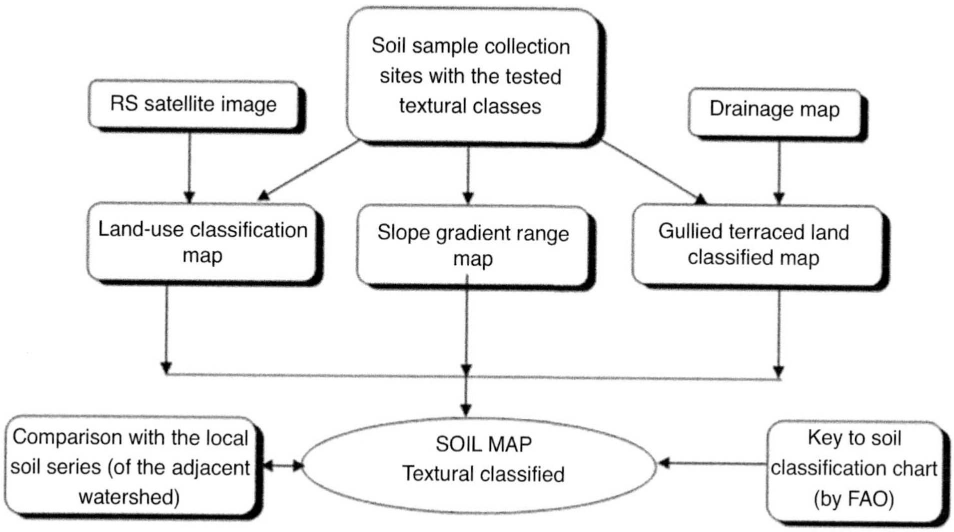

Fig. 16.4. Forested watershed soil map development procedure using geospatial technology and the Food and Agriculture Organization of the United Nations (FAO) soil classification key. RS, remote sensing. From Panda *et al.* (2004a).

map development process, forest soils differ from soils within other land-cover types. For example, a Himalayan overland flow study under forested versus degraded land cover showed that though Hortonian overland flow generation was dominant in both systems, hydrological characteristics varied in terms of runoff coefficients and soil physical properties. Geospatial technologies, including ground penetrating radar, can help study the soil water phenomena of the forest in a non-intrusive and efficient manner. For soil map development in areas where no maps are available/developed, forest land cover (using RS data) can be used as the area of a specific type of soil (Panda *et al.*, 2004a). Fig. 16.4 shows a GT-based procedure to create soil maps using forest hydrology information and FAO-suggested processes.

Understanding the dynamics of soil moisture and its measurements and modeling is critical for broad environmental areas such as agricultural and silvicultural crop management, water cycle and climate dynamics, flooding and forest fires, including hydrological processes. Although many methods are available to measure soil moisture, *in situ* measurement of the spatial distribution of soil moisture on a watershed/landscape scale is not typically possible. International efforts have been under way for decades to reliably measure soil moisture with an acceptable spatial resolution using a satellite-based RS technique.

Active microwave RS observations of backscattering, such as C-band vertically polarized synthetic aperture radar (SAR) observations from the second European RS (ERS-2) satellite, have the potential to measure moisture content in a near-surface layer of soil (Walker *et al.*, 2004). However, SAR backscattering observations are highly dependent on topography, soil texture, surface roughness and soil moisture, meaning that soil moisture inversion from single-frequency and polarization SAR observations is difficult. The authors reported some improvements in measurements of near-surface soil moisture with the ERS-2 satellite over Landsat. Microwave RS-based soil moisture estimates are limited to bare soil or low to moderate amounts of vegetation cover. Passive microwave sensors have the advantage of collecting soil moisture remote data with high vegetation cover like forest land cover but with a trade-off in the spectral resolution range. While the most useful frequency range for soil moisture sensing is 1–5 GHz, passive microwave RS is in a range of 10–20 km (Njoku and Entekhabi, 1996). Njoku and Entekhabi (1996) outlined the basic principles of the passive microwave technique for soil moisture sensing and how to optimally assimilate passive microwave data into hydrological models. Schmugge *et al.* (2002) remotely estimated forest surface soil moisture from passive microwave data.

Nolan and Fatland (2003) reported that recent advancements in making soil moisture models may act as the 'Rosetta Stone' that allows interferometric synthetic aperture radar (InSAR) measurement of soil moisture using existing satellites. Lu *et al.* (2005) demonstrated the feasibility of measuring changes in water level beneath tree cover more accurately using C-band InSAR images from the ERS-1 and ERS-2 satellites than the L-band for swamp forests in Louisiana. The ability to measure water-level changes in wetlands, and consequently in water-storage capacity, using RS may provide a required input for hydrological models and flood hazard assessments. Panda *et al.* (2015) used Band 5 (near infrared) and Band 7 (mid-infrared) to estimate plant moisture (stomatal conductance) and soil moisture in the forest cover with matured and young pine, switchgrass and pine understory with more than 70% accuracy.

Gessner *et al.* (2023) observed that forested watershed moisture content can be estimated directly as volumetric soil moisture or relative soil moisture and indirectly using indices that reflect moisture content conditions at various spatial scales through satellite RS to support the development of the soil water index, normalized difference moisture index, normalized difference water index, NDVI and normalized difference drought index. Wałęga *et al.* (2024), in their study of satellite data use for total direct runoff prediction using the curve number-based Modified Soil Moisture Equation model, used NOAA AVHRR-based NDVI to estimate soil moisture in larger forested watersheds in Europe and the USA.

Singh *et al.* (2021) evaluated the ability of using the Indian Space Research Organisation's Radar Imaging Satellite-1 (RISAT-1) C-band SAR observations in the SMAP active–passive algorithm to obtain soil moisture at 1, 3 and 9 km over the 'paddy-field' agricultural region that experiences seasonal flooding in the Mahanadi River Basin. They observed that the combination (RISAT-1 and SMAP) had the potential to estimate soil moisture in the forest landscape. Fig. 16.5 is a representation of the global SMAP L4 band map that contains the digital information of root zone (5 cm) soil moisture, surface soil moisture and average soil moisture at a crude 9 km spatial resolution scale. Vergopolan *et al.* (2021) developed improved spatial and temporal resolution SMAP-HydroBlocks (2021, https://

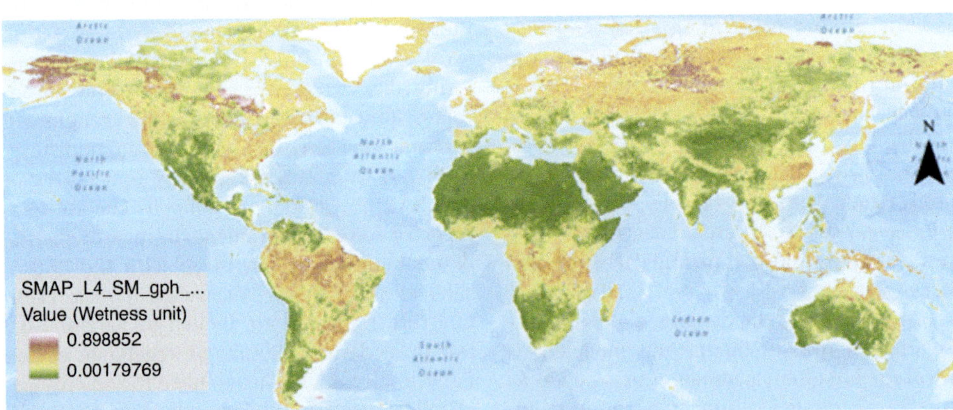

Fig. 16.5. Global SMAP L4 band map containing digital information on root zone (5 cm) soil moisture, surface soil moisture and average soil moisture. Note that SMAP root zone soil moisture is typically measured as dimensionless 'wetness units' with values ranging from 0 to 1. It represents the relative saturation of the soil, with 0 being completely dry and 1 being completely saturated. In this date image constructed by NASA, the green areas are very dry (e.g. northern Africa (Saharah), Australia, northern China), while the brownish to white areas are saturated areas (e.g. Greenland, rainforest areas including the Amazon rainforest)

waterai.earth/smaphb; accessed 14 May 2025) at 30 cm and 6 hour, respectively, through an 'image fusion' approach. This could assist forest hydrology managers in their proactive management decision support. Users should replicate the procedure by downloading the open-source code for the image fusion approach from Dr. Vergopolan's GitHub page (https://github.com/NoemiVergopolan/SMAP-HydroBlocks_postprocessing; accessed 14 May 2025). Mani and Varghese (2018) provided a detailed review of the integrated use of remotely sensed data, GPS and GIS to enable natural resource managers and researchers to develop management plans for a variety of agriculture and forest resource management applications in India. This detailed the procedure of using satellite RS data to forecast crop and forest yield through soil moisture, biomass and coverage area analyses. They explained how RS is being used to estimate spatially suitable locations for orchard development, drought assessment and monitoring, and soil moisture estimation for crop and forest growth management decision support over a study area encompassing the entire state of Maharastra, India (part of the Western Ghat Mountain Ranges). Studies in many other parts of the world have shown the potential of RS, GIS and GPS in monitoring forest hydrology for its sustainable good health.

16.2.3 Forest vegetation and biomass mapping

Forest vegetation has an apparent influence on microclimate (air temperature, humidity and wind speed) under the canopy compared with open area land covers. It is the 'active' surface for the absorption of solar energy and CO_2 and the release of O_2 and water vapor through ET, and has a localized effect. In the context of global warming and a climate change scenario, understanding the forest microclimate with respect to forest vegetation or forest biomass is a necessity. High-resolution orthoimagery along with an advanced image-processing approach are successful in forest vegetation speciation. Fig. 16.6 depicts the use of an object-based image analysis (OBIA) image segmentation approach in forest tree speciation in the Elachee Nature Center in Georgia with the use of very high-resolution (30 cm) orthoimagery and LiDAR data (to determine tree height) along with Visual Basic for Application (VBA) coding in ArcObjects platform.

Panda *et al.* (2016b) developed a GT-based procedure to estimate the biomass of pine, pine intermixed with understory, switchgrass and pine intercropped with switchgrass using: (i) LiDAR point cloud analyses to obtain tree/plant/crop heights: (ii) ultra-high spatial resolution (15 cm) orthoimagery analysis using the OBIA image segmentation approach as described above to map the canopy coverage of the plants; and (iii) an allometric equation to estimate the plant biomass with plant heights and canopy coverage area. They conducted the study in Carteret (North Carolina), Calhoun County (Mississippi) and Greene County (Alabama). Average matured pine biomass yield was approximately 140 Mg/ha, compared with other studies (obtained through literature review) that estimated it to be approximately 70–80 Mg/ha. The switchgrass biomass yield was around 40 Mg/ha in the three researched watersheds, compared with other USA studies showing yields of around 20–30 Mg/ha. The study by Panda *et al.* (2016b) was in commercially managed forests, which possibly explains the higher biomass yield. Fig. 16.7 is an example of LiDAR point cloud data analysis graphics of one of the three study areas described.

In Europe in general (and in France specifically), forest mapping and species identification were developed using RS data such as aerial infrared photography in the 1970s (Touzet and Lecordix, 2010) and more recently (since the 1990s) SPOT imagery at 10–20m resolution (i.e. panchromatic and visible and near infrared bands). The Soil and Water Assessment Tool (SWAT) model ArcGIS add-in was used to conduct ecohydrological modeling. Bärlund *et al.* (2007) assessed SWAT model performance in the evaluation of hydrology management actions for implementation of the Water Framework Directive in a Finnish forested catchment. They suggested that GT is an efficient tool to differentiate individual trees in the forest, determine the forest biomass and subsequently define the loss of water from the forest land via ET. Panda *et al.* (2015) developed a procedure to assess the water loss through ET in plots

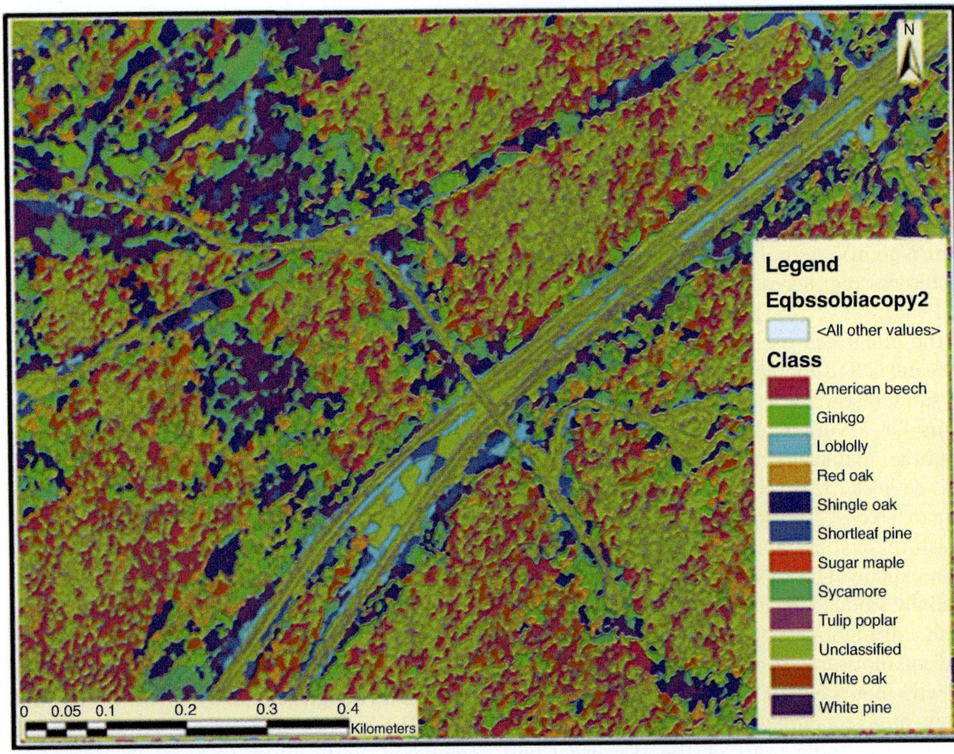

Fig. 16.6. Object-based image analysis of forest tree speciation using high-resolution orthoimagery and Visual Basic for Application (VBA) coding in the ArcObjects platform of ArcGIS 9.3.

with pine only, pine plus understory, pine and switchgrass intercropping, and switchgrass only using 30 cm LiDAR, 15 cm orthoimagery and expert knowledge of the field. Riegel (2012) used LiDAR data to develop a forest biomass quantification model, which provided insight on the overall forest health and helped in forest ET modeling. Forest net primary production is a field of RS research that includes hyperspectral data from airborne or satellite platforms such as AVIRIS or Hyperion (Ollinger and Smith, 2005). 'Primary production' is the accumulation of organic material produced by a plant (biomass), while 'net primary production' is the remaining biomass after subtracting energy used (respiration) for plant growth and development. In the coming years, the HYPXIM project (Michel *et al.*, 2011) aims at providing researchers, including forest RS topics, with hyperspectral satellite data, including visible and near infrared, and short-wave infrared sensors, at a high resolution (8–15 m). This project will be of great interest for forest research that needs the spatial resolution of airborne data with the global coverage facility of a satellite mission.

LiDAR usage as a combination of high-resolution orthoimagery to map forest biomass has gained ground globally, especially in Europe. Brovkina *et al.* (2017) mapped forest aboveground biomass (AGB) using an airborne hyperspectral and LiDAR data combination in a central Europe mountainous landscape. They developed three methods to estimate forest AGB at tree and plot levels and obtained good correlation with field biomass values for spruce, beech and a mixture of these species at tree and plot levels with the fused HS and LiDAR data analysis method. Puletti *et al.* (2020) analyzed the efficiency of different types of LiDAR. These included terrestrial laser scanning (TLS), aerial laser scanning (ALS) and spaceborne LiDAR in the form of GEDI data. These entailed different spatial scales for AGB mapping in the Mediterranean Forests in Europe. They

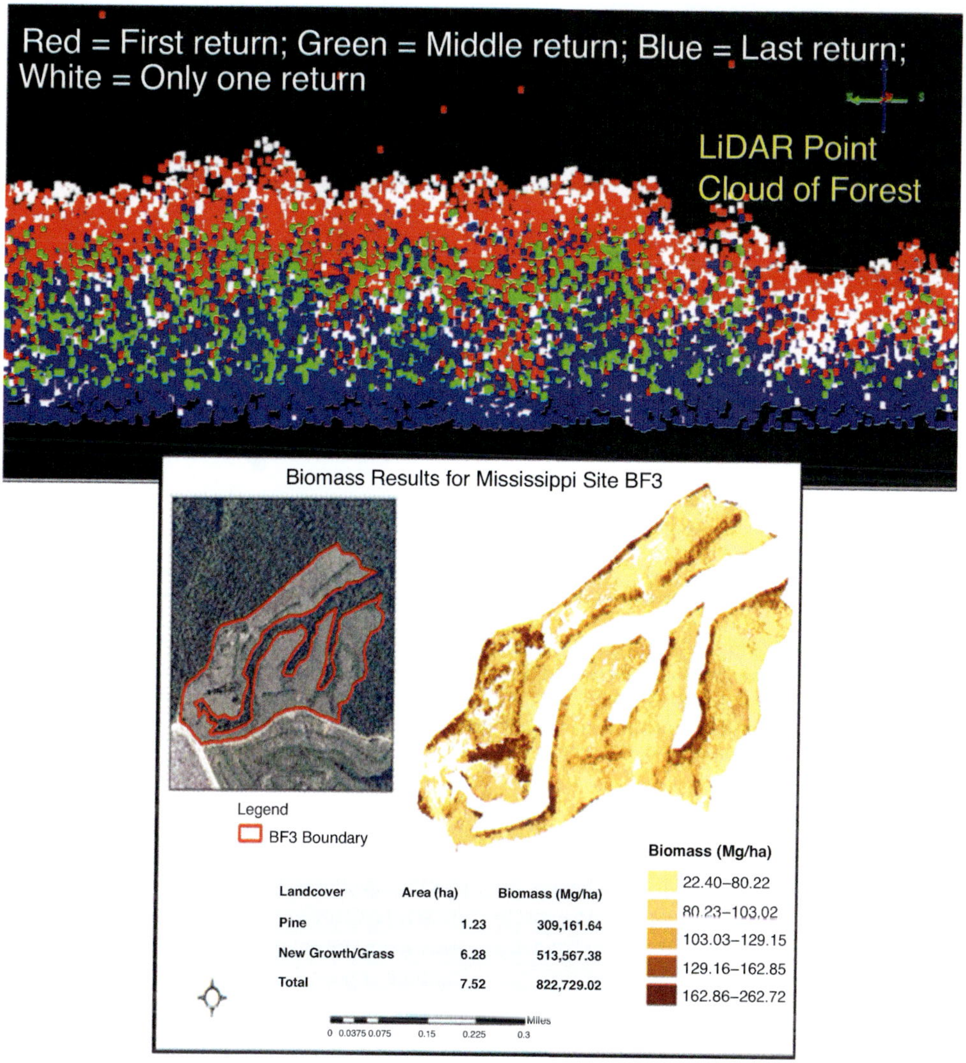

Fig. 16.7. LiDAR point cloud data analysis (top) and biomass estimated result (Mg/ha) of pine and switchgrass in the Calhoun County (Mississippi) site (bottom).

followed a three-step procedure: (i) validation of AGB estimates obtained from TLS against reference values obtained from conventional field inventory; (ii) calibration and validation of AGB estimates derived from ALS against TLS measurements, and (iii) calibration and validation of AGB estimates derived from GEDI data against mapped AGB values obtained from ALS. They obtained better correlation (R^2 range of 0.6–0.9 and root-mean-square error range of 29–49%) compared with field measurements with TLS. Panda *et al.* (2024a) used GEDI data to estimate the biomass of pine plantations in several Experimental Forest Research Stations in south-eastern USA, and are currently validating the results with field data. Fig. 16.8 shows the 2023 global biomass coverage map as created in ArcGIS Pro software after being downloaded from the Oak Ridge National Laboratory (ORNL) Distributed Active Archive Center (DAAC) site (https://daac.ornl.gov/cgi-bin/theme_dataset_lister.pl?theme_id=5; accessed 14 May 2025)).

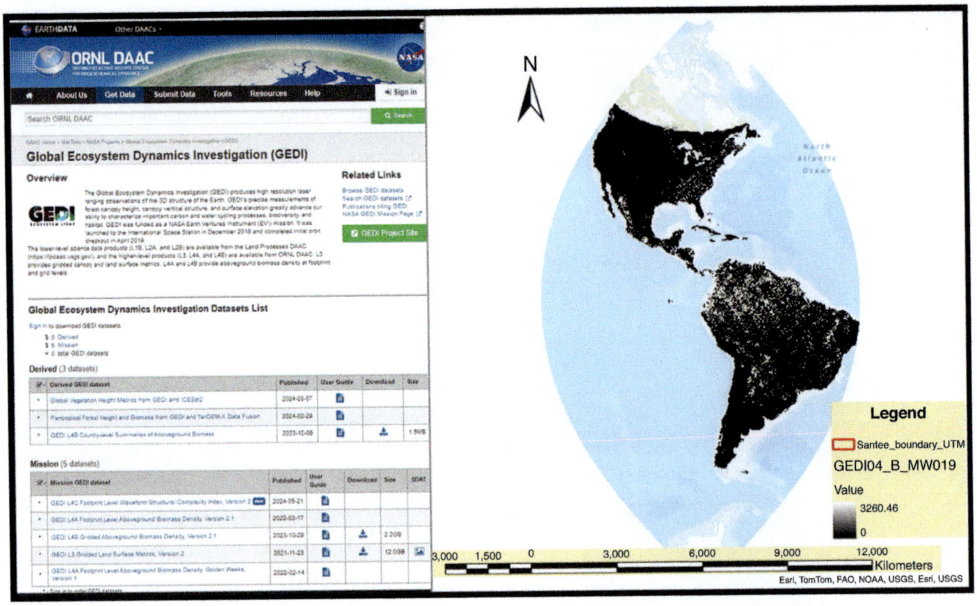

Fig. 16.8. Global biomass coverage raster (snapshot of the Americas) created in ArcGIS Pro software and shown alongside the Oak Ridge National Laboratory (ORNL) Distributed Active Archive Center (DAAC) site download page preview. Note that for the Global Ecosystem Dynamics Investigation (GEDI) digital data downloading links as shown on the left side of the figure, one can follow the site at https://daac.ornl.gov/cgi-bin/dataset_lister.pl?p=40 (accessed 14 May 2025).

The DAAC site contains 250 different types of GEDI-supported biomass spatial data (raster/grid format) and is updated periodically. It provides an amazing geospatial resource (well worth exploring) for remote forest management decision support.

16.2.4 Forest ET estimation

Different plant species compete for water at different rates in a forest due to the dense and complex composition of vegetation. The forest ET rate depends on many factors such as forest soils, vegetation, and climatic conditions such as air and canopy temperature, solar radiation, vapor pressure, wind velocity, and the nature and type of the evaporating surface in the forest range (Viessman and Lewis, 2002). Plant evaporation occurs mostly from the above-canopy interception and understory/litter evaporation. Transpiration encompasses the withdrawal and transport of water from the soil/aquifer system from plant roots and stems, and eventually from plant leaves into the atmosphere (Senay *et al.*, 2013). According to Viessman and Lewis (2002), available heat energy (radiation and air temperature), capacity to transport vapor away from the evaporative surface by wind and humidity, and soil water content availability are the guiding factors for ET. LAI, canopy temperature, canopy or stomatal conductance, wind velocity, and soil moisture or volumetric water content are the most important parameters of ET estimation (Panda *et al.*, 2014).

In recent years, RS-based GT has increasingly been used for development and application of ET models for determining and assessing ET rates compared with the field-measured data for agricultural and irrigated crop ecosystems (Cammalleri *et al.*, 2014). These novel approaches have recently been tested for individual forest species (see Panda *et al.*, 2014; Panda *et al.*, 2016a). These ET-related parameters (albedo, conductance, canopy temperature, soil moisture and LAI) are estimated with RS imagery data (Narasimhan *et al.*, 2003; Mu *et al.*, 2007; Chen *et al.*, 2014; Panda

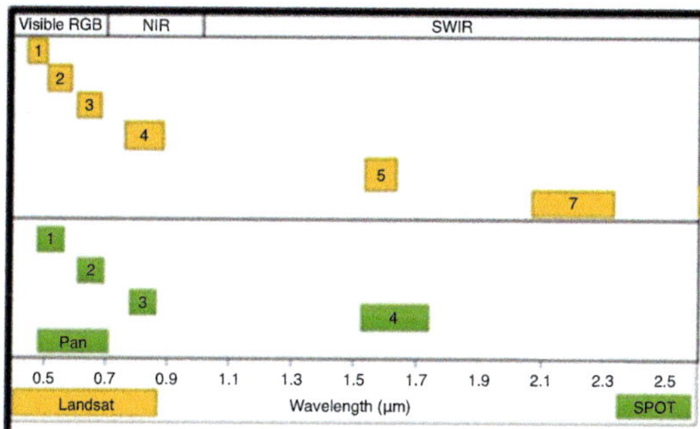

Landsat 7 ETM+ band (μm)	Applications	SPOT 4 band (μm)
Band 1 – Blue (0.45–0.52)	Coastal water mapping, soil vegetation differentiation, deciduous coniferous differentiation	
Band 2 – Green (0.52–0.60)	Green reflectance from healthy vegetation, iron content in rocks and soil	Band 1 (0.50–0.59) (green)
Band 3 – Red (0.63–0.69)	Chlorophyll absorption for plant differentiation	Band 2 (0.61–0.68) (red)
Band 4 – Near IR (0.76–0.90)	Biomass survey, water body delineation	Band 3 (0.79–0.89) (NIR)
Band 5 – Mid IR (1.55–1.75)	Plant moisture content, cloud/snow differentiation	Band 4 (1.58–1.75) (MIR)
Band 6 – Thermal IR (10.4–12.5)	Thermal mapping, soil moisture	
Band 7 – Mid IR (2.08–2.35)	Soil analysis	

Landsat 8 OLI and TIRS bands (μm)	Applications
Band 1 – Coastal aerosol (0.43–0.45)	Coastal and aerosol studies
Band2 – Blue (0.45–0.51)	Bathymetric mapping, distinguishing soil from vegetation and deciduous from coniferous vegetation
Band 3 – Green (0.53–0.59)	Emphasizes peak vegetation, which is useful for assessing plant vigor
Band 4 – Red (0.64–0.67)	Discriminates vegetation slopes
Band 5 – NIR (0.88–0.85)	Emphasizes biomass content and shorelines
Band 6 – SWIR 1 (1.57–1.65)	Discriminates moisture content of soil and vegetation penetrates thin clouds
Band 7 – SWIR 2 (2.11–2.29)	Improved moisture content of soil and vegetation and thin cloud penetration
Band 8 – Panchromatic (0.50–0.68)	15 m resolution, sharper image definition
Band 9 – Cimus (1.36–1.38)	Improved detection of cirrus cloud contamination
Band 10 – TIRS 1 (10.60–11.19)	100 m resolution, thermal mapping and estimated soil moisture
Band 11 – TIRS 2 (11.5–12.51)	100 m resolution, improved thermal mapping and estimated soil moisture

Fig. 16.9. General comparison of Landsat and SPOT spectral bands for earth observation application. B, blue; ETM+, Enhanced Thematic Mapper Plus; G, green; MIR, mid-infrared; NIR, near-infrared; OLI, Operational Land Imager; R, red; SWIR, shortwave infrared; TIR, thermal infrared; TIRS, thermal infrared sensor.

Table 16.1. Input–output correlation relationship for model development.

Models (with 2006–2012 data)	Input parameters (remote-sensing Landsat 7 ETM+ based)	Output parameters (field data)
ET	Plot SAVI meansPlot NDVI meansPlot VVI meansIndividual band 5, 6 and 7 DN value averages	Plot averages of calculated ET values from FLUX instrument (average of 12.00–14.00 hour) (W/m^2)
Soil moisture	Band 7 means	Plot averages of 30 cm depth soil moisture value (%)
Canopy temperature	Band 6 means	Plot averages of 12.00–14.00 hour (°C)
Canopy conductance	Band 5 means	Plot averages of 12.00–14.00 hour (m/s)

DN, digital number; ET, evapotranspiration; NDVI, normalized difference vegetation index; SAVI, soil-adjusted vegetation index; VVI, vegetation vigor index.

et al., 2016a). The spectral information is particularly useful in applications dealing with mapping and modeling biophysical properties of ecosystems such as water quality, plant vigor and soil nutrients (i.e. Landsat individual bands cater to very specific earth observation applications; see Panda *et al.*, 2016a). As shown in Fig. 16.9, Landsat individual bands or a combination of bands through ratio development can estimate ecohydrological parameters. Panda *et al.* (2016a) used Landsat 7 and Landsat 8 images to develop ET and ET parameter models for homogenous pine forest in coastal North Carolina. Table 16.1 provides a geospatial-based input and an ET and ET parameter output correlation chart. Owe *et al.* (2001) included many individual research articles describing the use of RS in ET and ET parameter estimation along with other hydrological processes.

Panda *et al.* (2024a) developed correlation analytics using remotely sensed ET data obtained from ECOSTRESS (L2 and L3 bands), forest floor and plant soil moisture data from SMAP, and an ET distribution raster created with a Landsat soil-adjusted vegetation index (SAVI)/LAI/albedo-based algorithm (Huete, 1988; Querino *et al.*, 2016; Chen *et al.*, 2020) and compared them with eddy flux tower-provided *in situ* ET data from 15 flux towers in Virginia, North Carolina, South Carolina and Georgia, encompassing much of the historic range of long-leaf pine. The Landsat SAVI/LAI/albedo-based algorithm estimated daily ET (12 June 2023) of 5.38 mm (WS77) and 5.35 mm (WS80) was found to be close to the actual flux tower-based measurement of 2.94 mm (WS77) and 4.35 mm (WS80) in the Santee Experimental Forest region, South Carolina. The 15 flux towers sites ECOSTRESS, SMAP and GEDI digital values had a strong correlation with R^2 values greater than 0.77, demonstrating that biomass, soil moisture, land surface temperature, precipitable water vapor and emissivity had strong correlations, although they were acquired through different satellites (Panda *et al.*, 2024a).

16.2.5 Forest hydrology-attributed geohazards analysis

Different geohazards are directly or indirectly related to forest and forest hydrology. Geohazards such as wildfires, landslides, drought and flooding may be harmful to humans and the biodiversity directly associated with forest cover. These hazards can be monitored, managed and even pre-warned with the use of GT. The following are a few applications analyzing the susceptibility or vulnerability of such geohazards related to forest hydrology.

Drought and forest fires

Forest fire management is always an issue. A persistent La Niña effect over the last two decades has created severe drought conditions in the USA. These were particularly evident in the west coast states of the USA from 2011 to 2015 and at many locations in the country over the years due to erratic rainfall patterns (Lenihan and Bachelet, 2015). Ault (2020) predicted that, globally, droughts in the future are likely to be more frequent, severe and longer lasting than they have been in recent decades. Drought hazard over Europe during 2018–2020 was exceptional according to a comprehensive spatiotemporal assessment benchmarking past exceptional events during the period 1766–2020, according to Rakovec (2022). They identified the 2018–2020 drought event as a new benchmark that showed unprecedented intensity persisting for more than 2 years. This is a longer period than older benchmark events and similar to the US West Coast drought in 2011–2015 (Rakovec, 2022). It exhibited a mean areal coverage of 35.6% over Europe and an average duration of 12.2 months in most spatial locations with its near-surface air temperature anomaly reaching +2.8 K compared with past events. These persistent and benchmark global drought events not only damage crops but also pose a significant threat to global forests. These events can have far-reaching consequences, including increased wildfire risk, decreased forest health and altered hydrological cycles. Drought reduces soil moisture, making forests more susceptible to ignition and rapid fire spread with more available fuel load. This dry vegetation becomes highly flammable, and this can create wildfires that become intense, unpredictable and difficult to contain.

The drought in California and other West Coast forests led to increased wildfires in 2015; since that date, wildfires in the USA and globally have increased. The increasing frequency and intensity of wildfires globally is driven by factors including climate change, deforestation and land-use change/human activities that are interconnected. Forest fires or wildfires are regulated by many environmental features of forests, including soil water content, forest topography, forest infrastructure, forest cover microclimate and especially forest species. Dudley *et al.* (2015) developed a geospatial model for determining the locations of forest fire susceptibility in Sumter National Forest in South Carolina. The authors used slope, aspect, slope rate of spread, slope suppression difficulty, NDVI, road buffer, fuel biomass density, urban fuel load and lightning strike frequency rasters to develop a fully automated geospatial model that predicts wildfire-vulnerable locations on a scale of low to high. Perry and Panda (2022) used SMAP soil moisture data and North American Land Data Assimilation System (NLDAS) Noah Land Surface Model L4 Monthly 0.125 × 0.125 degree V002 (NLDAS_NOAH0125_M) at GES DISC (https://worldview.earthdata.nasa.gov/; accessed 14 May 2025), along with geospatially distributed rasters such as slope, ET, lightning flash density, soil available water storage, soil drainage, detailed land use derived from the National Agriculture Imagery Program (www.fsa.usda.gov/programs-and-services/aerial-photography/imagery-programs/naip-imagery/index; accessed 14 May 2025) imagery, normalized difference water index derived from Landsat imageries (MIR and red band indexing), precipitation, temperature and soil hydrological group to predict the wildfire vulnerability using the Delphi method of weight selection for each parameter vulnerability scale combination. Figures 16.10 and 16.11 are the comprehensive GeoAI-supported automated geospatial model schematic graphics and the final wildfire spatial vulnerability map of Okefenokee Swamp (wetlands) in Georgia, USA, respectively. Comparison of historical a wildfire location map with our results showed that more than 76% of the area that has been under wildfire in the swamp is estimated/predicted to be under the highest category of wildfire spatial vulnerability scale (Perry and Panda, 2022). This modeling approach can be replicated for other study areas for proactive management decision support to safeguard lives and biodiversity.

Remote sensing applications for studying watershed-scale fires, their measurement, their effects on biogeochemistry and the atmosphere, and their ecohydrological effects have been studied extensively by Riggan *et al.* (2004) and Riggan and Tissell (2009). This group also led the development of the FireMapper thermal-imaging radiometer. This was applied to measurement and monitoring of large wildland fires and forest drought stress and mortality in mixed conifer forest (Riggan *et al.*, 2003). A study on tracking the MODIS NDVI time series to estimate fuel accumulation was conducted by Uyeda *et al.* (2015).

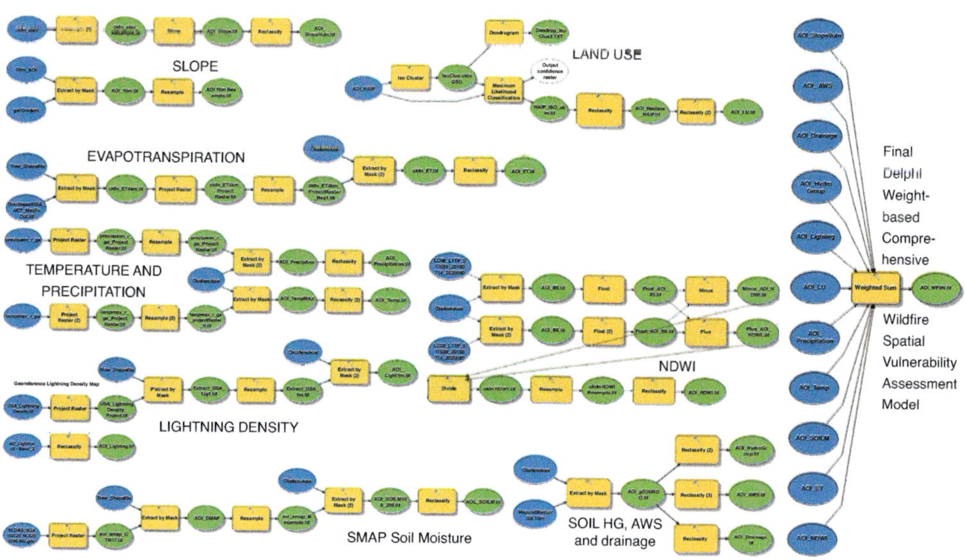

Fig. 16.10. Wildfire spatial vulnerability assessment comprehensive automated geospatial model schematic developed with geospacial artificial intelligence for proactive management decision support (Perry and Panda, 2022).

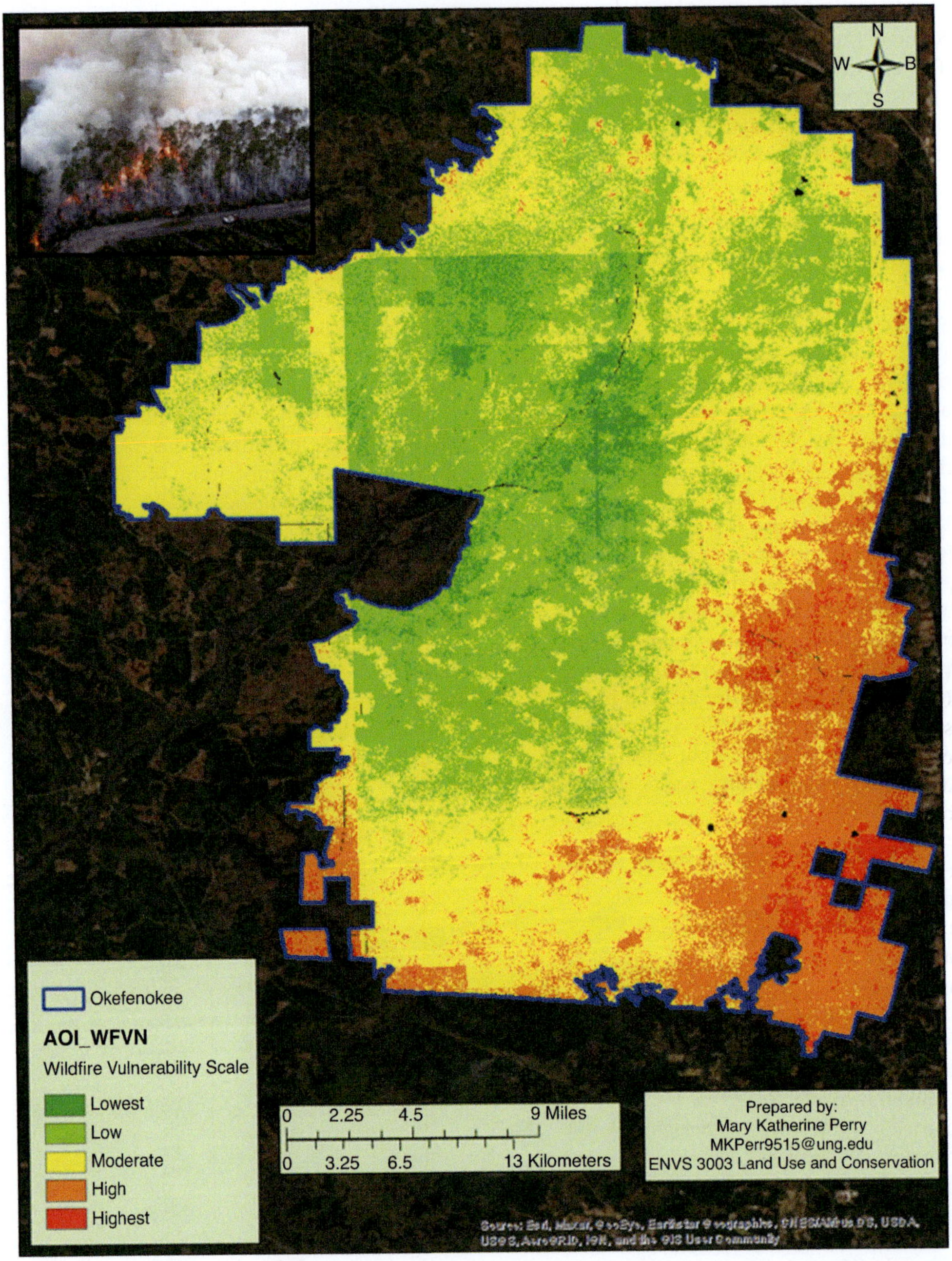

Fig. 16.11. Wildfire susceptibility/vulnerability map of Okefenokee Swamp in Georgia, USA (Perry and Panda, 2022).

Forest fires may affect the hydrological cycle and thus rainfall–runoff modeling (Eisenbies *et al.*, 2007; Folton *et al.*, 2015). Recently, Chen *et al.* (2013) analyzed satellite observations of terrestrial water storage from the Gravity Recovery and Climate Experiment (GRACE) mission, along with satellite observations of fire activity from the MODIS mission for the Amazon region. Based on the contrasting analysis of data for high- and low-fire years from 2002 to 2011, the authors suggested that, at least qualitatively, water storage as measured by GRACE can provide information to help predict the severity of a fire season in the region several months in advance.

Landslides

Landslides may be associated with drought and large wildfires. Large wildfires after a persistent drought decrease the forest plant density, and hence the plant root and soil-binding power diminishes. Forest soils are looser as a result of drought conditions and hence are more vulnerable to erosion. Therefore, heavy precipitation may lead to a large downslope movement of soil in the form of a landslide. The geology of the forest area plays a large role in landslides. Nolan *et al.* (2011) showed the advantage of GT in determining susceptibility to landslides in the Coosawattee watershed in the Chattahoochee National Forest of north Georgia. They used geospatial data, such as soil texture, soil drainage, maximum water capacity, bulk density, lithology, basement depth, slope, storm surge and land use/land cover.

Due to its mountainous nature and extreme precipitation events (attributable to global warming and climate change), increasingly numerous landslides are being observed at the base of the Himalayan Mountain ranges. Rana (2024) developed a landslide and geospatial database and an automated geospatial model (Fig. 16.12) with GeoAI to determine landslide susceptibility in the Himachal Pradesh, northern India. He used contributing factors such as bulk density, clay percentage, sand percentage, silt percentage, soil texture, land use/land cover, curvature, sand:clay ratio, lineament density, normal annual rainfall, slope, rainfall intensity, distance from rivers and distance from roads on a GeoAI-supported ArcGIS ModelBuilder platform. He classified each contributing factor on a scale of 1–8 (highest to lowest) in terms of landslide susceptibility and assigned each factor a weight according to its correlation with landslide probability/susceptibility using the Delphi process (radar chart in Fig. 16.12, top-right inset). The comprehensive automated geospatial model provided a landslide spatial vulnerability map (Fig. 16.12, bottom-right inset) to help stakeholders make proactive management decisions. This methodology provides a framework for identifying landslide-prone areas with satisfactory outcomes, especially in ungauged areas. The study results can assist local governments and researchers in landslide management.

Floods

Forest cover may reduce flooding due to its infiltrating soil (Booth *et al.*, 2002; van Dijk and Keenan, 2007). Deforestation or forest degradation may change the soil dynamics and lead to higher runoff from such areas. Often forests include areas of steeper topography, and flooding vulnerability increases in these areas. Several methods have been used to model flood potential sites throughout the world, but GT usage is preferred, because all flooding parameters are considered to be spatial in nature. Choi and Liang (2010), in South Korea, used the digital elevation model (DEM) hydrological soil group in their models to study the mostly mountainous watershed for flood vulnerability analysis. Ramsey *et al.* (2013) reported that SAR inundation mapping could provide an improved representation of coastal flooding, including flooding in the mangrove areas.

Yellowstone National Park (YNP) experienced a ‘500-year event’ in June 2022 due to ‘unprecedented’ rainfall and snowfall. Perry *et al.* (2022) made this a case study and noted that GeoAI-supported geospatial modeling automation could accurately assess the spatial flooding possibility. They used snowmelt algorithms with a spatial snow-depth raster layer along with other contributing temporal parameters to develop years 2000 and 2020 spatial flood potential assessment maps (Fig. 16.13) showing the increased potential of flooding in Yellowstone National Park over the years. The 2020 result (highest potential) matched the actual flooding areas of Yellowstone National Park by over 91%. These models help forest hydrologists and other managers for proactive management decision support.

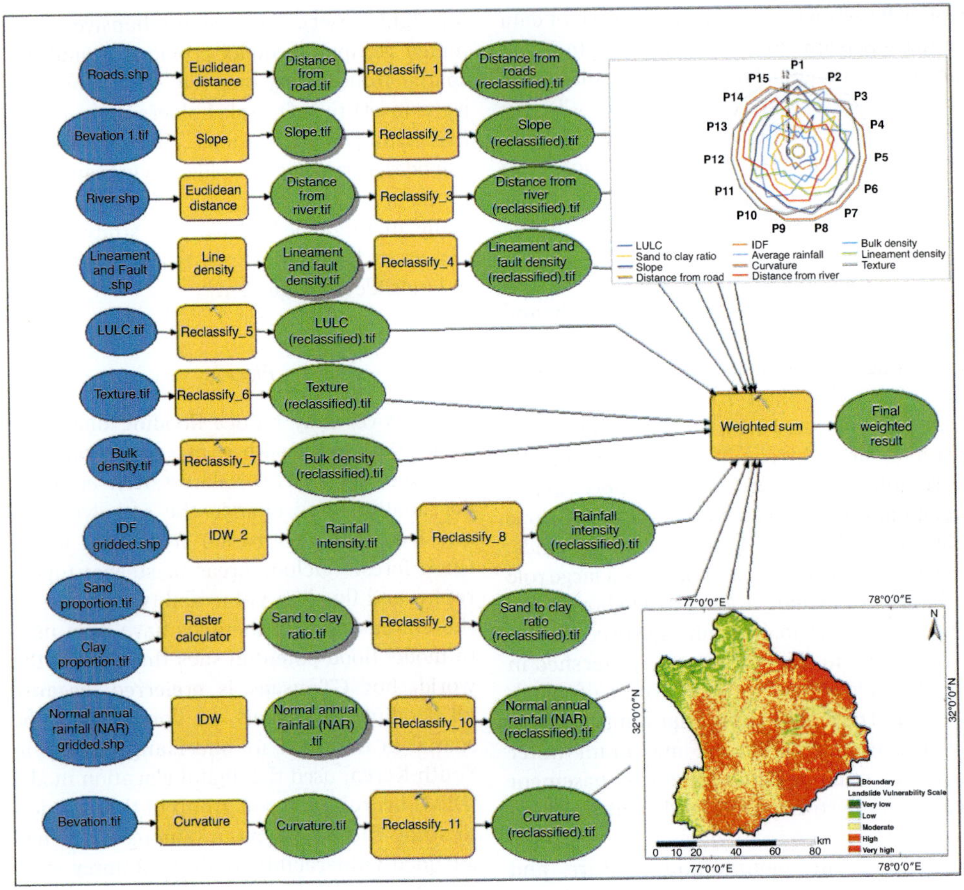

Fig. 16.12. Landslide spatial vulnerability assessment comprehensive automated geospatial model schematic developed with geospacial artificial intelligence for proactive management decision support with (top-right inset) radar chart depicting the weights of each contributing factor in the model design as determined using the Delphi method, and the final landslide vulnerability map (bottom-right inset) of the Beas River catchment of Himachal Pradesh, India (Rana, 2024).

16.2.6 Forest stream water quality management

Stream water quality may be a consequence of forest hydrology management. The riparian forest cover along streams is the transition or ecotone between terrestrial and aquatic ecosystems. It supports a host of essential functions (Naiman and Décamps, 1997), including filtering runoff nutrients, providing shade that influences water temperature and dissolved oxygen concentration in waterbodies, depositing leaf litter into the water as a carbon source for microbes and invertebrates at the base of the food web, supporting the stream banks structurally, supporting channels with large woody debris, diversifying stream habitats, and providing essential cover for flood flows and sediment transport. RS technology is efficiently being used to delineate the riparian forest cover (or lack of it) along streams. The 'Watershed Habitat Evaluation and Biotic Integrity Protocol' developed by Dr. Reuben Goforth (Carlsen *et al.*, 2004) and similar protocols developed with the USDA Forest Service use stream riparian forest cover and stream channel attributes as major parameters to determine stream health.

Fig. 16.13. Flooding potential temporal maps (2000 and 2020) of Yellowstone National Park in Wyoming (Perry *et al*., 2022). Note that in both 2000 and 2020, most of Yellowstone National Park showed a similar flooding potential. However, the bottom central areas with green polygons are the actual areas where the flood potential increased over the 20 years. The green polygon areas (highest flooding potential) dramatically increased from 2000 to 2020 based on geomorphological and enviro-climatic modeling analyses.

Zhang and Barten (2008) developed the Watershed Forest Management Information System (WFMIS) to help protect the water resources from watershed/forest degradation. WFMIS was developed as an extension of ArcGIS with three submodules to address non-point-source pollution mitigation, road system management and silvicultural operations (Zhang and Barten, 2008). Panda *et al.* (2004b) developed a GIS-based watershed management DSS for determining water quality and quantity variability due to annual land cover changes. The study area, the Beaver Lake watershed, was a forested watershed with more than 61% forest cover. This DSS is very important for forest hydrology management decision support in water quality monitoring of forested streams (Panda *et al.*, 2004b). Zhang and Barten (2008) also developed a stand-alone interface in VBA. A user can input the forest cover loss area in acres, and the software will predict the water quality change (total phosphorus, total nitrogen, PO_4, NO_3 and total suspended solid) in kg/ha/year. The GT-based forest biomass studies discussed earlier would help quantify the water quality dynamics of forest streams. Forest trails, nutrient-rich forest soils and unique forest hydrological cycles are the causes of different forest stream water quality dynamics (Lowrance *et al.*, 1997). The Water Erosion Prediction Project (WEPP) is a process-based model that allows continuous simulation in small watersheds and hillslope profiles to estimate soil erosion and subsequent water quality dynamics in forests (Flanagan *et al.*, 1995). Geospatial interface for WEPP (GeoWEPP) has the potential to predict soil and water erosion-based forest stream water quality monitoring and management using PRISM climate data, burn severity data, distributed WEPP land-use data, distributed WEPP soil parameters and DEM. The model will accurately and efficiently predict the forest soil erosion rate to support forest managers.

16.3 Modelling Forest Hydrological Processes with Geospatial Technology Support

Distributed models such as the MIKE SHE, SWAT, TOPMODEL (a topographic model) and others are widely used to simulate ecohydrological processes in a large watershed-scale landscape, which generally contains a forest land use (Amatya *et al.*, 2011). Such distributed models use parameters directly related to the physical characteristics of the catchment (watershed), including topography, soil, land use/land cover, geology, spatial variability in physical characteristics and meteorological conditions (Pietroniro and Leconte, 2000). Such models provide the possibility of deriving their inputs from remotely sensed data (Gupta *et al.*, 2008). The RS technique is useful in deriving high-resolution information in spatial and temporal domains about necessary hydrological parameters. These provide a new means for calibration and validation of distributed hydrological models (Fortin *et al.*, 2001).

A decade in hydrological research on ungauged basins (Hrachowitz *et al.*, 2013) has demonstrated interest in using RS for collecting data to predict water flows including topographical (i.e. DEM) and land-cover layers for spatially distributed hydrological models (Doten *et al.*, 2006; Khan *et al.*, 2010). In France, the impacts of the Mediterranean forest basin have been studied (Cosandey, 1993; Cosandey *et al.*, 2005) using IFN (National Forest Inventory, France) forest land-cover data derived from aerial photographic mapping. Using an existing hydrological model that includes lateral groundwater flow, Sutanudjaja *et al.* (2014) showed that remotely sensed soil moisture data can be valuable for accurately predicting groundwater dynamics at a local level and could be scaled up to provide more accurate information about groundwater variability and availability, and reserves across the globe.

Troch *et al.* (2007) investigated the potential use of GRACE data to detect the monthly terrestrial water storage changes in the Colorado River basin using *in situ* data from 2003 to 2006 and compared that data against the basin-scale water balance (BSWB)-based models. The authors found that the GRACE results agreed with the BSWB model that the winter of 2005 was generally wet, but the GRACE results disagreed with the exact timing of the rainfall events. With respect to BSWB, GRACE underestimates the severity of the subsequent dry period. Scanlon *et al.* (2012) reported that general correspondence between GRACE and groundwater level data found in the California Central Valley validates the methodology and increases confidence in the use of GRACE satellites to monitor groundwater storage changes. Van Griensven *et al.* (2012) evaluated LAI and ET simulated by the SWAT model with corresponding values obtained using remotely sensed data. The authors' evaluation showed that values for ET tended to be slightly underestimated, while those for LAI were visibly overpredicted. At the same time, the satellite images clearly followed the land-use pattern of the basin and showed uniform values for the different types of vegetation. This suggests that, for the SWAT model, the forest species input parameter development process needs updating to provide 'correct' results in forest ET estimation.

16.4 New Technology in Forest Hydrology Management

Higgins *et al.* (2014) used a satellite-based interferometry technique to map the subsidence of the Ganges–Brahmaputra river delta covering a 10,000 km^2 area over 4 years. The authors found that the delta is subsiding at a rate of about 10 mm/year around Dhaka (Bangladesh's capital) and about 18 mm/year outside the city. They indicated that satellite interferometry can be a useful method in accurately gauging the subsidence in deltas. Such techniques may be useful in large deltas with mangrove forests in Asia and Africa. Groundwater is the last component of the hydrological cycle to be quantified by RS (Becker, 2006). The author explored the potential for RS of groundwater in the context of active and planned satellite-based sensors. Again, these methods may well be applicable for large groundwater-dominated forested landscapes around the world.

Ongoing efforts under the planned NASA/Center National d'Etudes Spatiales Surface Water and Ocean Topography (SWOT) satellite mission, including the planned new algorithm using AirSWOT (an airborne platform approximating SWOT's capabilities), will provide an enhanced tool to accurately characterize river discharge from space by providing concurrent observations of water surface elevation, surface slope and inundated area for wide rivers (Pavelsky, 2012). Efforts are also under way to develop and expand space techniques to measure changes in terrestrial waters (Alsdorf *et al.*, 2003; Cazenave *et al.*, 2004). Such techniques will be useful for large forest landscapes such as the Amazon River basin and streams/rivers draining long-term USDA Forest Service experimental forests and ranges in the conterminous USA.

16.5 Conclusions

GTs including RS, GIS, GNSS and IT have tremendous potential for better decision support in forest hydrology management. More and more hydrology models/software, such as GeoWEPP https://fargo.nserl.purdue.edu/geowepp/; accessed 14 May 2025), the Automated Geospatial Watershed Assessment (AGWA) tool (www.epa.gov/water-research/automated-geospatial-watershed-assessment-agwa-tool; accessed 14 May 2025) and the USDA Forest Service database tools, Natural Resource Manager (NRM) (https://data.fs.usda.gov/geodata/edw/datasets.php?xmlKeyword=nrm; accessed 29 May 2025), are being developed for forest hydrology management that uses GT. Comprehensive complete automated geospatial models are being developed in the ArcGIS ModelBuilder platform, which can use any type of RS and GIS data to analyze forest hydrological behavior. Most importantly, GPS technology is getting better and more efficient with the introduction of more satellites into space by Europe, Russia, India and China. GNSS – the advanced version of GPS – is being used as a major tool in fighting forest fires, landslides and other forest-related geohazards in all parts of the world. Image spatial and spectral resolutions are getting better, in part due to the participation of private entrepreneurs in real-time image data collection, and also with the introduction of large-scale hyperspectral imaging.

Similar to SMAP and ECOSTRESS, a new age satellite, TRISHNA (Thermal InfraRed Imaging Satellite for High-resolution Natural Resource Assessment) is being launched as a sun-synchronous satellite this year (2025) through a joint satellite mission between the Indian Space Research Organization (ISRO) (India) and CNES (France). Unlike SMAP and ECOSTRESS satellites, which are placed on International Space Station, TRISHNA is geosynchronous and will revisit any location on the globe in every 3 days due to its large field of view and it has a better spatial resolution, i.e., 57 m on land and coastal areas and 1 km on the ocean. Trishna will have four Thermal InfraRed (TIR) and suplemented with six optical Visible and Near Infrared (VNIR)/Short Wave InfraRed (SWIR) sensor (VSWIR), making it an excellent data juxtaposition platform to analyze remotely for proactive and interactive management decision support in several fields of investigation including ecosystem water stress and water resource management, hydrology of coastal strips and land surfaces, urban, heat islands, the cryosphere and atmosphere (https://www.eoportal.org/satellite-missions/trishna#vswirinstrument). Forest Hydrology study will be most benefitted through TRISHNA.

Erratic weather conditions due to global warming and climate change and the consequential El Niño and La Niña effects are creating severe disruption in forest management. Therefore, freely available MODIS and Landsat 8 data and the subsequently generated NDVI and Enhanced Vegetation Index, along with open-source (free) GIS software such as Map Window (www.mapwindow.org/; accessed 14 May 2025), QGIS (www.qgis.org/; accessed 14 May 2025) and GRASS (https://grass.osgeo.org/; accessed 14 May 2025), would help develop forest hydrology management decision support to save forest land cover from degradation. It is expected that with the advent of UAVs and UASs, there will be opportunities for improved forest management. GT is getting easier, and the working procedures are becoming available in the public domain for the on-specialist use. Stakeholders should take advantage of these advanced technologies to take prudent steps toward forest hydrology management decision support.

References

Alsdorf, D., Lettenmaier, D.D. and Vörösmarty, C. (2003) The need for global, satellite-based observations of terrestrial surface waters. *Eos* 84, 269–280.

Amatya, D., Trettin, C., Panda, S.S. and Ssegane, H. (2013) Application of LiDAR data for hydrologic assessments of low-gradient coastal watershed drainage characteristics. *Journal of Geographic Information System* 5, 175–191.

Amatya, D.M., Douglas-Mankin, K.R., Williams, T.M., Skaggs, R.W. and Nettles, J.E. (2011) Advances in forest hydrology: Challenges and opportunities. *Transactions of the ASABE* 54, 2049–2056.

Amatya, D.M., Sun, G., Green Rossi, C., Ssegane, H.S., Nettles, J.E. *et al.* (2015) Forests, land use change, and water. In: Zolin, C.A. and Rodrigues, R. d. A.R. (eds) *Impact of Climate Change on Water Resources in Agriculture*. CRC Press, Boca Raton, Florida, pp. 116–153.

Ault, T.R. (2020) On the essentials of drought in a changing climate. *Science* 368, 256–260.

Austin, M.P. and Meyers, J.A. (1996) Current approaches to modelling the environmental niche of eucalypts: Implication for management of forest biodiversity. *Forest Ecology and Management* 85, 95–106.

Bärlund, I., Kirkkala, T., Malve, O. and Kämäri, J. (2007) Assessing SWAT model performance in the evaluation of management actions for the implementation of the water framework directive in a finnish catchment. *Environmental Modelling and Software* 22, 719–724.

Barten, P.K. (2006) Overview of forest hydrology and forest management effects. Massachusetts Envirothon, Westford, Massachusetts. Available at: https://portal.ct.gov/-/media/deep/water/wetlands/seg32009foresthydrologyoverviewpbartenpdf.pdf (accessed 2 May 2025).

Becker, M.W. (2006) Potential for satellite remote sensing of ground water. *Ground Water* 44, 306–318.

Booth, D.B., Hartley, D. and Jackson, R. (2002) Forest cover, impervious surface area, and the mitigation of stormwater impacts. *Journal of the American Water Resources Association* 38, 835–845.

Brovkina, O., Novotny, J., Cienciala, E., Zemek, F. and Russ, R. (2017) Mapping forest above-ground biomass using airborne hyperspectral and LiDAR data in the mountainous conditions of Central Europe. *Ecological Engineering* 100, 219–230.

Cammalleri, C., Anderson, M.C., Gao, F., Hain, C.R. and Kustas, W.P. (2014) Mapping daily evapotranspiration at field scales over rainfed and irrigated agricultural areas using remote sensing data fusion. *Agricultural and Forest Meteorology* 186, 1–11.

Carlsen, W.S., Trautmann, N.M., Cunningham, C.M., Krasny, M.E., Welman, A. *et al.* (2004) *Watershed Dynamics*. Cornell Scientific Inquiry Series. NSTA Press, Arlington, Virginia.

Cazenave, A., Milly, P.C.D., Douville, H., Benvineste, J., Kosuth, P. *et al.* (2004) Space techniques used to measure change in terrestrial waters. *Eos* 85, 106–107.

Chen, H., Zhu, G., Zhang, K., Bi, J., Jia, X. *et al.* (2020) Evaluation of evapotranspiration models using different LAI and meteorological forcing data from 1982 to 2017. *Remote Sensing* 12, 2473.

Chen, Y., Velicogna, I., Famiglietti, J.S. and Randerson, J.T. (2013) Satellite observations of terrestrial water storage provide early warning information about drought and fire season severity in the Amazon. *Journal of Geophysical Research: Biogeosciences* 118, 495–504.

Chen, Y., Xia, J., Liang, S., Feng, J., Fisher, J.B. *et al.* (2014) Comparison of satellite-based evapotranspiration models over terrestrial ecosystems in China. *Remote Sensing of the Environment* 140, 279–293.

Choi, H.I. and Liang, X.Z. (2010) Improved terrestrial hydrologic representation in mesoscale land surface models. *Journal of Hydrometeorology* 11, 797–809.

Cosandey, C. (1993) *Conséquences hydrologiques d'une coupe forestière. Le cas du bassin de la Latte (Mont-Lozère, France)*. In: Gresilen, M. (ed.) *L'eau, La Terre et Les Hommes, Hommage à René Frécaut*. Presses Universitaires de Nancy, Nancy, France, pp. 355–363.

Cosandey, C., Andréassian, V., Martin, C., Didon-Lescot, J., Lavabre, J. *et al.* (2005) The hydrological impact of the Mediterranean forest: A review of French research. *Journal of Hydrology* 301, 235–249.

Dassot, M., Constant, T. and Fournier, M. (2011) The use of terrestrial LiDAR technology in forest science: Application fields, benefits and challenges. *Annals of Forest Science* 68, 959–974.

Doten, C.O., Bowling, L.C., Lanini, J.S., Maurer, E.P. and Lettenmaier, D.P. (2006) A spatially distributed model for the dynamic prediction of sediment erosion and transport in mountainous forested watersheds. *Water Resources Research* 42, W04417.

Dudley, A., Panda, S.S., Amatya, D. and Kim, Y. (2015) Hydro-climatology based wildfire susceptibility automated geospatial model development for forest management. In: *Proceedings of the 2015 Georgia Water Resources Conference*, Athens, Georgia, April 28–30, 2025.

d'Urso, G. and Minacapilli, M. (2006) Semi-empirical approach for surface soil water content estimation from radar data without a-priori information on surface roughness. *Journal of Hydrology* 321, 297–310.

EEA (2006) *CLC2006 Technical Guidelines*. Technical Report no.17/2007. European Environmental Agency, Copenhagen, Denmark.

Eisenbies, M.A., Aust, W.M., Burger, J.A. and Adams, M.B. (2007) Forest operations, extreme flooding events, and considerations for hydrologic modeling in the Appalachians – a review. *Forest Ecology and Management* 242, 77–98.

Flanagan, D.C., Ascough, J.C., Nicks, A.D., Nearing, M.A. and Laflen, J.M. (1995) Overview of the WEPP erosion prediction model. In: *USDA-Water Erosion Prediction Project: Hillslope Profile and Watershed Model Documentation*. National Soil Erosion Research Laboratory, US Department of Agriculture, West Lafayette, Indiana, p. 1.

Folton, N., Andréassian, V. and Duperray, R. (2015) Hydrological impact of forest-fire from paired-catchment and rainfall–runoff modelling perspectives. *Hydrological Sciences Journal* 60, 1213–1224.

Fortin, J.P., Turcotte, R., Massicotte, S., Moussa, R., Fitzback, J. *et al.* (2001) Distributed watershed model compatible with remote sensing and GIS data. I: Description of Model. *Journal of Hydrologic Engineering* 6, 91–99.

Gessner, U., Reinermann, S., Asam, S. and Kuenzer, C. (2023) Vegetation stress monitor – assessment of drought and temperature-related effects on vegetation in Germany analyzing MODIS time series over 23 years. *Remote Sensing* 15, 5428.

Giri, C., Pengra, B., Zhu, Z., Singh, A. and Tieszen, L.L. (2007) Monitoring mangrove forest dynamics of the Sundarbans in Bangladesh and India using multi-temporal satellite data from 1973 to 2000. *Estuarine, Coastal and Shelf Science* 73, 91–100.

Goward, S.N., Masek, J.G., Cohen, W., Moisen, G., Collatz, G.J. *et al.* (2008) Forest disturbance and North American carbon flux. *Eos* 89, 106–107.

Grenzdörffer, G.J., Engel, A. and Teichert, B. (2008) The photogrammetric potential of low-cost UAVs in forestry and agriculture. *International Archives of the Photogrammetry, Remote Sensing and Spatial Information Sciences* 31, 1207–1214.

Gupta, H.V., Wagener, T. and Liu, Y. (2008) Reconciling theory with observations: Elements of a diagnostic approach to model evaluation. *Hydrological Processes* 22, 3802–3813.

Hansen, M.C., Potapov, P.V., Moore, R., Hancher, M., Turubanova, S.A. *et al.* (2013) High-resolution global maps of 21st-century forest cover change. *Science* 342, 850–853.

Higgins, S.A., Overeem, I., Steckler, M.S., Syvitski, J.P.M., Seeber, L. *et al.* (2014) InSAR measurements of compaction and subsidence in the Ganges-Brahmaputra Delta, Bangladesh. *Journal of Geophysical Research, Earth Surface* 119, 1768–1781.

Hrachowitz, M., Savenije, H.H.G., Blöschl, G., McDonnell, J.J., Sivapalan, M. *et al.* (2013) A decade of predictions in ungauged basins (PUB) – a review. *Hydrological Sciences Journal* 58, 1198–1255.

Huete, A.R. (1988) A soil-adjusted vegetation index (SAVI). *Remote Sensing of the Environment* 25, 295–309.

Jipp, P.H., Nepstad, D.C., Cassel, D.K. and Carvalho, C.R. (1998) Deep soil moisture storage and transpiration in forests and pastures of seasonally-dry Amazonia. In: Markham, A. (ed.) *Potential Impacts of Climate Change on Tropical Forest Ecosystems*. Springer, Dordrecht, Netherlands, pp. 255–272.

Johnson, D.W., Cheng, W. and Burke, I.C. (2000) Biotic and abiotic nitrogen retention in a variety of forest soils. *Soil Science Society of America Journal* 64, 1503–1514.

Khan, S., Hong, Y., Wang, J., Yilmaz, K.K., Gourley, J.J. *et al.* (2010) Satellite remote sensing and hydrologic modeling for flood inundation mapping in Lake Victoria Basin: Implications for hydrologic prediction in ungauged basins. *IEEE Transactions on Geoscience and Remote Sensing* 49, 85–95.

Khosravipour, A., Skidmore, A.K., Wang, T., Isenburg, M. and Khoshelham, K. (2015) Effect of slope on treetop detection using a LiDAR canopy height model. *ISPRS Journal of Photogrammetry and Remote Sensing* 104, 44–52.

Kim, R.H., Kwon, J.H., Amatya, D.M. and Kim, H.W. (2015) Monitoring forest degradation in North Korea using satellite imagery. In: *Proceedings of the 70th Annual Meeting of the Korean Association of Biological Sciences*, Seoul National University, Seoul, Korea, August 12–13, 2015.

Kittredge, J. (1948) *Forest Influences: The Effects of Woody Vegetation on Climate, Water, and Soil, with Applications to the Conservation of Water and the Control of Floods and Erosion*. McGraw-Hill, New York.

Krishnaswamy, J., Robert, J. and Shijo, J. (2014) Consistent response of vegetation dynamics to recent climate change in tropical mountain regions. *Global Change Biology* 20, 203–215.

Lenihan, J.M. and Bachelet, D. (2015) Historical climate and suppression effects on simulated fire and carbon dynamics in the conterminous United States. In: Bachelet, D. and Turner, D. (eds) *Global Vegetation Dynamics: Concepts and Applications in the MC1 Model*. American Geophysical Union, Washington, DC, pp. 17–30.

Lepers, E., Lambin, E.F., Janetos, A.C., DeFries, R., Achard, R. *et al.* (2005) A synthesis of information on rapid land-cover change for the period 1981–2000. *BioScience* 55, 115–124.

Liu, N., Oishi, A.C., Miniat, C.F. and Bolstad, P. (2021) An evaluation of ECOSTRESS products of a temperate montane humid forest in a complex terrain environment. *Remote Sensing of the Environment* 265, 112662.

Lowrance, R. (1992) Groundwater nitrate and denitrification in a coastal plain riparian forest. *Journal of Environmental Quality* 21, 401–405.

Lowrance, R., Altier, L.S., Newbold, J.D., Schnabel, R.R., Groffman, P.M. *et al.* (1997) Water quality functions of riparian forest buffers in Chesapeake Bay watersheds. *Environmental Management* 21, 687–712.

Lu, Z., Crane, M., Kwoun, O.-I., Wells, C., Swarzenski, C. *et al.* (2005) C-band radar observes water level change in swamp forests. *Eos* 86, 141–144.

Mani, J.K. and Varghese, A.O. (2018) Remote sensing and GIS in agriculture and forest resource monitoring. In: Reddy, G. and Singh, S. (eds) *Geospatial Technologies in Land Resources Mapping, Monitoring and Management. Geotechnologies and the Environment*, Vol. 21. Springer, Cham, Switzerland, pp. 377–400.

Michel, S., Gamet, P. and Lefevre-Fonollosa, M.J. (2011) HYPXIM – A hyperspectral satellite defined for science, security and defense users. In: *Proceedings of 2011 3rd Workshop on Hyperspectral Image and Signal Processing: Evolution in Remote Sensing (WHISPERS)*, IEEE, London, Lisbon, Portugal, June 6–9, 2011, pp. 1–4. Available at: https://ieeexplore.ieee.org/document/6080864 (accessed 2 May 2025).

Mu, Q., Heinsch, F.A., Zhao, M. and Running, S.W. (2007) Development of a global evapotranspiration algorithm based on MODIS and global meteorology data. *Remote Sensing of the Environment* 111, 519–536.

Muwamba, A., Rau, B., Trettin, C., Amatya, D.M., Tollner, E.W. *et al.* (2019) Regional differences in stream water nitrogen, phosphorus, and sediment responses to forest harvesting in conterminous USA. *Journal of Environmental Quality* 48, 634–644.

Naiman, R.J. and Décamps, H. (1997) The ecology of interfaces: Riparian zones. *Annual Review of Ecology and Systematics* 28, 621–658.

Narasimhan, B., Srinivasan, R. and Whittaker, A.D. (2003) Estimation of potential evapotranspiration from NOAA-AVHRR satellite. *Applied Engineering in Agriculture* 19, 309–318.

Njoku, E.G. and Entekhabi, D. (1996) Passive microwave remote sensing of soil moisture. *Journal of Hydrology* 184, 101–129.

Nolan, M. and Fatland, D.R. (2003) New DEMs may stimulate significant advancements in remote sensing of soil moisture. *Eos* 84, 233–237.

Nolan, J., Panda, S.S. and Mobasher, K. (2011) *Landslide Probability Study of Coosawhatchee 8-Digit HUC Watershed using Geospatial Technology*. Georgia Urban and Regional Information Systems Association (GA-URISA), Atlanta, Georgia. USA.

Ollinger, S.V. and Smith, M.L. (2005) Net primary production and canopy nitrogen in a temperate forest landscape: An analysis using imaging spectroscopy, modeling and field data. *Ecosystems* 8, 760–778.

Osman, K.T. (2013) Forest soils. In: Osman, K.T. (ed.) *Soils: Principles, Properties and Management*. Springer, Dordrecht, Netherlands, pp. 229–251.

Owe, M., Brubaker, K., Ritchie, J. and Rango, A. (eds) (2001) *Remote Sensing and Hydrology 2000*. International Association of Hydrologic Sciences, Wallingford, UK.

Panda, S.S., Andrianasolo, H., Murty, V.V.N. and Nualchawee, K. (2004a) Forest management planning for soil conservation using satellite images, GIS mapping, and soil erosion modeling. *Journal of Environmental Hydrology* 12, 1–16.

Panda, S.S., Chaubey, I., Matlock, M.D., Haggard, B.E. and White, K.L. (2004b) Development of a GIS-based decision support system for Beaver Lake watershed management. In: *Proceedings of the American Water Resources Association (AWRA) Spring Specialty Conference*, Nashville, Tennessee, May 17–19, 2004.

Panda, S.S., Andrianasolo, H. and Steele, D. (2005) Application of geotechnology to watershed soil conservation planning at the field scale. *Journal of Environmental Hydrology* 13, 1–22.

Panda, S.S., Ames, D.P. and Panigrahi, S. (2010) Application of vegetation indices for agricultural crop yield prediction using neural network. *Remote Sensing* 2, 673–696.

Panda, S.S., Amatya, D.M. and Hoogenboom, G. (2014) Stomatal conductance, canopy temperature, and leaf area index estimation using remote sensing and OBIA techniques. *Journal of Spatial Hydrology* 12, 4.

Panda, S.S., Rao, M., Fitzgerald, J. and Thenkabail, P.S. (2015) Remote sensing satellites and sensors: Optical, radar, LiDAR, microwave, hyperspectral, and UAVs. In: Thenkabail, P.S. (ed.) *Remote Sensing Handbook: Vols I, II, and III*. CRC Press, New York.

Panda, S.S., Amatya, D.M., Sun, G. and Bowman, A. (2016a) Remote estimation of a managed pine forest evapotranspiration with geospatial technology. *Transactions of the ASABE* 59, 1695–1705.

Panda, S.S., Amatya, D., Walters, B. and Nettles, J. (2016b) Forest biomass estimation and ET correlation study with advanced geospatial technology for environmental and climate modeling. In: *International American Society of Agricultural and Biological Engineers (ASABE) Conference 2016*, Orlando, Florida, July 17–20, 2016. Paper no. 2455782.

Panda, S.S., Amatya, D.M., Jackson, R., Sun, G. and Noormets, A. (2018) Automated geospatial models of varying complexities for pine forest evapotranspiration estimation with advanced data mining. *Water* 10, 1687.

Panda, S.S., Amatya, D.M., Oishi, A.C., Sun, G., O'Halloran, T. *et al*. (2024a) *Forest Evapotranspiration–Soil Moisture-Biomass Correlation Analysis in Southeastern United States with ECOSTRESS, SMAP, and GEDI based Remotely Sensed Information*. In: *NASA-JPL SBG Science and Applications Technical Interchange Meeting*, May 29–31, 2024.

Panda, S.S., Rao, M., Thenkabail, P.S., Misra, D. and Fitzgerald, J. (2024b) Remote sensing satellites and sensors: Optical, radar, LiDAR, microwave, hyperspectral, and UAVs. In: Thenkabail, P.S. (ed.) *Remote Sensing Handbook: Vols I, II, III, IV, V and VI*, 2nd edn. CRC Press, New York.

Pavelsky, T. (2012) Developing new algorithms for estimating river discharge from space. A meeting note. *Eos* 93, 457.

Perry, M.K. and Panda, S.S. (2022) Even wetlands can burn: Developing a wildfire spatial vulnerability model for the Okefenokee swamp for extreme geohazard mitigation decision support. In: *Annual International ASABE Conference*, Houston, Texas, July 17–20, 2022.

Perry, D.A., Oren, R. and Hart, S.C. (2008) *Forest Ecosystems*, 2nd edition. Johns Hopkins University Press, Baltimore, Maryland.

Perry, M.K., Panda, S.S., Amatya, D.M., Oyuyang, Y., Grace, J. *et al*. (2022) Comprehensive geospatial modeling for flood potential analysis in present day climate change condition: Yellowstone National Park case study. In: *American Geophysical Union (AGU) conference*, Chicago, Illinois, December 12–16, 2022.

Pietroniro, A. and Leconte, R. (2000) A review of canadian remote sensing applications in hydrology, 1995–1999. *Hydrological Processes* 14, 1641–1666.

Pinay, G., Roques, L. and Fabre, A. (1993) Spatial and temporal patterns of denitrification in a riparian forest. *Journal of Applied Ecology* 30, 581–591.

Potter, C., Genovese, V., Gross, P., Boriah, S., Steinbach, M. *et al.* (2007) Revealing land cover change in California with satellite data. *Eos* 88, 269–274.

Pritchett, W.L. (1979) *Properties and Management of Forest Soils*. Wiley, New York.

Puletti, N., Grotti, M., Ferrara, C. and Chianucci, F. (2020) LiDAR-based estimates of above-ground biomass through ground, aerial, and satellite observation: A case study in a Mediterranean forest. *Journal of Applied Remote Sensing* 14, 044501.

Querino, C.A.S., Beneditti, C.A., Machado, N.G., da Silva, M.J.G. and da Silva Querino, J. (2016) Spatiotemporal NDVI, LAI, albedo, and surface temperature dynamics in the southwest of the Brazilian Amazon forest. *Journal of Applied Remote Sensing* 10, 026007.

Rakovec, O. (2022) The 2018–2020 multi-year drought sets a new benchmark in Europe. *Earth's Future* 10, e2021EF002394.

Ramsey, E.I., Ragoonwala, A. and Bannister, T. (2013) Coastal flood inundation monitoring with satellite C-band and L-band synthetic aperture radar data. *Journal of the American Water Resources Association* 49, 1239–1260.

Rana, M.S. (2024) Integration of delphi method and geospatial modelling automation to determine landslide susceptibility for proactive management decision support: Beas river catchment, Himachal Pradesh, a case study. PhD thesis, Civil Engineering Department, Indian Institute of Technology Guwahati, Guwahati, India.

Richmond, E. (2007) Forest hydrology. In: Fierro, P.Jr. and Nyer, E.K. (eds) *The Water Encyclopedia: Science and Issues*, 3rd edn. CRC Press, Boca Raton, Florida.

Riegel, J.B. (2012) A comparison of remote sensing methods for estimating above-ground carbon biomass at a wetland restoration area in the Southeastern Coastal Plain. MSc thesis, Nicholas School of the Environment of Duke University, Durham, North Carolina.

Riggan, P.J., Tissell, R.G., Lockwood, R.N., Brass, J.A., Pereira, J.A.R. *et al.* (2004) Remote measurement of energy and carbon flux from wildfires in Brazil. *Ecological Applications* 14, 855–872.

Riggan, P.J. and Tissell, R.G. (2009) Airborne remote sensing of wildland fires. In: Bytnerowicz, A., Arbaugh, M., Anderson, C. and Riebau, A. (eds) *Wildland Fires and Air Pollution. Developments in Environmental Science 8*. Elsevier, Amsterdam, Netherlands, pp. 139–168.

Riggan, P.J., Tissell, R.G. and Hoffman, J.W. (2003) Application of the firemapper™ thermal-imaging radiometer for wildfire suppression. In: *2003 IEEE Aerospace Conference Proceedings*, IEEE, Piscataway, New Jersey, March 8–15, 2003, pp. 8–15 (Vol. 4).

Samec, P. (ed.) (2023) *Forest Degradation Under Global Change*. IntechOpen, London. Available at: www.intechopen.com/books/11457 (accessed 29 May 2025).

San Juan, R.F.D.V. and Domingo-Santos, J.M. (2018) The role of GIS and LiDAR as tools for sustainable forest management. In: Teodoro, A.C. (ed.) *Frontiers in Information Systems, Vol. 1. GIS – An Overview of Applications*. Bentham Science Publishers, Sharjah, UAE, pp. 124–148.

Sankey, T. and Tatum, J. (2022) Thinning increases forest resiliency during unprecedented drought. *Scientific Reports* 12, 9041.

Scanlon, B.R., Longuevergne, L. and Long, D. (2012) Ground referencing GRACE satellite estimates of groundwater storage changes in the California Central Valley, USA. *Water Resources Research* 48, W04520.

Schmugge, T.J., Kustas, W.P., Ritchie, J.C., Jackson, T.J. and Rango, A. (2002) Remote sensing in hydrology. *Advances in Water Resources* 25, 1367–1385.

Schumacher, J. and Christiansen, J.R. (2020) LiDAR applications to forest–water interactions. In: Levia, D.F., Carlyle-Moses, D.E., Iida, S., Michalzik, B., Nanko, K. *et al.* (eds) *Forest–Water Interactions*, Vol. 240. Ecological Studies, Springer, Cham, Switzerland, pp. 87–112.

Senay, G.B., Bohms, S., Singh, R.K., Gowda, P.H., Velpuri, N.M. *et al.* (2013) Operational evapotranspiration mapping using remote sensing and weather datasets: A new parameterization for the SSEB approach. *Journal of the American Water Resources Association* 49, 577–597.

Singh, G., Das, N.N., Panda, R.K., Mohanty, B.P., Entekhabi, D. *et al.* (2021) Soil moisture retrieval using SMAP L-band radiometer and RISAT-1 C-band SAR data in the paddy dominated tropical region of India. *IEEE Journal of Selected Topics in Applied Earth Observations and Remote Sensing* 14, 10644.

Sutanudjaja, E.H., Beek, L.P.H., Jong, S.M., Geer, V.C. and Bierkens, M.R.P. (2014) Calibrating a large-extent high-resolution coupled groundwater-land surface model using soil moisture and discharge data. *Water Resources Research* 50, 687–705.

Szott, L.T., Palm, C.A. and Buresh, R.J. (1999) Ecosystem fertility and fallow function in the humid and subhumid tropics. *Agroforestry Systems* 47, 163–196.

Touzet, T. and Lecordix, F. (2010) *La Carte Forestière Sans Papier*. No. 206. CFC, Paris, pp. 53–62.

Troch, P., Durcik, M., Seneviratne, S., Hirschi, M., Teuling, A. *et al.* (2007) New data sets to estimate terrestrial water storage change. *Eos* 88, 469–470.

USGS Land Cover Institute (2015) *NLCD 92 Land Cover Class Definitions*. US Geological Survey, Reston, Virginia.

Uyeda, K.A., Stow, D.A. and Riggan, P.J. (2015) Tracking MODIS NDVI time series to estimate fuel accumulation. *Remote Sensing Letters* 6, 587–596.

van Dijk, A.I. and Keenan, R.J. (2007) Planted forests and water in perspective. *Forest Ecology and Management* 251, 1–9.

van Griensven, A., Maskey, S. and Stefanova, A. (2012) The use of satellite images for evaluating a SWAT model: Application on the Vit Basin, Bulgaria. In: Seppelt, R., Voinov, A.A., Lange, S. and D, B. (eds) *Proceedings of 2012 International Congress on Environmental Modelling and Software, Managing Resources of a Limited Planet, Sixth Biennial Meeting Leipzig, Germany*. Available at: https://scholarsarchive.byu.edu/cgi/viewcontent.cgi?article=1874&context=iemssconference (accessed 2 May 2025).

Vergopolan, N., Chaney, N.W., Pan, M., Sheffield, J., Beck, H.R. *et al.* (2021) SMAP-HydroBlocks, a 30-m satellite-based soil moisture dataset for the conterminous US. *Scientific Data* 8, 264.

Viessman, W. and Lewis, G.L. (2002) *Introduction to Hydrology*. Prentice Hill, Upper Saddle River, New Jersey.

von Gadow, K. and Bredenkamp, B. (1992) *Forest Management*. Academica, Pretoria.

Wałęga, A., Wojkowski, J., Sojka, M., Amatya, D., Młyński, D. *et al.* (2024) Exploiting satellite data for total direct runoff prediction using CN-based MSME model. *Science of the Total Environment* 908, 168391.

Walker, J.P., Houser, P.R. and Willgoose, G.R. (2004) Active microwave remote sensing for soil moisture measurement: A field evaluation using ERS2. *Hydrological Processes* 18, 1975–1997.

Zarco-Tejada, P.J., Diaz-Varela, R., Angileri, V. and Loudjani, P. (2014) Tree height quantification using very high resolution imagery acquired from an unmanned aerial vehicle (UAV) and automatic 3D photo-reconstruction methods. *European Journal of Agronomy* 55, 89–99.

Zhang, Y. and Barten, P.K. (2008) Watershed forest management information system (WFMIS). *Environmental Modelling and Software* 24, 569–575.

Zinke, P.J. (1967) Forest interception studies in the United States. In: Sopper, W.E. and Lull, H.W. (eds) *International Symposium on Forest Hydrology*. Pergamon Press, New York, pp. 137–161.

17 Future Directions in Forest Hydrology

J. Nettles[1]*, L.J. Bren[2], D.M. Amatya[3], G. Sun[4], T.M. Williams[5] and C. De Jong[6]

[1]*Weyerhaeuser Company, Columbus, Mississippi, USA;* [2]*The University of Melbourne, Creswick, Victoria, Australia;* [3]*Santee Experimental Forest, Southern Research Station, USDA Forest Service, Cordesville, South Carolina, USA;* [4]*Eastern Forest Environmental Threat Assessment Center, USDA Forest Service, Research Triangle Park, North Carolina, USA;* [5]*Clemson University, Clemson, South Carolina, USA;* [6]*University of Strasbourg, Strasbourg, France*

Abstract

The chapter highlights the similarities and differences in hydrological characteristics and research needs among boreal, temperate, subtropical and tropical forests. Forest hydrology began as an attempt to understand the variations in stream behavior induced both naturally and by humankind's management of forests. This interest is now being pushed along by the realities of climate change, with new demands. The major volumetric component of the hydrological cycle is evapotranspiration, and its quantification (and management) presents key challenges. Difficulties in measurement at both tree and forest scales are still present, but there are whole new sources of information from flux towers to satellite measurements now publicly available. These will allow insights into ecohydrology, viewed as the basis of modern watershed ecosystem management.

17.1 Forest Hydrology: What Have We Learned?

17.1.1 The hydrological cycle in forests

Forest hydrology is a distinct branch of hydrology that addresses the unique conditions created by trees and the understory within forests. Understanding forests, with trees that can grow over 100m tall, have crowns up to 20–30m in diameter, roots extending 5–10m deep and lifespans ranging from 50 to 5000 years, presents significant scientific challenges. Forests cover approximately one-third of the world, with over half of that in just five countries – Russia, Brazil, Canada, the USA and China. Tropical forests make up 45% of area by climatic region, followed by boreal (27%), temperate (16%) and subtropical (11%) forests (FAO, 2020). The wide distribution of forests makes it difficult to generalize about the role of trees and forest ecosystems in the global hydrological cycle, and although *in situ* data collection has been decreasing for several decades (Tourian *et al.*, 2022), satellite data, models and public data sets are increasingly available to study the global hydrological cycle (e.g. Koutsoyiannis, 2020).

The chapters in this book have dealt with major hydrological processes such as runoff, drainage and evapotranspiration (ET) in forests across scales, climatic regimes and dominant processes – from snow-dominated temperate mountain forest watersheds to low-gradient humid coastal plain forests, and most other forest

*Corresponding author: jami.nettles@weyerhaeuser.com

 Forest Hydrology: Processes, Management and Assessment (eds Leon J. Bren *et al.*)

DOI: 10.1079/9781800625310.0017

types including flooded and wetland forests. In most forests, the largest component of water loss is evaporation of precipitation (see Chapter 3, Fig 3.1, this volume), with losses greatest for conifers in regions of frequent low-intensity rainfall separated by dry periods (see Chapter 11). Yet in some tropical montane forests, water condenses from the atmosphere on to tree leaves and the resulting drip may increase precipitation by up to 20% (see Chapter 9). In very cold continental climates, open areas are more likely to lose water (snow) by sublimation and wind than forests, but in warmer regions, openings have greater melt and produce more water than forests. Transpiration is a dominant process in the forest hydrological cycle but is generally difficult to measure directly (see Chapter 3), except on a single tree basis. The word 'evapotranspiration' was coined less than 100 years ago to reflect the difficulty in separating water loss processes, but recent authors suggest that use of ET undermines our recognition of these as separate processes (Savenije, 2004; Miralles *et al.*, 2020). Although the abovementioned issues are only a few of the challenges in understanding the many hydrological processes that depend upon temperature, rainfall, species, tree age, slope, drainage and soil type, the summary provided in the first 16 chapters of this book strives to outline some of the major findings on the most recent forest hydrological science and applications across the forests of the world.

Figure 16.1 shows the distribution of forests spread over much of the earth. Tropical forests are concentrated in Africa, South America and South-East Asia, and subtropical forests in south-eastern North America and south-eastern Asia. Temperate forests are concentrated in Europe, coastal Australia, eastern North America and far eastern Asia, while the vast northern boreal forests circle the Earth in a band across Europe, Asia and North America. Temperature limits alpine forests with treeline elevations varying from 700 m in Sweden (68°N) to 4000 m in Mexico (19°N) and Ecuador (0°), but all with mean annual surface temperatures of 6–7°C (Körner and Paulsen, 2004). Likewise, forests are absent in regions of permafrost that occur with mean annual air temperatures below about −6°C (Smith and Riseborough, 2002). The other major limit to forest distribution is the balance of annual precipitation and potential evaporation. In the tropics, forests are absent when precipitation is below 1000 mm (Staver *et al.*, 2011), but in the arctic (61°N), forests are present with 260 mm of annual precipitation (Carey and Woo, 2001). In both subtropical and temperate zones, forests are also constrained by precipitation and evaporation, but patterns of forest are highly fragmented due to human land uses. Analysis of tropical forests shows forest-rainfall feedback at a region scale, with loss of precipitation following deforestation (Smith *et al.*, 2023), increasing the vulnerability of areas already facing drought due to climate change.

Below are highlights of learnings, applied to the major forest climatic regions, or various aspects of forest hydrological science, including its applications, limitations, challenges and opportunities for a future of changing land use and climate.

Boreal forests

Boreal forests, found in high latitudes of the northern hemisphere, are the second most widespread forest in the world (FAO, 2020). Chapters 7 and 11, on mountains and taiga, discuss how forests cause snow hydrology to vary with temperature and winds during and shortly after snow falls. In colder regions, snow does not adhere to tree crowns as well and is easily dislodged by wind, and on the ground, snow is subject to further wind distribution and sublimation (Chapter 11). In warmer and more maritime regions, interception of snow is greater due to the tendency of snow near 0°C to adhere to foliage, with greater likelihood of partial melting and refreezing. These effects reverse the normal impact of forest removal in far northern regions of Siberia, causing a postharvest reduction of water available for streamflow (Chapter 11). While snowmelt is the most important factor in streamflow, in more southern areas, rain-on-snow events are often associated with the largest flows. Streamflow may cease in small watersheds due to freezing in winter and/or increased ET rates in the summer, while flow beneath the ice in larger rivers is difficult to measure.

Temperate forests

The vast majority of temperate forests lie in the northern hemisphere, with other areas found in the South Island of New Zealand, south-eastern Australia and Chile. These forest types have been studied the most extensively, and forest hydrology as a science originated in the temperate forests of central Europe in the 18th century. Most of the long-term US forest hydrology data originated in temperate forest types (see Chapter 5). Nearly all the runoff process studies cited in Chapter 2 occurred in the temperate forests of North America, Europe, New Zealand and Australia.

Distinctive characteristics of this forest region are a long dormant season due to low temperatures, and extensive deciduous forests in the northern hemisphere. High ET rates during late spring and summer generally result in considerably more ET demand than precipitation in late summer and early autumn. Precipitation during the winter may be stored as snow or soil moisture, when deciduous forest vegetation is dormant and evergreen vegetation is semi-dormant. Streamflow is generally seasonal, with highest flows during the spring (Wasko *et al.*, 2020), due to snowmelt, rain on snow or high soil moisture.

Subtropical forests

Forests in the subtropical climatic zones have two distinct climatic patterns: (i) rainfall well distributed throughout the year; and (ii) winter rainfall with hot dry summers. Forests in the well-distributed rainfall zone occur in the south-eastern USA, south-eastern China, southern Japan, north-eastern Australia and the North Island of New Zealand. Winter rainfall and hot dry summers are typical around the Mediterranean Sea, which gives this climatic type its common name. Mediterranean climates are also common in parts of south-western North America and southern Australia.

The higher latitudes of the subtropical zone, as in the temperate zone, have a winter season of plant dormancy and lower evaporative demands. In the Mediterranean climate, lower winter evaporative demand combines with higher rainfall for a highly seasonal runoff pattern. Although forest fires are common in all climate types, the dry summers of the Mediterranean type make the hydrological impacts of forest fire (see Chapter 10) an important aspect in this region.

In eastern North America and Asia, the subtropical climatic zone is also influenced by the maritime climate from the Northern Tropical Convergence Zone during summer and fall. Tropical cyclones (hurricanes and typhoons) are the most spectacular aspect of this influence, but higher summer rainfall is common. Summer runoff is generally small or non-existent. However, rainfalls of 100–600 mm associated with tropical cyclones produce large areas (approaching 100% of the catchment) of saturation-excess overland flow throughout coastal low-gradient watersheds. For example, Figure 8.4 (see Chapter 8) illustrates a very large rainfall event associated with a hurricane that produced 500 mm of rainfall in 48 hours, flooding almost the entire forest. This extreme precipitation-induced flood may have had severe impacts on both forest stands and road infrastructure, disrupting both traffic and aquatic organism passage and causing economic losses.

Tropical forests

One important aspect of tropical forest hydrology is the lack of cold-temperature-induced dormancy. Trees are evergreen and transpire the entire year, as long there is adequate rainfall. Tropical forests are divided into rainforests, with high year-round rainfall, and seasonal monsoonal forests, with a pronounced dry season. High energy associated with direct solar angles causes high rates of evaporation and intense thunderstorms where high atmospheric moisture is available. High energy results in extreme rates of hydrological processes, which may not follow the principles tested primarily in the temperate zone.

Tropical rainforests are located primarily in the Amazon and Congo Basins and south-east Asia, as well as insular and montane forests where prevailing winds cause advection of coastal moisture. These forests are generally close to the equator and have relatively constant daily temperature fluctuations throughout the year. Despite multi-layered evergreen forests, interception losses can be 9–11.6% of rainfall in the Amazonian rainforest (Lloyd *et al.*, 1988; Shuttleworth and Pereira, 1997; Czikowsky and Fitzjarrald, 2009).

Seasonal monsoons are most typically associated with India and South-East Asia, but seasonality is fairly high between 15° and 20° latitudes, associated with seasonal movement of the Intertropical Convergence Zone. During the rainy season, these regions may have high-intensity rainfall over sustained periods. Bonell and Gilmour (1978) found that surface runoff and rainfall intensity were factors in the forest hydrology of northern Australia, in contradiction to contention that rainfall intensity did not explain streamflow volume or peak discharge in humid temperate forests of the Appalachian Mountains in the USA (Hewlett *et al.*, 1977, 1984). Chapter 9 discussed the importance of understanding stream generation pathways in tropical forests, along with noting that, while field data collection lags behind that of temperate regions, the large fluxes significantly impact regional and global climate and are increasingly important.

17.2 Where Do We Go from Here?

17.2.1 What will forest hydrology become?

Forest hydrology, the study of water dynamics within forest catchments, emerged in an effort to understand how human interference to forests alters the amount of water in our rivers during floods and droughts. Now humans are changing both landscape and climate, while depending on forests for resources and sustainability of both ecosystems and societies. Farooqi *et al.* (2024) also stated that the intensified changes in global climate, land use and cover, and pollution that occurred over last the century have brought about adverse impacts on forests and their services in terms of water regulation, signifying the importance of forest hydrological research as a re-emerging topic of scientific interest. Accordingly, there is an urgent need for a better understanding of the linkages between trees, forests and water to raise awareness and increase capacity building in forest hydrology, and for embedding this knowledge and research findings in policies and management practices. In addition, Farooqi *et al.* (2024) emphasized that there is also an urgent need to include more under-represented nations and regions in future investigations to comprehensively grasp how interactions between forest ecosystems and water can impact the entire planet.

Many of the challenges of the coming decades discussed in the context of Europe and the south-eastern USA (see Chapters 5 and 15) are equally applicable to many other parts of the world. Forest hydrology over the last century has been concerned primarily with the effects of various forms of forest management on water quantity and quality, but the coming years will see a greater understanding of the important role of forests, both managed and unmanaged, as a natural infrastructure for water supporting ecosystem health – a 'nature-based solution' – across the globe (Springgay, 2019).

Over the next century, the role of forests in mitigating climate change may become the greatest challenge. Forests currently sequester a substantial amount of anthropogenic carbon, but there is potential for much more, either through improved management of existing forests or through afforestation (Mo *et al.*, 2023). Rapidly growing forests can provide sustainable resources and carbon-neutral energy, but as we saw in Chapter 3, forest carbon assimilation and transpiration are controlled by the same physiological mechanism, stomatal opening. While we do understand the correlation between carbon intake and transpiration, we have a limited understanding of the integration of tree genetics and larger-scale factors that influence the variation in that relationship (i.e. water-use efficiency) or rates. Zhang *et al.* (2023) outlined many of the factors involved in this relationship, including forest composition, climate and forest management, along with the future research needed to understand and manage forests to balance growth for wood products, energy production and carbon sequestration. The stakes are high: the Global Commission on the Economics of Water stated: 'We will fail on climate change if we fail on water' (Mazzucato *et al.*, 2023).

Streamflow from forested watersheds produces high-quality water requiring minimal treatment for drinking water (Neary *et al.*, 2009), an increasingly valuable and scarce resource. In the USA, 46% of the total surface water supply comes from forests (Liu *et al.*, 2022). Forests play a role in aquifer recharge by affecting the processes by which rain is partitioned into subsurface flow and surface runoff. Although these processes have been well defined (see Chapter 2), understanding how climate, forest characteristics, soils and geology determine the pathways and quantities of water movement, from crowns to streams or aquifers, is still far from our grasp. Because forests make up a relatively large portion of many of our watersheds, it is important to understand the hydrology, processes and pathways of the water cycle on both natural and managed forests, while considering the contribution of other land uses.

Much of our present understanding of forest hydrology is limited mainly to research on temperate forests, so even the most well-established tenets do not apply universally. We have given examples of contrasting situations; for example, cutting forests in parts of Siberia may decrease streamflow rather than increase it. In another example, forest floor infiltration may not exceed rainfall intensity during intense tropical thunderstorms. To extend forest hydrology, we must understand the large variations of hydrological responses and their drivers, and extend research into all forested regions. Farooqi *et al.* (2024) conducted a thorough review of global research, trends and future directions in advancing forest hydrological research.

17.2.2 ET and transpiration

The studies of ET reveal how little we really know about the principles that drive movement of water from forests into the atmosphere. ET accounts for the greatest loss of precipitation in most forested ecosystems (see Chapter 3), but in traditional paired-watershed studies, it is estimated as the residual water balance, assuming no loss of water from the watershed other than that measured as stream outflow. While these relationships have taught us much about individual watersheds and led to general principles (Zhang *et al.*, 2001), more direct measurement of ET will help us close gaps in the water balance.

Although many studies have been done on forest interception, our understanding of how interception varies as a function of forest structure is limited due to a lack of sufficient observational data and thus remains poorly constrained, with estimates ranging from 10% to 50% of precipitation, depending on canopy characteristics (Grunicke *et al.*, 2020). Lian *et al.* (2022) reported a recent decline in global interception due to altered rainfall regimes (e.g. rainfall events are becoming less frequent and more intense since 2000), suggesting that ongoing rainfall changes favor a partitioning toward more soil moisture and runoff, benefiting ecosystem functions but simultaneously increasing flood risks.

Sapflow measurements have been a part of forest hydrology research for almost 100 years (Swanson, 1994), although of course the technology has advanced in accuracy and deployability. Wide differences in sapflow are evident among different tree species and sizes as can be seen in Fig. 3.3 in Chapter 3. Upscaling to the stand level has long been recognized as a challenge (Čermák *et al.*, 2004), with upscaled sap flux measurements not capturing the variability in forest characteristics across the landscape (Ford *et al.*, 2007).

Fortunately, ET measurement and modeling may be the most rapidly expanding part of forest hydrology, with programs developed to aggregate ground-based data, such as Ameriflux, Fluxnet and Sapfluxnet with satellite measurements, and models, such as in OpenET in a 30m spatial scale. Integration of site-specific data with satellite data is giving scientists and land managers the ability to estimate site ET at a tract scale, although currently OpenET shows higher model error and variability for forested landscapes than for agricultural land (Volk *et al.*, 2024), consistent with Rashid and Tian (2024). OpenET provides data in timesteps as small as daily, and ultimately plans to provide retrospective ET data, giving scientists the ability to compare satellite data with historical watershed study data. This is currently being developed in the USA by a non-profit consortium, expanding across the Amazon Basin (OpenET-Brazil), and has the goal of being available in other regions across the globe. Similar other efforts by NASA's ECOSTRESS (Ecosystem Spaceborn Thermal Radiometric instrumentation on Space Satellite) program are under way to quantify surface temperature and its derived products such as ET and its components on a 70m scale in 3–4 days interval (Fisher *et al.*, 2020). However, strong divergences exist among remote-sensing products when partitioning ET, and thus the individual components remain largely unvalidated (Talsma *et al.*, 2018). Therefore, future studies should also focus on obtaining field data for the ET components, particularly the canopy evaporation and the understory ET on varying forest types of ecoregions for ground-truthing of remote sensing products applied to large-scale assessments.

While ET measurements will be invaluable, research is still needed to improve models to predict the effect of forest management on water resources (Komatsu and Kume, 2020; Sun *et al.*, 2023). New remote-sensing techniques of spaceborne, airborne or ground-based LiDAR (light detection and ranging) will produce the better estimates of forest structure needed for hydrological and ecosystem models (Massey *et al.*, 2023). LiDAR data from GEDI (Global Ecosystem Dynamics Investigation), a project to map vertical structures in the world's forests (Dubayah *et al.*, 2020), will be combined with other remote-sensing products for more accurate and extensive data (e.g. Silva *et al.*, 2021) and Landsat data including Landsat Operational Land Imager and ICESAT-2 (ice, cloud and elevation Satellite) for change estimates. New approaches can be explored for forest systems, such as the one studied by Good *et al.* (2015), who combined two distinct stable-isotope flux partitioning techniques to quantify ET subcomponents (interception, transpiration, soil evaporation and surface water evaporation) and the hydrological connectivity of bound, plant-available soil waters with more mobile surface waters. Stoy *et al.* (2019) conducted a detailed review and synthesis of studies for partitioning ecosystem ET, including the challenges and opportunities. Others are using machine learning techniques to improve ET model accuracy (e.g. Jiang *et al.*, 2021; Elbeltagi *et al.*, 2024). While there is established work on partitioning ET into evaporation and transpiration (Kool *et al.*, 2014), there is much less on partitioning transpiration by vegetation, despite evidence that in temperate and boreal forests, understory vegetation averages one-third of growing-season ET (Balandier *et al.*, 2022). Below-canopy eddy flux towers, while still relatively rare, are producing valuable insights into vegetative water use (e.g. Constantin *et al.*, 1999; Kassuelke *et al.*, 2022; Wolf *et al.*, 2024). Short time-scale data may also improve our understanding of processes such as hydraulic redistribution of soil water (Domec *et al.*, 2010) and upscaling instantaneous ET to daily time scales (Jiang *et al.*, 2021).

17.2.3 Atmospheric redistribution of ET

One of the most important factors in the water cycle, the atmospheric redistribution of water, has possibly been the least pursued in forest hydrology, even though it may be the most critical for the future. This process of ET, condensation and precipitation drives downwind water availability (Ellison *et al.*, 2017), and the cooling effect also extends beyond the watershed scale, allowing reforesting regions to better weather temperature increases due to climate change (Barnes *et al.*, 2024). Historically, people associated forests and rainfall, assuming that forests increased rainfall instead of understanding that forests thrived in areas of higher rainfall (van Meerveld and Seibert, 2024). However, the relationship of forest to airshed is not as well understood as forest to watershed, and predictions of the atmospheric response to forest management are limited. A study by Hoek van Dijke *et al.* (2022) based on models and moisture recycling data showed that potential reforestation efforts could increase water availability by 6% in some regions while decreasing other regions by up to 38%, with streamflow also responding regionally. Another model of moisture recycling also showed large variability between regions in the amount of precipitation from vegetative ET, but also noted how sensitive some areas were to even small losses of precipitation (Keys *et al.*, 2016).

As regions improve sustainable forest-management practices, increase forest area and density to sequester carbon, and respond to the demand for wood and wood products, atmospheric effects will transcend borders. There is a critical need to understand and predict these effects of forest policy and practice, as they have the potential to affect water availability at local to global scales (Creed *et al.*, 2019).

17.2.4 Runoff processes

Chapter 2 discussed runoff concepts, with interesting observations of how early research assumed site-specific conditions to be more generalized than they are, biasing results. More recently, we have come to understand how multiple thresholds of runoff generation influence response, and that interactions of weather, climate, surface and subsurface connections vary with location and time. This includes the 'fill and spill' concept introduced by Spence and Woo (2003), and recently generalized by McDonnell *et al.* (2021), that allow representation of processes common to forested systems. However, quantitative estimates of flow generation mechanisms are dependent on the climate and physiographic characteristics of the site (Wu *et al.*, 2021). Most of our quantitative understanding has come from isotope or geochemical tracer analysis of streamflow. An outstanding discussion of the use and limitations of isotopes can be found in Klaus and McDonnell (2013), and a call for standardization of isotope methods includes a thorough discussion of challenges in their use in ecohydrological research (Millar *et al.*, 2022). While end-member mixing has been a common technique using geochemical tracers, the recent ability to differentiate dissolved organic carbon fractions of stream natural organic matter may provide alternatives to examine flow through the forest floor (Yang *et al.*, 2015).

17.2.5 Merging forest hydrology and ecosystems

Ecohydrology includes forest hydrology as the study of climate, soil and vegetation dynamics through space and time, and the intersection with ecological dynamics (Rodriguez-Iturbe, 2000), with recent scientists including the microbiome, atmosphere, climate and humans (Porporato and Yin, 2022) and emphasizing the importance of the catchment as the fundamental scale (Rinaldo and Rodriguez-Iturbe, 2022).

A recent concept that integrates sustainability and ecology is nature-based solutions, a concept especially applicable to forests and water (Springgay, 2019). While forests have been recognized as providing a wide range of ecosystem services that include water quality and quantity, nature-based solutions are actions that use, improve or protect ecosystems to provide multiple benefits for humans and biodiversity (Cohen-Shacham *et al.*, 2016). Springgay (2019) called forests a 'natural infrastructure for water,' and recognition of the central role of forests in natural and optimized solutions will only increase the need for accurate analysis and prediction of the effects of forest management on water resources.

17.2.6 Broader dimensions of forest hydrology

The scope of forest hydrological science must be expanded from understanding the meteorological and hydrological influences of forests based on small watershed research of the 20th century (Hewlett, 1982) to quantifying the ecohydrological impacts of global changes today (Amatya *et al.*, 2011; Vose *et al.*, 2011). It must also advance to address current complex issues, including urbanization, climate change, wildfires, invasive species, instream flow, floods, droughts, beneficial water uses, changing patterns of development and ownership, and changing societal values. In this context, there is a critical need for continued monitoring of existing long-term forest watersheds worldwide, as they are well suited for documenting and detailing baseline hydrological conditions and serving as valuable benchmarks for advancing forest hydrological science and addressing emerging forest and water issues. In addition, such research should be undertaken as an integrated approach in collaboration with a multidisciplinary team in future forest hydrological research that emphasizes not only the long-term data but also utilizes various innovative methodologies including the ground-based and remote-sensing-based data backed by process-based models.

Our ability to model large-scale processes has been limited by our ability to collect data that appropriately reflects and integrates parameter variability. Intelligent, field-based, real-time monitoring of forest hydrological processes will improve data collection at a much finer spatial and temporal scale than traditional research methods (Sun *et al.*, 2016, 2023). Recent advancements in monitoring with high-resolution sensors and mapping technology using LiDAR, satellite imageries, stable isotopes for partitioning water flux sources (Good *et al.*, 2015; Klaus *et al.*, 2015) and other sensor technologies, together with rapidly advancing computational techniques, should also be used as opportunities to address these complex processes. Global water models are being used not just for policy and science but also to drive publicly available information services, although predictions still show considerable differences between models (Gnann *et al.*, 2023).

17.3 Meeting Forest Management Needs

UNESCO's World Water Development Report (UN Water, 2024) noted that water demand continues to grow, even as half of the world's population is already experiencing water stress for at least part of the year. Global water demand has doubled since 1960 and is expected to increase by an additional 20–25% by 2050, with demand in sub-Saharan Africa expected to increase a staggering 163% (Kuzma *et al.*, 2023a; b). Climate change and natural variability may reduce water availability, even in areas unaccustomed to drought. These conditions may put strong pressure on forest managers to sustain and even increase water yields of forested watersheds for municipal and other downstream uses, while water stress leaves the forest itself more vulnerable to dieback, pests and fire (Grant *et al.*, 2013). At the same time, carbon sequestration initiatives typically favor afforestation, longer rotations and increased stocking (Kaarakka *et al.*, 2021) that will increase ET and forest water use (Farley *et al.*, 2005; Jackson *et al.*, 2005).

To develop tools to manage complex and sometimes competing interactions, we should not only advance forest hydrological science for understanding complex interactions but also learn to scale currently available research and model results to define reasonably achievable benchmarks of water quantity and quality. As the world demands an increased clean water supply and wood, energy and carbon storage from forests, forest hydrology becomes equally critical to sustainably providing services while protecting water resources.

From a recent list of 23 unsolved problems in hydrology (Blöschl *et al.*, 2019), one could get discouraged at the vast number of processes we do not understand and research questions to address within forest hydrology. What hydrological laws control catchments and how do they vary with scale? What causes spatial heterogeneity in hydrological processes? Why is runoff more sensitive to land use/land cover in some watersheds than others? How do we adapt models to changing vegetation dynamics? However, there is room to be hopeful that the coming years will bring many advances as our ability to collect and share data at relevant scales has greatly increased, and analysis of these large data sets will follow. The next decades promise a confluence of advanced instrumentation, computational power and analysis tools to put better understanding within our reach.

17.4 Conclusions

In its century of existence, forest hydrology has come a long way in quantifying the relationship between the world's forests and its water. The approaches have used field studies, modeling and remote sensing. There is a rapid escalation of information available due to both satellite measurement of hydrological variables and new ground-based approaches, which may change our basis approach to research. The issues of climate change and rising Earth temperatures are creating new research issues.

References

Amatya, D., Douglas-Mankin, K., Williams, T., Skaggs, R. and Nettles, J. (2011) Advances in forest hydrology: Challenges and opportunities. *Transactions of the ASABE* 54, 2049–2056.

Balandier, P., Gobin, R., Prévosto, B. and Korboulewsky, N. (2022) The contribution of understorey vegetation to ecosystem evapotranspiration in boreal and temperate forests: A literature review and analysis. *European Journal of Forest Research* 141, 979–997.

Barnes, M.L., Zhang, Q., Robeson, S.M., Young, L., Burakowski, E.A. *et al.* (2024) A century of reforestation reduced anthropogenic warming in the eastern United States. *Earth's Future* 12, e2023EF003663.

Blöschl, G., Bierkens, M.F.P., Chambel, A., Cudennec, C., Destouni, G. *et al.* (2019) Twenty-three unsolved problems in hydrology (UPH) – a community perspective. *Hydrological Sciences Journal* 64, 1141–1158.

Bonell, M. and Gilmour, D.A. (1978) The development of overland flow in a tropical rainforest catchment. *Journal of Hydrology* 39, 365–382.

Carey, S.K. and Woo, M. (2001) Slope runoff processes and flow generation in a subarctic, subalpine catchment. *Journal of Hydrology* 253, 110–129.

Cohen-Shacham, E., Walters, G., Janzen, C. and Maginnis, S. (2016) *Nature-based Solutions to Address Global Societal Challenges*. International Union for Conservation of Nature, Gland, Switzerland.

Constantin, J., Grelle, A., Ibrom, A. and Morgenstern, K. (1999) Flux partitioning between understorey and overstorey in a boreal spruce/pine forest determined by the eddy covariance method. *Agricultural and Forest Meteorology* 98–99, 629–643.

Creed, I.F., Jones, J.A., Archer, E., Claassen, M., Ellison, D. *et al.* (2019) Managing forests for both downstream and downwind water. *Frontiers in Forests and Global Change* 2, 64.

Czikowsky, M.J. and Fitzjarrald, D.R. (2009) Detecting rainfall interception in an Amazonian rain forest with eddy flux measurements. *Journal of Hydrology* 77, 92–105.

Domec, J.-C., King, J.S., Noormets, A., Treasure, E., Gavazzi, M.J. *et al.* (2010) Hydraulic redistribution of soil water by roots affects whole-stand evapotranspiration and net ecosystem carbon exchange. *New Phytologist* 187, 171–183.

Dubayah, R., Blair, J.B., Goetz, S., Fatoyinbo, L., Hansen, M. *et al.* (2020) The global ecosystem dynamics investigation: High-resolution laser ranging of the earth's forests and topography. *Science of Remote Sensing* 1, 100002.

Elbeltagi, A., Heddam, S., Katipoğlu, O.M., Alsumaiei, A.A. and Al-Mukhtar, M. (2024) Advanced long-term actual evapotranspiration estimation in humid climates for 1958–2021 based on machine learning models enhanced by the RReliefF algorithm. *Journal of Hydrology: Regional Studies* 56, 102043.

Ellison, D., Morris, C.E., Locatelli, B., Sheil, D. and Cohen, J. (2017) Trees, forests and water: Cool insights for a hot world. *Global Environmental Change* 43, 51–61.

FAO (2020) *Global Forest Resources Assessment 2020: Main Report*. Food and Agriculture Organization of the United Nations, Rome.

Farley, K.A., Jobbágy, E.G. and Jackson, R.B. (2005) Effects of afforestation on water yield: A global synthesis with implications for policy. *Global Change Biology* 11, 1565–1576.

Farooqi, T.J.A., Portela, R., Xu, Z., Pan, S., Irfan, M. *et al.* (2024) Advancing forest hydrological research: Exploring global research trends and future directions through scientometric analysis. *Journal of Forestry Research* 35, 128.

Fisher, J.B., Lee, B., Purdy, A.J., Halverson, G.H., Dohlen, M.B. *et al.* (2020) ECOSTRESS: NASA's next generation mission to measure evapotranspiration from the International Space station. *Water Resources Research* 56, e2019WR026058.

Ford, C.R., Hubbard, R.M., Kloeppel, B.D. and Vose, J.M. (2007) A comparison of sap flux-based evapotranspiration estimates with catchment-scale water balance. *Agricultural and Forest Meteorology* 145, 176–185.

Gnann, S., Reinecke, R., Stein, L., Wada, Y., Thiery, W. *et al.* (2023) Functional relationships reveal differences in the water cycle representation of global water models. *Nature Water* 1, 1079–1090.

Good, S.P., Noone, D. and Bowen, G. (2015) Hydrologic connectivity constrains partitioning of global terrestrial water fluxes. *Science* 349, 175–177.

Grant, G.E., Tague, C.L. and Allen, C.D. (2013) Watering the forest for the trees: An emerging priority for managing water in forest landscapes. *Frontiers in Ecology and the Environment* 11, 314–321.

Grunicke, S., Queck, R. and Bernhofer, C. (2020) Long-term investigation of forest canopy rainfall interception for a spruce stand. *Agricultural and Forest Meteorology* 292, 108125.

Hewlett, J.D. (1982) *Principles of Forest Hydrology*. University of Georgia Press, Athens, Georgia.

Hewlett, J., Fortson, J. and Cunningham, G. (1977) The effect of rainfall intensity on storm flow and peak discharge from forest land. *Water Resources Research* 13, 259–266.

Hewlett, J., Fortson, J. and Cunningham, G. (1984) Additional tests on the effect of rainfall intensity on storm flow and peak flow from wild-land basins. *Water Resources Research* 20, 985–989.
Hoek van Dijke, A.J., Herold, M., Mallick, K., Benedict, I., Machwitz, M. *et al.* (2022) Shifts in regional water availability due to global tree restoration. *Nature Geoscience* 15, 363–368.
Jackson, R.B., Jobbágy, E.G., Avissar, R., Roy, S.B., Barrett, D.J. *et al.* (2005) Trading water for carbon with biological carbon sequestration. *Science* 310, 1944–1947.
Jiang, L., Zhang, B., Han, S., Chen, H. and Wei, Z. (2021) Upscaling evapotranspiration from the instantaneous to the daily time scale: Assessing six methods including an optimized coefficient based on worldwide eddy covariance flux network. *Journal of Hydrology* 596, 126135.
Kaarakka, L., Cornett, M., Domke, G., Ontl, T. and Dee, L.E. (2021) Improved forest management as a natural climate solution: A review. *Ecological Solutions and Evidence* 2, e12090.
Kassuelke, S.R., Dymond, S.F., Feng, X., Savage, J.A. and Wagenbrenner, J.W. (2022) Understory evapotranspiration rates in a coast redwood forest. *Ecohydrology* 15, e2404.
Keys, P.W., Wang-Erlandsson, L. and Gordon, L.J. (2016) Revealing invisible water: Moisture recycling as an ecosystem service. *PLOS ONE* 11, e0151993.
Klaus, J. and McDonnell, J.J. (2013) Hydrograph separation using stable isotopes: Review and evaluation. *Journal of Hydrology* 505, 47–64.
Klaus, J., McDonnell, J., Jackson, C.R., Enhao Du, E. and Griffiths, N.A. (2015) Where does streamwater come from in low-relief forested watersheds? A dual-isotope approach. *Hydrology and Earth System Sciences* 19, 125–135.
Komatsu, H. and Kume, T. (2020) Modeling of evapotranspiration changes with forest management practices: A genealogical review. *Journal of Hydrology* 585, 124835.
Kool, D., Agam, N., Lazarovitch, N., Heitman, J.L. and Sauer, T.J. (2014) A review of approaches for evapotranspiration partitioning. *Agricultural and Forest Meteorology* 184, 56–70.
Körner, C. and Paulsen, J. (2004) A world-wide study of high altitude treeline temperatures. *Journal of Biogeography* 31, 713–732.
Koutsoyiannis, D. (2020) Revisiting the global hydrological cycle: Is it intensifying? *Hydrology and Earth System Sciences* 24, 3899–3932.
Kuzma, S., Bierkens, M.F., Lakshman, S., Luo, T. and Saccoccia, L. (2023a) *Aqueduct 4.0: Updated Decision-Relevant Global Water Risk Indicators*. World Resources Institute, Washington, DC.
Kuzma, S., Saccoccia, L. and Chertock, M. (2023b) *25 Countries, Housing One-Quarter of the Population, Face Extremely High Water Stress*. World Resources Institute, Washington, DC.
Lian, X., Zhao, W. and Gentine, P. (2022) Recent global decline in rainfall interception loss due to altered rainfall regimes. *Nature Communications* 13, 7642.
Liu, N., Dobbs, G.R., Caldwell, P.V., Miniat, C.F. and Sun, G. (2022) *Quantifying the Role of National Forest System and Other Forested Lands in Providing Surface Drinking Water Supply for the Conterminous United States*. General Technical Report WO-100. Washington, DC: Forest Service, US Department of Agriculture.
Lloyd, C.R., Gash, J.H.C., Shuttleworth, W.J. and de O. Marques, F.A. (1988) The measurement and modelling of rainfall interception by Amazonian rain forest. *Agricultural and Forest Meteorology* 43, 277–294.
Massey, R., Berner, L.T., Foster, A.C., Goetz, S.J. and Vepakomma, U. (2023) Remote sensing tools for monitoring forests and tracking their dynamics. In: Girona, M.M., Morin, H., Gauthier, S. and Bergeron, Y. (eds) *Boreal Forests in the Face of Climate Change: Sustainable Management*. Springer International Publishing, Cham, pp. 637–655.
Mazzucato, M., Okonjo-Iewala, N., Rockström, J. and Shanmugaratnam, T. (2023) *Turning the Tide: A Call to Collective Action*. Global Commission on the Economics of Water, Colombo, Sri Lanka.
McDonnell, J.J., Spence, C., Karran, D.J., van Meerveld, H.J. and Harman, C.J. (2021) Fill-and-spill: A process description of runoff generation at the scale of the beholder. *Water Resources Research* 57, e2020WR027514.
Millar, C., Janzen, K., Nehemy, M.F., Koehler, G., Hervé-Fernández, P. *et al.* (2022) On the urgent need for standardization in isotope-based ecohydrological investigations. *Hydrological Processes* 36, e14698.
Miralles, D., Brutsaert, W., Dolman, H. and Gash, J. (2020) On the use of the term "evapotranspiration". *Water Resources Research* 56, e2020WR028055.
Mo, L., Zohner, C.M., Reich, P.B., Liang, J., Miguel, S. *et al.* (2023) Integrated global assessment of the natural forest carbon potential. *Nature* 624, 92–101.

Neary, D.G., Ice, G.G. and Jackson, C.R. (2009) Linkages between forest soils and water quality and quantity. *Forest Ecology and Management* 258, 2269–2281.

Porporato, A. and Yin, J. (2022) *Ecohydrology: Dynamics of Life and Water in the Critical Zone*. Cambridge University Press, Cambridge, UK.

Rashid, T. and Tian, D. (2024) Improved 30-m evapotranspiration estimates over 145 eddy covariance sites in the contiguous United States: The role of ECOSTRESS, harmonized Landsat Sentinel-2 imagery, climate reanalysis, and deep neural network postprocessing. *Water Resources Research* 60, e2023WR036313.

Rinaldo, A. and Rodriguez-Iturbe, I. (2022) Ecohydrology 2.0. *Rendiconti Lincei. Scienze Fisiche e Naturali* 33, 245–270.

Rodriguez-Iturbe, I. (2000) Ecohydrology: A hydrologic perspective of climate–soil–vegetation dynamics. *Water Resources Research* 36, 3–9.

Savenije, H.H. (2004) The importance of interception and why we should delete the term evapotranspiration from our vocabulary. *Hydrological Processes* 18, 1507–1511.

Shuttleworth, W.J. and Pereira, H.C. (1997) Evaporation from Amazonian rainforest. *Proceedings of the Royal Society of London B: Biological Sciences* 233, 321–346.

Silva, C.A., Duncanson, L., Hancock, S., Neuenschwander, A., Thomas, N. *et al.* (2021) Fusing simulated GEDI, ICESat-2 and NISAR data for regional aboveground biomass mapping. *Remote Sensing of the Environment* 253, 112234.

Smith, M. and Riseborough, D. (2002) Climate and the limits of permafrost: A zonal analysis. *Permafrost and Periglacial Processes* 13, 1–15.

Smith, C., Baker, J.C.A. and Spracklen, D.V. (2023) Tropical deforestation causes large reductions in observed precipitation. *Nature* 615, 270–275.

Spence, C. and Woo, M. (2003) Hydrology of subarctic Canadian shield: Soil-filled valleys. *Journal of Hydrology* 279, 151–166.

Springgay, E. (2019) Forests as nature-based solutions for water. In: *Unasylva 251: Forests: Nature-Based Solutions for Water*. Food and Agriculture Organization of the United Nations, Rome, pp. 3–13.

Staver, A.C., Archibald, S. and Levin, S.A. (2011) The global extent and determinants of savanna and forest as alternative biome states. *Science* 334, 230–232.

Stoy, P.C., El-Madany, T.S., Fisher, J.B., Gentine, P., Gerken, T. *et al.* (2019) Reviews and syntheses: Turning the challenges of partitioning ecosystem evaporation and transpiration into opportunities. *Biogeosciences* 16, 3747–3775.

Sun, G., Amatya, D. and McNulty, S. (2016) Forest hydrology. In: Singh, V.V. (ed.) *Handbook of Applied Hydrology*. CAB International, Wallingford, UK.

Sun, G., Wei, X., Hao, L., Sanchis, M.G., Hou, Y. *et al.* (2023) Forest hydrology modeling tools for watershed management: A review. *Forest Ecology and Management* 530, 120755.

Swanson, R.H. (1994) Significant historical developments in thermal methods for measuring sap flow in trees. *Agricultural and Forest Meteorology* 72, 113–132.

Talsma, C.J., Good, S.P., Jimenez, C., Martens, B., Fisher, J.B. *et al.* (2018) Partitioning of evapotranspiration in remote sensing-based models. *Agricultural and Forest Meteorology* 260–261, 131–143.

Tourian, M.J., Elmi, O., Shafaghi, Y., Behnia, S. and Saemian, P. (2022) HydroSat: Geometric quantities of the global water cycle from geodetic satellites. *Earth System Science Data* 14, 2463–2486.

UN Water (2024) *Water for Prosperity and Peace*, United Nations World Water Development Report. UNESCO, Paris.

van Meerveld, I. and Seibert, J. (2024) Reforestation effects on low flows: Review of public perceptions and scientific evidence. *WIREs Water* 12, e1760.

Volk, J.M., Huntington, J.L., Melton, F.S., Allen, R., Anderson, M. *et al.* (2024) Assessing the accuracy of openET satellite-based evapotranspiration data to support water resource and land management applications. *Nature Water* 2, 193–205.

Vose, J.M., Sun, G., Ford, C.R., Bredemeier, M., Otsuki, K. *et al.* (2011) Forest ecohydrological research in the 21st century: What are the critical needs? *Ecohydrology* 4, 146–158.

Wasko, C., Nathan, R. and Peel, M.C. (2020) Trends in global flood and streamflow timing based on local water year. *Water Resources Research* 56, e2020WR027233.

Wolf, S., Paul-Limoges, E., Sayler, D. and Kirchner, J.W. (2024) Dynamics of evapotranspiration from concurrent above- and below-canopy flux measurements in a montane Sierra Nevada forest. *Agricultural and Forest Meteorology* 346, 109864.

Wu, S., Jianshi, Z., Wang, H. and Sivapalan, M. (2021) Regional patterns and physical controls of streamflow generation across the conterminous United States. *Water Resources Research* 57, e2020WR028086.

Yang, L., Chang, S.-W., Shin, H.S. and Hur, J. (2015) Tracking the evolution of stream DOM source during storm events using end member mixing analysis based on DOM quality. *Journal of Hydrology* 523, 333–341.

Zhang, L., Dawes, W.R. and Walker, G.R. (2001) Response of mean annual evapotranspiration to vegetation changes at catchment scale. *Water Resources Research* 37, 701–708.

Zhang, Z., Zhang, L., Xu, H., Creed, I.F., Blanco, J.A. *et al.* (2023) Forest water-use efficiency: Effects of climate change and management on the coupling of carbon and water processes. *Forest Ecology and Management* 534, 120853.

Čermák, J., Kučera, J. and Nadezhdina, N. (2004) Sap flow measurements with some thermodynamic methods, flow integration within trees and scaling up from sample trees to entire forest stands. *Trees* 18, 529–546.

Index

Note: **bold** page numbers indicate figures and tables.

Acacia spp. 27
Acer spp.
    A. rubrum **39**
    A. saccharum **39**
adaptive forest management (AFM) 206, 210
afforestation 39, 41, 198, 199, 205, 210, 212
AFM. *see* adaptive forest management
Africa 145, 152, 315
    see also specific countries
Afro-tropical ecozone (AFR) 145, 147, **148**
agriculture 7, 8, 45, **46**, 47
AGWA (Automated Geospatial Watershed Assessment) 171
air pollution 38, 206
air temperature 37, 43, 103
    mean maximum/minimum 47, 144
    measurement of 3, 43
alpine/subalpine forests 197, 199, 201, 209
Alps 197, 198, 209, 211
Amazon region 145, 148, 152, 316
anisotropy 8
annual maximum PIs (ANMAXPI) 91
annual water yield (AWY) **4**, 99, **100**, 166
Appalachians (USA) 39
aquifers 7, 203, 317
    water-table 8
Arctic/subarctic 197, 315
    see also taiga forests
Ardennes forest (Belgium) 203
Arizona (USA) **167**
Asia 316
    South-East 145–146, 152, 315
    see also Eurasia; Indo-Malay-Australalasia ecozone
Australia 2, 4, **5**, **9**, 11, 145, **150**, 315–317
    forest hydrology research in 197, 316
    wildfire in 170
Australian flooding forest 124, **125**
Austria 206, 208
Automated Geospatial Watershed Assessment (AGWA) 171
AWY. *see* annual water yield

BAERTOOLS website 171
BARC. *see* Burned Area Reflectance Classification
baseflows 7, **161**
    and wildfire 160, **161**, 166
baseflow–stormflow separation 10
basin-scale water balance (BSWB)-based models 306
beech (Fagus) 204
Belgium 203
Bernese Emmental (Switzerland) 3
Betula spp 60, 64, 127
    B. lenta 39
biodiversity 198, 203–204, 210
Biomass Action Plan (BAP) 203
Bol'shoy Pit River 185, **185**
boreal forests 199, 212, 315
Borneo (Indonesia) 146, 148, 152
Bowen ratio methods 43
Brazil **150**, 151, 153
Britain (UK) 197
British Columbia (Canada) 98, 100, 105
Budyko formulations 12–13, 103, 147
Budyko framework 224
Burned Area Reflectance Classification (BARC) 170

California (USA) 170
Canada 11, 314
canopy density 164
canopy interception 146, 166
  and evapotranspiration 37, 41, 43, 146
  species variations in 37
  in tropical forests 146, 154
CAP. *see* Common Agricultural Policy
carbon balance 35, **36**
carbon sequestration 36, 317
Cascade Range (USA) 165
cavity flow systems 23
CECA. *see* cumulative equivalent clearcut area
Central Siberia
  forest cover 180–181
  hydrological regimes 181, **182, 183,** 183–184
Chadobets River basin 190, **192**
China 145, 316
China, forest hydrology
  canopy interception 221–222
  discharge monitoring 226
  evapotranspiration 222–223
  forest floor interception 222
  forest–water yield relationship 223–224
  hydrometric observations 219–221, **220, 221**
  massive forestation effort 227–228, **228**
  modelling tools 224–225
  National ecosystem research 219
  overview 218–219
  remote sensing of precipitation 225
  soil erosion 224
  stable isotope method 225–226
  UAV technology 226–227
Chinese Ecosystem Research Network (CERN) 219
Chinese Forest Ecosystem Research Network (CFERN) 219
City of Ballarat 246
City of Melbourne 246
clearcutting 206
climate change 5, 11–13, 315, 317, 320, 321
  adaptation 210–211
  and carbon sequestration 317
  direct impact 60–61
  and El Niño–Southern Oscillation 152
  in Europe 198, 199, 201, 205, 206, 209–211
  evapotranspiration 58
  hydrological impact 62, **63–64**
  indirect impact 60–61
  tidal freshwater 138–139
  on tree-level transpiration 58, **59**
  in tropical forests 144, 152
ClimWatAdapt 210
closed-canopy forests 37
closed/open watersheds 248–249
Coalburn catchment (UK) 198
cold-region montane forested watersheds
  continental climates 112
  forests and avalanches 112–114
  high-elevation steep watersheds 114–120
  maritime climates 111
  overview 109–110
  predictions of future climates 120–121
  snow processes 110–111
Colorado (USA) 162, 316
Common Agricultural Policy (CAP) 203
Congo Basin (Africa) 145, 316
coniferous forests 168
contactless technology 226
continental climates 315
Cotter River watershed 245, **245**
coupled surface–subsurface hydrology models 264, 272
covariance analysis 4, **4**, **40**, 43, 44
Coweeta Hydrologic Laboratory (CHL, North Carolina, USA) 10
cumulative equivalent clearcut area (CECA) 98
cyclones 145, 316

dams 2, 166, 212
Danube River, Europe 129–130
Dausse, Law of 2
deciduous forests 145, 164, 210
  evapotranspiration in 37, **46**
deforestation 4, 97, 99, 197, 198
  in tropical forests 146, 153
DHSVM model 98
digital elevation model (DEM) 23
dipterocarps *(Dipterocarpaceae)* 146
disease 11
DMC (double mass curve) method 102, **102**
downslope pipes 22
drainage 41, 149, **149**
DRAINMOD-FOREST model 272
drinking water supply (DWS) 197, 199, 203, 204, 206, 317–318
drought/water scarcity 212, 320, 321
  and climate change 211
  in Europe 197, 198, 201–203, 210
  and forest management 206, 210–211
  tolerance 210
  in tropical forests 144, 145–146
dryness index (DI) 103, 147
duff. *see* forest floor
DWS. *see* drinking water supply

EAFRD. *see* European Agricultural Fund for Rural Development
Eberswalde (Germany) 197
ECA. *see* equivalent clearcut area
ecohydrology 320
ecosystem productivity 35, 36, **36**
ecosystem services 47, 201, 206
  payment for (PES) 203
  in tropical forests 143–144

Ecuador **150**, 315
eddy covariance method **40**, 43, 44
EFI. *see* European Forest Institute
El Niño–Southern Oscillation (ENSO) 146, 152
end-member mixing 320
energy balance 35, **36**, 44
epikarst zone 22, 23
equifinality 101
equivalent clearcut area (ECA) 98, 100
ERA-Interim data 148
ERMiT (Erosion Risk Management Tool) 170
EU Biodiversity Action Plan 203
eucalypt forest **5**, 7, 164
Eucalyptus camaldulensis 124
EU Forest Action Plan (FAP) 203
Europe 196–212, 315, 316
 adaptation/water-sensitive management in 210–211
 Central 201, 202, 210, 316
 challenges/future research needs for 211–212
 climate change in 198, 199, 201, 205, 206, 209–211
 deforestation/afforestation in 198
 drinking-water/groundwater protection in 199, 203
 drought/water scarcity in 197, 198, 201, 202, 204, 209, 210
 flood prevention in 197–198, 199–201
 forest hydrological hypothesis for 211–212, **211**
 forest hydrology literature for 316
 future research needs in 211–212
 paired watersheds in 197
 runoff from ski runs/resorts in 206–209
 see also specific countries
European Agricultural Fund for Rural Development (EAFRD) 203
European Flood Directive 205
European Forest Fire Information System (EFFIS) 202
European Forest Institute (EFI) 198, 205, 209–210
European Union (EU) 196
 2020 Strategy 204
 and drinking-water/groundwater protection 205–206
 on droughts/water scarcity 202–204
 forest policies 197–198, 203–205
 Forest Strategy 198, 201, 202
 Green Paper on forest protection 203
 Resolution on Forests and Water (2007) 204
EU Water Scarcity and Drought Policy 203
evaporation 12, 315
evapotranspiration (ET) 12, **13**, 35–47, 164, 211, 314
 actual 103
 and canopy/litter interception 37, 146, 164
 and empirical models 45–47, **46**
 formulae for 35–36
 future study of 317
 global 35, 38, 47
 indirect estimation of **40**, 44–47
 and latitude 146–147, **147**
 and mathematical models 44–47
 measurement of **40**, 43–44, **148**, 317
 potential. *see* PET
 and precipitation 147, **148**
 rates 41–43, 316
 and satellite/remote sensing data 44, 316
 species/age variations in 38
 taiga forests 185–187
 and temperature 146–147, **147**
 in tropical forests 146–149, **147**, **148**, 154
 and water balance **36**, 149
evergreen broadleaf forest **46**
evergreen needle-leaf forest **46**

FDCs. *see* flow duration curves
fire, wild. *see* wildfire
flooded and wetland forests
 blackwater events 133–134
 climate change and tidal freshwater 138–139
 groundwater-fed forested wetlands 127–128
 homogeneous ecosystems 130
 hydrology of flooded forests 131–133, **132**, **133**
 intertidal forests 134–137
 northern bogs and fens 138
 overview 124–126
 precipitation-excess wetlands 126–127
 red gum forests 128–129
 river development and flood modification 130–131
 surplus surface water flow 128
 swamp cypress forests 129
 wetlands and stored carbon **137**, 137–138
 willow forests 129–130
floods 2, 201, 205, **211**, 303
 and protective role of forests 197–201
flow duration curves (FDCs) **198**
FLUXNET 43, 45
fog 146
Food and Agriculture Organisation (FAO) 45, **46**
Forest Action Plan (FAP) 203
forest age 11
forestation 62
forest fires. *see* wildfire
forest floor 318
 litter layer 163
forest fragmentation 146, 198, 209, 315
forest hydrological hypothesis 211–212, **211**
forest hydrological process
 biomass mapping 295–298, **296**, **297**
 drought and forest fires 300–303, **301**, **302**
 ET estimation 298–300, **299**
 floods 303
 forest vegetation 295–298, **296**, **297**
 geohazards analysis 298–300
 geospatial technology support 306

forest hydrological process (*Continued*)
landslides 303, **304**
mapping and change analysis 289–292, **290, 291**
soil water/moisture estimation 292–295, **293**
SPOT spectral bands 299, **299**
stream water quality 304–306, **305**
forest hydrology 29
adaptive forest management 65–66
characterizing and quantifying interactions 258
coupled surface–subsurface hydrology models 264, 272
diagnostics and evaluation 263–264
direct and indirect climate change 55, **55**
emerging modeling methods 272, 275, 276
evapotranspiration 58, **59**
evapotranspiration functions 261
forest–water–climate interactions 61–62
groundwater models 264, 271–272, **273–274**
model functionality and complexity 262
model parameterization 261
model selection 262–263
offsetting and amplifying effects 61
overview 54–56
plot and watershed models 264, 265, **266–269,** 270, 271
rainfall–runoff functions 260–261
silvicultural options 62, **63–64,** 64
simulation tools 259, **260**
streamflow 59–61
transpiration 56–58
uncertainty and sensitivity analysis 264
watershed-level forest 64–65
forest management 210, 321
catchment scale 205, 210
and climate change 203, 210–211
and national policies 203
and paired catchments. *see* paired catchment studies
and water yield 11, 198
forest management plans (FMPs) 204
forest ownership 197
forest regrowth 211, **211**
see also reforestation
forest runoff process
controversy over process 20
current knowledge 19–20
fill-and- spill watershed 24–29, **25, 26**
generalized forested cross-section 24, **25**
infiltration excess *vs.* variable source area 20
macropore flow 21–22
proposed and discarded mechanisms 20–21
saturated zone process 23
soil piping 21–22
thresholds and connectivity 23–24
forests and avalanches
effects on avalanches 113–114
hydrology 114
snow avalanche 112–113, **113**
Forest Service Water Erosion Prediction Project (FSWEPP) 170
forests, global distribution of 314
forest/tree mortality 152, 201
forest vegetation 295–298, **296, 297**
forest–water–climate interactions 61–62
France 2, 166, 199
freshwater forests 134–135, **135**
future of forest hydrology 211–212, 314–321

Ganges–Brahmaputra river delta 306
geographic information systems. *see* GIS
Geographic Information Systems (GIS) 170–171
geomorphology 150, 154
Geospatial interface for WEPP (GeoWEPP) 305
geospatial technology
distributed models 306
forest hydrological processes management 289–306
land-cover mapping classifications 287
new technology 306–307
overview 287–289
satellite-based interferometry technique 306–307
see also specific geospatial applications
Germany 197, 200–202, 204, 205, 210
GIS (Geographic Information Systems) 170–171
global eddy flux 46, **46**
Global Terrestrial Observation System (GTOS) 219
global warming. *see* climate change
GPP (gross primer productivity) 35–36, **36**
grassland **13**, **46**, 47, 198
green/blue water 206
groundwater 7–9, 12, 204–205
flows/fluxes 44, 149
models 264, 271–272, **273–274**
protection 205–206
recharge 166
ridging 20
GSFLOW model 272, **275**

Hadley circulation 144
Halley, Edmond 2
Hamon PET method 45
hardwood forests 37, 210
harvesting. *see* timber production/logging
Harz Mountains (Germany) 197, 205
hemlock 37
Hewlett separation procedure **9**
high- and low-flow frequency analysis 80–81
high-elevation forests. *see* mountainous areas
high-elevation steep watersheds
bedrock permeability 118
forest access 119–120
infiltration and runoff process 116

rain- *vs.* snow-dominated hydrological systems 115–116
schematic representation **115**
soil water 116–118
topography 118–119, **119**
upland forested watersheds 115
high-flow quantiles 86–89, **87, 88**
history of forest hydrology 1–8
Hubbard Brook Experimental Forest (HBEF, New Hampshire, USA) **316**
Hugo, Hurricane **42**
humidity 43, 45
relative 103–104
hurricanes **42**
hydraulic conductivity, saturated 151
hydraulic redistribution (HR) 41
hydroclimatic extremes
characteristics of precipitation extremes 81, **82**
high- and low-flow frequency analysis 80–81
high-flow characteristics 81–82, **83**
high-flow quantiles 86–89, **87, 88**
low-flow characteristics 83–85, **84, 85**
low-flow quantiles 89, **90, 91**
NOAA PIDF values 74
overview 73–75
PIDF estimation 75, 80, 85–86
site description 75, **76–79**
hydrograph analysis 4, **4**, 8–10, **9**, **11**
hydrograph separation 10
hydrological cycle 2, 210, 314–317
hydrological cycling 154
hydrological modelling 201, 205
DEM. *see* digital elevation models
and equifinality 101
for ET estimation 44–45
and large watersheds 97, 98, 101–102
see also specific models
hyetograph 9
hypothesized runoff process 26, **26**

India 145, 316
Indo-Malay-Australasia ecozone (IMA) 145, **148**
Indonesia 153
inferred dominant pathways **150**, 151–152, **152**
infiltration 7–8, 10, **36**, 97, 206
capacity 8, 151, 206
and wildfire 160, 161, **161**, 163–164, 165
infiltration excess. *see under* overland flow
infiltration excess *vs.* variable source area 20
insects 11
Intergovernmental Panel on Climate Change (IPCC) 152
intertidal forests
freshwater forests 134–135, **135**
mangrove forests 135–136
Inter-Tropical Convergence Zone (ITCZ) 144, 146
invasive species 320
irrigation 41, 201
isotope measurement techniques 43–44
Ivory Coast **150**

Japan 197, 316
Jelanisca catchment (Serbia) **202**

Karabula River 190, **192**
KINEROS2 (Kinematic Runoff and Erosion Model) 171
Köppen climate classification 144

La Cuenca basin (Peru) **150**, 151, **152**
LAI. *see* leaf area index
landscape scale 43, 44
landslides 303, **304**
land use 3–4, 205, 212, 315
land-use changes 4, 38, 205, 212, 315
subsidies for 205
in tropical forests 144, 153
La Niña 145
large-scale clear-cutting 187–191, **188–191**
large watersheds 11, 97–105, 149
and breakpoints/double mass curves 102, **102**
and deforestation 97
and hydrological models 97, 98, 101–102
and paired watershed experiments 97–98
and sensitivity-based approach 102–103
and simple water balance 103–104
three research challenges for 98
and time trend method 104
and Tomer-Schilling framework 104–105
and wildfire 98, 99
lateral flows 151
shallow **150**
leaf area index (LAI) 37, 39, 43, 45, **46**
LiDAR (light detection and ranging) 226
LiDAR technology 319, 321
Liriodendron tulipifera **39**
litter interception 35, 37
litter layer 163
local water use ratio (LEUR) 147, **148**
logging. *see* timber production/logging
Lower Mississippi River, USA 129
low-flow quantiles 89, **90, 91**
low flows. *see* baseflow
lysimeters **40**, 197

macropore flow 21–22
macropores 7–8, 10
Maimai watersheds 20
Malaysia **150**, 151
Man and the Biosphere (MAB) program 219
mangrove forests 135–136

maritime climates 315, 316
Marsh, George Perkins 3–4
MDMC. *see* modified double mass curves
Mediterranean 202, 206, 316
Melbourne's population 239–241, **240**
Melbourne's water catchments 247, **247**
Mexico 44, 315
microclimates 39
MIKE SHE model 98, 272
mineral soil 8, **150**
 and wildfire 161, 163
mires. *see* wetland forests
mixed forests **46**, 204, 210
Moderate Resolution Imaging Spectroradiometer (MODIS) 290
modified double mass curves (MDMC) 102, **102**
MODIS data 187
monsoon 143, 144–145, 146
 see also rainforest climate
monsoon climate (Am) 143–146
montane cloud forests 146
mountain ash *(Eucalyptus regnans)* 11
mountainous areas 205, 314
 snow processes in. *see* snowmelt; snow processes
 steep watersheds in. *see* steep watersheds
 and water supply 199
mountain pine beetle 98
MOUNTFOR 205
municipal water supply watersheds
 agricultural watersheds 242–243
 benchmarking watershed forests 250–251
 caveats 253
 closed/open watersheds 248–249
 forested municipal watersheds 251
 forested watersheds 243–244, **244**
 future-gazing 253
 historical view 235–236
 hydrological measurements 252
 management issues **245**, 245–248
 Melbourne's population 239–241, **240**
 natural watersheds 251–252
 overview 234–235
 ownership 252–253
 paired watershed 236
 Quabbin Reservoir and watershed development 249–250
 quantitative properties 236–237
 recognizing the effects on time 253
 runoff elasticity 238–239, **239**
 slope water storage 241–242
 watershed efficiency **237**, 237–238, **238**
Mura River 190, **192**

National Oceanic and Atmospheric Administration (NOAA) 74
NDVI. *see* normalized difference vegetation index
neo-tropical ecozone (NEO) 145, 148, **148**
NEP (net ecosystem productivity) 36, **36**
New Guinea 145, 148
new/old water 10
New Zealand 316
NOAA. *see* National Oceanic and Atmospheric Administration
normalized difference vegetation index (NDVI) 98
North America 315, 316
North Carolina (USA) 44
northern bogs and fens 138
Norway 196

oak. *see Quercus* spp.
old/new water 10
one-factor-at-a-time (OFAT) approach 101
Ontario (Canada) 99
overland flow 7–8, 10, 151–152, **161**
 infiltration-excess 151, **152**, 154, 165
 saturation-excess 151, 165, 316

Pacific Northwest (USA) 98
paired catchment studies 4–9, **4**, **5**, 197
 and spatial scale 97–98
 strengths/weaknesses of 7
paired-watershed experiments 19
peak flows 97, **161**, **200**
 in large watersheds 99–100
 and wildfire 166, 169
Penman–Monteith model 44
permafrost 212, 315
Perrault, Pierre 2
Peru **150**
pests/disease 209
 pine beetle 98
PET (potential evapotranspiration) 37, 44–45, 47
 and large watersheds 103–105
photosynthesis 36–38, 47
pine beetle 98
pine (*Pinus* spp.) 41, 44, 164, 204
 eastern white 37
 loblolly *(P. taeda)* 37
 longleaf *(P. palustus)* 37
 ponderosa *(P. ponderosa)* 37, 163, 165
 radiata *(P. radiata)* 4, **5**
 red *(P. resinosa)* 37
plantation forests
 canopy interception in 37
 ET rates in 41, **42**
 water use by 146
plant–water–atmosphere relationship 14
plot and watershed models 264, 265, **266–269**, 270, 271
Plynlimon catchment (Wales) 197
population growth 198
post-wildfire flood vulnerabilities 89
precipitation intensity–duration–frequency (PIDF) 73
Puerto Rico **150**

Quabbin Reservoir and watershed development 249–250
Quercus spp
Q. prinus 166
Q. rubra 38
Q. suber 166
quickflow separation technique 9–10, **9**

radiant energy **36**
rainfall 2–3, **9**, 11, 27, 197, 315
and air temperature 2
and canopy interception. *see under* canopy interception
infiltration 10, **11**, 151
intensity/duration of 8–10
mean annual 11–12, **13**
measurement of 2, 6–7, 8, 9
return period **200**
and streamflow 3–4, 7
in tropical forests 144, 145
rainforest climate (Af) 143, 144–145
see also monsoon climate
red gum forests 128–129
reforestation 39, 41, 99, 100, 204, 206
remote sensing (RS) 98, 318, 320
and evapotranspiration 44
Reserva Ducke basin (Brazil) **150**, 151, **152**
reservoirs 205, 206
return flow 151
River Murray, Australia 128–129
roads 206
ROS (rain-on-snow) events 165
runoff 10, **161**, 315
and macropores. *see* macropores
pathways 150, **150**, 152, **152**, **153**
and roads 199
from ski runs/mountain resorts 206–208
see also overland flow
runoff elasticity 238–239, **239**
Russia 127, 138, 177, 187, 191, 196
see also Siberia

sandy soil 205
Santee Experimental Forest (SEF, South Carolina, USA) **42**
sapflow measurement **40**, 43, 318
sapping 23
saturated zone process 23
saturation excess. *see under* overland flow
savannah **46**
savannah climate (Aw) 144–145
see also monsoon climate; rainforest climate
seasonal mean crop coefficient (K) 45, **46**
sediment load 1
sediment/sedimentation 2, 4, 197, 206
and wildfire 166–170, **168**, **169**
sensitivity-based approach 102–103
Serbia 208
shallow lateral flow **150**
shrubs/shrubland 38, **46**, 47
Shui Yuan Han Yang 218
Siberia (Russia) 315, 318
silvicultural options 62, **63–64,** 64
silviculture 39, 210
Silvistrat project 210
site-specific forest management decision support (SSFMDS) 288
ski resorts 206
slope 22
slope-parallel flow 201
snow-dominated watersheds. *see* mountainous areas; taiga forests
snowmelt **4**, 120, 164, 165, 199, 315, 316
and ski resorts 208
snow processes 314, 315
rain-on-snow (ROS) events 165
in taiga forests. *see under* taiga forests
soil
composition/structure of 7, 8
degradation **211**
erosion 154, 168–169
evaporation 37, 38, 44, 166
infiltration. *see* infiltration
mineral. *see* mineral soil
permeability 208
and spatial variability 163–164
voids. *see* macropores
water-holding capacity **161**, 162–163
and wildfire 160–164, **162**, 168–170
Soil and Water Assessment Tool (SWAT) 101
soil heat flux **36**, 43
soil piping 21–22
soil water
availability 36, 37, 41
storage 12
and transpiration 8
solar radiation, measurement of 3
South Creek basin (Australia) **150**, 151, **152**
South-East Asia 145–146, 152, 315
Soviet Union. *see* Russia
Sperbelgraben and Rappengraben (Switzerland) 197
spike hydrographs 11
SPOT spectral bands 299, **299**
spruce–fir–hemlock forest 37
spruce *(Picea)* 37
SSFMDS. *see* site-specific forest management decision support
steep watersheds 166
stemflow 27, 204
stochastic variation 7, 8
stormflow 7, 9, 165
separation procedure 9–10, **9**
subsurface. *see* subsurface stormflow
streamflow 1, 2–4, **9**, 11, **161**
and afforestation 212
direct impact of climate change 59–60

streamflow (*Continued*)
 and evapotranspiration. *see* evapotranspiration
 and experimental basins/watersheds 149, **150**, 151, **152**
 generation pathways 150, **150**, 152, **152**, **153**
 indirect impact of climate change 60–61
 inferred dominant pathways **150**, **152**
 and latitude 149, **149**
 measurement of 6–7, 8, 9
 and storms 151
 in tropical forests 149–152, **149**, **150**, **152**, **153**
 and wildfire 165–166
 see also baseflows; peak flows
stream water quality 304–306, **305**
subalpine forests. *see* alpine/subalpine forests
subsurface drainage. *see under* drainage
subsurface flow 151, 317
 return 151
subtropical forests 315, 316
Sukhoy Pit River basin 185, **185**, 186
Surface Water and Ocean Topography (SWOT) 307
sustainable forestry 204, 210
swamp cypress forests 129
SWAT. *see* Soil and Water Assessment Tool
Sweden 210, 315
Swiss Alps 199
Switzerland 3, 196
 Sperbelgraben and Rappengraben 197

Tai Forest (Ivory Coast) **150**, 151
taiga forests
 evapotranspiration 185–187, **185–187**
 forest cover, Central Siberia 180–181
 hydrological regimes, Central Siberia 181, **182**, **183**, 183–184
 large-scale clear-cutting 187–191, **188–191**
 literature review 177–178
 overview 176–177
 study sites and methods 178, **179–180**
Taxodium distichum 129
temperate forests 315–316, 317
Thailand 99
thinning 210, 212, 239, 242, 246
 and evapotranspiration 43
3PG model 14
throughflow/throughfall 151
throughflow trough system **150**, 151
Thuja occidentalis 127
timber production/logging 205, 206
 and streamflow 318
time trend method 104
Tomer-Schilling framework 104–105
transpiration (T) 3, 8, 14, 35, 38–41, 43, 44, 47, 211
 and climate change 209
 measurement of 43, 314, 318–319
 species/age variations in 38
 in tropical forests 146
 and wildfire 163, 164, 166
 see also evapotranspiration
tree transpiration
 CO_2 concentration 57
 droughts 57
 increasing VPD 56–57
 to warming 56
 water-use efficiency (WUE) 57–58, **58**
tropical forests 10, 143–154, 315, 317
 canopy interception in 146, 314
 characteristics of 143
 and climate change 144
 climatic regime/forest types 144–146
 deforestation in 146
 and drought/water stress 144–145
 evapotranspiration in 146–149, **147**, **148**, 152–154
 experimental basins in **150**, 152
 land-use changes in 144, 153
 rainfall distribution in 145
 research needs for 152–154
 streamflow generation in 150, **150**, 152, **152**, **153**
tropical storms 145
 see also monsoon
Tsuga. *see* hemlock

UAV. *see* unmanned aerial vehicles
understorey vegetation 8, 160
 evapotranspiration by 41
United States (USA) 11, 314
 forest hydrology literature/research in 197, 206
 Pacific Northwest 98
 post-fire rehabilitation in 170
 south-eastern 37
 southern 39, 316
unmanned aerial vehicle (UAV) technology 226
urban areas 198
urbanization 320
USDA Forest Service experimental forests 73, **74**
USDA (United States Department of Agriculture) 170
 Forest Service 170
US Geological Survey 170

Variable Infiltration Capacity (VIC) model 98
variable source area (VSA) 10, **11**, 19
VIC (Variable Infiltration Capacity) model 98

Wagon Wheel Gap experiment (Colorado) 4, **4**
Wales 197
WaReLa project 201
Washington (USA) 163, 165
water availability 41, 47, 205, 206, 321
 for evapotranspiration 147

water balance 6–7, 103–104, 318
and evapotranspiration 149
formula for 6, 35
water budget 197, 210
water–carbon coupling 36
Water Erosion Prediction Project (WEPP) 170, 305
Water Framework Directive (WFD, EU) 203, 204
water levels, measurement of 3–4
water pollution 198, 205, 206
water quality 196, 197, 204, 205
and droughts 210
and wildfire 166–168
water scarcity. *see* drought/water scarcity
Water Scarcity and Drought Policy (EU) 203
watershed
flows 7–8
hydrology 7–8, 10–11
watershed-level forest 64–65
watershed-scale research 149–150
water-storage capacity of forests 201, 205
water stress 41, 145, 201, 210, 321
water supply 12, 205
constraints/deficiencies 201
and mountainous areas 206
see also drinking water supply
water-table aquifers 8
water-use efficiency (WUE) 36, 57–58, **58**
water year 12
water yield 11, 198
wet canopy evaporation. *see* canopy interception
wetland forests 44, 205
wetlands and stored carbon **137**, 137–138
wildfire 160–171
and baseflow 166
and climate change 11, 211
and debris flows 169–170
and drought/water scarcity 202, 205, 206, 210
effects of, overview **161**
and fuel loads 11
and large watersheds 98, 99
and peak flows 166, 169
and precipitation 165
and sediment 166–170, **168**, **169**
and soil 160–164, **162**, 168–170
stabilization/rehabilitation after 170–171
and streamflow 165–166
and surface runoff/overland flow 165
and vegetation 164
and water quality 166–168
and watershed hydrology 11
willow forests 129–130
wind 315
and evapotranspiration 41

xylem anatomy **39**

Printed and bound by CPI Group (UK) Ltd, Croydon, CR0 4YY
13/09/2026
14961113-0003